5G for the Connected World

5G for the Connected World

Edited by

Devaki Chandramouli
Nokia
Texas
USA

Rainer Liebhart
Nokia
Munich
Germany

Juho Pirskanen
Wirepas
Tampere
Finland

This edition first published 2019

Registered Offices
John Wiley & Sons, Inc., 111 River Street, Hoboken, NJ 07030, USA
John Wiley & Sons Ltd, The Atrium, Southern Gate, Chichester, West Sussex, PO19 8SQ, UK

Editorial Office
The Atrium, Southern Gate, Chichester, West Sussex, PO19 8SQ, UK

For details of our global editorial offices, customer services, and more information about Wiley products visit us at www.wiley.com.

Wiley also publishes its books in a variety of electronic formats and by print-on-demand. Some content that appears in standard print versions of this book may not be available in other formats.

Library of Congress Cataloging-in-Publication Data

Names: Chandramouli, Devaki, editor. | Liebhart, Rainer, editor. | Pirskanen, Juho, editor.
Title: 5G for the connected world / edited by Devaki Chandramouli, Nokia, Texas, USA, Rainer Liebhart, Nokia, Munich, Germany, Juho Pirskanen, Wirepas, Tampere, Finland.
Description: First edition. | Hoboken, NJ : John Wiley & Sons, Inc., 2019. | Includes bibliographical references and index. |
Identifiers: LCCN 2018051891 (print) | LCCN 2018056369 (ebook) | ISBN 9781119247074 (AdobePDF) | ISBN 9781119247135 (ePub) | ISBN 9781119247081 (hardcover)
Subjects: LCSH: Mobile communication systems–Technological innovations. | Broadband communication systems–Technological innovations. | Wireless sensor networks–Technological innovations.
Classification: LCC TK5103.2 (ebook) | LCC TK5103.2 .A147 2019 (print) | DDC 621.3845/6–dc23
LC record available at https://lccn.loc.gov/2018051891

Cover Design: Wiley
Cover Image: Iaremenko Sergii/Shutterstock

Set in 10/12pt WarnockPro by SPi Global, Chennai, India
Printed in the UK

Contents

About the Editors

DEVAKI CHANDRAMOULI has over 18 years of experience in the telecommunication industry. She spent the early part of her career with Nortel Networks and is currently with Nokia. At Nortel, her focus was on the design and development of embedded software solutions for CDMA networks. Later, she represented Nortel on the Worldwide Interoperability for Microwave Access (WiMAX) Forum with a focus on WiMAX architecture and protocol development. At Nokia, her focus areas include architecture and protocol development of 5G System and EPS related topics. She has been instrumental in developing Nokia's vision for 5G System Architecture and she has also been instrumental in developing Nokia's strategy for radio and architecture standardization (phased) approach in the 3rd Generation Partnership Project (3GPP). She leads 5G System Architecture specification in 3GPP SA2 and continues to focus on active contribution toward evolution of 5G System work in SA2. She is now Head of North American Standardization in Nokia. She has co-authored IEEE papers on 5G, co-authored a book on "LTE for Public Safety" published by Wiley in 2015. She has (co-)authored over 100 patents in wireless communications. Devaki received her B.E. in Computer Science from Madras University (India) and M.S. in Computer Science from University of Texas at Arlington (USA).

RAINER LIEBHART has 25 years of experience in the telecommunication industry. He held several positions within the former Siemens Fixed and Mobile Networks divisions and now in Nokia Mobile Networks. He started his career as SW Engineer, worked later as standardization expert in 3GPP and the European Telecommunications Standards Institute (ETSI) in the area of Internet Protocol Multimedia Subsystem (IMS), took over responsibilities as WiMAX and Mobile Packet Core System Architect and was head of the Mobile Core Network standardization team in Nokia Networks for more than eight years. He was also the Nokia Networks main delegate in 3GPP SA2 with the focus on Long-Term Evolution/System Architecture Evolution (LTE/SAE). After working as Research Project Manager within Nokia Bell Labs with a focus on 5G, he is now Head of 5G Solution Architecture in the Mobile Networks Global Product Sales department of Nokia. He is (co-)author of over 70 patents in the telecommunication area and co-editor of the book "LTE for Public Safety" published at Wiley. Rainer Liebhart holds an M.S. in Mathematics from the Ludwig-Maximilians University in Munich, Germany.

JUHO PIRSKANEN has 18 years of experience on technology development on wireless radio technologies such as 3G, HSPA, LTE and WLAN and most recently on 5G. He has held several positions in Nokia Networks, Nokia Wireless Modem, Renesas Mobile Corporation and Broadcom Corporation and then again at Nokia Networks for 5G research and standardization. He has participated actively for several years on different standardization forums such as 3GPP and IEEE802.11 by doing numerous technical presentations, being rapporteur of technical specifications and leading different delegations. His research work has resulted in more than 40 (co-)authored patent families on different wireless technologies and several publications on radio interface solutions including 5G. On 5G his research focused on physical layer and radio protocol layer concepts and first implementations of the 5G radio solutions. In late 2017, he joined Wirepas having headquarters in Tampere, Finland. Wirepas develops de-centralized wireless IoT mesh networks that can be used to connect, locate and identify lights, sensors, beacons, assets, machines and meters with unprecedented scale, density, flexibility and reliability. Juho Pirskanen holds a Master of Science in Engineering, from Tampere University of Technology, Finland.

List of Contributors

Subramanya Chandrashekar
Nokia
Bangalore
India

Betsy Covell
Nokia
Naperville
USA

Sami Hakola
Nokia
Oulu
Finland

Volker Held
Nokia
Munich
Germany

Hannu Hietalahti
Nokia
Oulu
Finland

Jürgen Hofmann
Nokia
Munich
Germany

Keeth Jayasinghe
Nokia
Espoo
Finland

Toni Levanen
Tampere University
Tampere
Finland

Zexian Li
Nokia
Espoo
Finland

Andreas Maeder
Nokia
Munich
Germany

Jarmo Makinen
Nokia
Espoo
Finland

Tuomas Niemela
Nokia
Espoo
Finland

Karri Ranta-aho
Nokia
Espoo
Finland

Rapeepat Ratasuk
Nokia
Naperville
USA

Rauno Ruismäki
Nokia
Espoo
Finland

Peter Schneider
Nokia
Munich
Germany

Mikko Säily
Nokia
Espoo
Finland

Thomas Theimer
Nokia
Munich
Germany

Laurent Thiebaut
Nokia
Paris-Saclay
France

Samuli Turtinen
Nokia
Oulu
Finland

Mikko Uusitalo
Nokia
Espoo
Finland

Fred Vook
Nokia
Naperville
USA

Sung Hwan Won
Nokia
Seoul
South Korea

Foreword by Tommi Uitto

It has been said that hyper-successes often happen in business when three distinct and major inflection points or disruptions coincide. Not just one, not two, but three. With that theory, we can explain the hyper-success of GSM. First, there was a new global and open standard for mobile communication technology, driving up volumes and allowing for an ecosystem to emerge. Second, deregulation took place, allowing for competition with the incumbent, previously monopolistic operators. Third, electronics had evolved to a point, where mobile devices have become affordable for a bigger mass of the population. And sure enough, mobile telephony proliferated throughout the world and made people's lives more enjoyable, secure, efficient and effective. Obviously, businesses also benefited from the ability of their employees to stay connected with one another and Internet regardless of their location. With the advent of 5G, we have the ingredients for something equally profound to happen. You see, the introduction of 5G as a wireless technology is more or less coinciding with adoption of both cloud computing and Artificial Intelligence (AI)/Machine Learning (ML) by operators. Looking back, we can certainly be satisfied with the improvements that 3G and 4G brought as mobile communication technologies, as well as the new device paradigms, pricing models and business models that were introduced in parallel with them. But 5G can become a much bigger step for the humankind, relatively speaking, than 3G and 4G did at their time of introduction and adoption. With the extreme mobile broadband aspects of 5G, we have so much speed and capacity that it is difficult to see how we could run out of it with applications and use cases known today. With design criterion of one million connected objects per square kilometer, we can say that 5G has been designed for the Internet of Things (IoT) from the outset. It will be more affordable and technically feasible than ever before to embed radio sensors and transceivers in physical objects. With ultra-reliable low-latency communication (URLLC) performance criteria and functionalities in 5G, we open a host of new use cases and business potential as we can be certain enough about a reliable and robust connection to a physical object accurately in space and time. In addition to such superior wireless connectivity and capacity, we then have virtually infinite computing capacity thanks to cloud computing. And not just any computing, not linear or simplistically deterministic computing, but rather computing that learns with AI, ML, including deep learning. A self-improving machine that can collect data and command objects in a wireless manner. Furthermore, a technology called network slicing will allow us to logically segregate different performance sets by using the common cloud infrastructure end-to-end, rather than building separate physical networks for separate use cases. It is difficult to see why just about any physical or physical/digital business process could not be automated with such technology. Therefore, we can expect 5G, together with Cloud and AI/ML, to have, relatively speaking, a bigger, more profound impact on enterprise than previous generations of mobile communications have had. For consumers, we are lifting and eradicating many barriers to use mobile technology to its full potential. To put it another way, we are creating ubiquitous embedded computing, not just islands of computing, not just communication networks. We are seamlessly interweaving physical and digital worlds. We will have a perfect, programmable model of the physical world in digital space. Welcome to the 5G future. The authors are deeply involved in the work on 5G – network slicing technology as an example – and are in a prime position to provide valuable first-hand insights on 5G-related 3GPP activities and all relevant technical details.

Tommi Uitto
President
Mobile Networks
Nokia

Foreword by Karri Kuoppamaki

Wireless connectivity touches almost every aspect of our daily lives, and LTE has delivered the ubiquitous high-speed wireless broadband experience for us. This has unlocked the potential of mobile video and the mobile innovation that rides on those networks that we interact with every day. Lyft, Uber, Snapchat, Venmo, Square, Instagram ... these are companies that simply would not exist without LTE. In addition to new innovations, global leaders such as Facebook, Alphabet, Amazon, and Netflix adapted their businesses to benefit from mobile broadband and their growth exploded!

As a result, mobile data use keeps growing, and there seems to be no end for consumer demand for more. Additionally, the need for more sophisticated mobile broadband services as well as new industries and users adopting the power of mobile broadband will push the limit on LTE technology creating a need for the next generation of mobile technology – 5G.

The next-generation of wireless technology, 5G, will not only enhance and improve the services we enjoy today, but will also transform entire industries, from agriculture to transportation and manufacturing to become more capable, efficient, and intelligent. In other words, the evolution toward 5G is a key component in the digital transformation of almost every industry as well as of society. As such, 5G is an integral component of our continued Un-Carrier journey into the future.

Although the promise and vision of 5G is well described, the 5G system behind it is somewhat covered in a veil of mystery. This book explaining the new 5G system from an end-to-end perspective, from vision and business motivation to spectrum considerations and then the technology from Radio to Core Network and service architecture demystifies what the 5G system is about. I would like to thank the authors for doing an excellent job in translating this complex topic into a book, and I hope it will serve as a useful tool for anyone wanting to understand what 5G really is all about.

Karri Kuoppamaki
Vice President
Network Technology
Development and Strategy
T-Mobile USA

Preface

After the considerable success of LTE, why do we need a new system with a new radio and a new core? First, 5G will boost some of the LTE key performance indicators to a new horizon: capacity, latency, energy efficiency, spectral efficiency, and reliability. We will describe the relevant radio and core features to enable optimizations (5G to be 10, 100, or 1000 times better than LTE) in these areas in respective chapters of the book. But this is only half of the 5G story. With the service-based architecture 5G Core supports natively a cloud-based architecture, the higher layer split of the radio protocol as specified by 3GPP paves the path for a cloud-based implementation of the radio network or parts of it and network slicing will open totally new revenue streams for operators by offering their network as a service to vertical industries. Slicing will enable operators implementing logical networks for diverse use cases (e.g. industrial applications) in an optimal way on their physical hardware. The 5G System also supports compute and storage separation natively, and introduces enablers to support the ability to perform dynamic run time load (re-)balancing with no impact to user's services. In addition, open interfaces and the possibility to expose data from network functions to third parties via open APIs enables operators to monetize these contextual real-time and non-real-time data or simply use the data to optimize network deployment and configuration. Operators can broker information to different industries like providers of augmented reality services, traffic steering systems, factories, logistical systems, and utilities. Real-time big data analytics will play a crucial role for this brokering model.

In a nutshell, the main intention of this book is to explain what 5G is from a technical point of view (considering 3GPP Release 15 content), how it can be used to enable new services, and how it differs from LTE. The book covers potential 5G use cases, radio and core aspects, and deals also with spectrum considerations and new services seen as drivers for 5G. Although 5G will not support IoT and massive machine-type-communication from the very beginning in 3GPP Rel-15, we felt that this topic is extremely important and thus we provided a detailed description about IoT, M2M, and technical enablers like LTE-M, NB-IoT, GSM, 5G in this book.

Some familiarity of the reader with the basics of 3GPP networks, especially with LTE/EPC, would be helpful, but this is not a pre-requisite to understand the main parts of this book. The reader can find a more detailed description of the book's content in the introduction section.

This book is intended for a variety of readers such as telecommunication professionals, standardization experts, network operators, analysts, and students. It is also intended for infrastructure and device vendors planning to implement 5G in their products, and regulators who want to learn more about 5G and its future applicability for a large variety of use cases. We hope everyone interested in the subject of this book benefits from the content provided.

The race towards 5G has already begun in major markets round the world like North America, China, Japan, and South Korea. To be first on the market with 5G is a key differentiator between operators and vertical industries. The year 2018 will be the first where 5G commercial networks are deployed in some key markets, even on a medium or large scale. People will start benefitting from the huge and broad step forward 5G is bringing to them individually with faster traffic downloads, faster setup times, lower latency, higher connection reliability and to the whole society with smarter cities, factories and traffic control coping with the challenges of the future. This brings us close to the vision of a truly connected world where everyone and everything are potentially connected with each other.

The Editors
December 2018

Acknowledgements

The book has benefited from the extensive contribution and review of many subject matter experts and their proposals for improvements. The editors would like to thank in particular the following people for their extensive contribution and review that helped to complete this book:

Subramanya Chandrashekar, Betsy Covell, Sami Hakola, Volker Held, Hannu Hietalahti, Jürgen Hofmann, Günther Horn, Keeth Jayasinghe, Toni Levanen, Zexian Li, Andreas Maeder, Jarmo Makinen, Tuomas Niemela, Karri Ranta-aho, Rapeepat Ratasuk, Rauno Ruismäki, Mikko Säily, Peter Schneider, Thomas Theimer, Laurent Thiebaut, Samuli Turtinen, Mikko Uusitalo, Fred Vook, Sung Hwan Won.

We would also like to thank Sandra Grayson, Louis Vasanth Manoharan, Rajitha Selvarajan and Mary Malin from Wiley for their continuous support during the editing process.

Finally, we thank our families for their patience and cooperation during the writing of the book.

The editors appreciate any comments and proposals for enhancements and corrections in future editions of the book. Feedback can be sent directly to devaki.chandramouli@nokia.com, rainer.liebhart@nokia.com, juho.pirskanen@gmail.com.

Introduction

This book explains the new 5G system from an end-to-end perspective, starting from the 5G vision and business drivers, deployment and spectrum options going to the radio and core network architectures and fundamental features including topics like QoS, mobility and session management, network slicing and 5G security. The book also contains an extensive discussion of IoT-related features in GSM, LTE, and 5G. As indicated in the preface, some familiarity of the reader with basic concepts of mobile networks and especially with LTE/EPC is beneficial, although not a must for all chapters.

Chapter 1 will address the drivers and motivation for 5G. It will also provide insights into 5G use cases, requirements from various sources, like NGMN, ITU-R and 5GPPP, and its ability to support new services. In addition, it will touch on the business models enabled by the new radio and core architecture, and on possible deployment strategies. Furthermore, it will provide insights into organizations involved in defining use cases, requirements and developing the 5G eco system. It also provides an overview of the 3GPP timeline and content of Release 15 and Release 16. This chapter does not require detailed technical knowledge about mobile networks.

Chapter 2 provides insights into spectrum considerations for the new 5G radio regarding all possible bands. Additionally, readers will get a good understanding of the characteristics of available new spectrum for 5G that sets fundamental requirements for radio design deployment and how spectrum is used, based on available channel models and measurements. It will also provide information about regional demands for licensed and unlicensed spectrum and about new regulatory approaches for spectrum licensing in the 5G era.

Chapter 3 describes the new 5G radio access technology. It includes the evolution of LTE access towards 5G, description of new waveforms, massive MIMO, and beamforming technologies, which are key features of the new 5G radio. This chapter will also explain the physical layer frame structure with its necessary features and functionalities. Furthermore, the chapter will explain the complete physical layer design and procedures in both downlink and uplink. Finally, the chapter explains the radio protocols operating on top of the 5G physical layer and procedures required to build the complete 5G radio access system. This chapter will also discuss solutions on how 5G caters to the extreme bandwidth challenge, considering challenges providing broadband access in indoor, rural, sub-urban, and urban areas.

Chapter 4 provides detailed insights into the drivers and motivations for the new 5G system. It gives an overview of the System Architecture, RAN architecture, architectural requirements, basic principles for the new architecture, and the role of technology enablers in developing this new architecture. It provides a comparison with EPS, describes the essence of the 5G system, newly introduced features, and explains how interworking between EPS and the 5G system will work in detail. This chapter details the key features including network slicing, data storage principles for improved network resiliency and information exposure, generic exposure framework, architectural enablers for mobile edge computing, support for non-3GPP access, fixed-mobile convergence, support for IMS, SMS, location services, public warning system, and charging. It also includes a summary of control and user plane protocol stacks.

Chapter 5 describes the 5G mobility management principles followed in radio and core network. It will also provide a comparison of 5G mobility management with existing mobility management in LTE/EPC. This will include a description of 5G mobility states, connected and idle mode mobility for standalone and non-standalone deployments. It will also include procedures for interworking towards LTE/EPC. Furthermore, it will provide insights into how mobility support for ultra-high reliability applications or highly mobile devices is achieved in 5G, considering single connectivity and multi-connectivity features.

Chapter 6 provides an overview of Session Management and QoS principles in 5G. It defines the data connectivity provided by the 5GS (PDU sessions, PDU session continuity modes, traffic offloading, etc.). It describes the 5GS QoS framework (QoS Flows, parameters of 5GS QoS, reflective QoS, etc.). Finally, it gives an overview of how applications can influence traffic routing and policy control for PDU sessions.

Chapter 7 provides insights into the 5G security vision and architecture. It explains device and network domain security principles and procedures based on 3GPP standards. This also includes a description of the key hierarchy used within the 5G security framework. In addition, this chapter provides an overview of NFV, SDN, and network slicing security challenges and corresponding solutions.

Chapter 8 provides an overview of ultra-low latency and high reliability use cases, their challenges and requirements, e.g. remote control, industrial automation, public safety, and V2X communication. It also provides an overview of radio and core network related solutions enabling low latency and high reliability from an end-to-end perspective.

Chapter 9 provides a description of 5G solutions and features supporting massive machine type communication and IoT devices. The chapter outlines the requirements and challenges imposed by a massive number of devices connected to cellular networks. In addition, the chapter also gives a detailed overview of how M2M and IoT communication is supported with technologies like LTE-M, NB-IoT, and GSM along with System Architecture enhancements supported for M2M and IoT devices.

Chapter 10 is meant as a summary and wrap-up of the whole book, highlighting the most important facts about 5G and providing an outlook of new features that can be expected in future 3GPP releases.

Terminology

2G	2nd Generation
3G	3rd Generation
3GPP	3rd Generation Partnership Project
4G	4th Generation
5G	5th Generation
5G NR	5G New Radio
5G RG	5G Residential Gateway
5GAA	5G Automotive Association
5GACIA	5G for Connected Industries and Automation
5GC	5G Core
5G-GUTI	5G Globally Unique Temporary Identifier
5GIA	5G Infrastructure Association
5GPPP	5G Public-Private Partnership
5QI	5G QoS Identifier
5G-SIG	5G Special Interest Group
5GS	5G System
5G-S-TMSI	5G S-Temporary Mobile Subscription Identifier
5GTF	5G Task Force
6G	6th Generation
AAA	Authentication, Authorization and Accounting
AC	Access Class
ACC	Adaptive Cruise Control
ACK	Acknowledgement
ADSL	Asymmetric Digital Subscriber Line
AES	Advanced Encryption Standard
AF	Application Function
AGCH	Access Grant Channel
AI	Artificial Intelligence
AKA	Authentication and Key Agreement
AL	Aggregation Level
AM	Acknowledged Mode
AMF	Access and Mobility Management Function
AOA	Angle of Arrival
AOD	Angle of Departure
API	Application Programming Interface

APN	Access Point Name
AR	Augmented Reality
ARQ	Automatic Repeat Request
ARIB	Association of Radio Industries and Businesses
ARP	Allocation and Retention Policy
ARPC	Average Revenue Per Customer
ARPD	Average Revenue Per Device
ARPF	Authentication Credential Repository and Processing Function
ARPU	Average Revenue Per User
ARQ	Automatic Repeat Request
AS	Access Stratum
AS	Application Server
ASN.1	Abstract Syntax Notation Number 1
ATIS	Alliance for Telecommunications Industry Solutions
AUSF	Authentication Server Function
AV	Authentication Vector
BBF	Broad-Band Forum
BCCH	Broadcast Control Channel
BCH	Broadcast Channel
BFD	Beam Failure Detection
BFI	Beam Failure Instance
BFR	Beam Failure Recovery
BG	Base Graph
BLE	Bluetooth Low Energy
BLER	Block Error Rate
BNG	Broadband Network Gateway
BPSK	Binary Phase Sift Keying BS
BS	Base Station
BSD	Bucket Size Duration
BSR	Buffer Status Report
BSIC	Base Station Identity Code
BSS	Base Station Subsystem
BTS	Base Transceiver Station
BWP	Bandwidth Part
CAPEX	Capital Expenditures
CBC	Cell Broadcast Centre
CBE	Cell Broadcast Entity
CBRA	Contention Based Random Access
CBS	Cell Broadcast Service
CC	Coverage Class
CCCH	Common Control Channel
CCE	Control Channel Element
CCSA	China Communication Standards Association
CDF	Charging Data Function
CDMA	Code Division Multiple Access
CDN	Content Delivery Network
CDR	Charging Data Record

CE	Coverage Enhancement
CE	Control Elements
CFRA	Contention Free Random Access
CGF	Charging Gateway Function
CHF	Charging Function
CIoT	Cellular IoT
C-ITS	Cooperative Intelligent Transportation Systems
CK	Cyphering Key
CM	Connection Management
cmWave	centimeter Wave frequencies
cMTC	critical Machine Type Communication
CN	Core Network
CoMP	Coordinated Multipoint Transmission
CORESET	Control Resource Set
CP	Control Plane
CP	Cyclic Prefix
CPE	Customer Premises Equipment
CP-OFDMA	Cyclic Prefix Orthogonal Frequency-Division Multiple Access
CPRI	Common Public Radio Interface
CQI	Channel Quality Indicator
C-RAN	Centralized RAN
CRC	Cyclic Redundancy Check
C-RNTI	Cell Radio Network Temporary Identifier
CS	Circuit Switched
CSI	Channel State Information
CSI-RS	Channel State Information Reference Signal
CSFB	Circuit-Switched Fallback
CSS	Common Search Space
CU	Central Unit
CUPS	Control/User Plane Separation
DCCH	Dedicated Control Channel
DCI	Downlink Control Information
DECOR	Dedicated Core Network
DEI	Drop Eligible Indicator
DFT-s-OFDM	Discrete Fourier Transform-spread-OFDM
DHCP	Dynamic Host Configuration Protocol
DL	Downlink (Network to UE)
DMRS	Demodulation Reference Signal
DN	Data Network
DNAI	Data Network Access Identifier
DNN	Data Network Name
DoS	Denial of Service
DRB	Data Radio Bearer
DRX	Discontinuous Reception
DTCH	Dedicated Traffic Channel
DSL	Digital Subscriber Line
DU	Distributed Unit

E2E	End-to-End
EAB	Extended Access Barring
EAP	Extensible Authentication Protocol
EASE	EGPRS Access Security Enhancements
EC	European Commission
EC	Extended Coverage
EC-GSM-IoT	Extended Coverage for GSM based Internet of Things
eCPRI	enhanced CPRI
EC SI	Extended Coverage System Information
eDECOR	Enhanced Dedicated Core Network
EDGE	Enhanced Data Rates for GSM Evolution
eDRX	Extended idle mode DRX
EGPRS	Enhanced GPRS
EIR	Equipment Identity Register
EIRP	Equivalent Isotropic Radiated Power
eMBB	enhanced Mobile Broadband
EMM	EPS Mobility Management
EMSK	Extended Master Session Key
EN-DC	E-UTRAN - New Radio - Dual-Connectivity
EPC	Evolved Packet Core
ePDG	Enhanced Packet Data Gateway
EPS	Evolved Packet System
E-SIM	Embedded SIM
ESM	EPS Session Management
E-SMLC	Enhanced SMLC
ETH	Ethernet
ETSI	European Telecommunications Standards Institute
EU	European Union
E-UTRAN	Evolved Universal Mobile Telecommunications System Terrestrial RAN
eUTRAN	Evolved Universal Mobile Telecommunications System Terrestrial RAN
FAR	Forwarding Action Rule
FAR	False Alarm Rate
FCC	Federal Communications Commission
FCCH	Frequency Correction Channel
FDD	Frequency Division Duplex
FFT	Fast Fourier Transform
FH	Frequency Hopping
FMC	Fixed Mobile Convergence
FN	Frame Number
F-OFDM	Filtered OFDM
FQDN	Fully Qualified Domain Name
FR	Frequency Range
FUA	Fixed Uplink Allocation
FWA	Fixed Wireless Access
GBR	Guaranteed Bitrate

GERAN	GSM / EDGE RAN
GFBR	Guaranteed Flow Bitrate
GGSN	Gateway GPRS Support Node
GMLC	Gateway Mobile Location Centre
gNB	Gigabit NodeB, Next Generation NodeB
GPRS	General Packet Radio Service
GPSI	Generic Public Subscription Identifier
GSM	Global System for Mobile communications
GSMA	GSM Association
GTP	GPRS Tunneling Protocol
GUAMI	Globally Unique AMF Identifier
GUTI	Globally unique temporary UE identity
HARQ	Hybrid Automatic Repeat Request
HBRT	Hardware-Based Root of Trust
HD	High Definition
HLcom	High Latency Communication
HLR	Home Location Register
HLS	High-Layer Split
HPLMN	Home PLMN
HR	Home Routed
HSDPA	High Speed Downlink Packet Access
HSPA	High Speed Packet Access
HSS	Home Subscriber Server
HSUPA	High Speed Uplink Packet Access
HTTP	Hypertext Transfer Protocol
HW	Hardware
I 4.0	Industry 4.0
IaaS	Infrastructure as a Service
ICT	Information and Communication Technology
ID	Identity
IETF	Internet Engineering Task Force
IK	Integrity Key
IKEv2	Internet Key Exchange Version 2
IMS	IP Multimedia Subsystem
IMSI	International Mobile Subscriber Identity
IMT	International Mobile Telecommunications
IMT-2020	ITU-R process for defining 5G IMT technologies
InH	Indoor Hotspot
IoT	Internet of Things
IP	Internet Protocol
IPR	Intellectual Property Rights
IPsec	Internet Protocol Security
IR	Incremental Redundancy
ISDN	Integrated Services Digital Network
ISG	Industry Specification Group
ISI	Inter Symbol Interference
ISM	Industrial Scientific and Medical

ITS	Intelligent Traffic Systems
ITU	International Telecommunications Union
ITU-R	ITU Radiocommunication Sector
IWF	Interworking Function
IWK	Interworking
KPI	Key Performance Indicator
LADN	Local Area Data Network
LAI	Location Area Identity
LAN	Local Area Network
LBO	Local Breakout
LCG	Logical Channel Group
LCH	Logical Channel
LCP	Logical Channel Prioritization
LCS	Location Services
LDPC	Low-Density Parity Check coding
LLR	Log-Likelihood Ratio
LLS	Low-Layer Split
LMF	Location Management Function
LOS	Line of Sight
LPWA	Low Power Wide Area
LRF	Location Retrieval Function
LSA	Licensed Shared Access
LSB	Least Significant Bit
LTE	Long Term Evolution
LTE-M	LTE category M1
M2M	Machine to Machine
MAC	Media Access Control
MC	Multi-Connectivity
MCS	Modulation and Coding Scheme
MBMS	Multimedia Broadcast / Multicast Service
MCC	Mobile Country Code
MCG	Master Cell Group
MCL	Maximum Coupling Loss
MDBV	Maximum Data Burst Volume
ME	Mobile Equipment
MEC	Multi-Access Edge Computing
MeNB	Master eNB
METIS	Mobile and wireless communications Enablers for the Twenty-twenty Information Society
MFBR	Maximum Flow Bitrate
MFCN	Mobile/Fixed Communication Network
MIB	Master Information Block
MIB-NB	Master Information Block - Narrowband
MICO	Mobile Initiated Communication Only
ML	Machine Learning
MM	Mobility Management
MME	Mobility Management Entity

mMIMO	massive Multiple-Input-Multiple-Output
mmWave	millimeter Wave frequencies
mMTC	massive Machine Type Communication
MN	Master Node
MNC	Mobile Network Code
MNO	Mobile Network Operator
MOTD	Multilateration Observed Time Difference
MPDCCH	MTC Physical Downlink Control Channel
MR-DC	Multi RAT Dual Connectivity
MS	Millisecond
MSB	Most Significant Bit
MSG1…4	Message 1 to 4 in RACH procedure
MSI	Minimum System Information
MSC	Mobile Switching Centre
MSISDN	Mobile Subscriber ISDN Number
MSK	Master Session Key
MTA	Multilateration Timing Advance
MTC	Machine Type Communication
N3IWF	Non-3GPP Interworking Function
NaaS	Network as a Service
NACK	Negative Acknowledgement
NAI	Network Access Identifier
NAPS	Northbound API for SCEF - SCS/AS Interworking
NAS	Non-Access Stratum
NAT	Network Address Translation
NB-IoT	Narrow Band IoT
NCC	Network Color Code
NE-DC	NR E-UTRAN - Dual Connectivity
NEF	Network Exposure Function
NF	Network Function
NFV	Network Function Virtualization
NGAP	Next Generation Application Protocol
NGEN-DC	Next Generation EN-DC
NGMN	Next Generation Mobile Networks
NG-RAN	Next Generation RAN
NIDD	Non-IP Data Delivery
NLOS	Non-Line of Sight
NPBCH	Narrowband Physical Broadcast Channel
NPDCCH	Narrowband Physical Downlink Control Channel
NPDSCH	Narrowband Physical Downlink Shared Channel
NPRACH	Narrowband Physical Random Access Channel
NPSS	Narrowband Primary Synchronization Signal
NPUSCH	Narrowband Physical Uplink Shared Channel
NPV	Net Present Value
NR	New Radio
NRF	Network Repository Function
NRT	Non-Real Time

NS	Network Slice
NSA	Non-Standalone
NSSAI	Network Slice Selection Assistance Information
NSSF	Network Slice Selection Function
NSSS	Narrowband Secondary Synchronization Signal
O&M	Operation and Maintenance
OA&M	Operations, Administration & Maintenance
OAM	Orbital Angular Momentum
OAuth	Open Authorization
OECD	Organization for Economic Co-operation and Development
OFDM	Orthogonal Frequency-Division Multiplexing
ONF	Open Networking Foundation
OPEX	Operating Expenses
OSI	Other System Information
OTT	Over-the-Top
PA	Power Amplifier
PaaS	Platform as a Service
PACCH	Packet Associated Control Channel
PAPR	Peak to Average Power Ratio
PBCH	Physical Broadcast Channel
PC	Personal Computer
PCA	Packet Control Acknowledgement
PCC	Policy and Charging Control
PCCH	Paging Control Channel
PCell	Primary Cell
PCF	Policy Control Function
PCH	Paging Channel
PCI	Physical Cell Identity
PCO	Protocol Configuration Options
PCP	Priority Code Point
PDB	Packet Delay Budget
PDCCH	Physical Downlink Control Channel
PDCP	Packet Data Convergence Protocol
PDN	Packet Data Network
PDP	Packet Data Protocol
PDSCH	Physical Downlink Shared Channel
PDTCH	Packet Data Traffic Channel
PDU	Protocol Data Unit
PEI	Permanent Equipment Identifier
PEO	Power Efficient Operation
PER	Packet Error Rate
PFCP	Packet Forwarding Control Protocol
PFD	Packet Flow Descriptor
P-GW	PDN Gateway
PGW-C	P-GW Control Plane Function
PGW-U	P-GW User Plane Function
PH	Power Headroom

PHY	Physical Layer
PHR	Power Headroom Report
PICH	Paging Indication Channel
PLMN	Public Land Mobile Network
PMI	Precoding Matrix Indicator
PoC	Proof of Concept
PON	Passive Optical Network
PPDR	Public Protection and Disaster Relief
PRACH	Physical Random Access Channel
PRB	Physical Resource Blocks
P-RNTI	Paging RNTI
PS	Packet Switched
PSA	PDU Session Anchor
PSCell	Primary Secondary Cell
PSD	Power Spectral Density
PSM	Power Save Mode
PSTN	Public Switched Telephone Network
PSS	Primary Synchronization Signal
PTW	Paging Time Window
PUSCH	Physical Uplink Shared Channel
PWS	Public Warning System
QAM	Quadrature Amplitude Modulation
QC-LDPC	Quasi-Cyclic LDPC
QFI	QoS Flow Identifier
QoE	Quality of Experience
QoS	Quality of Service
QPSK	Quaternary Phase-Shift Keying
R	Code Rates in channel coding
RA	Routing Area
RA	Random Access
RACH	Random Access Channel
RAN	Radio Access Network
RAR	Random Access Response
RAT	Radio Access Technology
RAU	Routing Area Updating
RB	Radio Bearer
RCC	Radio Frequency Color Code
RDI	Reflective QoS flow to DRB mapping Indication
RDS	Reliable Data Service
RF	Radio Frequency
RFC	Request for Comments
RG	Residential Gateway
RI	Rank Indicator
RIT	Radio Interface Technology
RLC	Radio Link Control Protocol
RLF	Radio Link Failure
RM	Registration Management

RM	Reed-Muller
RMSI	Remaining Minimum System Information
RNTI	Radio Network Temporary Identifier
RO	RACH Occasion
RoHC	Robust Header Compression
RQI	Reflective QoS Attribute
RQoS	Reflective QoS
RRC	Radio Resource Control Protocol
RSRP	Reference Signal Received Power
RSRQ	Reference Signal Received Quality
RT	Real Time
RTT	Round Trip Time
RV	Redundancy Version
RX	Receiver Exchange
SA	Standalone
SAE	System Architecture Evolution
SAS	Spectrum Access System
SBA	Service Based Architecture
SC	Successive-Cancelation decoding
SCell	Secondary Cell
SCEF	Service Capability Exposure Function
SCH	Synchronization Channel
SCS	Services Capability Server
SCS	Sub-Carrier Spacing
SDAP	Service Data Adaptation Protocol
SDF	Service Data Flow
SDN	Software Defined Networking
SDO	Standards Development Organization
SDU	Service Data Unit
SEAF	Security Anchor Functionality
SEPP	Secure Edge Protection Proxy
SFN	System Frame Number
SgNB	Secondary gNB
SGSN	Serving GPRS Support Node
S-GW	Serving Gateway
S-GW-C	Serving Gateway Control Plane Function
S-GW-U	Serving Gateway User Plane Function
SIM	Subscriber Identity Module
SINR	Signal to Interference plus Noise Ratio
SIP	Session Initiation Protocol
SLA	Service Level Agreement
SLAAC	IPv6 Stateless Address Autoconfiguration
SM	Session Management
SME	Small and Medium Enterprises
SMF	Session Management Function
SMLC	Serving Mobile Location Centre
SMS	Short Message Service

SN	Secondary Node
SN id	Serving Network Identifier
SNDCP	Subnetwork Dependent Convergence Protocol
SNR	Signal to Noise Ratio
S-NSSAI	Single NSSAI
SON	Self-Organizing Networks
SR	Scheduling Request
SRB	Signaling Radio Bearer
SRVCC	Single Radio Voice Call Continuity
SS	Synchronization Signal
SSB	SS Block
SSC	Session and Service Continuity Mode
SSH	Secure Shell
SSP	Smart Secure Platform
SSS	Secondary Synchronization Signal Standalone
S-TMSI	S-Temporary Mobile Subscriber Identity
SUCI	Subscription Concealed Identifier
SUL	Supplementary / Supplemental Uplink
SUPI	Subscription Permanent Identifier
TA	Timing Advance
TAU	Tracking Area Updating
TBC	To Be Clarified
TBS	Transport Block Size
TCH	Traffic Channel
TCP	Transmission Control Protocol
TDD	Time Division Duplex
TDF	Traffic Detection Function
TDMA	Time Division Multiple Access
TEID	Tunnel Endpoint ID
TLS	Transport Layer Security
TM	Transparent Mode
TMSI	Temporary Mobile Subscriber Identity
TN	Timeslot Number
TPM	Trusted Platform Module
TR	Technical Report
TRP	Total Radiated Power
TRP	Transmission/Reception Point
TS	Technical Specification
TSC	Training Sequence Code
TSDSI	Telecommunications Standards Development Society India
TSG	Technical Specification Group
TTA	Telecommunications Technology Association
TTC	Telecommunication Technology Committee
TTI	Transmission Time Interval
TWAG	Trusted Wireless Lan Access Gateway
TX	Transmitter Exchange
TXRU	Transmitter Receiver Unit

UCI	Uplink Control Information
UDM	Unified Data Management
UDR	Unified Data Repository
UDSF	Unstructured Data Storage Function
UE	User Equipment
UFMC	Universally Filtered Multicarrier
UICC	Universal Integrated Circuit Card
UL	Uplink (UE to Network)
UL CL	Uplink Classifier
UM	Unacknowledged Mode
Uma	Urban Macro
Umbi	Urban Micro
UDP	User Datagram Protocol
UMTS	Universal Mobile Telecommunications System
UP	User Plane
UPF	User Plane Function
URI	Universal Resource Identifier
URLLC	Ultra-Reliable Low Latency Communication
URN	Universal Resource Name
USF	Uplink State Flag
USIM	Universal Subscriber Identity Module
UTRAN	Universal Terrestrial Radio Access Network (3G RAN)
V2I	Vehicle-to-Infrastructure
V2N	Vehicle-to-Network
V2V	Vehicle-to-Vehicle
V2X	Vehicle-to-X, Vehicle to Everything
VAMOS	Voice services over Adaptive Multi-user channels on One Slot
VCC	Voice Call Continuity
VID	VLAN Identifier
VLAN	Virtual Local Area Network
VNF	Virtual Network Function
VoIP	Voice over IP
VoLTE	Voice over LTE
VoNR	Voice over New Radio
VPLMN	Visited PLMN
VPN	Virtual Private Network
VR	Virtual Reality
WAP	Wireless Application Protocol
WB-E-UTRAN	Wide Band E-UTRAN
WCDMA	Wideband Code Division Multiple Access
WG	Working Group
Wi-Fi	Wireless Fidelity
WLAN	Wireless Local Area Network
WLCP	Wireless Local Area Network Control Plane Protocol
WOLA	Windowed Overlap-and-Add
WRC	World Radio Conference
ZSM	Zero touch network and Service Management

1

Drivers and Motivation for 5G

Betsy Covell[1] *and Rainer Liebhart*[2]

[1] *Nokia, Naperville, USA*
[2] *Nokia, Munich, Germany*

1.1 Drivers for 5G

Main drivers for the evolution of mobile networks in the past were mobile voice (2G/3G/Voice over Long Term Evolution [VoLTE]), messaging (Short Messaging Service [SMS], WhatsApp) and Internet access (Wideband Code Division Multiple Access [WCDMA], High Speed Packet Access [HSPA], Long Term Evolution [LTE]) whenever and wherever needed. Focus was on end consumers equipped with traditional handsets or smartphones.

Consumer demand continues to be insatiable with an ever growing appetite for the bandwidth that is needed for 4K and 8K video streaming, augmented reality (AR) and virtual reality (VR), among other use cases. On the same token, operators want the network to be "better, faster and cheaper" without compromising any of these three elements.

The biggest difference between 5G and previous "Gs" is the diversity of applications that 5G networks need to support. Objects ranging from cars and factory machines, appliances to watches and apparel, will learn to organize themselves to fulfill our needs by automatically adapting to our behavior, environment or business processes. New use cases will arise, many not yet conceived, creating novel business models. 5G connectivity will impact the following areas:

- *Real world mobility.* The way we travel and experience our environment;
- *Virtual mobility.* The way we can control remote environments;
- *High performance infrastructure.* The way the infrastructure supports us;
- *4th Industrial Revolution.* The way we produce and provide goods.

We already have indicators about these long-term trends and disruptions and they are not only driven by the Internet and the telecommunication industry but by a multitude of different industries. 5G will be the platform enabling growth in many of these industries; the IT, car, entertainment, agriculture, tourism and manufacturing industries. 5G will connect the factory of the future and help to create a fully automated and flexible production system. It will also be the enabler of a superefficient infrastructure that saves resources.

5G for the Connected World, First Edition.
Edited by Devaki Chandramouli, Rainer Liebhart and Juho Pirskanen.

Smartphones are becoming more and more a commodity which means that consumers will differentiate themselves increasingly with new gadgets such as VR devices, connected cars and devices for connected health.

With the decline or at least flattened Average Revenue Per User/Average Revenue Per Device (ARPU/ARPD) in many markets worldwide (typical ARPU is around $30 per month, while ARPD is not more than $3 per month) the telecom industry is more and more considering new revenue streams besides traditional business models. This is where we see new business drivers and requirements, from vertical industries, the Internet of Things (IoT) and the digital society. In the past, the main driver for the mobile industry was connecting people, in the future it is about the "connected world" meaning connecting everyone and everything at any time. This will create new and additional revenue streams for Mobile Network Operators (MNOs). Connecting "things" like cars (e.g. 125 million vehicles to be connected by 2022), robots, meters, medical machines bring new challenges to the next generation of mobile networks. To name only a some of them:

- Ubiquitous, pervasively ultra-fast connectivity;
- Resilient and secure networks;
- Massive radio resources and ultra-dense networks; and
- Instantaneous connectivity.

Vertical industries and applications do have very diverse requirements with regards to throughput, latency, reliability, number of connections, security and revenue (see Figure 1.1).

This requires a highly flexible architecture of the new generation of mobile networks, on radio and core network side, as well as at the transport. Flexibility also includes a high degree of automation in deploying and maintaining networks, parts of a network or single resources (e.g. network slices). Flexible architecture is achieved by different means, from flexible frame structures and intelligent radio schedulers to edge computing, slicing, Software Defined Networking (SDN) and fully automated Orchestration capabilities. We will explain the most important technical solutions throughout this book in the chapters to follow (see also Section 1.6).

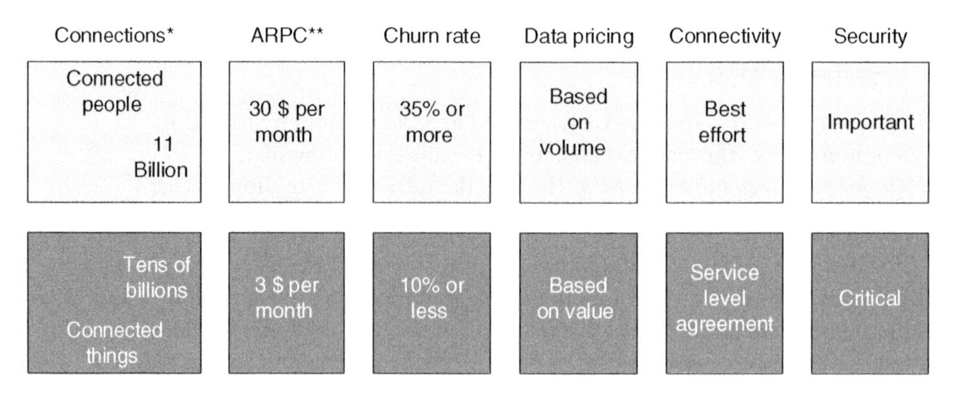

Figure 1.1 Market characteristics (people and things).

What are the concrete use cases that will drive the market? Nobody knows the answer today, but we can have a look at the industries that are forced into digital transformation by changes in their markets. There is an extensive list of possible use cases for various industry sectors where 5G may play a crucial role to interconnect people and things, to name only a few of them:

- *Healthcare*. Bioelectronic medicine, Personal health systems, telemedicine, connected ambulance including Augmented/Virtual Reality (AR/VR) applications.
- *Manufacturing*. Remote/motion control and monitoring of devices like robots, machine-to-machine communication, AR/VR in design (e.g. for designing machines, houses, etc.).
- *Entertainment*. Immersive experience, stadium experience, cooperative media production (e.g. production of songs, movies from various locations).
- *Automotive*. Platooning, infotainment, autonomous vehicles, high-definition (HD) map updates, remote maintenance and SW updates.
- *Energy*. Grid control and monitoring, connecting windfarms, smart electric vehicle charging.
- *Public transport*. Infotainment, train/bus operations, platooning for buses.
- *Agriculture*. Connecting sensors and farming machines, drone control.
- *Public Safety*. Threat detection, facial recognition, drones.
- *Fixed Wireless Access* (FWA). Replacing fixed access technologies like fiber at the last mile by wireless access.
- *Megacities*. Applications around mission control for public safety, video surveillance, connected mobility across all means of transport including public parking and traffic steering, and environment/pollution monitoring.

Among the listed use cases motion control appears the most challenging and demanding one. Such a system is responsible for controlling, moving and rotating parts of machines in a well-defined manner. Such a use case has very stringent requirements in terms of low latency, reliability, and determinism. Augmented Reality requires high data rates for transmitting (high-definition) video streams from and to a device. Process automation is between the two, and focuses on monitoring and controlling chemical, biological or other processes in a plant, involving both a wide range of different sensors (e.g. for measuring temperatures, pressures, flows, etc.) and actuators (e.g. valves or heaters).

1.2 ITU-R and IMT 2020 Vision

The International Telecommunication Union – Radio Sector (ITU-R) manages the international radio-frequency spectrum and satellite orbit resources. In September 2015 ITU-R published its recommendation M.2083 [1] constituting a vision for the International Mobile Telecommunications (IMT) 2020 and beyond. In this document ITU-R describes user and application trends, growth in traffic, technological trends and spectrum implications, and provides guidelines on the framework and the capabilities for IMT 2020. The following trends were identified, leading in a later phase to concrete requirements for the new 5G system defined by 3rd Generation Partnership Project (3GPP):

- Support for very low latency and high reliability communication;
- Support of high user density (in one area, one cell, etc.);
- Support of high accurate positioning methods;
- Support of the IoT;
- Support of high-quality communication at high speeds; and
- Support of enhanced multimedia services and converged applications.

Regarding the growth of traffic rates, it was estimated (based on various available forecasts) that the global IMT traffic will grow in the range of 10–100 times from 2020 to 2030 with an increasing asymmetry between downlink (DL) and uplink (UL) data rates.

ITU-R is not defining a new radio system itself but has listed some technology trends for both the radio and network side they deemed necessary to fulfill the new requirements and cope with new application trends and increased traffic rates. Technologies enhancing the radio interface capabilities mentioned in M.2083 are, e.g. new waveforms, modulation and coding techniques, as well as multiple access schemes. Spectrum efficiency enhancements and higher data rates can be achieved by techniques such as 3D-beamforming, an active antenna system, and massive Multiple-Input-Multiple-Output (mMIMO). Dual connectivity and dynamic Time Division Duplex (TDD) can enhance spectrum flexibility. On the network side features like SDN, Network Function Virtualization (NFV), Cloud Radio Access Network (C-RAN) and Self-Organizing Networks (SON) are mentioned.

One key item to allow for (much) higher data rates in future is the utilization of new spectrum, especially in higher frequency bands (above 6 GHz). ITU-R Report M.2376 [2] provides information on the technical feasibility of IMT in the frequencies between 6 and 100 GHz. The report includes measurement data on propagation in this frequency range in several different environments. Both line-of-sight and non-line-of-sight measurement results for stationary and mobile cases as well as outdoor-to-indoor results are included. Thus, ITU-R Report M.2083 is highlighting the need for spectrum harmonization and contiguous and wider spectrum bandwidth (above 6 GHz).

ITU-R is highlighting mainly three usage scenarios (rather use cases): enhanced mobile broadband (eMBB), ultra-reliable and low latency communication (sometimes called critical-machine type communication [cMTC]) and massive machine-type communication (mMTC). These three usage scenarios are the fundamental basis of 5G system specification. Figure 1.2 gives an illustrative overview of the three usage scenarios and how they relate to each other.

When it comes to concrete capabilities for IMT 2020 and beyond, Report M.2083 lists the following eight key items:

- *Peak data rate.* Maximum achievable data rate under ideal conditions per user.
- *User experienced data rate.* Achievable data rate per user.
- *Latency.* The contribution by the radio network to the time from when the source sends a packet to when the destination receives it (in milliseconds [ms]).
- Mobility maximum speed at which a defined Quality of Service (QoS) and seamless transfer between radio nodes can be achieved (in km/h).
- *Connection density.* Total number of connected devices per unit area.
- Energy efficiency
- *Spectrum efficiency.* Average data throughput per unit of spectrum resource and per cell.
- *Area traffic capacity.* Total traffic throughput served per geographic area.

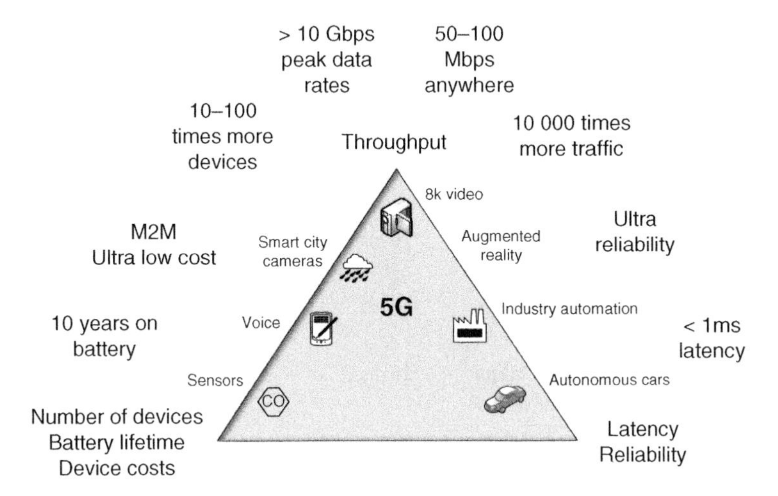

Figure 1.2 IMT 2020 usage scenarios.

Enhanced user experience will be realized by increased peak and user data rate, spectrum efficiency, reduced latency and enhanced mobility support.

1.3 NGMN (Next Generation Mobile Networks)

The Next Generation Mobile Networks (NGMN) Alliance is a mobile telecommunications association of mobile operators, vendors, manufacturers and research institutes. It was founded by major mobile operators in 2006 as an open forum to evaluate candidate technologies to develop a common view of solutions for the evolution of wireless networks. NGMN aims to establish clear functionality and performance targets as well as fundamental requirements for deployment scenarios and network operations. In February 2015 NGMN published a 5G White Paper [3] that contains requirements regarding user experience, system performance, device capabilities, new business models and network operation and deployment. The NGMN White Paper starts with a short introduction on use cases, new business models and value creation in the era of 5G, and is listing afterwards detailed requirements from operator perspective on user experience (data rates, latency, mobility), system performance (connection and traffic density, spectrum efficiency), devices (multi-band support, power and signaling efficiency), enhanced services (location, security, reliability), new business models (connectivity provider, XaaS, network sharing) and network deployment (cost and energy efficiency). New business models, technology and architecture options as well as spectrum and Intellectual Property Rights (IPR) aspects are also considered in the paper.

NGMN is categorizing 5G use cases in several sub-sets, e.g. broadband access in urban areas or indoor, 50+ Mbps anywhere, mobile broadband in vehicles, massive low-cost/long-range/low-power machine type communication, ultra-low latency and high reliability, broadcast like services.

For each of these use case categories NGMN has listed requirements for the user experience. User specific data rates of 1 Gbps are mentioned for special environments, and 10 millisecond latency in general while 1 millisecond must be achievable for selected use cases. Detailed requirements can be found in Table 1.1.

Table 1.1 User experience requirements.

Use case category	User experienced data rate	E2E latency	Mobility
Broadband access in dense areas	DL: 300 Mbps UL: 50 Mbps	10 ms	On demand, 0–100 km h^{-1}
Indoor ultra-high broadband access	DL: 1 Gbps UL: 500 Mbps	10 ms	Pedestrian
Broadband access in a crowd	DL: 25 Mbps UL: 50 Mbps	10 ms	Pedestrian
50+ Mbps everywhere	DL: 50 Mbps UL: 25 Mbps	10 ms	0–120 km h^{-1}
Ultra-low-cost broadband access for low ARPU areas	DL: 10 Mbps UL: 10 Mbps	50 ms	On demand: 0–50 km h^{-1}
Mobile broadband in vehicles (cars, trains)	DL: 50 Mbps UL: 25 Mbps	10 ms	on demand, up to 500 km h^{-1}
Airplane connectivity	DL: 15 Mbps per user UL: 7.5 Mbps per user	10 ms	Up to 1 000 km h^{-1}
Broadband Machine Type Communication (MTC)	See the requirements for the Broadband access in dense areas and 50+ Mbps everywhere categories		
Ultra-low latency	DL: 50 Mbps UL: 25 Mbps	<1 ms	Pedestrian
Resilience and traffic surge	DL: 0.1–1 Mbps UL: 0.1–1 Mbps	Regular communication: not critical	0–120 km h^{-1}
Ultra-high reliability and ultra-low latency	DL: From 50 kbps to 10 Mbps UL: From a few bps to 10 Mbps	1 ms	On demand: 0–500 km h^{-1}
Ultra-high availability and reliability	DL: 10 Mbps UL: 10 Mbps	10 ms	On demand: 0–500 km h^{-1}
Broadcast like services	DL: Up to 200 Mbps UL: Modest (e.g. 500 kbps)	<100 ms	On demand: 0–500 km h^{-1}

Regarding overall system performance requirements, use case specific requirements for connection and traffic density are provided (see Table 1.2). In general, it is assumed that 5G allows for several hundred thousand simultaneous active connections per square kilometer and data rates of several tens of Mbps for tens of thousands of users in hotspot areas. 1 Gbps shall be offered simultaneously to some tens of users in the same limited area. Spectral efficiency should be significantly better compared to 4G. 5G should allow

Table 1.2 System performance requirements.

Use case category	Connection density	Traffic density
Broadband access in dense areas	200–2 500 km^{-2}	DL: 750 Gbps km^{-2} UL: 125 Gbps km^{-2}
Indoor ultra-high broadband access	75 000 km^{-2} (i.e. 75/1 000 m^2 office)	DL: 15 Tbps km^{-2} (15 Gbps/1 000 m^2) UL: 2 Tbps km^{-2} (2 Gbps/1 000 m^2)
Broadband access in a crowd	150 000 km^{-2} (30 000/stadium)	DL: 3.75 Tbps km^{-2} (DL: 0.75 Tbps/stadium) UL: 7.5 Tbps km^{-2} (1.5 Tbps/stadium)
50+ Mbps everywhere	400 km^{-2} in suburban 100 km^{-2} in rural	DL: 20 Gbps km^{-2} in suburban UL: 10 Gbps km^{-2} in suburban DL: 5 Gbps km^{-2} in rural UL: 2.5 Gbps km^{-2} in rural
Ultra-low-cost broadband access for low ARPU areas	16 km^{-2}	16 Mbps km^{-2}
Mobile broadband in vehicles (cars, trains)	2 000 km^{-2} (500 active users per train × 4 trains, or 1 active user per car × 2 000 cars)	DL: 100 Gbps km^{-2} (25 Gbps per train, 50 Mbps per car) UL: 50 Gbps km^{-2} (12.5 Gbps per train, 25 Mbps per car)
Airplanes connectivity	80 per plane 60 airplanes per 18 000 km^2	DL: 1.2 Gbps/plane UL: 600 Mbps/plane
Massive low-cost/long-range/low-power MTC	Up to 200 000 km^{-2}	Non-critical
Broadband MTC	See the requirements for the Broadband access in dense areas and 50+ Mbps everywhere categories	
Ultra-low latency	Not critical	Potentially high
Resilience and traffic surge	10 000 km^{-2}	Potentially high
Ultra-high reliability and Ultra-low latency	Not critical	Potentially high
Ultra-high availability and reliability	Not critical	Potentially high
Broadcast like services	Not relevant	Not relevant

higher data rates to be achieved in rural areas based on the current grid of macro sites (depending on the frequency bands used).

Some general statements are made regarding expected device capabilities such as multi-band/multi-mode support (e.g. simultaneous support of TDD and Frequency Division Duplex [FDD] operation), support of LTE and 5G radio technology and the high degree of programmability of the device. However, this does not lead to concrete requirements for the device for modem manufacturers, but can be seen as high-level recommendations. More or the less the same applies for statements on subscriber security, privacy and network security, e.g. going beyond radio security to also consider end-to-end and higher-layer security solutions. With respect to network reliability and availability 5G should enable 99.999% network availability, including robustness against climatic events and guaranteed services at low energy consumption for critical infrastructures and high reliability rates of 99.999% or higher, for ultra-high reliability and ultra-low latency use cases. By design the 5G system should also allow for cost and energy efficient deployments and for enhanced flexibility and scalability, e.g. through decoupling Core and Radio Access Network (RAN) network domains (access agnostic core).

1.4 5GPPP (5G Public-Private Partnership)

The 5G Public-Private Partnership (5GPPP) is one of or even the world's biggest 5G research program. It is a joint initiative between the European Commission (EC) and the European Information and Communication Technology (ICT) industry and aims to deliver 5G solutions, architectures, technologies and standards. 5GPPP was initiated by the EU Commission and industry manufacturers, telecommunications operators, service providers, small and medium enterprises (SME) and research institutes.

Within the 5GPPP, the 5G Infrastructure Association (5GIA) represents the private side and the European Commission, the public side.

5GPPP is working on the document "5GPPP use cases and performance evaluation," with version 1 published in April 2016, version 2 is work in progress. This is a living document, i.e. it is constantly updated. The document provides an overview of use cases and models. It covers 5G scenarios, definitions of key performance indicators (KPIs) and models (e.g. of wireless channel, traffic or user's mobility), as well as corresponding assessment results. Developed use case families are mapped to corresponding business cases identified in vertical industries. Additionally, performance evaluation approaches are compared with the latest version of performance evaluation framework proposed in 3GPP.

5GPPP work is grouped round the three well-known 5G services extreme mobile broadband (xMBB), ultra-reliable machine-type communication (uMTC), and massive machine-type communication (mMTC).

5GPPP defines the following KPI values for clustering the different use cases:

- Device density:
  - *High*: $\geq$10 000 devices per km^2
  - *Medium*: 1000–10 000 devices per km^2
  - *Low*: <1000 devices per km^2
- Mobility:
  - *No*: static users
  - *Low*: pedestrians (0–3 km h^{-1})
  - *Medium*: slow moving vehicles (3–50 km h^{-1})
  - *High*: fast moving vehicles, e.g. cars and trains (>50 km h^{-1})

- Infrastructure:
  - *Limited*: no infrastructure available or only macro cell coverage
  - *Medium density*: Small number of small cells
  - *Highly available infrastructure*: Large number of small cells available
- Traffic type:
  - Continuous
  - Bursty
  - Event driven
  - Periodic
  - All types
- User data rate:
  - *Very high data rate*: ≥1 Gbps
  - *High*: 100 Mbps–1 Gbps
  - *Medium*: 50–100 Mbps
  - *Low*: <50 Mbps
- Latency:
  - *High*: >50 milliseconds
  - *Medium*: 10–50 milliseconds
  - *Low*: 1–10 milliseconds
- Reliability:
  - *Low*: <95%
  - *Medium*: 95–99%
  - *High*: >99%
- Availability (as related to coverage):
  - *Low*: <95%
  - *Medium*: 95–99%
  - *High*: >99%
- 5G service type, comprising of:
  - xMBB, where the mobile broadband is the key service requirement of the use case.
  - uMTC, where the reliability is the key service requirement of the use case.
  - mMTC, where the massive connectivity is the key service requirement of the use case.

In addition to these KPIs, localization and security requirements are important KPIs for vertical industries.

1.5 Requirements for Support of Known and New Services

5G use cases have been developed by a wide range of sources, including both the traditional telecommunications organizations such as Global System for Mobile Communications Association (GSMA), NGMN, and ITU, and the vertical industries such as automobile, gaming, and factory automation, that are considering how 5G can benefit them. In addition to the ITU-R, NGMN and 5GPPP use cases already discussed in this chapter, several other standards and industry organizations have provided significant input to the 5G vision being developed in 3GPP. These include the China IMT 2020 Promotion Group, the German Electrical and Electronic Manufacturers Association,

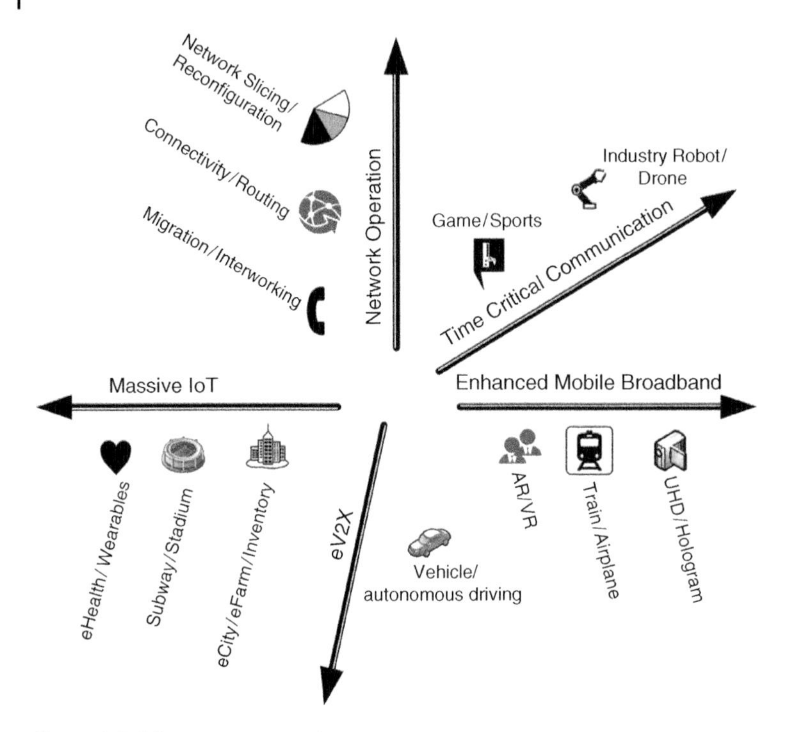

Figure 1.3 5G use case categories.

the METIS project, the ARIB 2020 and Beyond Ad Hoc Group. The 5G for Connected Industries and Automation (5G-ACIA) and the 5G Automotive Association (5GAA) are focusing on 5G use cases and their requirements from the perspective of specific verticals. 5G-ACIA serves as the central forum for addressing, discussing, and evaluating relevant technical, regulatory, and business aspects with respect to 5G for the industrial domain (see [4]). 5GAA on the other hand is considering requirements, architectures and solutions enabling 5G to become the ultimate platform for Cooperative Intelligent Transportation Systems (C-ITS) and the provision of Vehicle-to-X (V2X) services. 5G will be able to better carry mission-critical communications for safer driving and further support enhanced V2X communications and connected mobility solutions. Several white papers can be found at the 5GAA Internet page www.5gaa.org.

In general, 5G use cases can be grouped into five main categories (for details see [5]): massive IoT, time critical communication, eMBB, network operations, and enhanced V2X, as shown in Figure 1.3.

1.5.1 Massive IoT

As more and more connected devices, from home appliances and medical monitors to industrial robots and vehicles, are developed and deployed, the demand for efficient, reliable, secure communications between and with these devices has been increasing. While there are many existing technologies to support communication for IoT devices, such as Bluetooth™, Wi-Fi™, and LTE support for machine-type communication (MTC) and NB-IoT, 3GPP 5G technology is specifically designed to support the various IoT use cases in an efficient, reliable, and secure manner.

Within the massive IoT category, there is no single set of criteria that will meet all IoT needs in the future. A range of use cases cover a diverse set of sometimes conflicting requirements. For example, sensors entail a potentially enormous number of stationary, localized, devices sending infrequent small data bursts. Industrial robot controls require very high reliability and ultra-low latency within a constrained physical space. Wearables impose requirements for nomadic connectivity that may be over very wide areas (e.g. global roaming) and may occur at varying speeds, from pedestrian to high speed trains. Wearables also have a range of service requirements from voice and small data to streaming video. A 5G system therefore must be capable of being tailored to meet each of these diverse needs and use cases efficiently and reliably.

Security requirements are more common across all these cases. No matter the type of device or type of IoT service, the user expects the communications to be secure from access by unauthorized applications or users. 5G systems provide the high-level of security expected in each case, including support for secure system access, data integrity protection, and confidentiality.

1.5.2 Time Critical Communication

Many of the time critical communication use cases are based on 3rd party use of telecommunications systems. These 3rd parties may include industries such as utilities, factories, and public safety authorities. Utility use cases include support for smart grids, which require constant monitoring and immediate action to be taken when an outage or power surge occurs. Utilities may also use 5G drones to monitor and perform routine maintenance on remote equipment. In these cases, both drone controls as well as data transmitted from the drone rely on the time critical communications provided by 5G. Factories are considering 5G as an enhancement to existing wired robotic control technologies. Eliminating wires in a robotic factory improves safety as well as increasing flexibility in the factory floor configuration, but it must not come with any loss of quality in terms of speed and accuracy in controlling a robot. Public safety organizations around the globe are already migrating to LTE-based systems, while looking ahead to the additional enhancements supported in 5G for better video, faster data, and more reliable communications particularly when out of range of a public network.

Other time critical communications use cases address specific technology such as augmented reality, virtual reality, and tactile internet. These technologies are gaining ground in a variety of industries including health care, gaming, education, and real estate. Very precise location information and a highly reliable communications path with low latency are necessary to use these technologies without causing the user to feel discomfort in the process.

3GPP's initial analysis of the various industry needs resulted in the KPIs for each use case shown in Table 1.3 (see [6]).

At the time of writing, it should be noted that the KPIs in Table 1.3 are undergoing refinement as additional use cases from vertical industries are being explored in 3GPP. Much of the original input was based on numerous studies and white papers predating the 5G standards. Input from industrial automation verticals such as Siemens, ABB and

Table 1.3 Performance requirements for time critical communication.

Scenario	Max allowed end-to-end latency[a] (ms)		Survival time (ms)	Communication service availability[b] (%)	Reliability[b] (%)
Discrete automation	10		0	99.99	99.99
Process automation – remote control	60		100	99.999	99.999
Process automation – monitoring	60		100	99.9	99.9
Electricity distribution – medium voltage	40		25	99.9	99.9
Electricity distribution – high voltage[c]	5		10	99.999	99.999
Intelligent transport systems – infrastructure backhaul	30		100	99.999	99.999
Scenario	**User experienced data rate**	**Payload size[d]**	**Traffic density[e]**	**Connection density[f]**	**Service area dimension[g]**
Discrete automation	10 Mbps	Small to big	1 Tbps km^{-2}	100 000 km^{-2}	1 000 × 1 000 × 30 m
Process automation – remote control	1 Mbps up to 100 Mbps	Small to big	100 Gbps km^{-2}	1 000 km^{-2}	300 × 300 × 50 m
Process automation – monitoring	1 Mbps	Small	10 Gbps km^{-2}	10 000 km^{-2}	300 × 300 × 50
Electricity distribution – medium voltage	10 Mbps	Small to big	10 Gbps km^{-2}	1 000 km^{-2}	100 km along power line
Electricity distribution – high voltage[c]	10 Mbps	Small	100 Gbps km^{-2}	1 000 km^{-2} [h]	200 km along power line
Intelligent transport systems – infrastructure backhaul	10 Mbps	Small to big	10 Gbps km^{-2}	1 000 km^{-2}	2 km along a road

a) This is the maximum end-to-end latency allowed for the 5G system to deliver the service in the case the end-to-end latency is completely allocated to the 5G system from the UE to the interface to the Data Network.
b) Communication service availability relates to the service interfaces, reliability relates to a given system entity. One or more retransmissions of network layer packets may take place to satisfy the reliability requirements.
c) Currently realized via wired communication lines.
d) Small: payload typically ≤256 bytes.
e) Based on the assumption that all connected applications within the service volume require the user experienced data rate.
f) Under the assumption of 100% 5G penetration.
g) Estimates of maximum dimensions; the last figure is the vertical dimension.
h) In dense urban areas.
i) All the values in this table are targeted values and not strict requirements.

Bosch are providing a more detailed analysis of the KPIs needed in a factory automation setting, leading to more accurate KPIs as well as a closer look at how different deployment configurations can be used to meet the KPIs.

1.5.3 Enhanced Mobile Broadband (eMBB)

5G eMBB addresses the increasing use of data by providing for higher data rates, increased traffic density, and high-speed mobility scenarios such as use on a train or airplane, all with an improved Quality of Experience (QoE) for the end user. Use cases for higher data rates identify KPIs for peak, experienced, uplink and downlink data rates under varying traffic (e.g. urban, rural) and mobility (e.g. stationary, pedestrian, high speed train) conditions. Increased traffic density use cases identify KPIs considering both large volumes of traffic in a localized area as well as varying traffic volume in an area with a high connection density. Coverage areas are also addressed in these use cases, identifying KPIs considering different usage conditions such as indoor/outdoor, wide area/local area, and speed at which the User Equipment (UE) is moving (e.g. pedestrian vs automobile). Other use cases address KPIs considering the increased speed at which mobile service is being used, from stationary to high speed trains and automobiles. While technology enhancements to the 5G radio and core network facilitate meeting these KPIs, additional factors are also considered, including use of small cells and femto cells, and optimized deployment configurations that will also be needed to meet these KPIs. Table 1.4 provides a view of the KPIs identified for these eMBB use cases.

1.5.4 Enhanced Vehicular Communications

While many of the time critical communication requirements and KPIs apply for vehicular communications, there are also other specific use cases related to vehicles addressed in 5G. These include enhancements beyond the V2X support provided by 4G systems specifically for platooning, advanced driving, extended sensors and remote driving (for more details see [7]). The relative level of automation is factored into the use cases for 5G V2X. These include the following levels 0–5:

- *Level 0 (no automation).* Automated system issues warnings and may intervene but has no sustained vehicle control.
- *Level 1 (drive assistance).* The driver and the automated system share control of the vehicle. Examples are Adaptive Cruise Control (ACC), where the driver controls steering and the automated system controls speed, and the parking assistance where steering is automated while speed control is manual. The driver must be ready to retake full control at any time.
- *Level 2 (partial automation).* The automated system takes full control of the vehicle (accelerating, braking, and steering). The driver must monitor the driving and be prepared to intervene immediately at any time. Contact between hand and wheel is often mandatory during Level 2 driving to confirm that the driver is ready to intervene.
- *Level 3 (conditional automation).* The driver can safely turn attention away from the driving tasks. The vehicle will handle situations that call for an immediate response, like emergency braking. The driver must still be prepared to intervene within some

Table 1.4 Performance requirements for high data rate and traffic density scenarios.

Scenario	Experienced data rate (DL)	Experienced data rate (UL)	Area traffic capacity (DL)	Area traffic capacity (UL)	Overall user density	UE speed	Coverage
Urban macro	50 Mbps	25 Mbps	100 Gbps km^{-2}	50 Gbps km^{-2}	10 000 km^{-2}	Pedestrians and users in vehicles (up to 120 km h^{-1})	Full network
Rural macro	50 Mbps	25 Mbps	1 Gbps km^{-2}	500 Mbps km^{-2}	100 km^{-2}	Pedestrians and users in vehicles (up to 120 km h^{-1})	Full network
Indoor hotspot	1 Gbps	500 Mbps	15 Tbps km^{-2}	2 Tbps km^{-2}	250 000 km^{-2}	Pedestrians	Office and residential
Broadband access in a crowd	25 Mbps	50 Mbps	3, 75 Tbps km^{-2}	7, 5 Tbps km^{-2}	500 000 km^{-2}	Pedestrians	Confined area
Dense urban	300 Mbps	50 Mbps	750 Gbps km^{-2}	125 Gbps km^{-2}	25 000 km^{-2}	Pedestrians and users in vehicles (up to 60 km h^{-1})	Downtown
Broadcast-like services	Maximum 200 Mbps (per TV channel)	N/A or modest (e.g. 500 kbps per user)	N/A	N/A	15 TV channels of 20 Mbps on one carrier	Stationary users, pedestrians and users in vehicles (up to 500 km h^{-1})	Full network
High-speed train	50 Mbps	25 Mbps	15 Gbps/train	7,5 Gbps/train	1 000/train	Users in trains (up to 500 km h^{-1})	Along railways
High-speed vehicle	50 Mbps	25 Mbps	100 Gbps km^{-2}	50 Gbps km^{-2}	4 000 km^{-2}	Users in vehicles (up to 250 km h^{-1})	Along roads
Airplanes	15 Mbps	7, 5 Mbps	1, 2 Gbps/plane	600 Mbps/plane	400/plane	Users in airplanes (up to 1 000 km h^{-1})	

limited time, specified by the manufacturer, when called upon by the vehicle. Some manufacturers have announced that this time limit is around 10 seconds, others talking even about 30 seconds or more. Some modern cars already fulfill Level 3 requirements, at least these cars can take full driving control at speeds up to, e.g. 60 km h^{-1} and/or this level only works on highways.
- *Level 4 (high automation).* This is like Level 3 but no driver attention is required, i.e. the driver can even sleep. Self-driving is supported only in limited spatial areas or under exceptional circumstances such as traffic jams. Otherwise, the vehicle must be able to safely abort the trip, i.e. park the car, if the driver does not retake control.
- *Level 5 (full automation).* No human intervention is required anymore.

A 5G enhancement for platooning is also of interest to the shipping industry. Enabling several trucks traveling together to be managed as a group provides many safety and efficiency enhancements for a trucking company. Specific KPIs for platooning are shown in Table 1.5.

Increased support for semi- to fully automated driving also has many safety aspects related to collision prevention and traffic flow efficiency. KPIs for automated driving are shown in Table 1.6.

Extended sensor use cases bring increased environmental awareness into the mix, providing vehicles additional data on the surroundings, such as pedestrians, cyclists, animals. Table 1.7 shows the KPIs for extended sensors.

And finally, remote driving use cases bring new opportunities for safer traversal through dangerous terrain. Table 1.8 shows the KPIs for remote driving.

1.5.5 Network Operations

The need for increased resource efficiency is a common thread running throughout the use cases for network operation enhancements. A primary driver for 5G is the ability to offer communications services in a manner that minimizes network resource usage, network power consumption, and device power consumption. Various techniques have been developed to provide these efficiencies.

Network slicing provides a significant advantage in the ability to support widely varying use cases. A network slice can be designed with the resources needed to meet a specific set of requirements (e.g. stationary sensors sending infrequent small data), while another network slice in the same network can be designed to address a separate set of requirements (e.g. high quality streaming video). Network slicing allows a network operator to deploy network resources in configurations that maximize resource efficiencies.

New mobility management techniques provide efficient support for diverse devices that might be stationary, geographically limited (e.g. confined to a factory), nomadic, or capable of high-speed travel. Providing specific support for each of these classes of devices allows for more efficient resource usage, particularly for the less mobile devices.

Content delivery is made more efficient in 5G using in-network caching and service hosting environments that can be located close to the end user. This minimizes the resources needed to provide content to the end user, as well as enhancing the user experience by reducing the transmission delay between the content host and end user.

Beyond resource efficiency, other 5G network operations open doors to new business opportunities. Many additional network capabilities are exposed through Application

Table 1.5 Performance requirements for vehicles platooning.

Communication scenario		Payload (Bytes)	Tx rate (messages per second)	E2E latency (ms)	Reliability (%)	Data rate (Mbps)	Min range (m)
Scenario	Degree						
Cooperative driving for vehicle platooning	Lowest degree of automation	300–400	30	25	90		
Information exchange between a group of UEs supporting V2X application.	Low degree of automation	6500	50	20			350
	Highest degree of automation	50–1200	30	10	99.99		80
	High degree of automation			20		65	180
Reporting needed for platooning between UEs supporting V2X application and between a UE supporting V2X application and RSU.	N/A	50–1200	2	500			
Information sharing for platooning between UE supporting V2X application and RSU.	Lower degree of automation	6000	50	20			350
	Higher degree of automation			20		50	180

Table 1.6 Performance requirements for advanced driving.

Communication scenario description		Payload (Bytes)	Tx rate (messages per second)	E2E latency (ms)	Reliability (%)	Data rate (Mbps)	Min range (m)
Scenario	Degree						
Cooperative collision avoidance between UEs supporting V2X applications.		2 000	100	10	99.99	10	
Information sharing for automated driving between UEs supporting V2X application.	Lower degree of automation	6 500	10	100			700
	Higher degree of automation			100		53	360
Information sharing for automated driving between UE supporting V2X application and RSU	Lower degree of automation	6 000	10	100			700
	Higher degree of automation			100		50	360
Emergency trajectory alignment between UEs supporting V2X application.		2 000		3	99.999	30	500
Intersection safety information between an RSU and UEs supporting V2X application.		UL: 450	UL: 50			UL: 0.25 DL: 50	
Cooperative lane change between UEs supporting V2X applications.	Lower degree of automation	300–400		25	90		
	Higher degree of automation	12 000		10	99.99		
Video sharing between a UE supporting V2X application and a V2X application server.						UL: 10	

Table 1.7 Performance requirements for extended sensors.

Communication scenario		Payload (Bytes)	Tx rate (messages per second)	E2E latency (ms)	Reliability (%)	Data rate (Mbps)	Min range (m)
Scenario	**Degree**						
Sensor information sharing between UEs supporting V2X application	Lower degree of automation	1600	10	100	99		1000
	Higher degree of automation			10	95	25	
				3	99.999	50	200
				10	99.99	25	500
				50	99	10	1000
				10	99.99	1000	50
Video sharing between UEs supporting V2X application	Lower degree of automation			50	90	10	100
	Higher degree of automation			10	99.99	700	200
				10	99.99	90	400

Table 1.8 Performance requirements for remote driving.

Communication scenario	Max end-to-end latency (ms)	Reliability (%)	Data rate (Mbps)
Information exchange between a UE supporting V2X application and a V2X Application Server	5	99.999	UL: 25 DL: 1

Programming Interfaces (APIs) to allow greater control and flexibility to 3rd parties. In 5G, these include support for 3rd party creation, use, and management of network slices, enhancements for 3rd party broadcast capabilities, use and management of 3rd party service hosting environments, and accessibility to required QoE for 3rd party applications. These new APIs allow network operators to consider new business opportunities when authorizing the levels of control and network visibility granted to 3rd parties.

Enhancing the end user experience is also considered in 5G network operations. Because 5G is intended to support flexible and variable network configurations, a 5G network may be configured to support any number of specialized market requirements. These can range from addressing markets where the need is to provide a minimum level of service in a very resource and power efficient manner, such as in remote rural areas with few users and limited or unreliable power sources to addressing markets where the need is to provide highly reliable service with very low latency, such as within the confines of an urban financial firm. Other markets, such as a suburban area, may be somewhere in between these two outliers, with a mix of end user requirements for transmission speed, throughput, and latency. The 5G network provides the network operator with the flexibility to tailor the QoE experience for end users in this environment as well through dynamic priority, QoS, and policy controls.

1.6 5G Use Cases

While we have already mentioned briefly some 5G use cases in Section 1.1 we will go into more detail in this chapter and explain use cases and the underlying business models (for more information see [8]). Different use cases require different technological solutions, e.g. enabling high throughput at high speeds, ultra-low latency and extreme high reliability. Although some of the use cases can also be covered by LTE today, future use cases requiring extreme data throughput (e.g. 8K video) or low latency and high reliability (e.g. some industry applications and AR/VR applications) and a combination of these, can only be fulfilled by 5G. And 5G is promising to be the future-proof technology for many or even all use cases we can think of today and the ones coming.

5G will most likely bring first benefits for MNOs and their customers in the following into three areas:

- *5G immersive and interactive experience.* 5G will create life-changing experiences for consumers in their homes and on public and private transportation systems.
- *5G live experience.* 5G will provide new experiences for people attending large events and meet very high demand at traffic hot spots.
- *5G industry experience.* 5G can become the communications standard of the fourth industrial revolution.

1.6.1 5G to the Home

Consumers living in households without fiber access can benefit from 5G bringing fiber-like speeds to their houses, they can join the ultra-broadband party including the world of Virtual and Augmented Reality.

5G presents an opportunity for MNOs to offer massive broadband access to homes in areas where conventional fiber-to-the-home is difficult or expensive to deploy ("5G for the last mile"). Avoiding the need for time-consuming and high-cost civil works to lay fiber, 5G delivers faster time to market and opens the home broadband market by enabling new entrants to compete against fixed line operator.

Analysis have shown that solutions using mmWave spectrum beyond 6 GHz (e.g. 28 GHz) allow each base station to serve tens of households. The 5G short-range FWA is expected to sustain 1 Gbps per household in the downlink (DL). To achieve such a high speed and longer ranges for 5G fixed wireless service, for example beyond urban and more densely packed suburban markets as well as rural applications, cmWave and mmWave radio technology must support large bandwidths and Multiple Input Multiple Output (MIMO) antenna beam forming techniques (see Figure 1.4).

The business modeling is based on an addressable market of, e.g. 100 000 households with 6% served by fiber and a MNO market share of 35% with a 30% take rate for the service. The discounted cash flow (year on year) is yielded by the difference between the MNO costs (capital and operational expenses) and revenue (present value of the money) from connected households (domestic units, people who live together along with non-relatives). With these assumptions in mind, the analysis shows that the MNO business case mostly depends on the number of households served per site, the site capital expenditure and the ARPU. Simulation results for period ranging from 2019 to 2028 show that the business case appears quite sensitive to ARPU, which needs to be kept above 40 Euros (premium) and number of households per site, which should be at least 30 for a positive business case, under the above assumptions. The price erosion is insignificant, hence the sooner the service is launched, the better it is for a successful

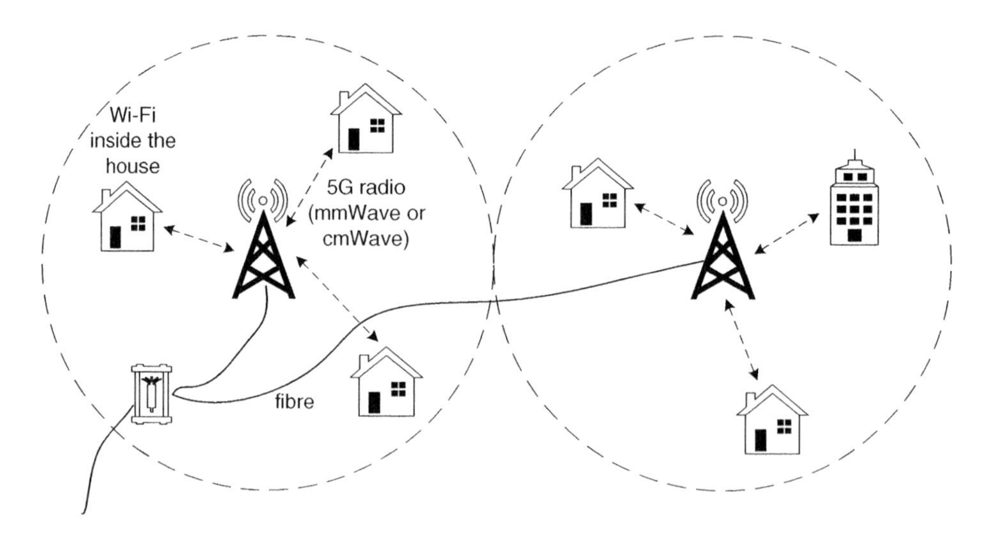

Figure 1.4 FWA using cmWave or mmWave 5G radio.

business. It should be noted that this business case is also considered by some MSOs (i.e. cable operators) for two main reasons: One reason is to leverage 5G for the last mile where fiber is not deployed yet. Another reason is to position cable technology as a backhaul for 5G.

1.6.2 In-Vehicle Infotainment

Public transport users with ultra-broadband connectivity can make more of their time. Watching streamed high-definition augmented reality or conducting business meetings via video calling, many passengers will welcome new experiences on the move enabled by 5G.

5G will enable MNOs to win revenue by delivering information and entertainment services to traveling subscribers, particularly in dense urban areas as depicted in Figure 1.5. Such services will include high-quality video streaming, augmented and virtual reality applications, online gaming and video calling. The network must be able to deliver services consistently across the area and with superior performance simultaneously to many users on public transport, traveling at relatively high speed.

High performance services could also be provided for users of non-5G devices by transmitting 5G to a vehicle and distributing bandwidth via Wi-Fi, for example on a train or bus. This gives an MNO the chance to use early 5G deployments to win new revenue ahead of the widespread availability of 5G devices.

Analysis of a use case for infotainment services on public transport as part of a 5G deployment across a dense city center area using 3.5, 3.7, and 25 GHz bands have shown that meeting the predicted rise in mobile and video traffic demand, cell capacity needs to be increased from 1 Gbps in LTE to 10 Gbps downlink and 3 Gbps uplink peak, with more users being supported by each cell. In addition, 5G ultra-low latency performance will be needed to support virtual reality, gaming and other delay-sensitive applications.

The MNO could win revenue from high value passengers and from governments for supplying 5G bandwidth to public transport. Charging could be per trip for public transport, per time or per data volume.

Assuming that 5G ARPU increases in an analogous way as for LTE, the business analysis for the city center area shows that an MNO could achieve several hundred million Euros of additional Net Present Value (NPV) over 10 years.

1.6.3 Hot Spots

5G enables visitors of stadium events to get close to the sporting or entertainment action. Using real-time virtual reality to experience being trackside at a critical moment in a race or in the middle of a pit stop is a powerful attraction.

The high throughput and low latency enabled by 5G is well-suited to deliver services that provide alternative live views of sporting action at a major event, and to do so simultaneously to thousands of spectators. Consumers can experience being at the heart of the action with live streaming virtual reality, or they can select from a choice of camera views to see what's happening from any angle they want, click to see instant replays or enhance their experience with insights provided through augmented reality.

Furthermore, with 5G capacity in place a range of other services can be offered to subscribers at the event, from betting online to buying merchandise, from pre-ordering refreshments to instantly sharing experiences on social media.

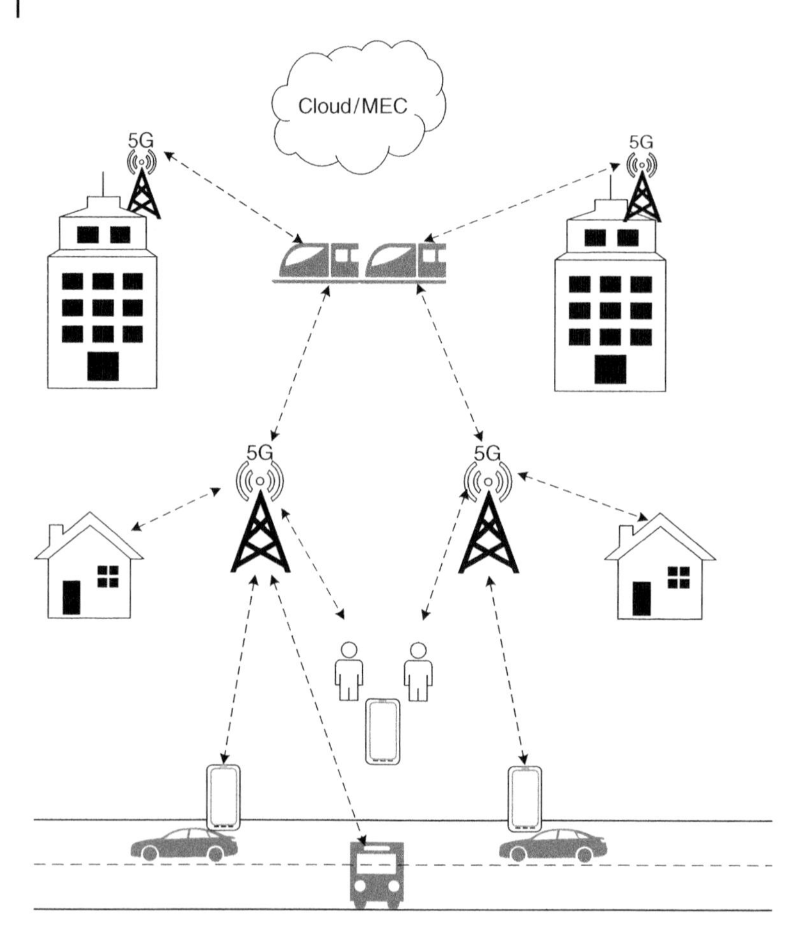

Figure 1.5 5G for in-vehicle infotainment.

The high density of users and extreme throughput and latency demands of these applications cannot realistically be met by Wi-Fi or LTE. Only 5G can support more than 500 users per cell, provide high cell edge performance with an acceptable QoE and deliver an end-to-end latency of less than 5 milliseconds to avoid virtual reality motion sickness.

A business case analysis has shown that an MNO providing services at five events per month at a major stadium could achieve several million dollars in NPV over 10 years from its investments in 5G. The business case viability is highly dependent on the number of events being held at a venue. At least five events per month seem to be needed to be profitable. Another analysis has shown that at a major stadium, an MNO could achieve breakeven in two years by supporting five events per month, while more than six events monthly would hit breakeven during the first year under certain assumptions on the service offering.

1.6.4 Truck Platooning

In many countries, roads are clogged with traffic, creating jams that are a frustrating the drivers. Truck platooning in which several trucks travel in a tightly-knit,

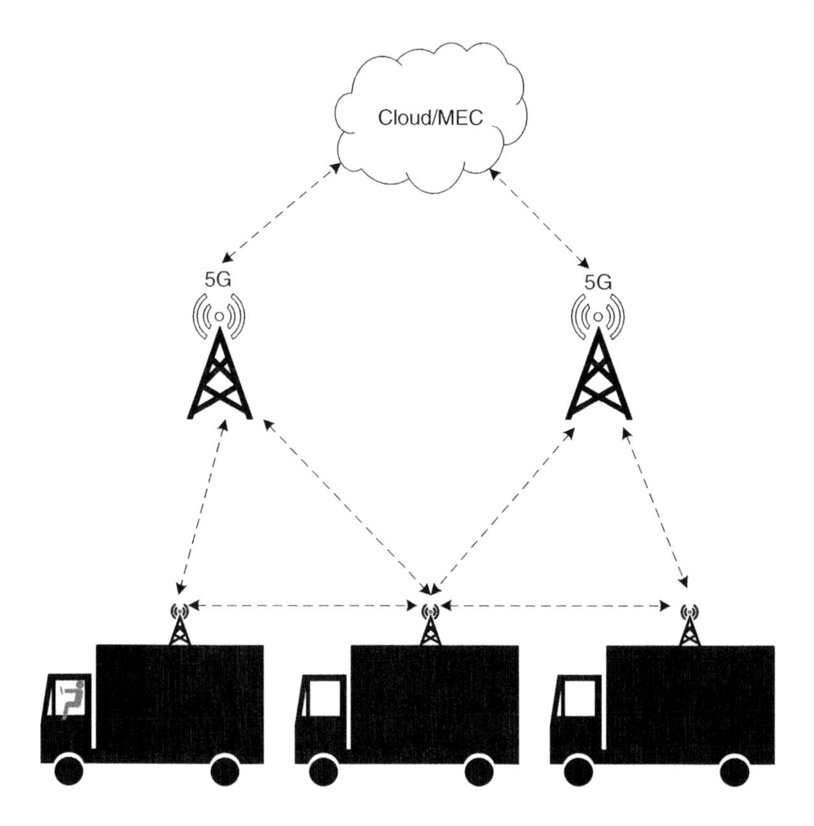

Figure 1.6 5G for truck platooning.

automatically-controlled convoy behind a lead human-driven vehicle, promises not only to reduce congestion and lower fuel consumption, but also to cut transport costs for logistics companies.

While the concept of truck platoons as a means of cutting operating costs and reducing road congestion has been around for many years, the advent of 5G looks to be finally bringing the idea to reality. Safe platooning depends critically on 5G's ultra-low latency and high reliability capabilities.

Small platoons are possible using LTE and Multi-access Edge Computing (MEC), however longer platoons are more cost-effective and require 5G (see Figure 1.6). Platooning will also become an integral part of the future of connected vehicles enabled by 5G, including infotainment, telematics and assisted driving. Ultimately, this will lead to autonomous driving.

Truck platooning should prove attractive to logistics companies by reducing their staff costs, fuel use and supporting more efficient use of the truck fleet. MNOs are also likely to partner with vehicle makers to provide an end-to-end solution.

Assuming average truck fuel consumption data and delivery distance statistics, considering fuel savings of 4% for a lead truck and 10% for following trucks a business case analysis has shown that an MNO could reach breakeven on its 5G investments in six years if it received a 12.5% share of the logistics company's cost savings.

It has been forecast that more than 7 million truck platooning systems could be shipped by 2025. With hundreds of private transportation companies running more than 100 trucks in their fleet, the overall revenue opportunity for an MNO could be immense.

1.6.5 Connected Health Care

Healthcare systems globally are under intense pressure as populations age and economic limitations are applied. 5G can address the issues in many ways, such as enabling skilled surgeons to work remotely and with the help of intelligent robotics to provide basic care needs. Main idea is to allow for efficient use of limited resources to improve access to healthcare services for people wherever they are. Two business cases illustrate the possibilities.

Wireless telesurgery brings telecommuting to the surgical world. Procedures are performed on remotely-located patients by surgeons with the aid of a robot. The target is to provide a remote surgeon, who could be located hundreds of kilometers from patients, with the same sense of touch (essential for localizing hard tissue or nodules) while substituting doctor's hands with robotic probes. To achieve such an experience, delay and stability are crucial in transmitting the haptic feedback (kinesthetic, as force or motion; and/or tactile, as vibration or heat), in addition to audio/video data, as substantial delays can seriously impair the stability of the feedback process and lead to cyber-sickness.

The second business case is wireless service robots, or personal assistant robots, that use artificial intelligence to help the elderly and other patients remain active and independent with an excellent quality of life, while also reducing care costs. Service robots for care are being primed to join the labor force in roles, such as logistics, cleaning and monitoring, which can be fully automated. Beyond these simple tasks, androgynous robots are anticipated to interpret human emotions, interact naturally with people and perform complex care or household jobs. They could also assist patients and elderly people in hospital and hospice campus areas, and at home to reduce care costs, and help aging people remain active and independent with an excellent quality of life.

The target with 5G wireless is to meet the required latency and throughput, with ultra-high reliability, between wireless robots and edge computing centers, where most of the intelligence is located, for example for object tracking, recognition and related application. Attention should be paid to the control loops, which cannot be executed locally. For instance, visual processing cannot be handled locally (because of the computational load/amount of data) and therefore is managed by a server remotely. This means a 5G wireless connectivity between peer points of Gbps, extremely low latency (below 5 milliseconds) for force control, and with extremely low failure rates.

Calculations have shown that a care provider working in partnership with an MNO would achieve breakeven in less than six years for its robot and backend server costs.

For care providers, the business case is quite sensitive to rising Operating Expenses (OPEX) and Capital Expenditures (CAPEX) per robot, including its replacement. The business case will unquestionably fly when more powerful robotic platforms will help save more costs of care, and their price will be much more accessible to consumers.

1.6.6 Industry 4.0

As part of the so-called fourth Industrial Revolution, smart (digital) factories (see Figure 1.7) will deploy greater automation, interconnecting all their areas and activities

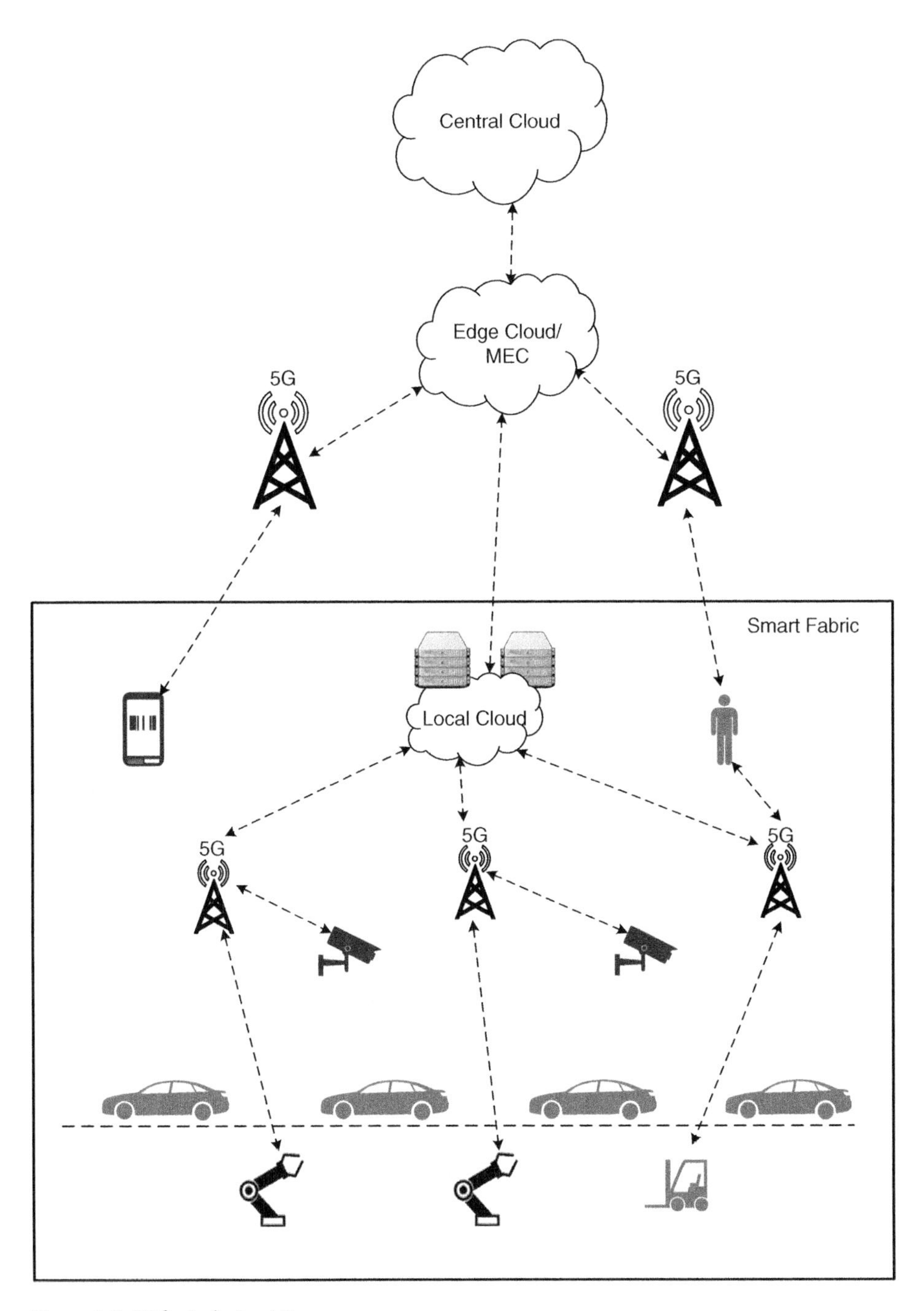

Figure 1.7 5G for Industry 4.0.

to vastly improve productivity, increase staff safety and become far more flexible to meet rapidly evolving market needs.

One of the most important enablers of the smart factory of the future will be vastly increased connectivity that will link machines, processes, robots and people to create more flexible and more dynamic production capabilities. About 90% of industrial connectivity today uses wired connections which provide the high performance and reliability needed for automation, but lack flexibility to be able to rapidly meet changing production demands.

5G is the first wireless technology with high throughput, low latency and extreme reliability that can replace wireline connectivity in the factory. Wireless connectivity allows additional machines to be connected by simply equipping them with wireless sensors and actuators and if required, scaling the network capacity to handle new traffic.

In factories 5G wireless connectivity has up to five times lower costs than wired connectivity. Wireless 5G connectivity can replace wired systems in an existing facility with a short payback period.

Functional safety is one of the most crucial aspects in the operation of industrial sites like factories (see also [4]). Accidents can harm people, machines and the environment. Safety measures must be applied to reduce risks to an acceptable level, particularly if the severity and likelihood of hazards are high. Like an industrial control system, the safety system also conveys specific information from and to the equipment under control. Thus, a 5G network used at industrial sites must be able to transport safety messages between entities in a fast, secure, and reliable manner. In addition, current industrial communication systems are often isolated from the Internet and not exposed to attacks from outside. This will change with wirelessly connected industry applications. Therefore, extreme high security measures must be applied to the used wireless technologies and the overall network architecture avoiding attacks from inside and outside.

1.6.7 Megacities

More than half of the world's population live in big cities and this portion is increasing fast. Traffic management, Public Safety, efficient supply with energy and water as well as waste management are becoming pressing issues for dense urban areas. Around 70% of the worldwide energy consumption and carbon emissions are done in cities. Until 2020 we will therefore see applications around mission control for Public Safety, video surveillance, connected mobility across all means of transport including public parking and traffic steering, and environment/pollution monitoring. These will evolve further into an intelligent traffic infrastructure, connected surveillance of drones for Public Safety scenarios. Also, tourism will benefit from AR and VR applications.

1.7 Business Models

With the introduction of 5G the traditional business model with the MNO as a pure connectivity provider offering voice and short message services is outdated. In fact, one of the biggest changes for the telco industry coming with 5G is most probably the change from this simple business model to new business models where the mobile operator provides not only pure connectivity plus voice but also highly sophisticated services such as Infrastructure-as-a-Service (IaaS). Network-as-a-Service (NaaS), Platform-as-a-Service (PaaS) to third parties, e.g. verticals, small/medium/large enterprises or tenants. These new models allow verticals to build their networks upon the mobile operator's infrastructure, optimized for their specific use cases. Consequently, many new players may be part of this game. In the past, it was mainly the MNO and the end consumer (individual people or enterprises). In future however, it is the MNO, various verticals, site (e.g. stadium or factory) owners, content owners, enterprises and their IT departments, authorities like public safety agencies, end consumers like "normal" smartphone users but also high-end gamers. Partnerships need to be established on multiple levels ranging from sharing the infrastructure, to exposing specific network capabilities as an end to end service, and to integrating partner services into the 5G system.

The NGMN white paper [3] differentiates three roles of parties engaged in the 5G game.

1.7.1 Asset Provider Role

One of the operator's key assets is infrastructure. Infrastructure is usually used by an operator to deliver own services to the end-customer. However, especially in the wholesale business it is common that parts of the infrastructure can be used by a third-party provider. Assets can be various parts of a network infrastructure that are operated for or on behalf of third parties resulting in a service proposition. Accordingly, one can distinguish between IaaS, NaaS or PaaS. Another dimension of asset provisioning is real-time network sharing that refers to an operator's ability to integrate 3rd party networks in the MNO network and vice versa.

1.7.2 Connectivity Provider Role

Another role of an operator is one of a connectivity provider. Basic connectivity involves best effort traffic for retail and wholesale customers. While this model is basically a projection of existing business models into the future, enhanced connectivity models will be added where Internet Protocol (IP) connectivity with QoS and differentiated feature sets (e.g. zero rating, latency, mobility) is possible. Furthermore, (self-) configuration options for the customer or the third party will enrich this proposition.

1.7.3 Partner Service Provider Role

Another role an operator can play in the future is one of a partner service provider, with two variants: The first variant directly addresses the end customers where the operator provides integrated service offerings based on operator capabilities enriched by partner content and specific applications. The second variant empowers partners to directly make offers to the end customers enriched by the operator's network or other value creation capabilities.

When collaborating with a 3rd party (tenant, vertical, Over-The-Top [OTT]) the MNO can, e.g. offer the following three basic business models to follow:

1. No control model
   The 3rd party has no control over deployed network services and functions, network deployment and operation is under full control of the MNO, 3rd party can monitor given KPIs (KPI).
2. Limited control model
   3rd party or MNO can deploy parts of the network, 3rd party can change configuration of deployed network functions and deploy own certified functions.
3. Extended control model
   Third party designs, deploys and operates its own network on the infrastructure of the MNO. Third party has tight control over own network functions and services, but limited control over MNO functions.
   Enabling these different new business models and modes of network control requires new technical solutions which 5G must build on. Some of these are e.g. NFV, SDN and network slicing. NFV allows the MNO to introduce new network sharing models: sharing of data center sites (buildings, rooms), infrastructure (switches, router, firewalls) and hardware (racks, blades), sharing of certain core or radio virtual network functions, sharing of frequencies. Network slicing will allow the MNO to offer its hardware, software and infrastructure to 3rd parties, i.e. offering NaaS kind of services to different service providers with the highest degree of security and isolation of slices.

1.8 Deployment Strategies

3GPP has defined different deployment options as to how 5G can be introduced into existing LTE networks (see [9]). We differentiate mainly between so-called non-standalone (NSA) and standalone (SA) deployment solutions. In the GSMA White Paper [10] these deployment options and their pros and cons are extensively discussed including support of voice via NSA and SA. For more information on this topic see also Chapter 4.

With an NSA solution one radio technology (5G or LTE) is "anchored" at the other one (LTE or 5G) by using a dual-connectivity mechanism. Only the master base station, the anchor point, maintains a signaling connection to the core network (either S1-C to Evolved Packet Core (EPC) or N2 to 5GC). The signaling connection state in the device, Radio Resource Control (RRC) state, is based on the RRC state of the master base station. In addition to the master, data can be received and transmitted from/to the device or the core network via the secondary (or slave) base station. The secondary base station can also decide to establish a signaling connection to the device, which is used to send reconfiguration messages and measurement reports. There is always a signaling connection between master and secondary node enabling the master to manage data radio bearers at the secondary base station. In a standalone deployment option, the 5G or enhanced LTE base station is directly connected to the 5G Core and maintains a signaling connection with the mobile device and the 5G Core. The NSA options 3, 4, and 7, as well as standalone options 2 and 5 with the 5G base station respectively the enhanced

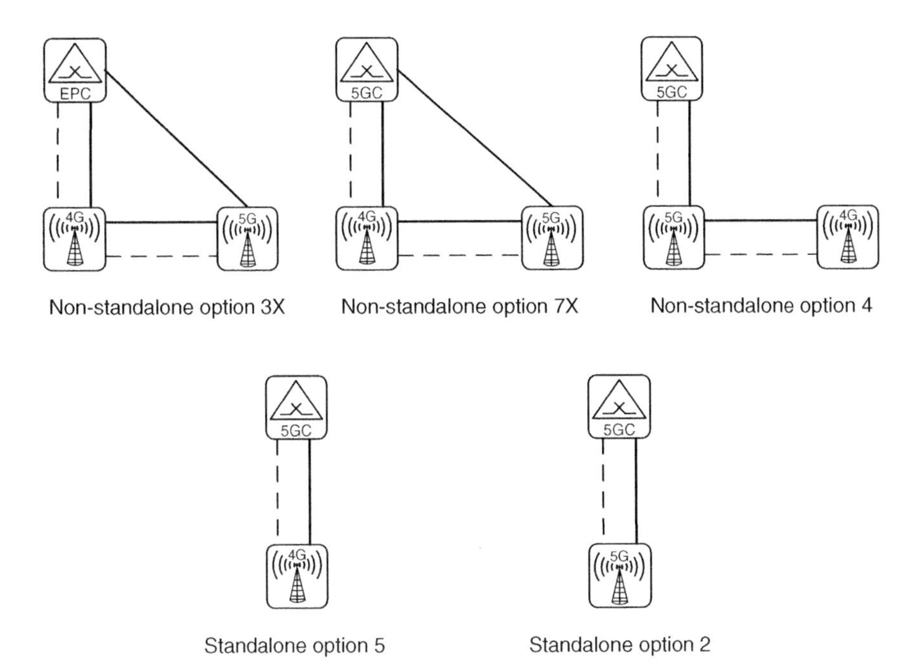

Figure 1.8 Non-standalone and standalone deployment options.

LTE base station directly connected to the new 5G Core are shown in Figure 1.8. Note that options 3, 4, and 7 have different flavors, depending on whether there exists a user plane path between LTE and 5G base station, respectively between secondary base station and core network. We focus on options 3X and 7X as these provide several benefits compared to other flavors of options 3 and 7 (e.g. less signaling between radio and core in case of mobility events). In NSA option 4 and its flavor 4a master and secondary roles are exchanged compared to options 3 and 7, i.e. 5G base station is now the master and LTE base station serves as secondary. Therefore, option 4 works well only if 5G radio can provide sufficient coverage to maintain a connection to the devices. In option 4, there is a user plane path between the two base stations, while in option 4a user plane path is directly between core network and each base station. Introducing NSA or SA deployment options with a new 5G Core requires interworking between existing EPC and new 5GC. Native interworking between 5G and legacy technologies such as 2G and 3G is not foreseen.

The solid lines indicate data paths (user plane) between two nodes while dotted lines indicate signaling connections (control plane).

Obviously, introducing 5G with NSA option 3X based on the existing LTE/EPC network deployment is the fastest and easiest way to provide 5G services such as low latency data connections and high user throughput to the end consumer. Option 3X requires only minor changes to the existing EPC (e.g. in Mobility Management Entity [MME] and HSS), it allows aggregation data rates on LTE and 5G and can leverage features like VoLTE already implemented in LTE. LTE provides wide area coverage, while 5G can boost the throughput in certain (limited) areas. In addition, 3GPP has accelerated option 3X, thus option 3X standard was completed several months earlier than option

2 and implementations can also start earlier. Option 3X requires an upgrade on the LTE base station to act as master for 5G, but this is mainly a SW upgrade at the sites where 5G is introduced, e.g. first in hot spot and urban areas where more capacity per user and cell is needed. However, it is not required that 5G and LTE base stations are deployed at the same site.

Most operators worldwide have decided to start their 5G deployment with one of the NSA option 3 versions and potentially evolve later to NSA option 7 (i.e. evolve EPC to 5GC and upgrade LTE to connect to 5GC and EPC in parallel) and/or to standalone option 2. NSA option 3 with EPC, NSA option 7 and standalone option 2 with 5GC are expected to co-exist for some time, together with the existing legacy LTE/EPC network (and potentially 2G/3G General Packet Radio System [GPRS] networks). In the long term, it is expected that operators migrate toward one of the options 2, 4, 5, 7 with 5G Core, so that they can smoothly shut down the EPC.

With standalone option 2 and other options with 5G Core, the operator can offer new 5G end-to-end services such as Network Slicing, Edge Computing, Ultra Reliable and Low Latency Communication (URLLC) services, lower latency without use of LTE, lower setup times and there is no need for LTE network upgrades. Also, operators can leverage cloud native enablers for improving availability and reliability of their Network Functions. One issue with options 2 and 4 is limited coverage of early deployments especially in high bands. This could lead to frequent handover events between 5G and LTE or requires use of a 5G coverage layer in low bands (e.g. below 1 GHz). Only few operators will start with standalone option 2 from the very beginning due to bigger investment costs.

1.9 3GPP Role and Timelines

In the context of 5G, the most important Standards Developing Organization (SDO), which is creating the 5G standards for the radio and system architecture, is the 3rd Generation Partnership Project 3GPP. 3GPP is a joint international standardization initiative between North American (Alliance for Telecommunications Industry Solutions [ATIS]), European (European Telecommunications Standards Institute [ETSI]) and Asian organizations (Telecommunications Standards Development Society India [TSDSI] in India, Association of Radio Industries and Businesses [ARIB] and Telecommunication Technology Committee [TTC] in Japan, Telecommunications Technology Association [TTA] in Korea and China Communication Standards Association [CCSA] in China) that was originally established in December 1998. The participating organizations are also called organizational partners. The scope of 3GPP was to specify a new worldwide mobile radio system. Global System for Mobile Communications (GSM) was a European initiative, while Code Division Multiple Access (CDMA) was initiated in North America, and these are not compatible with each other. The new system would be based on the evolved GSM techniques GPRS/Enhanced Data Rates for Global System for Mobile communications Evolution (EDGE). This activity has led to the standardization of the third generation (3G) Universal Mobile Telecommunications System (UMTS), which consists of WCDMA as radio technology and a core network supporting both circuit-based voice calls and packet-based data services. UMTS was meant as a universal standard that allows subscribers to use their UMTS capable mobile phones and subscriptions worldwide through roaming agreements between mobile operators. UMTS is a considerable success story.

But 3GPP did not stop work after UMTS; in the years that followed enhancements of UMTS like High Speed Packet Access [HSPA/HSPA+], new services like Multicast/Broadcast delivery, Location services and the Internet Protocol Multimedia Subsystem (IMS) were introduced. As a next step, LTE with a new Orthogonal Frequency-Division Multiplexing (OFDM) based radio technology and an All-IP core network architecture was developed to simplify the overall network architecture leading to increased operational efficiency and to cope with the rising demands on data throughput caused by new smartphone and tablet generations. The latest step in this evolution story is the development of the 5th Generation (5G) system comprising of a new radio and new core network that can fulfill the requirements on higher bandwidth and reliability, lower latency, increased operational efficiency and much higher network densification. 5G allows a wide variety of new use cases in the industrial ("digitization of industries or Industry 4.0"), enterprise and consumer market to be addressed.

3GPP is organized in different working groups (see Figure 1.9) that are responsible for various parts of the 3GPP system. The RAN groups define the radio parts of the GERAN/UMTS/LTE/5G system, the physical layer and radio protocols. The System Architecture (SA) and Core/Terminal (CT) groups specify all parts of the overall system (e.g. architecture, security, charging) and all other protocols (between the mobile device and network, within the network and between networks).

3GPP follows a phased approach; working output is delivered as a set of Technical Specifications (TS) in so-called System Releases. Technical Specifications contain normative requirements that must be implemented by chipset, device and network equipment vendors. Interim results of ongoing work in 3GPP are usually captured in non-normative Technical Reports (TR). Test specifications are also created by 3GPP (mainly test cases for UE to network communication). It must be noted that 3GPP defines only functions and protocols, how these functions are implemented in concrete network nodes or whether some functions are implemented in the same node is up to the network vendor. One basic design principle in 3GPP's standardization process is backward compatibility of new features with existing ones. This ensures that new features can be introduced in one network without the need to upgrade all inter-connected networks or all other nodes within this network at the same time. Also, it ensures that the upgraded network can provide services to legacy devices that are not upgraded yet.

Table 1.9 provides a brief overview of the official release dates and milestones of the 3GPP releases up to Release 16.

Figure 1.10 illustrates the given 5G timeline in 3GPP for completing the different 5G options. Options 3 and 2 are first to be specified by 3GPP, followed by options 4 and 7 completion in a late drop beginning of 2019. Option 5 (LTE base station connected to 5G Core) is completed by Q3 2018. Abstract Syntax Notation No. 1 (ASN.1) freeze follows usually three months after specifying the architecture and interfaces.

For more information on the history and structure of 3GPP, visit the official 3GPP site at http://www.3gpp.org/about-3gpp/about-3gpp.

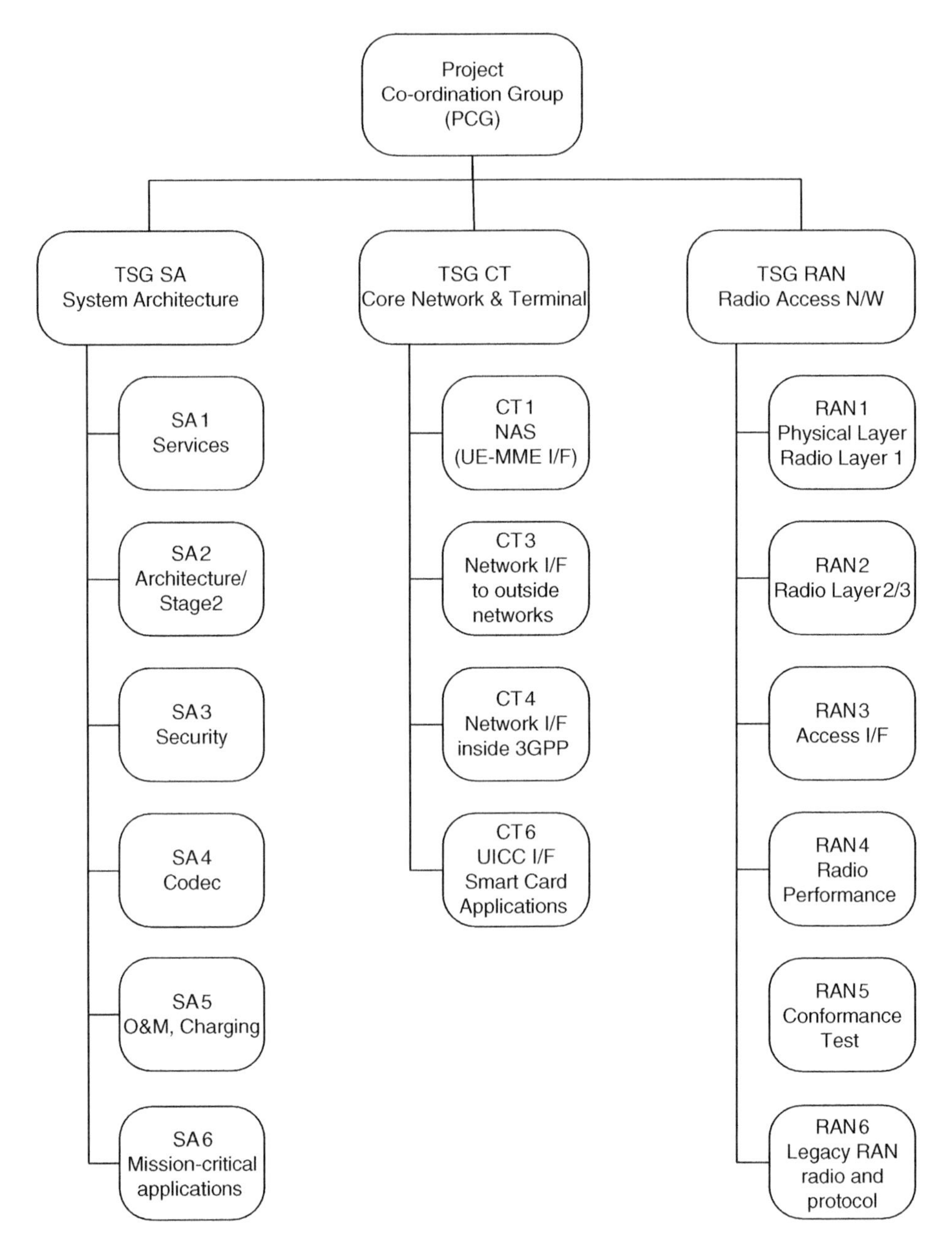

Figure 1.9 3GPP organizational structure.

Table 1.9 3GPP milestones up to Release 16.

Release	End date	Main content
Phase 1 and 2	1992 and 1995	Basic GSM Functions
Release 96, 97, 98, 99	1996, 1997, 1998, 1999	GPRS, HSCSD, EDGE, UMTS
Release 4	2001	MSC Server split architecture
Release 5	2002	HSDPA, IMS
Release 6	2004	HSUPA, MBMS, Push to Talk over Cellular (PoC)
Release 7	2007	HSPA, EDGE Evolution
Release 8	2008	LTE/SAE
Release 9	2009	LTE/ SAE Enhancements, Public Warning System (PWS), IMS emergency sessions
Release 10	2011	LTE Advanced, Local IP Access (LIPA), Selective IP Traffic Offload (SIPTO)
Release 11	2012	Heterogeneous Network Support (HetNet), Coordinated Multipoint Operation (CoMP)
Release 12	2014	Public Safety, Machine Type Communication, HSPA/LTE Carrier Aggregation
Release 13	2016	Cellular (Narrowband) Internet of Things (NB-IoT) Mission-critical Push-to-Talk (MCPTT) Small cell dual-connectivity Single cell point-to-multipoint (SC-PTM) Latency reduction for LTE
Release 14	2017	Mission Critical Video and Data over LTE (MCVIDEO, MCDATA) LTE support for V2X (LTE-V2X) Control and User Plane Separation of EPC nodes (CUPS) Latency reduction for LTE Channel model above 6 GHz Requirements for Next Generation Access Technologies
Release 15	2018	5G Phase 1 Phase 1 schedule is as follows (see Figure 1.10): • 5G Option 3 NSA completion Q4/17, ASN.1 freeze Q1/18 • 5G Option 2 SA completion Q2/18, ASN.1 freeze Q3/18 • Late drop 5G NSA architectures 4 and 7 Q4/18, ASN.1 freeze Q1/19 Shortened TTI for LTE EPC support for E-UTRAN Ultra Reliable Low Latency Communication Mobile Communication Systems for Railways
Release 16	2019+	5G Phase 2 Studies on various new 5G features and enhancements: • Support of waveforms above 52 GHz • Massive MTC support • 5G in shared and unlicensed spectrum • Interworking with trusted Non-3GPP access • Fixed Mobile Convergence • Multi-connectivity • Network Slicing in RAN • V2X communication with 5G • Broadcast support in 5G

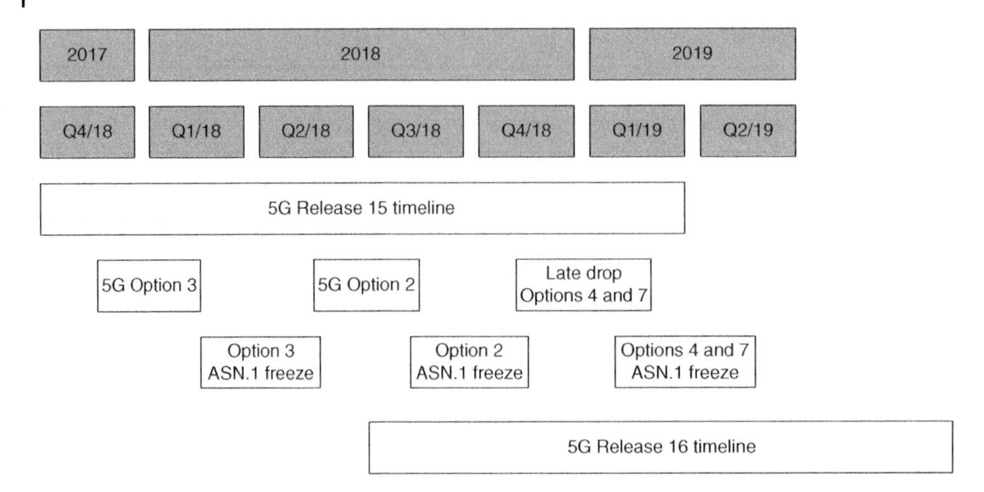

Figure 1.10 3GPP timeline for 5G.

References

All 3GPP specifications can be found under http://www.3gpp.org/ftp/Specs/latest. The acronym "TS" stands for Technical Specification, "TR" for Technical Report.

1 ITU-R M.2083-0: "IMT Vision – Framework and overall objectives of the future development of IMT for 2020 and beyond" (see https://www.itu.int/rec/R-REC-M.2083-0-201509-I/en), 2015.
2 ITU-R M.2376-0: "Technical feasibility of IMT in bands above 6 GHz" (see https://www.itu.int/pub/R-REP-M.2376), 2015.
3 NGMN Alliance: "5G White Paper" (see https://www.ngmn.org), 2015.
4 5GACIA: "5G for Connected Industries and Automation White Paper" (see https://www.5g-acia.org), 2018.
5 3GPP TR 22.891: "Feasibility Study on New Services and Markets Technology Enablers", 2016.
6 3GPP TR 22.261: "Service requirements for the 5G system", 2018.
7 3GPP TR 22.186: "Enhancement of 3GPP support for V2X scenarios", 2017.
8 Nokia White Paper: "Translating 5G use cases into viable business cases" (see https://networks.nokia.com/innovation/5g-use-cases/home), 2017.
9 3GPP TR 23.799: "Study on Architecture for Next Generation System", 2016.
10 GSMA White Paper: "Road to 5G: Introduction and Migration, April 2018" (see https://www.gsma.com/futurenetworks/5g/road-to-5g-introduction-and-migration-whitepaper), 2018.

2

Wireless Spectrum for 5G

Juho Pirskanen[1], Karri Ranta-aho[2], Rauno Ruismäki[2] and Mikko Uusitalo[2]

[1] *Wirepas, Tampere, Finland*
[2] *Nokia, Espoo, Finland*

2.1 Current Spectrum for Mobile Communication

Spectrum is a limited natural resource setting basic requirements for the wireless system operation. For commercially viable communication systems, the optimal spectrum should meet several requirements. The amount of spectrum needs to be sufficient to carry traffic generated by the expected use cases. The spectrum should be globally available so that product implementations (base station [BS] equipment, devices, chipsets) could be used in different countries. Additionally, lower carrier frequency and high allowed transmission power would enable to build constant network coverage with lower number of needed BSs (BS).

However, due to several reasons, such single spectrum does not exist for cellular communications. Countries like US, China, Japan and European countries have used spectrum for different purposes during previous decades (e.g. radio and TV broadcast, satellite communication). The single block of spectrum in lower frequencies is not wide enough to carry all traffic generated by today's cellular communication. On higher operating frequencies, more spectrum is available, but challenging propagation conditions have made it difficult to use those in cellular networks to provide wide area coverage with economically feasible number of BS.

As a result, today's cellular standards as well as state of the art (chipset) implementations support a high number of different operating bands for global operation as well as various band combination techniques, e.g. carrier aggregation (CA), to increase the simultaneously used bandwidth to achieve higher data rates in devices.

For 5G, the requirements for total transmitted traffic in a cell as well as user specific bit rates are ever increasing. As a consequence, completely new spectrum, new licensing methods, and advanced interplay of high and low spectrum to provide high capacity and throughput with ubiquitous coverage need to be utilized.

2.2 Spectrum Considerations for 5G

In the past, mobile systems like Global System for Mobile communications (GSM), Universal Mobile Telecommunications System (UMTS) and Long Term Evolution

5G for the Connected World, First Edition.
Edited by Devaki Chandramouli, Rainer Liebhart and Juho Pirskanen.

(LTE) typically utilized the lowest available spectrum to provide initial wide coverage for voice and mobile broadband services. The utilization of higher frequencies was considered when capacity demands required taking additional (currently not used) spectrum resources in use which also leads to network densification as higher spectrum comes with smaller cells and requires a higher number of cell sites. In the 5G era, the availability of spectrum in low frequency ranges remains vital to provide uniform network coverage for all different 5G services. Additionally, it must be possible in the future to refarm existing mobile cellular bands, currently used for GSM, UMTS or LTE operation, in an efficient manner. This aspect was addressed when defining operating bands for 5G New Radio (NR), see Section 2.6.

However, until World Radio Conference 2015 (WRC-15), only spectrum below 6 GHz was considered for cellular mobile systems limiting the amount of available spectrum and setting the limits to network densification. One consequence was that past generations of cellular standards did not consider support for higher spectrum bands. For WRC-19, bands above 24 GHz will be studied for mobile use, namely for 5G, referred to as International Mobile Telecommunications 2020 (IMT-2020) in International Telecommunication Union-Radio Sector (ITU-R) terminology (see [1]). These new higher bands are required to provide significantly higher (tens of Gigabits per second) data rates for mobile broadband users. There are also other new requirements, e.g. substantially lower communication latencies and billions of wireless devices that need to be connected.

Spectrum below 6 GHz is rather fragmented and composed of a mixture of bands for operation of Frequency Division Duplex (FDD) and Time Division Duplex (TDD), also referred to as paired and unpaired bands, respectively. Additionally, spectrum available in different regions and countries may have different frequency ranges and communication requirements, including unpaired unidirectional spectrum blocks. Depending on the region and country, roughly 1200 MHz of spectrum below 6 GHz is allocated for mobile service and identified for IMT (cellular) use but typically only half of it is made available. Bands for FDD are mainly below 3 GHz and TDD typically above 3 GHz, with some exceptions. Due to this fragmentation, efficient utilization of spectrum below 6 GHz calls for support for different bandwidths and carrier aggregation solutions. Spectrum available for wide and contiguous carrier bandwidths in the order of 100 MHz is only possible in the spectrum range 3–6 GHz, while narrower carrier bandwidths (of up to 40–100 MHz) for sub 3 GHz FDD deployments are only possible with carrier aggregation. Fragmented spectrum, different regional requirements, and the need to support different carrier aggregation techniques lead to the fact that current LTE specifications support more than 60 operational bands and hundreds of band combinations (see [2]).

For massive Machine-type-Communication (mMTC) services, licensed spectrum below 1 GHz is the most suitable spectrum range to be used due to good coverage and outdoor-to-indoor penetration properties. Nevertheless, the 5G mMTC design should also provision operation of Machine-type-Communication (MTC) devices in the local area and/or mesh type of deployments, and for these scenarios the licensed centimeter Wave (cmWave) or millimeter Wave (mmWave) bands are also suitable. The term cmWave refers to radio frequencies where the wave length is in centimeters, i.e. roughly 3–30 GHz. The term mmWave refers to radio frequencies where the wave length is in millimeters, i.e. roughly 30–300 GHz. However, typically spectrum around 20–30 GHz is also called mmWave. Operation in license-exempt frequency bands could be provisioned as well but is subject to Quality of Service (QoS) requirements of the use case, co-existence issues and requires additional radio protocol mechanisms to be supported in MTC devices (at higher cost).

From the reliability point of view, licensed and dedicated spectrum is preferred to safeguard QoS. Due to diverse requirements of multiple verticals regarding coverage, reliability and capacity (e.g. autonomous driving and industry automation), it is reasonable to use low frequencies, e.g. below 3 GHz for wide area communication, and higher frequencies for local communication, e.g. within a factory or campus.

2.3 Identified New Spectrum

The WRC-15 agreed an agenda item (AI 1.13) for WRC-19 to study frequency bands between 24.25 and 86 GHz for IMT/5G as shown in Tables 2.1 and 2.2 [1]. Table 2.1 defines frequency bands having an already mobile allocation on a primary basis and Table 2.2 lists bands for which mobile service allocation is needed. The WRC-19 study will consider:

- Technical and operational characteristics of terrestrial IMT systems that would operate in this frequency range;
- The deployment scenarios envisaged for IMT-2020 systems and the related requirements of high data traffic such as in dense urban areas and/or in peak times;
- The needs of developing countries; and
- The time-frame in which spectrum would be needed.

Before actual ITU-R sharing and compatibility studies can start, the suitability of those bands needs to be analysed to find out which band(s) have the best potential to avoid any undue limitations and restrictions, if identified for IMT and used for 5G, and, which band(s) have the best potential for global availability.

Despite the fact that no bands below 24 GHz were included in the ITU-R studies, it is still important to also continue 5G research activities in bands below 24 GHz (e.g. 5.9–7.7, 10–10.5, 14.8–15.35 GHz) as those bands have beneficial characteristics for some expected 5G use cases.

Table 2.1 Frequency bands studied for WRC-19 and allocated for mobile service on a primary basis.

Frequency range (GHz)
24.25–27.5
37–40.5
42.5–43.5
45.5–47
47.2–50.2
50.4–52.6,
66–76
81–86,

Table 2.2 Frequency bands studied for WRC-19 and not allocated for mobile service.

Frequency range (GHz)
31.8–33.4
40.5–42.5
47–47.2

Even though the 28 GHz band (27.5–29.5 GHz) was excluded from ITU-R studies, it has a good chance to become a de-facto 5G band in some of the 5G front-runner countries, such as the United States, South Korea and Japan. Furthermore, it overlaps with the European 5G pioneer band 24.25–27.5 GHz, which is the only ITU-R candidate for 5G bands below 30 GHz.

2.4 Spectrum Regulations

When considering different spectrum bands, the regional or national regulation authorities define by whom and how the band can be utilized. Typically, cellular mobile systems operate on licensed spectrum, whereas wireless local area networks (WLAN) such as Wi-Fi, and personal area networks such as Bluetooth operate on license-exempt (unlicensed) spectrum. The spectrum licensing policy has significant impact on how radio systems can operate as well as on business models of the industry. In addition to traditional licensing models new regulation approaches have also been considered. In the following sections we discuss these regulatory approaches.

2.4.1 Licensed Spectrum

Dedicated and licensed spectrum will continue to be preferred for many 5G key use cases. Licensed spectrum ensures a stable framework for investment with guaranteed coverage and QoS. This can be achieved as with a given licensed spectrum block, the licensee has the right to control all radio communication. The licensee, who is in the case of mobile communication, the mobile network operator, can thus perform accurate network planning including linking budget calculations, interference and system capacity modeling without the risk of interference from other users using the same spectrum.

The co-channel interference is caused by the operator's own mobile network operations, which can be controlled and mitigated by network algorithms and interference cancelation receivers on both the network and on the mobile side. Interference from neighboring channels is controlled by adjacent channel leakage requirements, set for all transmitters operating at the mobile spectrum. Additionally, for receivers' adjacent channel rejection and out of band blocking requirements the operator can define the minimum requirement for reducing adjacent channel and out of band interference. These requirements are defined in 3rd Generation Partnership Project (3GPP) Radio Access Network (RAN) Working Group (WG) 4, for each supported band to ensure that all mobile devices and BS support these minimum requirements, so that the regulatory requirements are met, and interference conditions in each carrier can be ensured. Therefore, licensed spectrum also remains the primary spectrum asset for mobile operators for 5G deployments.

Typically, as was the case with 2G, 3G and 4G deployments, these licenses are sold in spectrum auctions in a country or assigned to mobile operators under specific terms. Additionally, depending on the country the spectrum auction may set rules how the spectrum owner must utilize the spectrum, e.g. the spectrum may need to be taken into use within a specific time after the license has been granted and/or the deployment must meet certain coverage requirements (covering certain percentage of the country or population).

Mobile operators have been investing on spectrum since the addressable market has been significant, and utilization of auctioned spectrum is mainly market driven, i.e. operators can deploy their networks in areas where it is justified from a business perspective. Additionally, one should not underestimate the fact that by investing into the spectrum auction, existing mobile operators can avoid new competitors accessing the market.

In some smaller European countries like Finland, specific terms for auction has been applied that limits the amount of spectrum that each bidder can obtain, but also set requirements to when spectrum needs to be taken into use and to provide significant network coverage. An example of such terms is the frequency band allocation for telecom operating licenses for the spectrum 703–733 MHz and 758–788 MHz in Finland, which was defined as follows (see [3]): "License must be built so as to cover 99% of the population of mainland Finland within three years of the start of the license period. The coverage requirement should be structured so as to ensure reasonable indoor coverage within the coverage area. 'Reasonable indoor coverage' means that the telecom operator's services must be available without additional cost to users in normal circumstances of use in users' permanent residences or enterprises' places of business."

In addition, the rule defined that "No more than two 2 × 5 MHz frequency pairs will be allocated to any individual enterprise or organization." However, in coverage calculation, the broadband mobile communications networks previously built by the license holder using the 2.6 GHz, 1800 MHz and 800 MHz spectra are considered [3]. At the end, existing mobile operators bought the maximum frequency amount, almost with starting bid of 11 million Euros per frequency block (only TeliaSonera paid 11 330 000 Euros on other frequency blocks) as the cost of coverage build was prohibitive for new entrants and the existing operators could not compete against each other on the amount of spectrum [4].

However, such licensing rules lead to a situation where mobile telecommunication clearly flourishes in Finland, as mobile subscriptions prices are inexpensive, and utilization of mobile data is very high. Figure 2.1 presents data usage per mobile broadband subscription on average for OECD (Organisation for Economic Co-operation and Development) countries (for details see [5]). It can be observed that in Finland the average data usage was around 15.5 Gigabytes (GB) per month, which is over five times more than the OECD average and almost 9 times more than in Germany where the monthly costs of a mobile subscription are relatively high.

It is apparent that licensing rules have an impact on the utilization of the spectrum, thus regulators need to balance between several terms such as ensuring competition, auction income, spectrum utilization, license period and impact of auction prices to

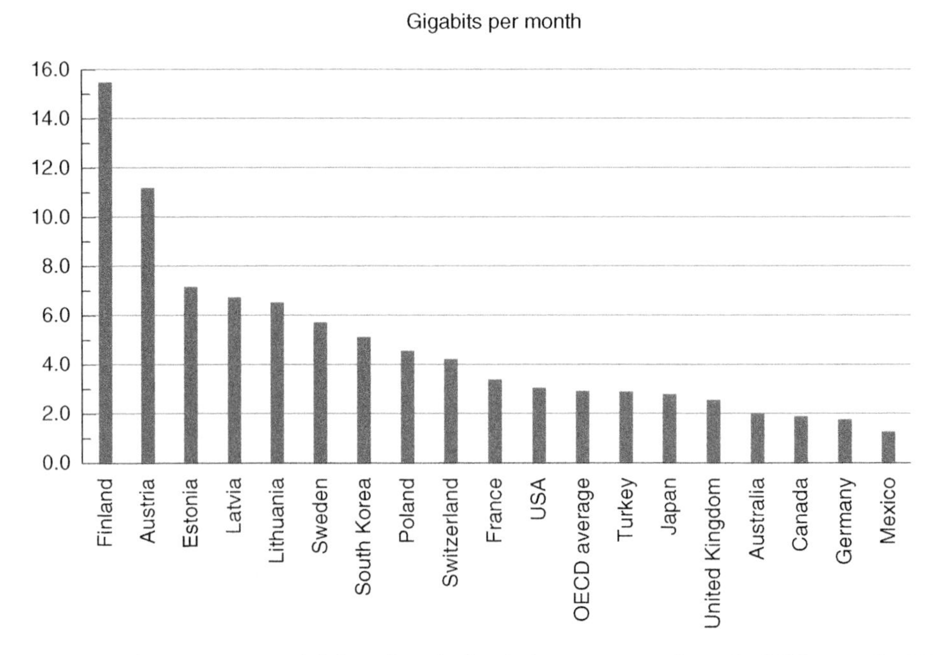

Figure 2.1 Data usage per mobile broadband subscription – average volume for OECD countries 2017.

mobile subscription prices, as well as to identify enough spectrum to be licensed. Both license-exempt (unlicensed) spectrum and new regulation approaches can provide additional tools in this process.

2.4.2 License-Exempt Spectrum

Instead of licensed spectrum, where radio operation is controlled by the licensee, the radio operator in license-exempt spectrum (Table 2.3) is free if operation fulfills

Table 2.3 Main license-exempt spectrum.

Spectrum	Region	Current use and technologies
863–870 MHz [6]	Europe (not completely harmonized in Europe)	Short range devices, IoT, Low Power Wide Area (LPWA)
902–928 MHz [7]	US, Canada, South Korea, Australia, etc.	IoT, short range devices
2400–2483.5 MHz [8]	Global, regional variations	Mobile broadband: short range devices (headsets, etc.), IoT Wi-Fi, Bluetooth, including other standards and proprietary radios
5150–5350 MHz and 5470–5725 MHz [9]	Global, regional variations	Mobile broadband: Wi-Fi, LTE, LAA, proprietary radios
57–66 GHz (Europe) [10] 59.3–64 GHz (US) [11]	Global, regional variations	Personal area networking, high data rate short range connections Outdoor point to point fixed links (US)

the regulation requirements. These requirements depend on the frequency band and are country specific. The best harmonization has been achieved for the 2.4 GHz Industrial Scientific and Medical (ISM) band which has only few global variations. Below 1 GHz frequency, there is the 915 MHz band in the US and other countries following the Federal Communications Commission (FCC) regulation in Canada, South Korea, and Australia. This frequency band is not available in Europe where frequency blocks between 863 and 870 MHz have been allocated for license-exempt use, however, this band is not completely harmonized inside Europe and thus it has variations on spectrum range between European countries. Spectrum below 1 GHz is used for the growing number of Internet of Things (IoT) devices and low power wide area networks (LPWA) but have not been addressed by 3GPP in previous license-exempt spectrum solutions as available spectrum and transmission powers are limited. It needs to be seen whether any of these bands will be addressed by 3GPP, e.g. for IoT purposes.

As has been said, the best harmonization on unlicensed license-exempt has been obtained for the 2.4 GHz band. Thus, it is tremendously utilized by Wi-Fi and Bluetooth. This spectrum block is also being more and more utilized by different IoT short range technologies, either with Bluetooth Low Energy (BLE) physical layer solution, other standard solutions such as ZigBee, or proprietary radio solutions. Due to the good propagation environment and the high number of available devices supporting this band, it will remain very important for existing and forthcoming Wi-Fi implementations. However, as interference is increasing in this band, bandwidth demanding applications will and have already moved to the less crowded 5 GHz band.

The 5 GHz band is also the primary band for 5G license-exempt solutions as it provides a significant amount of spectrum and is optimized for wideband transmissions for mobile broadband use cases. The requirement for wideband transmission is, e.g. defined in the European regulation via Power Spectral Density as shown in Table 2.4. For details see also [9]. Due to spectral power limitations the maximum transmission power can be only achieved, if the transmission bandwidth is 20 MHz, and narrower transmission need to reduce Transmitter Exchange (TX) power to meet the power spectral density limit. This is the case in LTE unlicensed spectrum operation when the eNB (enhamced Node-B, i.e. a BS) is transmitting the Broadcast Control Channel (BCCH) at 1.4 MHz bandwidth.

In addition to existing spectrum, it is expected that also unlicensed-exempt spectrum will be allocated during WRC-19 process or as part of it. The NR operation on unlicensed spectrum was not part of the first 3GPP NR release (Release 15), but it is planned to be addressed as part of Release 16.

Table 2.4 Mean EIRP (Equivalent Isotropic Radiated Power) limits for RF output power and power density at the highest power level.

Frequency range (MHz)	Mean EIRP limit for PH (dBm) With transmit power control in use	Mean EIRP density limit (dBm/MHz) With transmit power control in use
5150–5350	23	10
5470–5725	30	17

Slave devices without a radar interference detection function shall comply with the limits for the frequency range 5250–5350 MHz.

2.4.3 New Regulatory Approaches

As discussed above, dedicated and licensed spectrum will continue to be preferred for many key use cases of 5G especially where coverage and QoS needs to be guaranteed. Additionally, license-exempt spectrum can also play an important role for 5G. However, novel regulatory approaches and tools such as Licensed Shared Access (LSA) and Spectrum Access System (SAS) are expected to complement the existing dedicated and licensed spectrum access regimes. Future access to new spectrum for 5G in the 4–6 GHz range may depend on sharing possibilities with incumbent users in many regions where re-purposing of spectrum is not possible. This can provide means to establish dedicated or private networks for certain use cases in a limited area or region.

These innovative ways of allocating even licensed spectrum to users (not necessarily mobile operators) are under discussion in many countries. Low latency, high reliability use cases like collaborative robots and industry automation make it necessary to allow usage of licensed spectrum in local or regional areas for specific applications. In Germany, an auction of frequencies in the range of 3.6 GHz is planned for beginning of 2019. In parallel, frequencies in the range of 3.700–3.800 MHz and 26 GHz are made available for local or regional use based on request (i.e. no auction is planned). This allows deployment of local networks in factories, harbors, airports, campuses, cities and hot-spots like a stadium. Local frequencies need to be used within one year after allocation, otherwise the user must return the license to the regulator. Use of such local frequencies is also planned in other European countries like the Netherlands.

Additionally, operation in some of mmWave bands may initially be amenable to spectrum and network infrastructure sharing arrangements between operators to avoid installation of multiple network equipment (especially antennas) at each site and therefore reducing deployment costs. In general, 5G systems need to support different spectrum authorization modes and sharing scenarios in different frequency bands.

As illustrated in Figure 2.2, four different user modes can be defined under which 5G radio access systems are expected to operate, namely the "service dedicated user mode," the "exclusive user mode," the "LSA user mode" and the "unlicensed user mode" (see [12]). The use of radio spectrum can be authorized in two ways, first by individual authorization in the form of awarding licenses, and secondly by "general authorization," also referred to as license-exempt or unlicensed. The relationship between user modes and authorization schemes is visible in the upper part of Figure 2.2, named "regulatory framework domain."

Spectrum usage rights awarded by "individual authorization" are exclusive at a given location and/or time. The "service dedicated user mode" refers to spectrum designated to services other than Mobile/Fixed Communication Network (MFCN) operation, which are indented to be integrated into the 5G eco system, e.g. Intelligent Traffic Systems (ITS) or Public Protection and Disaster Relief (PPDR) services. This spectrum is only used for dedicated services and applications. Spectrum designated to MFCN falls into the "exclusive user mode." In the "LSA user mode" a non-MFCN license holder (incumbent) would share spectrum access rights with one or more mobile network operators (LSA licensee), which can use the spectrum under defined conditions, subject to individual agreement and permission by the relevant regulatory authority.

The user modes can occur in their basic form (continuous lines), or as evolution of current approaches in the form of "limited spectrum pool" or "mutual renting" (dashed lines), see Figure 2.2. Limited spectrum pool is used in spectrum usage scenarios where

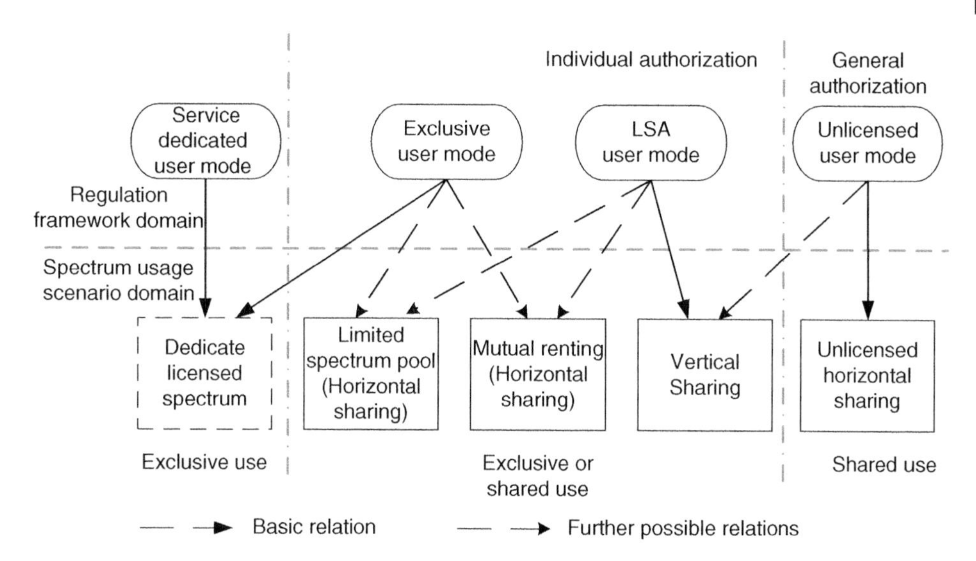

Figure 2.2 METIS-II concept for spectrum management and sharing for 5G mobile networks.

a limited number of known operators obtain authorizations to access a spectrum band dynamically. It is envisioned that mutual agreements between licensees are such that the long-term share of an individual operator has a predictable minimum value.

Mutual renting allows an operator to rent at least part of its licensed spectrum resources to another operator, based on mutually agreed rules. While the spectrum ownership stays unchanged, the rules may define spectrum usage restrictions and spectrum owner protections. Mutual renting can provide both static (i.e. like exclusive use), and/or dynamic shared spectrum.

Depending on the duration (static or dynamic) of the spectrum access, the spectrum usage scenarios "limited spectrum pool," "mutual renting" , and "vertical sharing" are considered as exclusive use or shared use.

In the "unlicensed user mode" spectrum access and usage rights are granted by general authorization, i.e. without an individual license, but subject to certain technical restrictions or conditions like limited transmit power or functional features like duty cycle or listen-before-talk as required in many license-exempt bands. In this mode, users cannot claim protection and may receive interference from other users.

Spectrum sharing between systems of different priority, e.g. if incumbent users must be protected in the "LSA user mode," is referred to as "vertical sharing," and sharing between systems of equal priority is called "horizontal sharing." For example, 5 GHz WLAN systems need to ensure protection of the incumbent radar systems (vertical sharing), and coexistence with other WLAN systems (horizontal sharing).

To achieve high spectrum usage efficiency, 5G systems may preferably support all spectrum usage scenarios shown in Figure 2.2, noting that several scenarios may occur simultaneously.

2.5 Characteristics of Spectrum Available for 5G

As spectrum considered for 5G is foreseen to be in higher bands (above 6 GHz) than are used in today's mobile networks, the propagation characteristics of this use of spectrum needs to be well understood. Industry and academia have performed a significant number of measurements to evaluate channel characteristics and propagation conditions in expected 5G network deployments. Measurement results and analysis reported significant new aspects related to frequencies above 6 GHz (see [13–15]).

Based on these measurements and proposed models, 3GPP developed channel models that are used for evaluating different solutions for 5G radio technologies and finally evaluating the agreed 5G radio technology (see [16]).

Channel models developed by 3GPP are statistical models. This means that they are not absolutely exact at any given location, city center, indoor office, etc. but are giving a good approximation of signal propagation in those environments. A more detailed modeling would require exact models of locations, including buildings, vegetation, BS location, etc. with ray tracing tools where the actual signal path including reflections is considered. Such methods would provide exact knowledge of signal propagation at that location but utilizing such models more broadly would be difficult. In the next section the main parts of the 3GPP models are explained.

2.5.1 Pathloss

A first required evaluation in system design is to estimate pathloss with different receiver distances (Figure 2.3). For details see also [16]. A good estimation of the pathloss can provide a very intuitive view to the system as part of link budged calculations. With given transmission powers and antenna configurations achievable data rates can be estimated on a given link distance. The benefit of such calculations is that these results can be obtained without any detailed link or system simulations. Additionally, the pathloss models are used in system simulations as well as in network planning tools when no detailed environment information is available.

The 3GPP channel models cover several different environments, each having different pathloss models separately for both Line of Sight (LOS) and Non-Line of Sight (NLOS) conditions. Supported environments are Rural Macro (RMa), Urban Macro (UMa), Urban Micro Street Canyon (UMi – SC) and Indoor Hotspot (InH). For example, the indoor NLOS pathloss is defined by following functions with additional random loss based on normal distribution $N(0,\sigma^2{}_P)$ probability function with an 8 dB standard deviation:

$$PL_{\text{InH-NLOS}} = \max(PL_{\text{InH-LOS}}, PL'_{\text{InH-NLOS}})$$

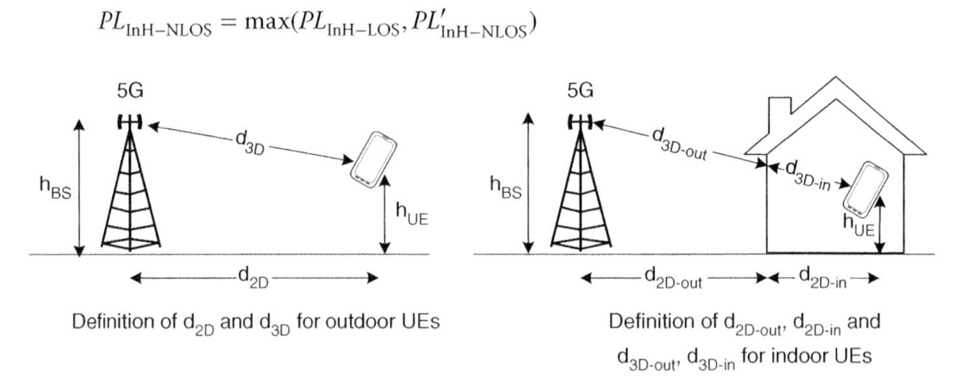

Figure 2.3 Definition of distances for pathloss models.

where $PL_{InH\text{-}LOS}$ is the pathloss for LOS conditions and defined as

$$PL_{\text{InH-LOS}} = 32.4 + 17.3\log_{10}(d_{3D}) + 20\log_{10}(f_c)$$

and $PL'_{InH\text{-}NLOS}$ is defined as

$$PL'_{\text{InH-NLOS}} = 38.3\log_{10}(d_{3D}) + 17.30 + 24.9\log_{10}(f_c)$$

The function shown above assumes that both transmitter and receiver are inside the building and there is a signal path from transmitter to receiver so that penetration through internal walls is not needed. By using similar kind of formulas, the pathloss for other environments can be calculated.

When considering an outdoor BS and indoor mobile terminals the total pathloss is the sum of outdoor pathloss PL_b, indoor pathloss PL_{in} and additional wall penetration loss PL_{tw} as shown in the following formula:

$$\text{PL} = \text{PL}_\text{b} + \text{PL}_\text{tw} + \text{PL}_\text{in} + N(0, \sigma_P^2)$$

The PL_{tw} is defined by the formula:

$$\text{PL}_{tw} = \text{PL}_{npi} - 10\log_{10} \sum_{i=1}^{N} \left(p_i \times 10^{\frac{L_{material\_i}}{-10}}\right)$$

where PL_npi is a constant value of 5 dB currently used in [16], and p_i is the proportion of the i-th material of the wall. The carrier frequency dependent values for each material $L_{materila\_i}$ is given in Table 2.5. This model allows e.g. calculating wall penetration loss for walls with certain percentage of concrete and standard multi-panel glass.

2.5.2 Multipath Propagation

After understanding pathloss to get a high-level view of the achievable link budget, it is essential for the system design to understand multipath propagation of the transmitted signal in different environments. The transmitted signal typically propagates through multiple paths, each path introducing different amplitude and phase to the received signal, causing frequency selective channel, i.e. frequency selective fading. This can be further used in frequency selective scheduling in Orthogonal Frequency-Division Multiplexing (OFDM) based systems where only strong frequency components, OFDM subcarriers, are used for transmitting data to the intended receiver. Fading frequency parts are left unused and assigned to other receivers, which have different multipath

Table 2.5 Material penetration losses.

Material	**Penetration loss (dB)**
Standard multi-pane glass	$L_{\text{glass}} = 2 + 0.2f$
IRR glass	$L_{\text{IIRglass}} = 23 + 0.3f$
Concrete	$L_{\text{concrete}} = 5 + 4f$
Wood	$L_{\text{wood}} = 4.85 + 0.12f$

Note: f is in GHz.

channel conditions. Thus, fading occurs to different OFDM subcarriers. Furthermore, in system design the frequency selectivity can be considered when defining reference symbols for channel estimation processes, so that the receiver's equalizer can compensate fading and phase distortion of the different OFDM subcarriers.

Additionally, the delay spread of these multipath signals is an important design criterion for physical layer numerology of the OFDM signal waveform. The OFDM signal utilizes Fast Fourier Transform (FFT) processing in the receiver, which can recover multipath signal components that are not delayed more than a cyclic prefix (CP) of the OFDM symbol. This enables the receiver to utilize the received energy of all multipath components in the demodulation process. However, multipath signal components that fall outside of a CP introduce inter symbol interference (ISI) from that portion of the symbol falling on the next symbol. To avoid ISI completely, the used CP should be longer than the delay of any multipath component received with any meaningful power. However, as CP is direct overhead to an OFDM signal, it cannot be made extensively long compared to the overall symbol duration, rather the CP length is a compromise between expected delay spread, used symbol length, and acceptable ISI and system overhead.

It is important to recognize that the deciding factor is the relative signal power of these different multipath components. Thus, even if there are multipath components that fall outside the CP, the ISI does not immediately destroy the data reception completely, but rather introduces noise floor to the signal that can become dominant in high Signal to Noise Ratio (SNR) conditions and can therefore be the limiting factor to utilize high modulation orders such as 256 QAM or 512 QAM (QAM stands for Quadrature Amplitude Modulation). Thus, longer symbols and lower modulation orders are more tolerant on ISI caused by delay spread, however longer symbols lead to narrower subcarrier spacing, which is more vulnerable for phase noise. The different waveform and OFDM numerology issues are further discussed in Chapter 3.

Finally, the understanding of angle of arrival (AOA) and angle of departure (AOD) properties in different environments allows exploding multi-antenna systems (see [17]). Multi-antenna systems with efficient beamforming solutions are essential on high millimeter operating frequencies to compensate high pathloss. Beamforming further increases the relative power of main multipath components, reducing the impact of delay spread and frequency selective fading. Beamforming and beam selection/steering are further discussed in Chapter 3.

2.6 NR Bands Defined by 3GPP

When defining operating bands for 5G NR in 3GPP, the approach was similar to LTE. In this approach, almost all existing bands defined for LTE or 3G are converted to be applicable to NR. As a result, the number of possible bands where NR can be deployed is high, even in the very first 3GPP release without the necessity of identifying completely new spectrum, if the bands already defined for LTE or 3G are available, or existing spectrum is refarmed to 5G. This can be seen from Table 2.6, where all 3GPP Rel-15 NR operating bands below 6 GHz are defined, notable band n1 is identical to Evolved Universal Mobile Telecommunications System Terrestrial Radio Access Network (E-UTRAN) band 1 as defined in [2]. The table also shows how fragmented the below 6 GHz spectrum is globally, as only some of the bands are available in single countries.

Table 2.6 Operating bands for NR at below 6 GHz defined in 3GPP Rel-15.

NR operating band	Uplink operating band BS receive/UE transmit F_{UL_low}–F_{UL_high} (MHz)	Downlink operating band BS transmit/UE receive F_{DL_low}–F_{DL_high} (MHz)	Duplex mode
n1	1920–1980	2110–2170	FDD
n2	1850–1910	1930–1990	FDD
n3	1710–1785	1805–1880	FDD
n5	824–849	869–894	FDD
n7	2500–2570	2620–2690	FDD
n8	880–915	925–960	FDD
n12	699–716	729–746	FDD
n20	832–862	791–821	FDD
n25	1850–19 515	1930–1995	FDD
n28	703–748	758–803	FDD
n34	2010–2025	2010–2025	TDD
n38	2570–2620	2570–2620	TDD
n39	1880–1920	1880–1920	TDD
n40	2300–2400	2300–2400	TDD
n41	2496–2690	2496–2690	TDD
n50	1432–1517	1432–1517	TDD
n51	1427–1432	1427–1432	TDD
n66	1710–1780	2110–2200	FDD
n70	1695–1710	1995–2020	FDD
n71	663–698	617–652	FDD
n74	1427–1470	1475–1518	FDD
n75	N/A	1432–1517	SDL
n76	N/A	1427–1432	SDL
n77	3300–4200	3300–4200	TDD
n78	3300–3800	3300–3800	TDD
n79	4400–5000	4400–5000	TDD
n80	1710–1785	N/A	SUL
n81	880–915	N/A	SUL
n82	832–862	N/A	SUL
n83	703–748	N/A	SUL
n84	1920–1980	N/A	SUL
n86	1710–1780	N/A	SUL

In addition to existing bands below 6 GHz (see [18]), also completely new bands and new ways of using bands have been introduced (see [19]). From new bands one should note bands n71, n77, n78, and n79. The n71 introduces 600 MHz support for NR that can be effectively used to provide wide area coverage in the US. The bands n77 and n78 are the bands for 3.5 GHz spectrum in different countries that is expected to be

Table 2.7 Operating bands for NR above 6 GHz defined in 3GPP Rel-15.

NR operating band	Uplink operating band BS receive/UE transmit F_{UL_low}–F_{UL_high} (MHz)	Downlink operating band BS transmit/UE receive F_{DL_low}–F_{DL_high} (MHz)	Duplex mode
n257	26 500–29 500	26 500–29 500	TDD
n258	24 250–27 500	24 250–27 500	TDD
n260	37 000–40 000	37 000–40 000	TDD
n261	27 500–283 500	27 500–283 500	TDD

available globally with few exceptions. Band n79 for 4.5 GHz is defined for Japan to provide NR operation comparable to the 3.5 GHz bands n77 and n78. These bands can provide continuous 100 MHz channel bandwidths support.

Regarding new type of utilization of spectrum, one should note bands n75, n76 and n80–n84. Bands n75 and n76 are so called Supplementary Downlink (SDL), which have only Downlink (DL) transmission without corresponding Uplink (UL) band as in FDD. The intention of these two bands is to be used as DL only channels coupled with other FDD or TDD bands together with Carrier Aggregation (CA) to boost DL data rates. Then bands n80–n84 are Supplementary Uplink (SUL) bands, to provide improved UL coverage for higher frequency FDD and TDD bands. Main use case for these SUL bands is to serve as additional UL for n77, n78 and n79 bands, which have higher pathloss due to higher carrier frequency and compensate limited transmission powers of the User Equipment (UE), although notably the same UL bands are also available as UL parts of regular FDD bands.

Above 6 GHz, several operating bands for NR are defined. Band n257 is matching with requirements of US, Republic of Korea, and Japan, however, this frequency area was not included as part of the International Telecommunications Union (ITU) studies as discussed above. The n258 is the band definition for the European pioneering band mentioned above, which may become more globally available at WRC-2019. Finally, n260 and n261 definitions are addressing the frequency bands available in US.

As 5G must support a high variety of operating bands, with different carrier frequencies and carrier bandwidths, 3GPP has defined a flexible radio interface that can be parameterized to allow most optimal operation at the given band and avoid unnecessary implementation complexities. Different physical layer configurations and the applicability of these bands are discussed in Chapter 3 with more details together with NR physical radio interface concepts.

In addition to bands in Tables 2.6 and 2.7, 3GPP has defined a set of band combinations that can be used either with CA or dual-connectivity (DC) to increase the operating bandwidth of the UE. Finally, the specification in [20] defines the bands that can be used together with other radio access technologies, namely LTE.

References

All 3GPP specifications can be found under http://www.3gpp.org/ftp/Specs/latest. The acronym "TS" stands for Technical Specification, "TR" for Technical Report.

1 The World Radiocommunication Conference (Geneva, 2015: RESOLUTION 238 (WRC-15) R0C0A00000C0014PDFE.pdf: Studies on frequency-related matters for International Mobile Telecommunications identification including possible additional allocations to the mobile services on a primary basis in portion(s) of the frequency range between 24.25 and 86 GHz for the future development of International Mobile Telecommunications for 2020 and beyond.
2 3GPP TS 36.101: "Technical Specification Group Radio Access Network; Evolved Universal Terrestrial Radio Access (E-UTRA); User Equipment (UE) radio transmission and reception", 2018.
3 Finnish Ministry of Transport and Communication INVITATION FOR APPLICATIONS FOR FREQUENCY BAND ALLOCATION Telecom operating licenses for the spectrum 703–733 MHz and 758–788 MHz, Finnish Ministry of Transport and communication. https://www.viestintavirasto.fi/attachments/maaraykset/LVM_hakuilmoitus_eng.pdf.
4 Finnish Communications Regulation Authority Spectrum auction concluded, The auction of six operating licenses within the 703 - 733 MHz and 758 - 788 MHz bands concluded on 24th November 2016, Finnish Communications Regulation Authority: https://www.viestintavirasto.fi/en/ficora/news/2016/spectrumauctionconcluded.html.
5 OECD Broadband Portal, Mobile data usage per mobile broadband subscription, 2017, last updated: 10-Oct-2018. http://www.oecd.org/sti/broadband/broadband-statistics/
6 ETSI EN 300 220-1 V3.1.1 (2017-02): "Short Range Devices (SRD) operating in the frequency range 25 MHz to 1000 MHz; Part 1: Technical characteristics and methods of measurement", 2017.
7 Electronic Code of Federal Regulations, e-CFR data is current as of May 10, 2018, §15.247 Operation within the bands 902-928 MHz, 2400-2483.5 MHz, and 5725-5850 MHz.
8 ETSI EN 300 328 V2.1.1 (2016-11): "Wideband transmission systems; Data transmission equipment operating in the 2.4 GHz ISM band and using wide band modulation techniques; Harmonised Standard covering the essential requirements of article 3.2 of Directive 2014/53/EU", 2016.
9 ETSI EN 301 893 V2.1.1 (2017-05): "5 GHz RLAN; Harmonised Standard covering the essential requirements of article 3.2 of Directive 2014/53/EU", 2017.
10 ETSI EN 302 567 V2.0.22 (2016-12): "Multiple-Gigabit/s radio equipment operating in the 60 GHz band; Harmonised Standard covering the essential requirements of article 3.2 of Directive 2014/53/EU", 2016.
11 Revision of Part 15 of the Commission's Rules Regarding Operation in the 57-64 GHz Band REPORT AND ORDER, Released: August 9, 2013.
12 METIS-II, Deliverable D3.2: "Roadmap to enable and secure sufficient access to adequate spectrum for 5G", Mobile and wireless communications Enablers for the Twenty-twenty Information Society-II".
13 Rappaport, T.S., MacCartney, George R., Samimi, Matthew K., et al.: "Wideband Millimeter-Wave Propagation Measurements and Channel Models for Future Wireless Communication System Design".
14 METIS: "Deliverable D1.2 Initial channel models based on measurements", Mobile and wireless communications Enablers for the Twenty-twenty Information Society".

15 Thomas, T.A., Nguyen, Huan Cong, MacCartney, George R., et al.: "3D mmWave Channel Model Proposal", Vehicular Technology Conference (VTC Fall), 2014 IEEE 80th, September 14 - 17, 2014.
16 3GPP TR 38.901: "Study on channel model for frequencies from 0.5 to 100 GHz", 2018.
17 Samimi M, Wang, Kevin, Azar, Yaniv, et al.: "28 GHz Angle of Arrival and Angle of Departure Analysis for Outdoor Cellular Communications using Steerable Beam Antennas in New York City", VTC2013.
18 3GPP TS 38.101-1 NR: "User Equipment (UE) radio transmission and reception; Part 1: Range 1 Standalone", 2018.
19 3GPP TS 38.101-2 NR: "User Equipment (UE) radio transmission and reception; Part 2: Range 2 Standalone", 2018.
20 3GPP TS 38.101-3 NR; User Equipment (UE) radio transmission and reception; Part 3: Range 1 and Range 2 Interworking operation with other radios, 2018.

3

Radio Access Technology

Sami Hakola[1], Toni Levanen[2], Juho Pirskanen[3], Karri Ranta-aho[4], Samuli Turtinen[1], Keeth Jayasinghe[4] and Fred Vook[5]

[1] *Nokia, Oulu, Finland*
[2] *Tampere University, Tampere, Finland*
[3] *Wirepas, Tampere, Finland*
[4] *Nokia, Espoo, Finland*
[5] *Nokia, Naperville, USA*

3.1 Evolution Toward 5G

3.1.1 Introduction

Cellular technologies have developed tremendously in past decades. The global system for mobile communications (GSM) technology as the best known 2G technology was highly successful for voice service, changing completely our way of communicating over telephone and being available everywhere and at any time. GSM also laid the foundation for cellular data services with the introduction of general packet radio service (GPRS). The 3G, i.e. Universal Mobile Telecommunications System (UMTS) with High Speed Downlink Packet Access (HSDPA) and High Speed Uplink Packet Access (HSUPA) features, started the real data boom with variety of different data applications in devices. At the same time different companies introduced new mobile phone types, more capable for data communication with different data application compared to earlier generation smartphones and without a real keypad like the iPhone. This data boom was further emphasized by introduction of Long-Term Evolution (LTE) and formally 4G with Long-Term Evolution Advanced (LTE-A) features. The development in cellular technologies as well as development in wireless local area networks (WLANs), i.e. WLAN802.11 changed our thinking on data connectivity, availability of the connectivity, as well as mobility and using mobile data services also in outside of the home country.

The first 3G UMTS Release '99 supported theoretical data rates of 2 Mbps in downlink (DL) and 768 kbps in uplink (UL), however, practical downlink data rates where around 384 kbps with uplink data rates limited many times between 64 and 128 kbps (see [1, 2]). Data rates in downlink increased with HSDPA to 7.2 Mbps and up to 14.4 Mbps in 3GPP Release 5 and in uplink by HSUPA from 1.4 Mbps to 5.7 Mbps in Release 6. Downlink data rates of High Speed Packet Access (HSPA), which is the combination of HSDPA and HSUPA, where further increased by dual carrier operation that doubled the available utilized bandwidth to 10 MHz and the corresponding data rates. The introduction of 64QAM modulation added another 50% to the peak data rates.

5G for the Connected World, First Edition.
Edited by Devaki Chandramouli, Rainer Liebhart and Juho Pirskanen.

Instead of 5 MHz channel bandwidth utilized by HSPA, LTE introduced a significant step upward in the maximum channel bandwidth by supporting 20 MHz channels, which was set as a minimum User Equipment (UE) capability already at the introduction of LTE in 3GPP Release 8. The change of the waveform and access technology from Direct Spread Code Division Multiple Access (DS-CDMA) to Cyclic Prefix Orthogonal Frequency-Division Multiple Access (CP-OFDMA) in downlink (DL) and Discrete Fourier Transform-spread-Orthogonal Frequency-Division Multiplexing (DFT-s-OFDM) in uplink, allowed better utilization of Multiple Input Multiple Output (MIMO) technique in downlink. The use of MIMO and increased bandwidth resulted in DL data rates from 100 to 150 Mbps in practical handheld Release 8 devices. Respectively uplink data rates were ranging from 25 to 50 Mbps. These data rates were further increased in 3GPP Release 10 mainly by the Carrier Aggregation (CA) feature adding more bandwidth to the radio link. The introduction of Release 10 is also known as LTE-A or 4G, as LTE Advanced was the official submission from 3GPP as a Radio Interface Technology (RIT) for 4G to the International Telecom Union Radio sector (ITU-R). After Release 10, next releases of LTE introduced new capabilities for the UE by extending the number of carriers that can be simultaneously aggregated to the UE. Thus, LTE data rates have been further increased, depicted in the Figure 3.1, which presents uplink and downlink data rates from 3GPP Release '99 to Release 12.

In addition to carrier aggregation, the future releases of LTE introduced several other features such as Dual Connectivity, full dimension MIMO, Coordinated Multipoint Transmission (CoMP), Machine Type Communication (MTC) and Narrow band Internet of Things (NB-IoT). Many of these features are conceptually used as part of 5G technology.

LTE originally focused on mobile broadband (MBB) use cases, MTC and narrow band IoT brought new uses cases into the picture by introducing optimized support for low power IoT devices connecting to cellular networks. In 5G, in addition to enhanced Mobile Broadband (eMBB) and massive Machine Type Communication (mMTC),

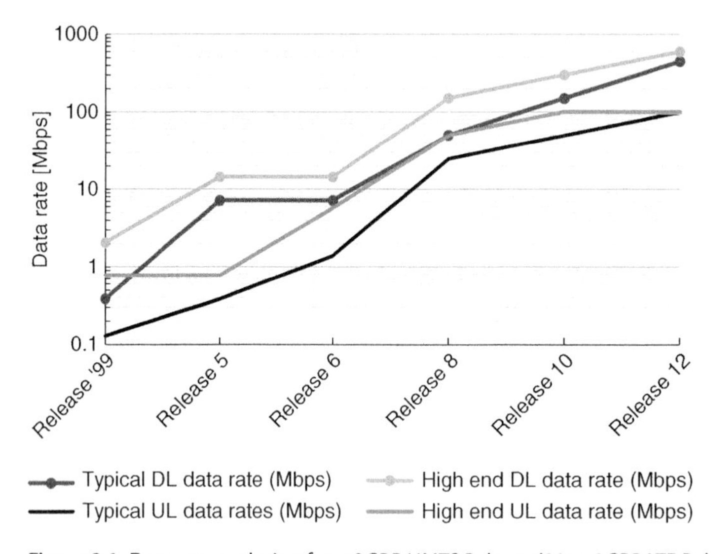

Figure 3.1 Data rate evolution from 3GPP UMTS Release '99 to 3GPP LTE Release 12.

the new Ultra Reliable and Low Latency Communication (URLLC) use case has been considered.

In eMBB, it is evident that for high quality of experience for the end user, low communication latency and high throughput are essential. However, eMBB services such as fast file transfer, video streaming in high-definition quality, and different mobile applications can tolerate long delays and errors in data transfers that are then recovered by re-transmissions. However, the URLLC service is characterized by requiring both very high reliability of the communication and latency that is at such a low level that possibilities of performing re-transmission is very limited or not possible at all in radio protocols, on TCP/IP or application level.

The utilization of cellular technologies for self-driving cars, robots, factory automation to name a few new applications, has raised significant interest in URLLC. However, it can be expected that many URLLC applications will be introduced in industrial environments where externally controlled machines, such as cranes, trucks, or forklifts, are operating in restricted areas. The break of communication could then lead to slow down of the operational speed of the machine, full stop or even emergency stop causing even damages to the machine. Depending on issues caused by the communication loss, such cases should be avoided especially when operation is required 24 hours a day and 7 days a week. The radio solutions designed for a URLLC communication area are discussed with overall system solutions in Chapter 8.

3.1.2 Pre-Standard Solutions

Before the 3GPP standardization process started to create a globally harmonized technology standard for 5G, several companies published proposals for a future radio interface design, such as presented in [3, 4]. Additionally, different industry players demonstrated 5G technology solutions and benefits of the features with proprietary Proof-of-Concept (PoC) implementations and test trials (see [5, 6]).

Two major telecom operators, Korean Telecom (KT) in South-Korea and Verizon Wireless in the US, called several network vendors and UE chipset vendors to define pre-standard 5G solutions. In the case of KT, the work was done in the KT PyeongChang 5G Special Interest Group (5G-SIG) to realize the world's first 5G trial service at PyeongChang 2018 Olympic Winter Games. In the US, the work was organized in the Verizon 5G Technology Forum (5GTF) targeting for a special commercial use case. As most companies working in KT 5G-SIG and 5GTF were the same, several physical layer (PHY) solutions were identical even though separate specifications were published (see [7, 8]).

The main difference between these two pre-standard solutions were that KT 5G-SIG aimed for seamless mobility between LTE and the pre-standard 5G trial service, while 5GTF was designed with a Fixed Wireless Access (FWA) use case in mind. The selected user protocol architecture depicted in Figure 3.2 was designed to support dual connectivity (DC) between commercial LTE radio access and the KT 5G-SIG radio solution. This new service would only be available for selected customers with new handheld devices able to support Korean Telecom-Special Interest Group (KT-SIG) defined pre-5G radio technology.

On the other hand, the work in 5GTF aimed to deliver a fiber over wireless type of service. In this service a very high data throughput radio link is used to provide last

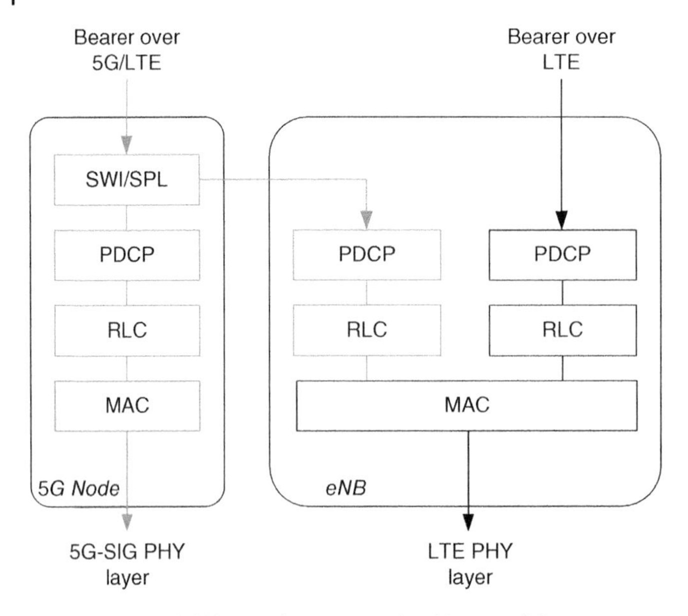

Figure 3.2 KT 5G-SIG user plane protocol architecture [9].

hop connection for home and offices, installing instead a fiber cable. In this solution fixed Customer Premises Equipment (CPE) devices, terminates the 5GTF radio connection and provides backhaul link to end users at home and office networks as shown in Figure 3.3. The benefit of this solution is that it allows end users to utilize their existing

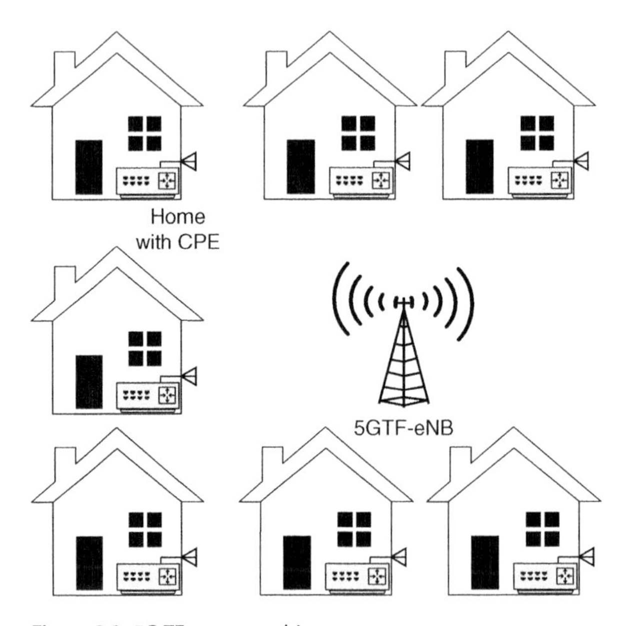

Figure 3.3 5G-TF system architecture.

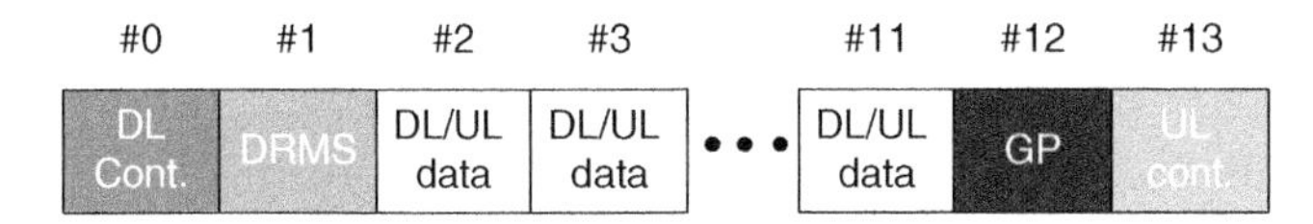

Figure 3.4 Bi-direction frame type in KT 5G-SIG and 5GTF.

devices with Wi-Fi radio connection to connect to the CPE and then to the Internet. Additionally, the installation costs in homes and offices is limited compared to fixed fiber cabling costs.

Due to the different uses cases, KT-SIG and 5GTF utilized different architecture options. However, to avoid extensive specification and implementation work, major parts of radio protocols and functions were harmonized between 5G-SIG and 5G-TF. The harmonization was done in the PHY, as well as higher layers whereever it was possible. Both systems supported:

- Operation at 28 GHz
- 100 MHz channel bandwidth, with maximum of eight aggregated carriers
- 5 Gbps as maximum DL data rate
- Massive-MIMO with support to hybrid beamforming Transmitter Exchange/ Receiver Exchange (TX/RX) architecture for both UE and Base Station (BS) receivers and transmitters.
- CP-OFDMA for both UL and DL direction.
- Subcarrier spacing (SCS) of 75 kHz with 2048 fast Fourier transform (FFT) size.
- 0.2 ms frame length with support of bi-directional frame as shown in Figure 3.4.
- low-density parity check (LDPC) channel coding for data channel.
- Tail-Biting Convolutional Coding (TBCC) for control channels.
- Support of TX and RX beamforming of all channels including synchronization channels.
- Beam management for beamformed data transmissions.
- Concatenation on Medium Access Control (MAC), for optimized L2 processing for high date rates.

When considering supported features defined for KT-SIG and 5GTF pre-standard solutions and features defined in 3GPP Rel15, one can observe that many features defined in these pre-standard solutions are also supported by 3GPP specifications. However, there are several differences in how these features are defined in detail.

Reasons for these differences, are manifold, for 5G New Radio (NR) standard in 3GPP needs to support carrier frequencies from 600 to 42 GHz, instead of a single 28 GHz band. Instead of single use case implementations supported by pre-standard solutions, cost and energy consumption optimized multimode NR and LTE device and BS implementations are expected. The pre-standard solutions also took many design choices as quick engineering decisions motivated by fast implementation rather than optimizing performance, energy consumption or generic adaptability. In the following sections, we go through the 3GPP Release 15 radio solution in more details, starting with basic building blocks of PHY and then discussing the actual realization of the 3GPP Release 15 radio interface covering also the radio protocols.

3.2 Basic Building Blocks

3.2.1 Waveforms for Downlink and Uplink

In the 5G research phase different waveform options were extensively evaluated. The main motivation for these studies was to find new waveform solutions for improving frequency localization, compared to CP-OFDM while keeping the benefits of CP-OFDM, such as simple channel estimation and equalization, good support for MIMO, and flexible multiplexing of users in the frequency domain (FDM). At the same time, the waveform should avoid high peak-to-average power ratio (PAPR), extensively complex solutions, and preferably maintain the good time resolution of CP-OFDM with low inter-symbol interference.

Additionally, studies considered NR propagation characteristics of between 6 and 52.6 GHz and frequencies above 52.6 GHz. In high frequencies, the radio environment is different compared to traditional cellular bands especially in terms of reduced delay spread, significantly higher attenuation of non-line of sight (NLOS) radio links.

The improved frequency localization is motivated by two main drivers. First, there was a clear desire to improve the spectral efficiency of the 5th generation telecommunication system by reducing required guard bands and improving spectral utilization from LTE where 90% spectral utilization was achieved in most of the channels. This implies that, in a LTE 20 MHz channel 2 MHz is dedicated for guard bands and only 18 MHz is used for data transmission. Secondly, improved spectral localization is especially beneficial when multiple different technologies, waveforms or OFDM numerologies are operating in parallel on the same band, as it reduces the guard bands needed between different technologies or numerologies. Furthermore, improved spectral localization would be beneficial when contention-based access with unsynchronized transmission is supported in the uplink. Improved spectral localization is needed to reduce interference caused by unsynchronized transmission occurring at adjacent OFDM subcarriers in the same subframe with synchronized uplink transmission.

The main candidate waveforms that gained interest after the research phase were conventional CP-OFDM, CP-OFDM with windowed overlap-and-add (WOLA) processing, universally filtered multicarrier (UFMC), and filtered orthogonal frequency division multiplexing (f-OFDM). These waveforms were mainly considered for below 6 GHz frequency bands, where channel bandwidths are narrower and spectrum availability is more limited compared to higher frequencies. The common nominator for considered waveforms is that they build on conventional CP-OFDM waveform, for which a transmitter is presented in Figure 3.5.

In Figure 3.6, the power spectral density (PSD) response for CP-OFDM, WOLA, universal filtered orthogonal frequency division multiplexing (UF-OFDM) and f-OFDM are given in the case of one physical resource block (PRB) allocation at the edge of a 10 MHz channel edge while assuming a maximum of 50 PRB allocation, following the LTE numerology.

WOLA, which is based on cyclic extension and time domain (TDM) windowing of CP-OFDM symbols combined with overlap-and-add processing, reduces the out-of-band spectral emissions by smoothing the transition from one CP-OFDM symbol to another. In Figure 3.6, the PSD response for WOLA is shown for two different window slope lengths, $N_{ws} = 18$ and $N_{ws} = 36$, where window slope length defines the rising or

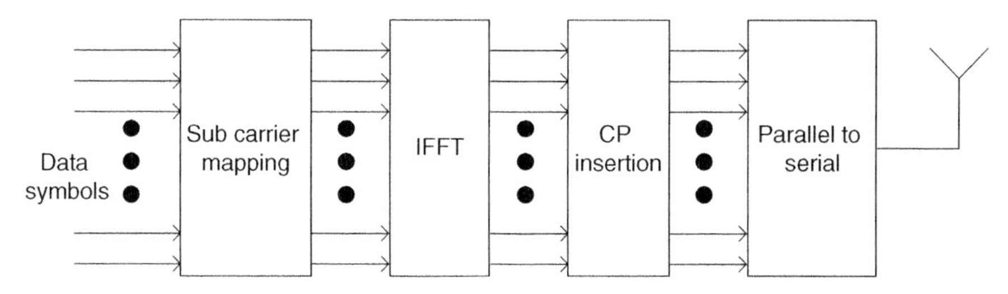

Figure 3.5 Conventional CP-OFDMA transmitter block diagram.

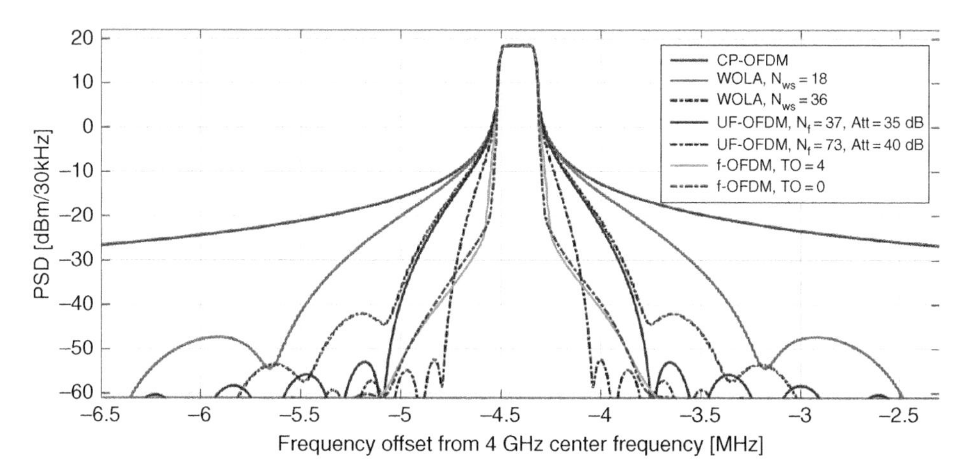

Figure 3.6 Example PSD realizations for different waveform candidates.

falling slope length of the used TDM raised cosine window. The window length can be also expressed as a roll off, which in this case would correspond to $r = 1.6\%$ and $r = 3.3\%$ for window slope lengths $N_{ws} = 18$ and $N_{ws} = 36$, respectively.

For UF-OFDM, Dolph-Chebyshev window is used to define the TDM filter response which is adapted by two different parameters: length of the filter, N_F, and minimum attenuation in the stopband, A_{min} (see [10, 11]). By tuning these two parameters a trade-off between filter length and filter's 3 dB bandwidth is obtained.

The f-OFDM waveform is a Hann windowed sinc-function-based sub-band filtered CP-OFDM waveform (see [12]). The sinc-function corresponds to the given allocation size and its TDM response is windowed with the well-known Hann window. The filter length is half of the OFDM symbol length, and in this example, corresponds to $N_F = 512$ samples. To adjust the in-band distortion caused by the filtering, a tone offset (TO) has been introduced. Tone offset defines the increase in the passband width in terms of integer multiple of SCSs. Thus, in this case TO = 4 corresponds to 4*15 = 60 kHz increase in the passband width.

Additionally, different single carrier waveforms such as Discrete Fourier Transform (DFT)-spread OFDM (DFT-s- with CP or zero tail [ZT]) have been considered for uplink access and for higher millimeter wave frequencies. Here we consider centimeter wave frequencies to cover frequencies from 3 to 30 GHz and wave frequencies to cover frequencies from 30 to 300 GHz.

For millimeter wave frequencies, it is important to have very low PAPR, because the power amplifier (PA) efficiencies tend to drop as the carrier frequency increases. Therefore, to obtain reasonable power efficiency and emitted powers from handheld devices, minimization of the PAPR is critical. When considering DFT-s-OFDM, using pulse shaping allows to further reduce the PAPR of the transmitted signal. The pulse shaping filter is typically a root-raised cosine filter, which is implemented in the frequency-domain of the ZT DFT-s-OFDM transmitter, this improvement in PAPR comes at a cost of spectral efficiency as increasing the pulse shaping roll off factor reduces PAPR, and hence a trade-off between PA efficiency and spectral efficiency needs to be carefully considered. On the other hand, at millimeter wave frequencies we can expect channel bandwidths up to 2 GHz, and therefore spectral efficiency is not that critical for achieving high throughputs. Furthermore, the ZT DFT-s-OFDM, allows to maintain symbol synchronization while allowing variable guard periods (GPs), e.g. for control and data signaling or for different users and is considered as a strong candidate as the waveform for beyond 52.6 GHz is studied in 3GPP Release 16.

During the 3GPP standardization process, different companies proposed different waveforms for evaluation, resulting in an industry agreement to relay on CP-OFDM-based waveforms in below 52.6 GHz communications (see [13]). Those were considered as the most suitable choice for the downlink, uplink, and device to device (D2D) transmissions in NR. For 5G MBB services, CP-OFDM is well suited due to the provided good time-localization properties enabling low latency and low-cost receivers with good MIMO and beamforming performance. Additionally, it was demonstrated that different filtering schemes, such as UF-OFDM and f-OFDMA, can be introduced separately in transmitter and receiver units and it is not necessary to utilize same signal processing method in both ends.

The simple reason for this is that all the considered methods try to manipulate the out-of-band signal component in transmitter so that the signal energy in adjacent frequencies is reduced while avoiding distorting the desired in-band signal. Similarly, in the receiver side, these techniques provide means to suppress the interference coming from adjacent channel and out-of-band signals without affecting desired in-band signal. Therefore, these transmitter and receiver characteristics can be tested with separate TX and RX tests as shown in Figures 3.7 and 3.8 for transmitter and receiver respectively (see [14]). Furthermore, this testing approach allows that transmitter and receiver signal processing techniques to be independent, enabling separate development for transmitter and receiver implementations, which can open completely new avenue for future technology development.

As different filtering solutions cannot totally avoid introduction of inter symbol interference (ISI) to an in-band desired signal, it is beneficial to perform filtering

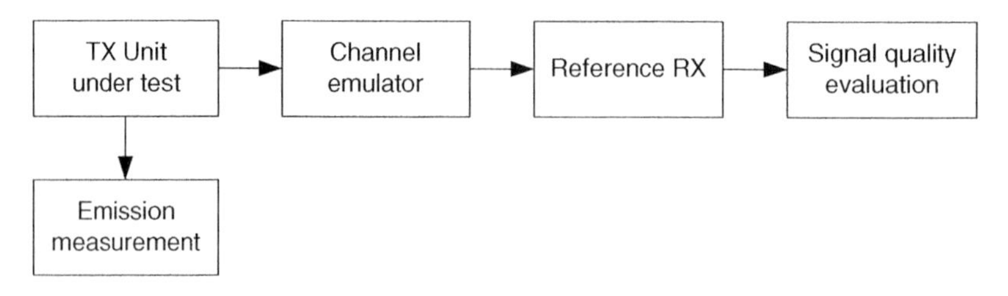

Figure 3.7 Transmitter unit test setup.

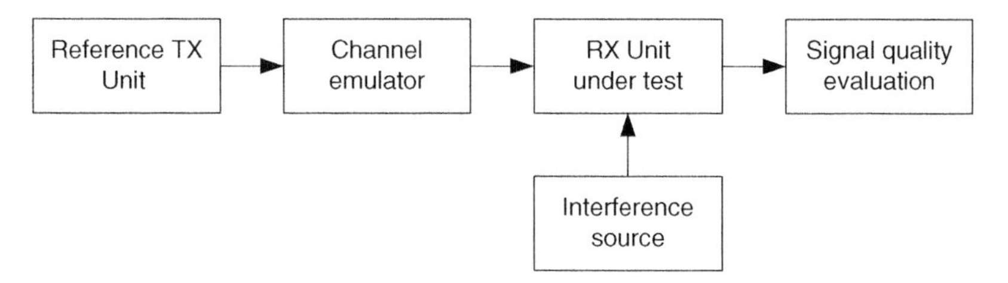

Figure 3.8 Receiver unit test setup.

only when needed. Especially in downlink, where transmissions for different UEs are synchronized, the transmitter should filter signal per numerology rather on than sub-band level (e.g. per UE allocation and per numerology). When high modulation and coding scheme (MCS) is used in high Signal to Noise Ratio (SNR) conditions, the ISI becomes a dominating factor and thus filtering may become the limiting factor on the maximum performance due to increased transmitter error vector magnitude (EVM) (see [14, 15]).

In previous mobile technology generations such as 2G and 3G, the waveform processing solutions were fixed by the standard, and the RX and TX processing solutions were limited to match the standard. In these cases, any changes to the standards would have required a new generation of mobile devices and network infrastructure operating on different bands than the originally used waveform. However, with the new transparent waveform processing approach both UE and BS implementations may introduce new TX solutions that improve signal spectrum containment and RX solutions to reduce adjacent carrier and out-of-band interference independently in fully backward compatible manner, with the requirement that transmitter and/or receiver may also be a conventional CP-OFDM transmitter and/or receiver.

Due to above reasons, it is difficult to find a globally optimal solution, when considering all the different frequency bands supported by 5G NR. Due to the wide variety of use cases having e.g. different spectral containment, latency, and TX and RX signal quality requirements, it can be expected that several different solutions on top of conventional CP-OFDM are applied in the network and in the UE. This is exemplifying the potential of transparent waveform processing in enabling 5G NR to support all the diverse use cases envisioned.

The 3GPP TSG-RAN WG1 specifications are not defining any filtering scheme or requirements. Rather filtering requirements are set in 3GPP TSG-RAN WG4 specification for below 52.6 GHz operation, when defining in-band blocking and emission requirements for adjacent PRBs and neighboring channels. Similarly, TX and RX side filtering processing can be considered when defining out-of-band blocking and emission requirements for receiver and transmitter respectively.

In addition, to the spectrum containment discussion, the CP-OFDMA's known limitation on having a considerably high PAPR, an additional support for single carrier waveform was considered to improve uplink coverage so that 5G uplink coverage would be comparable to LTE [16].

The conclusion was that DFT-s-OFDM was selected as an additional uplink waveform for single stream and low MCS transmission, targeted to be supported mainly in coverage limited scenarios. The support of DFT-s-OFDM is compulsory for UEs but optional for base stations. Therefore, the main difference compared to LTE uplink is that in NR CP-OFDMA is the main option for the uplink waveform and DFT-spread OFDM is an additional option, while being the only option for LTE uplink.

Release 15 covered communications in the frequency range up to 52.6 GHz [17]. Frequencies beyond 52.6 GHz are considered in Release 16. This study mainly considers operation with a single carrier like waveforms. The highest emphasis is on CP DFT-s-OFDM and ZT DFT-s-OFDM waveforms as they allow efficient (FDM) frequency domain processing and can share several TX and RX functions in the implementation with existing solutions.

As discussed, in high carrier frequencies the PA efficiencies tend to drop and low PAPR is a critical design target for waveforms evaluated for communications. CP-OFDM could also be supported as an optional, short distance and high throughput waveform as in WLAN 802.11ad (see [18]). As 3GPP is designing a global mobile communication system, the DL and UL waveforms must provide sufficient coverage to allow reasonably dimensioned network implementations, and therefore minimizing PAPR allowing to maximize emitted signal power and PA efficiency while simultaneously minimize energy consumption, and PA cost is clearly the most important design parameter. Other important aspects are computational complexity, good time resolution, and zero prefix.

As the targeted throughputs are gigabits per second, the computational complexity per data bit should be smaller than in traditional carrier frequencies. This is already alleviated by the assumptions of a lower degree of spatial streams and lower modulation schemes to be used beyond 52.6 GHz. The number of spatial streams is typically assumed to be limited to two, which is achieved in Line of Sight (LOS) communications using two different polarizations. The lower modulation schemes are sufficient to achieve high throughputs as the channel bandwidths may be up to 2 GHz, allowing ultra-high throughputs for end-users even with binary phase sift keying (BPSK) modulation.

Good time resolution is critical for the desired waveform as in millimeter wave communications the main use case is assumed to be beamformed Time Division Duplex (TDD). Beamforming is required to overcome the increased pathloss at higher carrier frequencies and TDD is the most likely duplexing scheme with very narrow beams, as it is difficult to achieve reasonable multiplexing gains with Frequency Division Duplex (FDD).

One promising solution is to utilize ZT instead of CP, in DTF-s-OFDM transmission. This allows the transmitted symbol energy to drop near to zero between symbols, allowing transmitter to switch TX beams between symbols, enabling a highly efficient and agile beamforming for millimeter wave communications. Zero prefix is also easy to combine with pulse shaping filtering allowing to further reduce the PAPR of the transmitted signal while keeping low-power symbol transitions. The above mentioned ZT DFT-s-OFDM is an excellent solution, as it allows to constantly maintain symbol synchronization in the system while fulfilling all the requirements and adding on top the possibility to adapt the ZT duration per symbol.

It is expected that $\pi/2$-BPSK modulation will also play an important role in millimeter wave communications. It has been accepted to be used with CP DFT-s-OFDM-based uplink in 5G NR. In frequency range (FR)1 (450–6 GHz) it plays a relatively small role, as devices can achieve the maximum allowed emitted power of 23 dBm even with Quaternary Phase-Shift Keying (QPSK) modulation using CP-OFDM. Although, the future high power (HP) UE classes, which can transmit with higher power using smaller duty cycles, may benefit from this modulation. On the other hand, in millimeter wave communications where the UE devices can achieve only small array gains in beamforming and use cheaper and lower efficiency PAs need to rely in most cases on $\pi/2$-BPSK modulation to achieve good UL coverage. As in waveform design, the pulse shaping of $\pi/2$-BPSK modulation was not explicitly defined in the specification. Instead, it is indirectly limited by the spectral flatness requirements imposed on the RX side.

3.2.2 Multiple Access

The 5G NR radio access is targeted to support both paired and unpaired spectrum with maximum commonalities of FDD used in paired spectrum and TDD in unpaired spectrum. The FDD transmission mode has been the dominant operational mode in previous cellular system, whereas TDD has been mainly used in China.

The FDD system requires separate frequency blocks for both DL and UL transmission with sufficient frequency gap allocated between them, so called duplex distance between DL and UL. The benefit of FDD is that with sufficient duplexing distance, the device's transmitter can be sufficiently isolated from the device receiver with a duplex filter allowing simultaneous UL transmission and DL reception. This is very suitable for URLLC as it minimizes both DL and UL latencies and Hybrid Automatic Repeat Request (HARQ) feedback loop delay is not bounded by DL to UL or UL to DL transmission direction switching.

The drawback of FDD is that separate spectrum allocation for DL and UL is needed with sufficient duplex distance between UL and DL portions on the same band. This can be very difficult due to the available spectrum band allocation. Additionally, the required duplex distance increases when carrier frequency increases with the resultthat above 3 GHz TDD starts to be more attractive and at high millimeter wave (mmWave) frequencies TDD is the only option. Finally, in FDD the spectrum resource allocation for UL and DL direction is fixed, as the allocation of the UL part for DL operation is not possible. Configurations, where distance between UL and DL part is not fixed, i.e. there are multiple possible downlink portions, or any downlink portion is possible for a single uplink band, can be considered as flexible duplexing approach, which sets new requirements for the duplex filtering.

In TDD, the transmission resources between UL and DL transmission are divided in time. The benefit of TDD is that only single spectrum allocation is needed as both uplink and downlink operate on the same frequency and no duplex distance definition is required, which is essential for frequencies above 6 GHz. This also allows TDD spectrum to be allocated between the FDD UL and DL frequency portions as shown in Figure 3.9. This is especially needed at spectrum allocations below 6 GHz trying to efficiently use the scarce spectral resources.

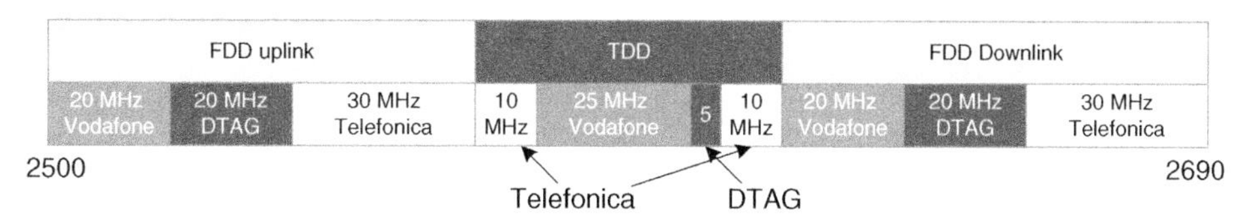

Figure 3.9 Spectrum allocation on 2.6 GHz in Germany.

In addition, in TDD operation the spectrum resource allocation is not fixed between UL and DL direction, as UL and DL frames can be dynamically allocated in NR by the packet scheduler of the Gigabit NodeB (gNB) based on capacity needs. This operation is often referred to as dynamic TDD. Utilizing spectrum between UL and DL frequency allocation of the FDD band and dynamic TDD has clear benefits, both have also some limitations and restrictions that set requirements for implementation and the system operation.

The TDD spectrum allocation between the FDD uplink and downlink portions is hindered as high power FDD downlink transmission can easily interfere significantly with UE DL reception and BS UL reception. Similarly, FDD uplink UE transmission may interfere with another UE TDD DL reception nearby. Finally, TDD UE and BS transmissions can cause interference to FDD UE reception. These interference conditions are most significant at the band edges, where out-of-band blocking requirements are toughest to meet. In LTE, these interference problems were relieved by reducing the actual operating bandwidth by not allocating resource blocks for UE transmission in UL. In LTE downlink transmission, the BS can either utilize additional proprietary TX filtering or non-schedule resource blocks for DL transmissions. Both methods reduce bandwidth utilization of the spectrum and therefore, advanced waveform processing solutions in both TX and RX, aiming to improve spectral containment, become interesting techniques for enable better utilization of such bands.

In dynamic TDD, the dynamic resource allocation between UL and DL can provide significant performance improvements, as the resources can be fully dynamically allocated either for DL or UL depending on traffic needs. Even though, the statistical ratio between uplink and downlink traffic is 1 : 10 in today's Internet traffic, the actual capacity demand can change significantly. In urban area network deployments, events such as music concerts, sport events, etc. are found to create significant uplink traffic with different social media picture and video uploads from the event. In office or hotspot deployments the number of simultaneously active users can be low, thus the immediate capacity need of actual user applications will dominate the UL/DL capacity split needed. Therefore, the capacity requirement at a given moment in time is not based on a statistical distribution; rather statistical distribution may only be achieved over a longer time period. In both cases, it is apparent that a fixed allocation between UL and DL resource can easily lead to the situation where either direction can be highly congested, limiting the achievable data rate. Additionally, as the traffic is bi-directional this may result in the under-utilization of resources in the other direction.

However, due to very similar UL-to-DL and DL-to-UL interference scenarios, as in TDD band allocations between FDD UL and DL portions, the TDD UL/DL allocation is expected to be quite fixed in NR TDD macro deployments. In such cases, all BS will operate synchronized with the same DL and UL frame pattern ensuring that cross-link interference does not occur. This will not only be needed between single operator BSs but also between operators in macro deployments without special antenna constellations. This results in the fact that DL and UL TDD configuration needs to be selected based on statistical distribution between DL and UL traffic. In higher frequencies, with small cell deployments and especially deployments with active antenna systems (AASs), the TDD configuration can more freely vary between different cells based on DL and UL traffic needs.

To enable dynamic TDD operation, several enabling design choices were made in the NR slot design and PHY operation. In the slot design, dynamic TDD is considered by defining such DL and UL control channel location options which do not suffer from cross-link interference between frame synchronized cells, even though user plane traffic is allocated to different directions. Additionally, to enable dynamic TDD operation, the UE does not make any assumption whether a certain frame contains DL and UL data portions. Rather, it operates based on scheduling information in the physical downlink control channel (PDCCH). Furthermore, the channel estimation and radio measurement design does not assume any continuous reference symbols in fixed locations mandating certain DL transmissions. These design choices allow gNBs to freely choose the frame format based on actual capacity needs. This way the fixed TDD operation is achieved by network implementation dependent configurations that mandate fixed scheduling for UL and DL frames.

When the transmission direction is not fixed to any uplink and downlink transmissions pattern, dynamic TDD can be further leveraged to support in-band backhaul in high mmWave deployments to simplify BS deployments as no fiber backhaul cabling would be needed. In this case, the time resources would be distributed between the data link and backhaul link, thus converting maximum throughput into deployment flexibility.

3.2.3 5G Numerology and Frame Structures

As discussed in Section 3.1.2, the pre-standard solutions applied 75 kHz SCS, which is five times higher than in LTE, for supporting 28 and 39 GHz frequencies with specific use cases. However, in 3GPP the target was to support a wider range of spectrum as well as different use cases and system bandwidths. The supported spectrum ranging from 600 to 100 GHz and possible system bandwidths from 5 to 1 GHz or even up to 2 GHz as for future extensions in above 52.6 GHz millimeter wave frequencies. Additionally, numerology must support excellent co-existence with LTE systems and allow economical implementation of multimode devices supporting both LTE and NR. Therefore, it become apparent that 75 kHz alone is not sufficient SCS even when considering spectrum below 40 GHz in the first phase of NR. Additionally, it was widely expressed by different UE vendors and gNB vendors to consider higher FFT/IFFT sizes than 2048 used in LTE for support of wider carrier bandwidths.

The requirement of supporting wide area cells and to have best possible co-existence between LTE resulted in 15 kHz SCS support in NR. To allow economical implementation of multimode devices, 3GPP adopted so called 2^N scaling of 15 kHz, where N is a positive integer, which results in SCS and nominal system bandwidth options in different frequency ranges as shown in Table 3.1.

As mentioned, the 15 kHz SCS is motivated by supporting excellent co-existence with LTE, as well as supporting very large cells in rural and sub-urban areas with large delay spreads. Additionally, 15 kHz SCS is needed for supporting bands with a maximum system bandwidth of only 5 MHz with high spectrum efficiency. The minimum operation bandwidth per numerology is defined by the Synchronization Signal (SS) Block as discussed in Section 3.3.1.

The 30 kHz SCS is mainly targeted to urban macro cells where typical inter-site distance (ISD) is below 1 km. Thus, it is anticipated that 30 kHz SCS would be directly

Table 3.1 Subcarrier spacing, nominal BW and frequency range.

N for $15*2^N$ Scaling	0	1	2	3	4
Subcarrier spacing (kHz)	15	30	60	120	240
Supported frequency range (GHz)	<1–6	<1–6	1.7–52.6	24.25–52.6	24.25–52.6
PRB bandwidth (kHz)	180	360	720	1440	—
Max number of resource blocks	270	273	264	264	—
Max BW (MHz)	48.6	99.280	190.08	380.16	—
Max FFT size	4096	4096	4096	4096	—
T symbol (μs)	66.7	33.33	16.6	8.33	4.17
CP (μs)	4.7	2.41	1.205	0.6	0.3
#symbols per slot	14	14	14	14	—
Slot duration (ms)	1	0.5	0.25	0.125	—
#slots in frame	10	20	40	80	160

usable in current urban macro sites in low carrier frequencies and is therefore the main deployment option in many below 6 GHz frequency bands.

The 60 kHz is SCS option in above 3 GHz carrier frequencies can be utilized even above 6 GHz. However, in high carrier frequencies the 60 kHz SCS sets sufficiently high requirements for oscillators to compensate carrier frequency and phase noise error. Therefore, 120 kHz SCS was also introduced and is the preferred numerology for above 24 GHz frequencies. Additionally, 120 kHz can provide 400 MHz system bandwidth with 4K FFT size, which is highly beneficial in above 24 GHz spectrum. To support a high number of synchronization signal and physical broadcast channel (PBCH) beams and a short sweeping procedure (see Section 3.3.1) the 240 kHz SCS was introduced for Synchronization Signal Block (SSBlock or SSB) transmissions (see Section 3.3.1).

A typical OFDM design is that cyclic prefix (CP) is approximately 5% of the symbol length introducing corresponding fixed overhead into the system. In addition to SCS and normal CP, an extended CP was considered. This was mainly proposed for 60 kHz SCS to extend possible coverage of the system operating with such short symbols in macro cell deployments, while maintaining the benefit of short slot length to minimize the delays of the system. However, the need for this option became significantly less important when the mini-slot concept with a two-symbol transmission length was introduced for frequencies below 6 GHz. Additionally, above 6 GHz a one symbol mini-slot concept is supported (see [13]).

The basic building unit of NR radio access is the Resource Element (RE), which is defined as one OFDM subcarrier in FDM and one OFDM symbol in TDM from a single antenna port. In the FDM, 12 RE's are grouped in a PRB. The PRB defines the minimum granularity of the transmission in FDM, being 180 kHz for 15 kHz SCS, and it is doubled when SCS is increased, as shown in Table 3.1. Due to this the maximum channel bandwidth is given in number of PRBs as shown in Table 3.1, and the remaining part of the carrier is used as guard band. As a result, with higher SCS the frequency utilization is slightly lower in narrower operating bands.

As the PRB defines the minimum granularity in FDM, it is also the basic unit for scheduling resources in FDM just like in LTE. There were no obvious benefits in using a different PRB size, and commonality with LTE was an enabler for LTE and NR frequency sharing. Thus, the LTE choice was carried forward to NR, even though, e.g. 16-subcarrier PRB was considered. Having frequency allocation in terms of PRBs of fixed number of subcarriers reduces the amount of signaling overhead compared to assigning subcarriers directly. Additionally, the same FDM scheduling implementation can be applied regardless of the used SCS. Furthermore, if multiple numerologies are used on a given carrier, there is always an integer number of lower SCS PRBs that match one higher SCS PRB, leading to a convenient nested PRB grid for all SCSs.

The TDM is organized in 10 ms frames, and slots of 14 symbols, with a varying number of slots in a frame depending on the used SCS as shown in Table 3.1. In addition, a 1 ms subframe is defined as in LTE but it has less meaning as all definitions are based on slot level operation.

For each slot length, the allowed slot structures have been defined in a flexible manner. However, all the different configurations follow three basic structures, which are downlink only slot, uplink only slot, and bi-directional slot. The uplink only and downlink only slots self-evidently contain only uplink or downlink symbols, which are used in FDD for UL and DL transmission, respectively. Additionally, uplink and downlink only slots can be used in TDD when longer transmission in either direction is desired by the network scheduler. Different bi-directional slot variants are introduced by different allocation of physical channels in the flexible symbols.

The flexible symbol can be allocated to be either a DL or UL symbol containing different DL or UL channels and the corresponding Demodulation Reference Symbol (DMRS) depending on the network scheduler decision. Between DL and UL symbols in actual configuration single symbol is used as switching gap, which also includes the necessary guard period (GP). Figure 3.10 depicts a frame containing first a DL symbol, then flexible symbols 1–12, i.e. either used for DL or UL direction, and finally symbol 13 used for UL direction.

For all slot structures the same principles are used to enable fast receiver processing and to maximize commonalities between different slot types. The DL only slot and bi-directional slot starts with a PDCCH portion, which can be from one to three symbols, followed by a DMRS, followed by the data symbols.

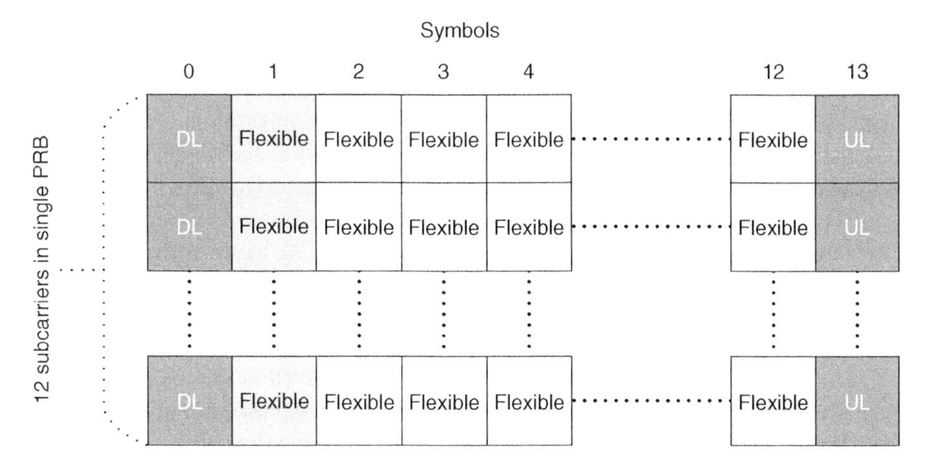

Figure 3.10 Bi-directional slot with DL symbol, flexible symbols and UL symbol.

In downlink only slots, all symbols 2–13 are DL data, and the Physical Downlink Shared Channel (PDSCH) portion can also use REs before the DMRS, if unoccupied by PDCCH. Containing the DMRS early in the structure enables efficient pipelined processing as the UE or gNB receiver can prepare the channel estimate early during the reception of the slot. This allows the receiver to start demodulating symbols and decoding the code blocks prior to having received all the symbols in the slot. Compared to LTE this allows for significantly reduced processing time. Processing time is defined as the time required before the UE can complete decoding of the received transport block and report Acknowledgement (ACK) or Negative Acknowledgement (NACK) back to the gNB. In difficult channel conditions additional DMRS symbols can be added to the frame structure to update the channel estimation. This comes with additional DMRS overhead as well as additional receiver processing requirements as the channel estimation must be updated during a single slot based on the additional DMRS symbol(s) before demodulation of the slot can be fully completed.

The time-domain resource allocation of PDSCH and Physical Uplink Shared Channel (PUSCH) and the placement of the scheduling PDCCH allows further flexibility to facilitate different slot structures and reduced scheduling and transmission latency. PDSCH allocations (including DMRS) can be of duration 2, 4, 7…14 symbols, and start at any DL symbol if the allocation does not span over the end of the 14-symbol slot, introducing mini-slots as discussed above. The PDCCH scheduling the PDSCH can be placed on any symbol in the slot, if the PDSCH scheduled by the PDCCH does not start earlier than the PDCCH. Similarly, in uplink the PUSCH allocation can be of any length, if the allocation does not span across the slot boundary (see [19]).

Such unlimited flexibility was designed to allow for constructing so-called mini-slot structures within slots for low latency traffic, as the data waiting for the start of the transmission as well as the data transmission duration take a shorter time. This comes with the cost of increased control and RS overhead. Thus, the basic scheduling of data can be expected to operate on a slot basis. With low-latency UE data processing it is also possible to construct self-contained slots, where the PDCCH and PDSCH are in the first part of the slot, the UE processes the data during the DL/UL switching gap and transmits the HARQ-ACK on Physical Uplink Control Channel (PUCCH) in the end of the same slot.

3.2.4 Bandwidth and Carrier Aggregation

As discussed above, the NR system supports many system bandwidths with exceptionally wide system bandwidths as depicted in Table 3.1. In previous cellular systems, 2G, 3G, LTE, the UE RF (Radio Frequency) bandwidth for reception and transmission was equal to the system bandwidth of the cell, i.e. 5 MHz in 3G, or 20 MHz in LTE respectively. In LTE Rel-8, also cell bandwidth narrower than 20 MHz were supported, which were enabled by network configuration. In such cases the UE RF bandwidth for RX and TX operation matched with the system bandwidth of the cell.

However, when the support for wider system bandwidth was increased significantly, mandating that UE RF bandwidth should always match the cell bandwidth for reception and transmission, was overwhelming and unnecessary. To allow UE to operate with narrower RX and TX bandwidth than the cell bandwidth, a definition of a bandwidth part (BWP) was introduced. The design of BWP allows UE specific control of each RF transmitter and receiver chain of the UE.

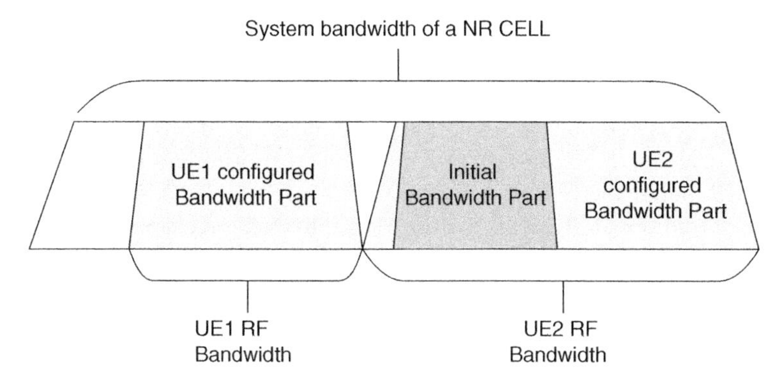

Figure 3.11 System bandwidth, initial BWP and configured BWP.

During initial access the UE utilizes the default BWP for receiving SSB and initializing connection via Random Access Channel (RACH). After the Radio Resource Control (RRC) connection is established, the gNB can configure the UE with UE specific set of BWP(s). In D and UL, a UE can be configured with up to four BWPs with a single BWP being active at a given time. Figure 3.11 illustrates the system bandwidth used by the gNB for transmitting and receiving data in the cell, the initial BWP for initial access, and separated dedicated BWP configured for two devices UE1 and UE2.

When a UE dedicated BWP is configured, the UE tunes its RF on the configured BWP, and therefore the UE is not expected to receive any physical channels or signals such as PDSCH, PDCCH, or Channel State Information Reference Signal (CSI-RS) outside an active BWP.

The carrier aggregation is a feature to aggregate transmissions from multiple cells. It was introduced in LTE Release 10. In NR, as the cell bandwidth is decoupled from UE RF bandwidth, NR carrier aggregation aggregates different BWP configured for the UE in different cells. In each cell, when BWP is configured for carrier aggregation operation of a UE, the same PHY definitions apply.

3.2.5 Massive MIMO (Massive Multiple Input Multiple Output)

Massive Multiple Input Multiple Output (mMIMO) is one of the essential features of 5G NR access. The utilization of mMIMO technology has been considered in all the aspects of the NR radio access design, and thus it is supported in all physical channels in both UL and DL directions. The primary purposes of mMIMO are to enhance system coverage and capacity. The benefits of mMIMO are highly dependent on many factors such as the deployment scenario, carrier frequency, channel characteristics, and the antenna array configuration.

The mMIMO is the extension of traditional MIMO technology to antenna arrays having a very large number of controllable antenna elements (AEs) for transmitting and receiving radio signals. The term MIMO is used in a rather broad manner to include any transmission scheme involving multiple transmit antennas and multiple receive antennas, and the term "massive" is intended to mean a number much greater than eight (8) transmit and receive antennas in the base station. Typically, the UE operates with a significantly lower number of antennas as small device size and low power consumption are desired. The term "controllable" antennas refers to antennas whose signals are adapted/modified by the PHY via both gain and phase control for adapting the overall response of the antenna array. Therefore, in addition to actual mMIMO transmission, mMIMO relies on appropriate mechanisms for obtaining information regarding the channel so that control of the antennas is possible and optimized. Methods for obtaining the necessary control information can vary depending on the transmission direction, antenna technology, UE mobility, etc.

The basic principle of traditional MIMO is to utilize separated uncorrelated channels between the transmitter and receiver antennas so as to transmit different data streams on the same physical resources. When multiple streams are transmitted between the gNB and a single UE on the same time-frequency resources, the term Single User Multiple Input Multiple Output (SU-MIMO) is used. However, as the number of uncorrelated antennas and transmitter receiver units (TXRUs) is often limited in practical UE implementations to 2 or 4, the number of SU-MIMO streams is typically limited to 2 or 4, in the downlink and to 2 or even single stream transmission in the uplink.

However, when the gNB has a higher number of TXRUs than the UE, the gNB capability can be utilized for multi-user Multiple Input Multiple Output (MU-MIMO). In MU-MIMO, multiple data streams are transmitted over the same PHY resources to multiple users simultaneously. At a high level, the principle of MU-MIMO is identical with SU-MIMO, namely the transmission of multiple data streams on the same PHY resources over uncorrelated channels between multiple transmit and multiple receive antennas. In MU-MIMO, as the distance between receiver antennas in different devices is typically much larger than the antennas on a single device, the channels in MU-MIMO are even less uncorrelated receiver antennas than in typical SU-MIMO transmission.

In SU-MIMO, multiple streams are transmitted between a single UE and the gNB for increasing single user throughput. However, since the practical maximum number of streams that can be achieved in SU-MIMO is limited to the number of antennas in the UE, the benefits of increasing the array size at the gNB tend to be limited unless MU-MIMO is leveraged. As a result, the primary system capacity gains from massive MIMO are achieved by leveraging MU-MIMO.

The coverage of the system can be improved by utilizing mMIMO with high gain adaptive beamforming, which focuses the transmitted energy toward the intended receiver. Coverage enhancement will be particularly important at higher carrier frequencies, where deployments tend to be coverage limited due to poor path loss conditions. Beamforming can also reduce the distribution of the interference seen in the system since the signals received at UEs other than the intended UE are typically combining non-coherently, which acts to increase the overall signal-to-interference-plus-noise experienced by UEs in the system.

Severe coverage limited situations at higher carrier frequencies pose two main difficulties that can be overcome with mMIMO. The first problem is that a cell wide broadcast control channel (BCCH) may not be feasible since the maximum pathloss may be too high to achieve a reasonable cell radius especially in high frequencies in mmWave bands. Therefore, to increase the cell radius, a grid-of-beams-based approach involving the sweeping of multiple narrow high-gain beams is supported for downlink synchronization and BCCHs, as discussed in Sections 3.3.1 and 3.3.2. The second problem is that it is difficult to acquire channel knowledge on a per-antenna-element basis when the individual antenna elements are low gain with high beam-width in severely path-loss limited channels.

To overcome these problems, a grid-of-beams type of approach is an appropriate solution for data channels as well, and the system can be configured to acquire channel knowledge on a per-scanned-beam basis rather than on a per-antenna basis. However, it is difficult to define precisely the conditions under which acquiring channel knowledge per antenna element is impractical and these limitations may and will change during different product generations when more advanced processing technologies are available.

Thus, one of the main design goals for NR mMIMO was to provide a framework that scales easily to handle any number of antenna elements and any of the gNB antenna array architectures depicted in Figures 3.12–3.14. Furthermore, NR-mMIMO provides solutions where the UE can be agnostic to the gNB array configuration.

Additionally, even though the number of antennas at an UE is expected to be significantly lower than the number at the gNB, similar antenna architecture considerations also apply to the UE design. Therefore, the NR mMIMO framework also supports a grid of beams strategy at the UE, which is mostly applicable for operation above 6 GHz. The number of RX and TX beams is left to UE implementation, but DL synchronization, uplink RACH as well as data and control channels transmission are compatible with UEs that have a hybrid beamforming architecture.

The mMIMO concept can be implemented with three different antenna array architectures, which are digital baseband beamforming, analog beamforming and hybrid beamforming. Each of these antenna array architectures have different characteristics and implications on system operation.

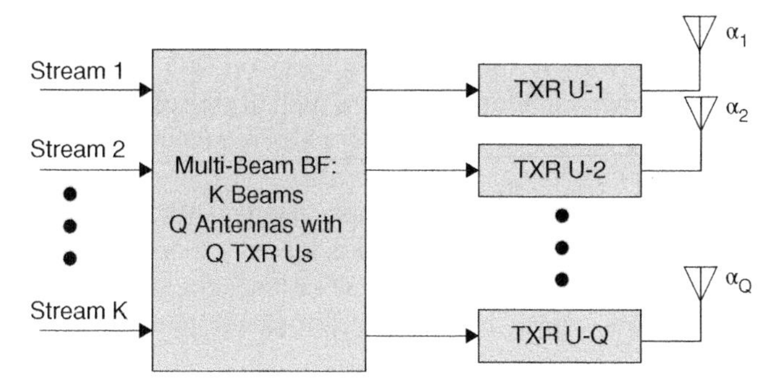

Figure 3.12 Digital baseband beamforming architecture, with K input streams and Q Transmitter-Receiver units and antennas.

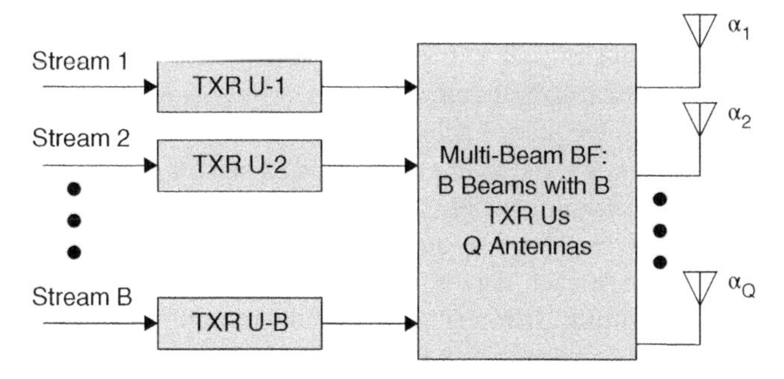

Figure 3.13 RF beamforming architecture, with B input streams with B Transmitter-Receiver units and Q antennas.

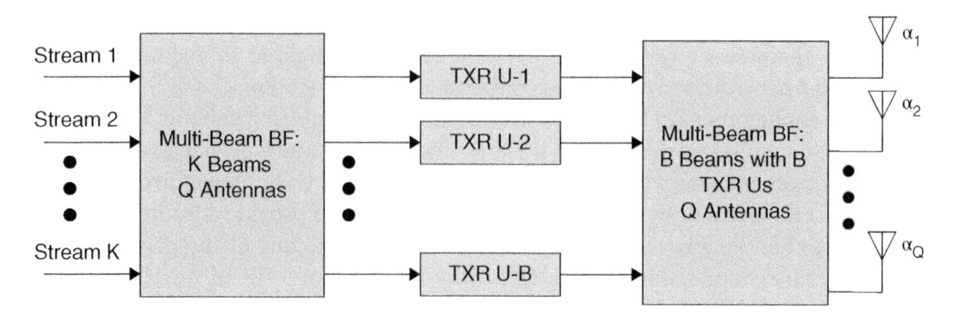

Figure 3.14 Hybrid beamforming architecture, with B input streams with B Transmitter-Receiver units and Q antennas.

The digital baseband beamforming architecture shown in Figure 3.12 is the architecture typically deployed by LTE macro-cells. In a digital baseband architecture, each antenna port is driven by a transceiver. The multi-antenna methods operate at the baseband in the digital domain, i.e. this is a baseband MIMO architecture. Extensions to multi-stream transmission and reception involve incorporating multiple receive and transmit weights in the baseband MIMO processing block, thus there is a full transceiver with Analog to Digital converter and Digital to Analog converter (ADC and DAC) behind every antenna. The digital baseband architectures are assumed for systems operating below 6 GHz frequency bands, where the deployments are expected to be mostly interference limited. The benefits of large-scale arrays will be realized by using high order spatial multiplexing, with an increasing emphasis on MU-MIMO as the base array size increases. Baseband architectures provide a high degree of flexibility such as frequency selective beamforming across Orthogonal Frequency-Division Multiple Access (OFDMA) subcarriers.

However, the complexity of baseband architectures increases significantly when the number of transceivers increase, as well as when the system bandwidth increases. The use of wider bandwidths can improve system capacity but necessitates the use of very high-speed ADC and DAC processors, which have significant power consumption requirements and increases the costs.

An alternative approach to the digital baseband beamforming architecture is the RF-beamforming architecture, also called the analog architecture, where control of MIMO and beamforming is performed at RF level with analog components. Figure 3.13 shows the RF MIMO architecture where a single transceiver drives the antenna array, and the transmit array processing is performed with RF components having phase shifting and potentially gain adjustment capabilities as well.

In contrast to baseband architectures, frequency selective beamforming with an RF architecture is not feasible as the transmit weights are applied at RF across the entire signal bandwidth. Additionally, as beamforming weights are applied over the entire signal bandwidth, multiplexing different UEs across frequency would require the use of multiple beams operating simultaneously, each driven by a separate transceiver unit. As a result, with RF architectures operating with a single transceiver, multiplexing of UEs is typically performed in the TDM instead of the FDM. Furthermore, as transmit weights are applied by using analog components, sufficient beam switching time should be enabled between symbols using different transmission weights. Therefore,

RF architectures are used mainly in millimeter frequencies, which are mainly pathloss rather than interference limited, and available bandwidths are relatively wide.

The hybrid beamforming architecture is an alternative to the full digital baseband and the RF beamforming architecture. The hybrid architecture tries to find a compromise between complexity and transmission flexibility. In hybrid beamforming control of MIMO and beamforming is split between RF and baseband. Figure 3.14 shows an example of the hybrid architecture, where multiple streams are beamformed at the RF in addition to the baseband MIMO processing. In the hybrid architecture, each RF beam is driven by a transceiver, and multi-stream beam weighting is applied at the baseband to the inputs of the transceivers. Figure 3.14 shows a hybrid architecture for a "fully connected" array configuration, where the multiple RF beamforming weight vectors are applied in parallel to all antenna elements of the array. In contrast, the alternative hybrid architecture is a "sub-array" configuration, where each RF weight vector is applied to a unique subset of the antenna elements. One advantage of the sub-array configuration is the lack of summation devices behind the antenna elements that are needed to form multiple parallel beams in the fully connected configuration. However, the beams in the sub-array configuration have reduced beamforming gain as not each TXRU is connected to all antennas. However, this reduction in the gain of the RF beams can be mitigated by the baseband MIMO operation. A hybrid architecture provides additional flexibility over an RF beamforming architecture as the baseband transmit portion can be adapted across the signal bandwidth to further optimize performance.

For NR mMIMO operation, several objectives and requirements were defined to meet the overall 5G system requirements:

- Support for different beamforming architectures, digital, hybrid, and RF beamforming as discussed above for both gNB and UE.
- Support for sector wide common channel transmission as in LTE as well as beam sweeping of the common control channels (CCCHs) to improve downlink coverage. The actual introduced schemes are discussed in Sections 3.3.1 and 3.3.2.
- Scalability in terms of the number of antenna ports, number of transceiver units and number of antenna elements, especially at the gNB.
- Support for UE operation with minimal assumptions on the MIMO operation at network side allowing network vendors to improve network implementation without requiring a new generation of UEs.
- Support for user-specific reference symbol designs while eliminating the use of common reference signals to enable network power savings when the number of UEs in the cell is low or when the cell is completely empty.

With these requirements, NR mMIMO transmissions can be divided into following different operational options.

The first option is to utilize precoding on CSI-RS. This technique involves the use of beamformed pilot signals where the UE sends feedback for one or more beamformed reference signals. Two main classes of pre-coded CSI-RS techniques are possible.

The first is the use of cell-specific beams with dynamic beam selection. As an example, the grid-of-beams concept involves the base station transmitting reference signals out of some number of fixed beams, and the UE can perform best-beam selection and/or feedback CSI for one or more beams. The feedback can include Channel Quality Indicator

(CQI), Rank Indicator (RI), and possibly codebook Precoding Matrix Indicator (PMI), corresponding to the beamformed channel measured by the UE.

The second is the use of UE-specific beams created by leveraging reciprocity. An example of this approach is where UL signals from the UE are leveraged to determine one or more beams over which the CSI-RS will be transmitted to the UE. The UE will feedback CSI for the one or more UE specific beams over which the CSI-RS was transmitted.

The use of precoding on CSI-RS, e.g. the grid-of-beams concept, is especially helpful in mmWave deployments where path loss limitations make it difficult to estimate the channel to each antenna port. Pre-coded CSI-RS techniques are appropriate for both baseband and hybrid array architectures and can support both SU-MIMO and MU-MIMO transmission. With baseband architectures, the precoding of the CSI-RS for forming the cell-specific or UE-specific beams would be applied at baseband level. In contrast, with hybrid architectures, the CSI-RS would be pre-coded in the RF/analog domain, and the number of CSI-RS ports would be limited by the number of transceivers in the array.

When precoding is not used on CSI-RS, the mMIMO technique involves the transmission of CSI-RS from all the transceiver units in the array and generally involve feedback from the UE. The feedback is obtained by PMI codebook signaling from the UE. NR defines two types of codebooks, a "standard resolution" codebook intended for SU-MIMO operation and a "high resolution" codebook intended to provide accurate channel knowledge suitable for MU-MIMO transmission. The non-pre-coded CSI-RS techniques are generally intended for digital baseband architectures operating in scenarios that allow for the acquisition of channel knowledge on a per-transceiver basis via CSI-RS. This operating methodology is possible on below 6 GHz frequency bands in deployments which are not path loss limited. Therefore, techniques without pre-coded CSI-RS are not well suited for operation in the mmWave bands where poor path-loss conditions make it difficult for a non-pre-coded reference signal to reach the cell edge.

In addition, the NR mMIMO framework leverages reciprocity of the propagation channel. Transmit weights for the downlink can be computed based on signals received on the uplink (and vice-versa) by leveraging the uplink/downlink reciprocity in the RF multipath channel response between a base station and a UE. As is well known, the instantaneous overall space-time-frequency RF multipath channel is reciprocal in a TDD system, but other aspects of the channel are also reciprocal even in FDD systems, e.g. parameters such as the multipath directions of arrival/departure and times of arrival. Various aspects of the uplink channel can be used for computing downlink transmit weights, e.g. the uplink spatial covariance matrix, directions of arrival, best uplink beam, or even the complete uplink channel response (TDD). However, leveraging reciprocity has its challenges. For example, link adaptation can be challenging given the non-reciprocal nature of the interference. Also, if UEs do not transmit with all antennas, then the full uplink matrix channel cannot be determined without antenna switching in the UE. Transmit power limitations in the UE may hinder the ability of the uplink signal to reach the base station with sufficient SNR unless antenna/beamforming gain is applied on one or both ends of the link, thereby making it difficult to acquire the full matrix uplink channel on a per-antenna basis. Leveraging reciprocity also requires antenna array calibration to remove the influence of transceiver hardware variations between uplink and downlink.

3.2.6 Channel Coding

Channel coding is an essential component of any telecommunication system with imperfect channel environment. Channel encoding is performed in a transmitter to enable a receiver to detect and to recover bit errors caused by imperfect channel. The selection of the channel coding method is a compromise between obtained channel coding gain, calculation complexity, and processing delay. All service scenarios in NR will have channel coding as a necessary functionality, but the exact coding scheme may be optimized depending on the use case. However, significant changes or adopting completely different coding schemes could cause complex hardware implementations which consequently lead to expensive NR operation. In general, the selection of channel code(s) which are quite flexible to match and satisfy different NR requirements was considered from the beginning of NR discussions in 3GPP. The selection of the channel coding scheme was mainly influenced by the requirements of the eMBB use case, given that hardware should be dimensioned mainly considering the case that supports very high data rate requirements. The other usage scenarios may preferably use the same channel coding scheme, and different schemes should only be introduced if compelling benefits are identified.

General requirements of eMBB scenarios in NR have a broader definition. Most of these requirements can be simplified into a smaller group of requirements. Performance of the coding scheme, implementation complexity, the latency of encoding and decoding, and flexibility (e.g. variable code length, code rate, HARQ) are identified as requirements that a code should satisfy. Turbo, LDPC and polar codes are the most promising candidates, which were identified in 3GPP for the eMBB data channel because they are capacity approaching codes.

Turbo coding is the existing coding scheme in LTE and capable of handling existing broadband scenarios. LTE turbo transmitter and receiver chain including interleaving, rate matching, and HARQ have also well developed over time. In LTE, turbo code also supports a wider range of block sizes. Their performances in lower code rates are competitive with many other coding candidates. However, turbo coding has limitations when achieving low decoding latencies due to its interleaving/deinterleaving stages and iterative decoding. Also, considering excessive energy consumptions and chip area of existing LTE turbo decoders, a substantial increase in energy consumption and chip area are expected with turbo codes when the data rates increase up to multi-gigabit range. Therefore, the implementation aspects, considering area-efficiency, i.e. encoded/decoded throughput per given chip area (Gbps mm^{-2}), and energy-efficiency, joules per bit in encoding/decoding (pJ/bit), played a significant role when deciding the coding candidate for NR. For example, to support 20 Gbps throughput with 1 W baseband power at the UE would require energy efficiencies around 50 pJ/bit, which is not possible with available turbo decoder implementations.

Polar coding is very promising in terms of theoretical performances. It was introduced with a simple decoding scheme, successive-cancellation (SC) decoding, which achieves high capacity when the block sizes are very large. However, such block sizes will not be used in NR, and the performance is lower for short to moderate block sizes. New algorithms have been proposed to improve the performance of polar codes for short to moderate block sizes by sacrificing its low complex decoding capability. For example, List-32 Cyclic Redundancy Check (CRC) assisted decoding performs better compared to available LDPC designs when the block sizes are less than 2000 bits. Incremental redundancy (IR) HARQ and implementation concerns when supporting high throughputs were identified as possible concerns for polar codes.

LDPC codes are the most common scheme used outside 3GPP and provide very good performance over a wider range of block sizes. This is also capable of achieving the performance close to the Shannon limit mostly for long block codes. In general, LDPC has superior performance when the code rates are closer to one. The flexibility of the implementation is one other key benefit of LDPC. For example, the latency associated with LDPC decoding is low due to parallelizable architecture. Like Turbo, LDPC codes are mature as they are already used in many other standards.

As implementation aspects are very important in NR eMBB data channel coding scheme, it is good to understand the capabilities of different codes. Tables 3.2 and 3.3 provide recent implementations considered for the Turbo, LDPC, and Polar codes.

In 3GPP discussions, it was understood that LDPC codes with limited flexibility provide the most attractive area and energy efficiency, and that the characteristics of LDPC codes in area and energy efficiency remain advantageous even when supporting full flexibility. For decoding hardware that can achieve acceptable latency, performance and flexibility, there are some concerns about the area efficiency and energy efficiency that are achievable with polar codes. Turbo codes are widely implemented in commercial hardware, supporting HARQ and flexibility comparable to NR requirements, but not at the high data rates or low latency as required for NR.

In addition, highly-parallelized LDPC decoders can help to reduce latency. Some concerns exist regarding Turbo and polar decoders as they incur longer latency than LDPC decoders. Moreover, it is understood that LDPC, Polar and Turbo codes can all deliver acceptable flexibility. Chase-combining (CC) and Incremental Redundancy-Hybrid Automatic Repeat Request (IR-HARQ) support is a concern that arose when discussing the polar code for eMBB data channel. On the other hand, LDPC schemes for support of both CC- and IR-HARQ and the ability of Turbo codes to support both CC- and IR-HARQ was well known.

Considering most of the above aspects, 3GPP decided to adopt LDPC codes for the NR eMBB data channel (see [35]).

Use of the same code for other use cases of NR is hardware efficient, if there are no real benefits of a different coding scheme. URLLC is the next scenario that has different requirements than eMBB. When it comes to URLLC coding, the most important requirements are support for low latency and very high reliability in encoding and decoding. This demands the channel coding scheme to have low latency in the encoding/decoding process and extremely low error floors. Low encoding/decoding often can be achieved by adopting small to moderate block length. In consequence, the system will work far away from the Shannon limit stated for very long codes. In the LDPC design details, these requirements were considered and URLLC will use the same coding scheme as eMBB.

Massive MTC requirements are quite different from the eMBB usage scenario. The key requirements for the mMTC use case are mainly to design low complex and low-cost solutions, which could operate for years while serving smaller throughput requirements. For many mMTC scenarios, the device might operate only with battery power and be required to communicate over a long period. Moreover, the cost of the device should be lower to deploy in massive numbers. Most capacity approaching coding schemes,

Table 3.2 Implementation for single code rate and block size.

Coding scheme	LDPC						Turbo		Polar			
Reference	[20]	[21]		[22]		[23]	[24]	[25]	[26]	[27]		[28]
Technology (nm)	65	65		65		65	45	65	90	65		40
Decoding algorithm	Split threshold min-sum	Offset min-sum		Split threshold min-sum		Partial parallel Sum-Product	Max-log-MAP	Max-log-MAP	SC	BP		Fast SSC
Code length	2048	2048		2048		672	6144	6144	1024	1024		1024
Code rate	0.84	0.84		0.84		0.8125	0.75	—	0.5	0.5		0.5
Clock (MHz)	195	700	100	185	40	500	1000	410	2.79	300	50	248
Chip area (mm^2)	4.84	5.35		5.10		0.16	11.1	109	3.21	1.48		—
Throughput (Gbps)	92.8	47.7	6.7	85.7	18.4	5.6	3.7	15.8	2.9	4.7	0.77	254.1
Area-efficiency (Gbps mm^{-2})	19.1	8.9	1.2	16.8	3.6	35	0.34	0.145	0.89	3.17	0.5	—
Energy-efficiency (pJ/bit)	15	58.7	21.5	13.6	3.9	17.65	2105	608	11.45	102.1	23.8	—
Maximum latency (ns)	56.4	137	960	81	375	—	—	—	358	—	—	1470

Table 3.3 Implementations for multiple code rates and block sizes.

Coding scheme	LDPC				Turbo	
Reference	[29]	[30]	[31]	[32]	[33]	[34]
Technology (nm)	90	28	65	65	65	45
Decoding algorithm	Stochastic	Min-sum	New	Partial layered BP	Max-log-map	Max-log-map
Code lengths (standard)	672 (802.15.3c)	672 (802.11ad)	672 (802.11ad)	2304	All block sizes in LTE	All block sizes in LTE
Code rates	1/2, 5/8, 3/4, 7/8	1/2, 5/8, 3/4, 13/16	1/2, 5/8, 3/4, 13/16	1/2–1	All code rates	All code rates
Clock(MHz)	768	260	400	1100	410	600
Chip area (mm^2)	2.67	0.63	0.575	1.96	2.46	2.004
Throughput (Gbps)	7.9	12	9.25	1.28	1.01	1.67
Area-efficiency (Gbps mm^{-2})	2.97	19	16.08	0.65	0.41	0.83
Energy-efficiency (pJ/bit)	55.2	30	29.4	709	1870	520

e.g. turbo, LDPC, and polar perform well when the block-length is larger. When block sizes are small their performances are not significantly better compared to simple coding schemes like convolutional codes. Considering decoder complexities associated with turbo, LDPC, and polar codes, it is likely that other codes must be considered for mMTC in future 3GPP releases.

3.2.6.1 Channel Coding for User Plane Data

The LDPC code adopted in NR is flexible and different from the LDPC codes standardized before. Moreover, good flexibility of the supported block sizes and code rates should be supported to handle the wider range of traffic requirements in eMBB. IR-HARQ support is not available in earlier LDPC standards, and 3GPP has taken major steps forward when optimizing LDPC codes for the eMBB data channel. In summary, NR LDPC design construction is supporting 1-bit granularity of block sizes, IR-HARQ support, and higher to lower code rate support by utilizing two base graphs. In the following section, we provide a quick overview of basic code construction details and coding chain, starting with basic LDPC operation.

An LDPC code is often defined by its $M \times N$ parity-check matrix **H**. The M rows in **H** specify the M constraints in the code. Different codes have different parity check matrices. For example, a parity matrix can be illustrated as below.

$$\mathrm{H} = \begin{bmatrix} \mathbf{1} & \mathbf{0} & \mathbf{0} & \mathbf{1} & \mathbf{0} & \mathbf{1} \\ \mathbf{0} & \mathbf{1} & \mathbf{0} & \mathbf{1} & \mathbf{1} & \mathbf{0} \\ \mathbf{0} & \mathbf{0} & \mathbf{1} & \mathbf{0} & \mathbf{1} & \mathbf{1} \end{bmatrix} \tag{3.1}$$

The N columns in **H** correspond to the total number of code bits within a code word. There are two types of LDPC codes, called regular and irregular LDPC codes.

The regular version code has exactly w_c ones (bit 1) per column (column weight) and exactly $w_r = w_c \times (N/M)$ ones per row (row weight), where w_c and w_r are both small compared to N. Each parity-check equation involves exactly w_r bits, and every bit in a code word is associated in exactly w_c parity check equations. In irregular LDPC codes, the number of ones per column or row is not a constant. Such irregular LDPC codes can perform better compared to the regular LDPC codes with similar dimensions.

The codeword x is constructed such that $\mathbf{H}x = 0 \pmod 2$. At the encoder side, generator matrix is required to encode the info bits. The codeword x can be written as info and parity check parts.

$$\mathbf{x}^T = [\mathbf{k}|\mathbf{p}] \tag{3.2}$$

The parity check matrix can be divided into two parts as

$$\mathbf{H} = [\mathbf{A}|\mathbf{B}] \tag{3.3}$$

Matrix multiplication gives the following

$$\mathbf{Ak} + \mathbf{Bp} = \mathbf{0} \tag{3.4}$$

When the matrix **B** is non-singular, parity bits p can be derived as

$$\mathbf{p} = \mathbf{B}^{-1}\mathbf{Ak} \tag{3.5}$$

In Eq. (3.5) the generator matrix **G** can be identified as $\mathbf{B}^{-1}\mathbf{A}$. At the decoder, LDPC use message passing algorithms and can be understood by the representation of a Tanner graph. Any LDPC code can be illustrated by a Tanner graph, as shown in Figure 3.15.

For LDPC codes, the Tanner graph can represent the parity check matrix with two nodes, known as check and variable nodes. In Figure 3.15, check nodes are illustrated with squares C1–C3 and bit nodes are shown with circles V1–V6. There are M check nodes (three in the example), and N variable nodes (six in the example) which correspond to the number of rows and columns in matrix **H**. The check nodes are connected to the variable nodes based on the ones in matrix **H**. The branches between nodes are considered in the message passing algorithms such that iterative computation of probabilistic quantities is possible. In the LDPC decoding process, likelihoods obtained from soft-decision components of a received vector r initialize the variable nodes and iteratively calculate relevant probabilistic values such that decoding of bits improve with the number of iterations.

The NR LDPC code design is based on quasi-cyclic low-density parity check (QC-LDPC), which has low complexity encoding/decoding compared to other variants.

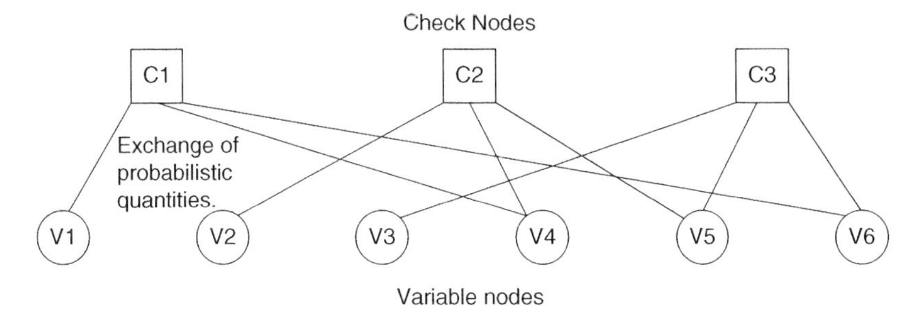

Figure 3.15 Tanner graph for parity check matrix in Eq. (3.1).

Table 3.4 NR LDPC base graphs.

Base graph	Maximum block size	Max code rate	Min code rate	LDPC lifting size	
				Min (Z_{min})	Max (Z_{max})
BG 1	8448	8/9	1/3	2	384
BG 2	3840	2/3	1/5	2	384

The parity-check matrix of a QC-LDPC is given as an array of sparse circulants of the same size. The circulant size, or the shift size, determines the overall complexity of the implementation together with the dimensions of the parity-check matrix. In NR LDPC design, two base graphs are introduced such that the code provides good performance at a broader range of block sizes and code rates and improves the latency and performance for lower block sizes and code rates. Parameters of LDPC designs are summarized in Table 3.4.

The representation of QC-LDPC base graph H can be represented as follows

$$\mathrm{H} = \begin{bmatrix} P_{1,1} & P_{1,2} & P_{1,3} & & P_{1,N} \\ P_{2,1} & P_{2,2} & P_{2,3} & & P_{2,N} \\ \cdot & \cdot & \cdot & & \cdot \\ \cdot & \cdot & \cdot & \cdots\cdots & \cdot \\ \cdot & \cdot & \cdot & & \cdot \\ \cdot & \cdot & \cdot & & \cdot \\ \cdot & \cdot & \cdot & & \cdot \\ P_{(N-K_b),1} & P_{(N-K_b),2} & P_{(N-K_b),3} & & P_{(N-K_b),N} \end{bmatrix} \tag{3.6}$$

where $P_{i,j}$ is a cyclic-permutation matrix obtained from the zero matrix and the z by z cyclically shifted identity matrix to the right. Also, $P_{i,j}$ is often represented as a numerical entry which is the value of the shift. All non-zero entries of H define the connections between check and variables nodes. This is generally known as the base graph. Two base graphs in NR have the following structure as show in Figure 3.16.

For BG #1, N = 68 and $K_b = 22$, while BG#2 has N = 52 and K_b depends on the supported block size. Matrix A corresponds to systematic bits, matrix B is square and corresponds to parity bits, has a dual diagonal structure (i.e. main diagonal and off-diagonal), matrix C is a zero matrix, matrix D corresponding to systematic and parity bits, and matrix E is an identity matrix.

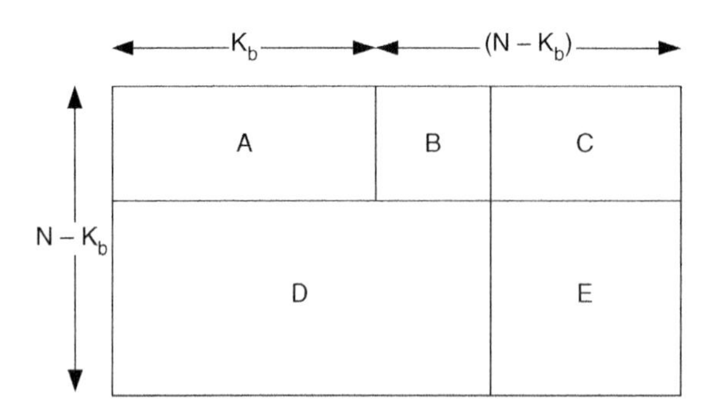

Figure 3.16 Dimensions of LDPC base graphs.

Table 3.5 Sets of LDPC lifting size.

Set number	Set of lifting sizes (Z)
1	{2, 4, 8, 16, 32, 64, 128, 256}
2	{3, 6, 12, 24, 48, 96, 192, 384}
3	{5, 10, 20, 40, 80, 160, 320}
4	{7, 14, 28, 56, 112, 224}
5	{9, 18, 36, 72, 144, 288}
6	{11, 22, 44, 88, 176, 352}
7	{13, 26, 52, 104, 208}
8	{15, 30, 60, 120, 240}

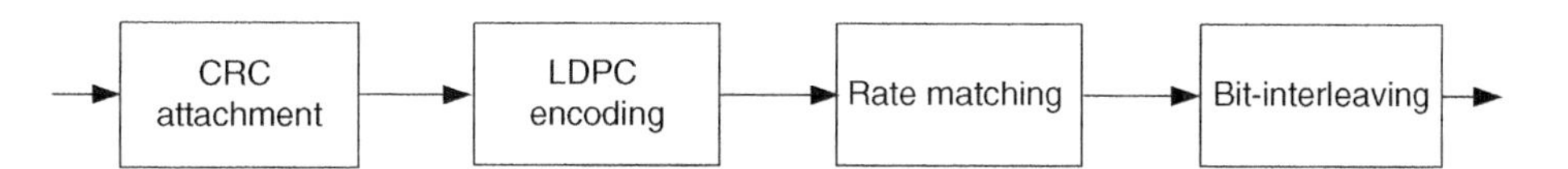

Figure 3.17 Coding chain for LDPC.

Each of the base graphs can have eight lifting coefficient designs sets as in Table 3.5, where values of $P_{i,j}$ can be different for the same i and j. Additionally, the maximum shift size of each shift coefficient design can be adjusted such that different code block sizes are supported with NR LDPC codes.

BG#1 and BG#2 have different operating region unlike in LTE turbo, which was used across all transport block sizes (TBSs) and code rates (R). The BG #2 is used when the TBS ≤ 292 for all code rates, or TBS ≤ 3824 and $R \leq 2/3$,or $R \leq 1/4$, otherwise LDPC BG #1 is used.

The NR LDPC utilizes coding chain steps like LTE as shown in Figure 3.17, however at each step there are differences.

The procedure starts with (CRC) attachment that can have one or two levels. First, CRC is attached to the transport block. Next CRC is appended on code block, if the code segmentation provides more than one code block. In NR, 16-bit CRC is used when the TBS is lower than 3824, whereas 24-bit CRC is used for all other TBS. Moreover, 24 CRC is appended per code block when there is more than one code block.

In the second step LDPC encoding is performed using the parity check matrices described in an earlier section. This has several other steps like selecting the lifting size, zero padding, encoding, and removal of the padding bits. The selection of the lifting size also depends on the base graph. For BG #1, $K_b = 22$, while K_b for BG#2 is decided per supported block size.

After channel coding, rate matching in NR is based on the circular buffer as in LTE. After encoding, coded bits are copied to the circular buffer without the first 2*Z bits. Redundancy versions (RVs) are defined as RV0, RV1, RV2, and RV3, where they are not uniformly separated like in LTE. For BG #1, starting positions of RVs are 0, 17/66, 33/66, and 56/66 in fractions of the circular buffer. For BG #2, they are 0, 13/50, 25/50, and 43/50 in portions.

Finally, bit-interleaving process is performed in channel coding, which is quite like the bit-interleaver LTE. In principle, both have block interleavers, and writing is row-wise left to right, reading from column-wise top to bottom. The only difference is that the number of rows in the block interleaver is defined by the modulation order used for the transmission.

After completing the coding chain, a code concatenation is done and the output bit-stream is send to the modulation mapper.

3.2.6.2 Channel Coding for Physical Control Channels

Polar coding is adopted as the coding scheme for both DL and UL control channels except for the very small block lengths considering the performance benefits observed with list decoding of polar codes (see [35]) For very short block sizes, LTE block codes are adopted in 3GPP. In particular, 1 bit, 2 bits, and 3–11 bits of payloads should be supported with repetition, simplex, and LTE RM codes. Polar coding is new to the standard bodies as it was invented in 2009.

Polar code is a channel coding scheme to approach communication channel capacity, and with the help of list decoders, polar codes have comparable and sometimes even better performance compared to the state-of-the-art codes like LDPC and turbo codes. Also, decoding complexity of polar codes also shows a lower number such as $O(L * N * log_2(N))$, where N is the encoded block length and L is the list size. These features made polar codes more attractive to control channels and polar was adopted as the main coding scheme in both downlink and uplink control.

Polar codes use the concept of polarization for error correction. The basic building block in polar codes is shown in Figure 3.18.

In Figure 3.18, u_1 and u_2 refer to the input bits, and y_1 and y_2 refer to the output/encoded bits of the encoder. In the information theoretic view, the mutual information $I(U_1; Y_1, Y_2)$ decreases compared to the pre-polarized pair, $I(U_1; Y_1)$, while $I(U_2; Y_1, Y_2, U_1)$ increases compared to $I(U_2; Y_2)$. In this way, one channel is degraded, and the other one is upgraded.

By duplicating and stacking the basic blocks, longer polar codes can be constructed. Figure 3.19 shows an example by a length-4 polar code.

When the number of blocks grow up, the polarization effect becomes visible, and when the block size is very large, some channels would have zero capacity and others would become error-free. This phenomenon is used in the data transmission, where the error-free channels can be used to transmit information bits and force the value of the bits transmitted in the zero-capacity channels to be some known value, e.g. 0, which are also called frozen bits.

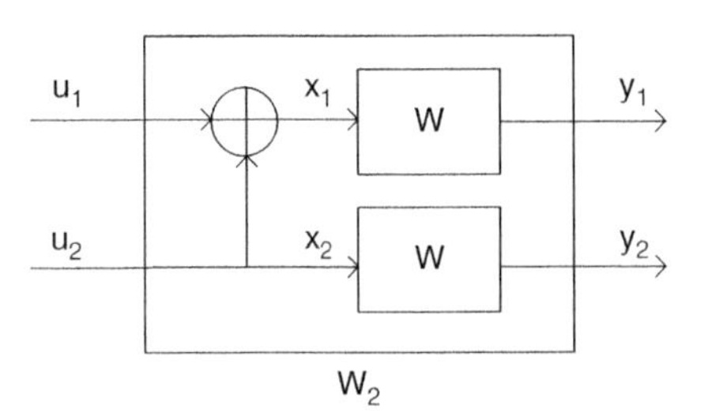

Figure 3.18 Basic building block of polar codes.

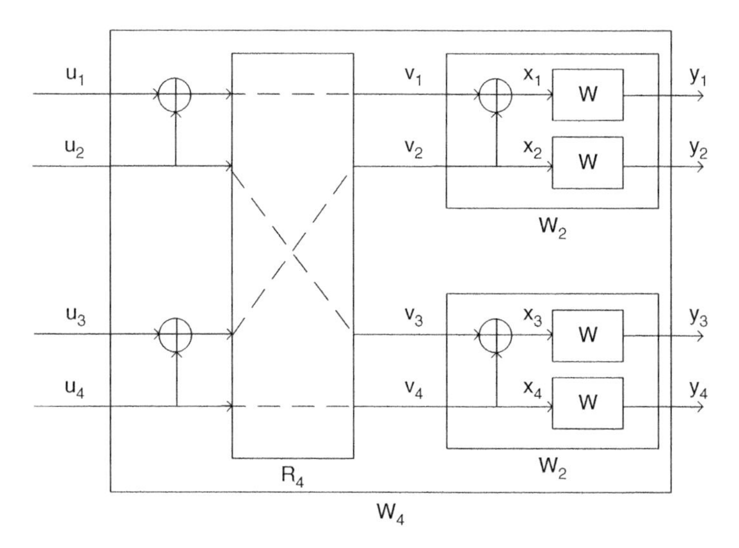

Figure 3.19 Encoding graph of length-4 polar codes.

In data transmissions, every channel is normally given a polarization weight or ranking, and best channels out of the total N polarized channels are used to transmit the data bits. As is visible in the stacking, polar codewords are always provided by the power of two as the output bits. However, rate matching schemes can be still applied without loss of significant performance. At the receiver side, polar coding can use many decoding algorithms just like many other error control coding schemes. SC list decoding has good performance with algorithmic complexity. In NR, most of the evaluations and design considerations were considered based on the assumption that SC list decoding is done at the receiver side.

NR control channels use concatenated polar coding schemes with CRC and parity check bits for Uplink Control Information (UCI), which is understood to give much better performance compared to traditional polar coding design. Two different designs were introduced for NR. The first design is for Downlink Control Information (DCI) transmitted in PDCCH, which uses distributed CRC polar code, with maximum polar codeword of 512 bits.

For uplink control the maximum polar codeword is 1024 bits. The uplink design uses parity and CRC concatenated polar code for UCI of 12–19 bits. Finally, CRC concatenated polar code is used for UCI message above 19 bits.

As DL control channel in NR is associated with blind decoding at the UEs, an optimized coding scheme is required to save UE energy and reduce latency. This is achieved by distribution of CRC bits inside the information bits. Overall, 8 CRC bits are distributed, and 16 bits are appended at the end. A nested interleaver is used to support any code block size, and a benefit of the distributed CRC polar code is early termination capability at the decoder. This saves UE energy consumption and reduces decoding latency. In addition, this supports the flexible decoding operation, where CRC bits can be used as error correction or error detection by a conventional CRC detector.

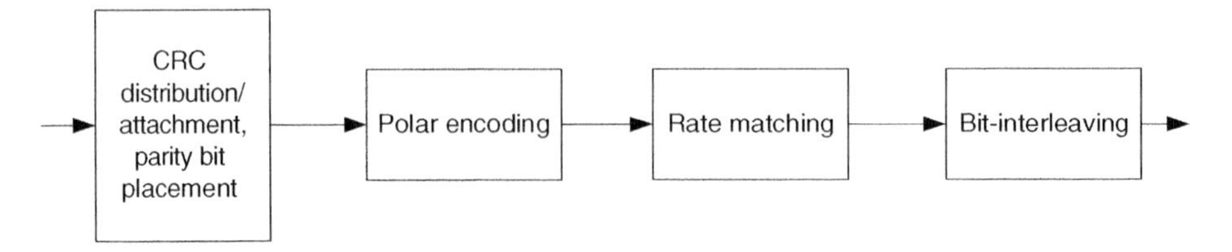

Figure 3.20 Coding chain of the NR polar coding.

Finally, the design reduces false alarm rate (FAR). The rate that incorrectly received messages are decoded as correct, FAR, is reduced to obtain targets as low as 2^{-21} by careful selection of the distribution pattern.

For the UL control channel, CRC bits are attached at the end of the information payload, and CRC lengths depend on the payload size. When the payload is between 12 and 19 bits, 6 CRC bits are appended with three parity check bits. Above 19 bits of information payload, 11 CRC bits are appended.

At the input of the encoder, these concatenated bits are mapped to the most reliable ones (from the ranked positions), and the remaining positions are set to zero. This ranking order is known as the polar sequence, which is a nested pattern supporting up-to 1024 bits long polar codeword.

Basic steps of the coding chain for control channels in NR are shown in Figure 3.20. As in the user plane coding chain, the coding chain for control channels starts with CRC attachment. As described before, the parity check bits are also applied for UCI when the payload is between 12 and 19 bits.

The polar encoding is done based on the encoding mechanism as discussed above. To ensure that the polar codeword size is selected efficiently, the selection of the codeword size depends on the information payload size, available resources for the control channel and the minimum threshold code rate such as 1/8. As highlighted before, the maximum codeword size for DCI encoding is 512 and for UCI 1024.

After polar encoding, rate matching is performed. In NR PDCCH transmission, rate matching takes several steps. First, sub-block interleaving is performed on the output bit stream of the polar encoded bits. In the second phase, bit stream bits are copied to the circular buffer. Finally, bit selection is done depending on the code rate supported in the control channel. If the rate matching output bits are larger than the polar codeword size, repetition is applied. Otherwise, puncturing and shortening are used depending on the code rate. When the code rate is lower than or equal to 7/16, puncturing is used, whereas shortening is used for other rates.

During the puncturing process the transmitter removes set of bits from coded bits to be transmitted in a way that the non-transmitted bits can be unknown to the receiver and the corresponding Log-Likelihood Ratio (LLR) of the bits can be set to zero.

During the shortening process the transmitter is setting input bits to a known value, and non-transmission of coded bits corresponding to those input bits, such that the corresponding LLRs can be set to a large value at the receiver.

Finally, for UL control channel transmission bit-interleaver is applied, however this step is not utilized in DL direction. The interleaver is known as the triangular interleaver, where the write-read operation is like the traditional block interleaving.

3.3 Downlink Physical Layer

In the design of downlink PHY, different beamforming architectures, discussed in 3.2.5, were considered for each downlink channel. Different architectures at gNB and UE side imposed several requirements for the design. In addition, the design was targeted to be independent of the used SCS and the frequency band of the cell.

3.3.1 Synchronization and Cell Detection

For initial cell search, downlink synchronization and cell level radio resource management (RRM) functions, a downlink synchronization signal block, called an SS Block, was defined. The SS Block comprises of primary synchronization signal (PSS), secondary synchronization signal (SSS) and PBCH, together with a demodulation reference signal (DMRS). The PSS and SSS are used for time and frequency synchronization acquisition and physical cell identity (PCI) determination. PBCH is used to broadcast the most essential system information of the cell.

The PSS and SSS occupy both one OFDM symbol in TDM, and PBCH occupies two OFDM symbols. PSS, SSS, and PBCH are multiplexed in time division manner as shown in Figure 3.21.

In total, the SS Block occupies 4 OFDM symbols in TDM and 20 PRBs in the FDM. PRBs around PSS and within SS Block bandwidth allocation are left unused to allow transmitting power boosting for the PSS.

The structure is the same for below 6 GHz and above 6 GHz carrier frequency ranges and for different numerologies. The SS Block can be transmitted using 15 or 30 kHz SCS at below 6 GHz frequency bands and using 120 or 240 kHz SCS at above 6 GHz carrier frequency bands. Different allowed SCS options are depicted in Table 3.6.

The SS Block used for initial cell search can be located also elsewhere than in the middle of the system bandwidth in FDM if PSS is in the predefined synchronization signal raster which can be sparser than the channel raster.

In addition, the gNB may configure an SS Block which is not located on the synchronization signal raster in FDM for measurements. Such configuration requires UE dedicated signaling to point to the location of the SS Block.

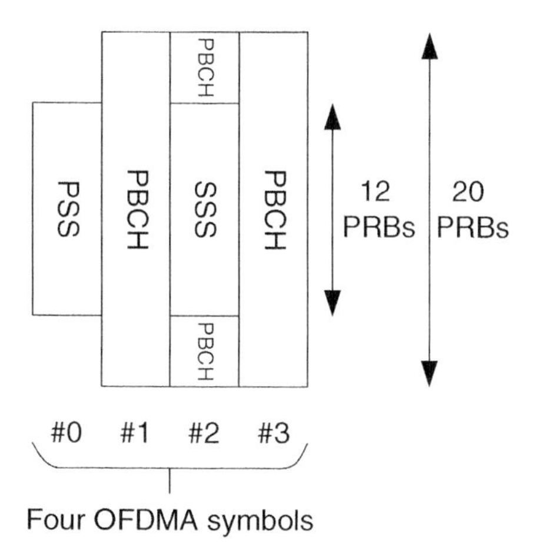

Figure 3.21 SS Block structure.

Table 3.6 SS Block subcarrier spacings in given bands.

SS block SCS (kHz)	NR operating band
15	n1, n2, n3, n7, n8, n20, n28, n38, n41, n50, n51, n70, n71, n74, n75, n76
15 or 30	n5, n66
30	n77, n78, n79
120	n258
120 or 240	n257, n260

3.3.1.1 Primary Synchronization Signal (PSS)

When the UE is performing cell search it searches first for the PSS. PSS is in a predefined synchronization signal raster in FDM. PSS is used for initial symbol boundary and coarse frequency synchronization to the NR cell. It is based on CP-OFDM waveform for below 52.6 GHz frequencies. For initial access, the UE can assume a specific SCS applied for the PSS in a given frequency band. This is defined in 3GPP specifications as depicted in Table 3.6, however, for frequency bands having two options, the UE needs to perform blind search with both options in initial cell search.

There are three PSS sequences like in LTE. Instead of using Zadoff-Chu sequences, NR has adopted an FDM-based BPSK m-sequence. M-sequence was selected because it has not a time and frequency offset ambiguity, present with Zadoff-Chu sequences. Ambiguity function plots of (a) LTE PSS and (b) NR PSS are presented in Figure 3.22. The LTE PSS sequence length is 62 and NR PSS sequence length is 127. Detection performance under initial frequency offset due to oscillator synchronization mismatch is improved and UE complexity is reduced since the UE does not need to try with that many different PSS hypothesis in NR SSS detection as would be the case with Zadoff-Chu sequence-based LTE PSS. Correspondingly, joint PSS and SSS detection performance is improved in NR compared to LTE.

An important design criterion was to improve one-shot PSS detection performance in NR compared to LTE. In addition, the selection of m-sequence provided better signal characteristics under time offset and frequency offset ambiguity. As a result, length-127 m-sequence was adopted to provide 3 dB larger processing gain and higher frequency diversity with the cost of increased UE complexity, since bandwidth and sampling rate would be doubled from that in LTE.

3.3.1.2 Secondary Synchronization Signal (SSS)

After a UE has detected the New Radio-Primary Synchronization Signal (NR-PSS) and acquired symbol timing and initial frequency synchronization, it tries to detect SSS which carries the PCI. Since the SSS is located on the same frequency location as NR-PSS and one OFDM symbol apart, the UE may perform either non-coherent or coherent detection using channel estimates based on NR-PSS.

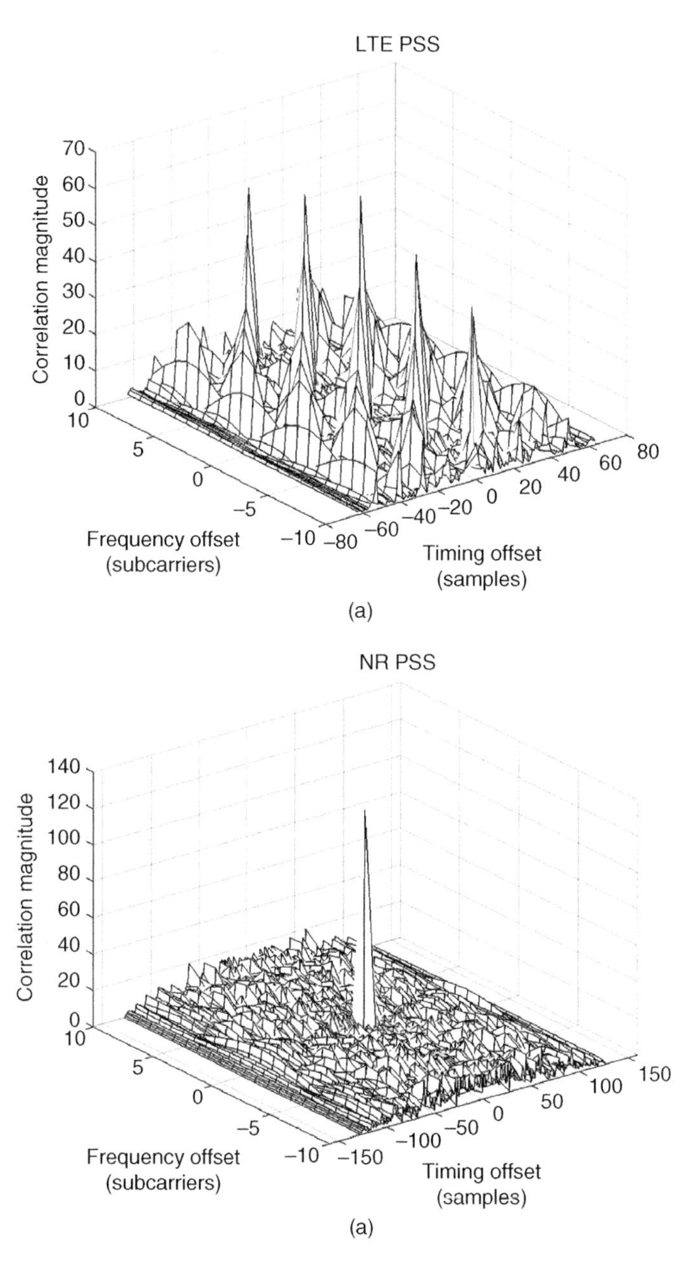

Figure 3.22 PSS time and frequency offset ambiguity of (a) LTE PSS sequence (left) and (b) NR PSS sequence (right).

SSS is a Gold sequence of length 127. There is one polynomial with 112 cyclic shifts and another polynomial with 9 cyclic shifts forming together 1008 different PCIs. Index of the detected NR-PSS sequence (0, 1, or 2) is used in the generation of nine cyclic shifts for the second polynomial. Gold sequence $d_{SSS}(n)$ is depicted in Eq. (3.7), where m_0 and m_1 are cyclic shifts for the first and second polynomial, $N_{ID}^{(1)}$ has values $0, 1, \ldots, 335$ and

$N_{ID}^{(2)}$ has values 0, 1, 2 corresponding to the index carried by the NR-PSS.

$$d_{SSS}(n) = [(1 - 2x_0(n + m_0)\text{mod}127)][1 - 2x_1((n + m_1)\text{mod}127)] \quad (3.7)$$

where $0 \leq n < 127$ and

$$m_0 = 15 \left\lfloor \frac{N_{ID}^{(1)}}{112} \right\rfloor + 5N_{ID}^{(2)} \quad (3.8)$$

and

$$m_1 = N_{ID}^{(1)}\text{mod}127 \quad (3.9)$$

This design provides enhancements compared to LTE SSS. Doubling the sequence length brings 3 dB processing gain enhancement and adopting long sequences instead of using two shorter sequences in interleaved manner as in LTE improves cell detection reliability especially at the cell edge. In LTE due to the use of two shorter sequences to deliver the cell ID, the cell-ambiguity issue may arise especially for UEs at the cell edge because performing piece-wise maximum-likelihood detection of each sub m-sequence will result in performance loss due to smaller spreading gain from the shorter sequence.

The improved overall design compared to LTE is illustrated in Figures 3.23 and 3.24. Figure 3.23 present performance when 5 ms transmission periodicity for PSS/SSS is used for both LTE and NR. Figure 3.24 depicts the performance when LTE has 5 ms and NR 20 ms PSS/SS transmission periodicity. Clearly, improved one-shot detection performance in NR allows using lower SS Block transmission periodicities without scarifying performance, while at the same time enables better energy savings in the network.

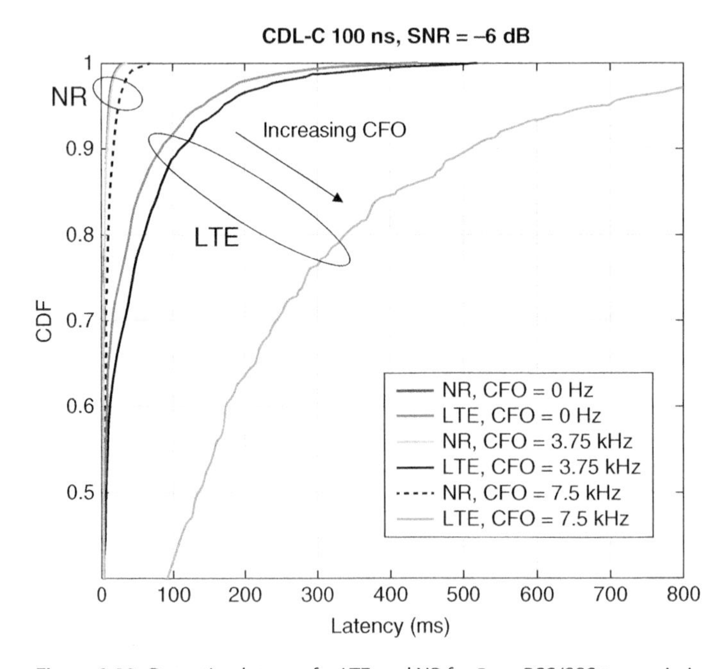

Figure 3.23 Detection latency for LTE and NR for 5 ms PSS/SSS transmission periodicity.

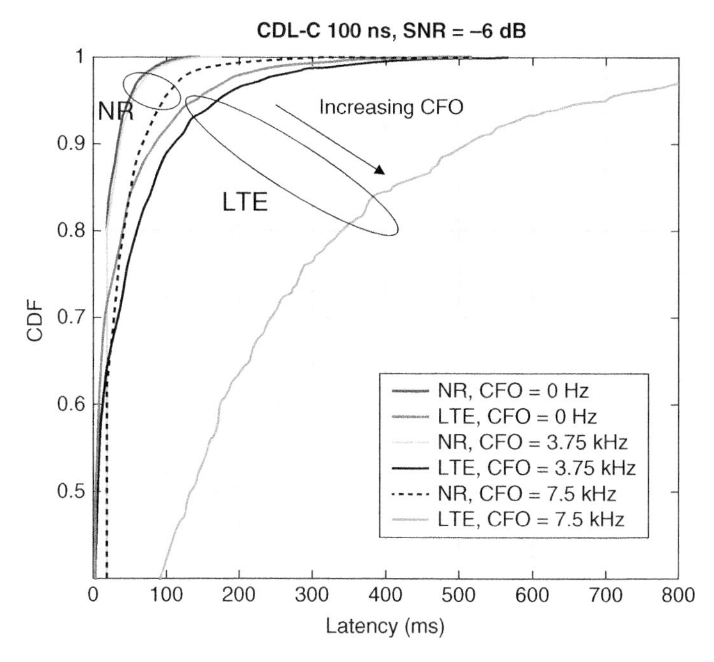

Figure 3.24 Detection latency for LTE and NR when LTE is having 5 ms and NR 20 ms PSS/SSS transmission periodicity.

3.3.1.3 Physical Broadcast Channel (PBCH)

PBCH, a part of the SS Block, is for signaling most essential system information, and shown in Table 3.7. The information includes timing info based on System Frame Number (SFN), half frame indicator and Most Significant Bits (MSBs) of the SS Block position within a half-frame, and information how to receive remaining minimum system information (RMSI). Parameters for receiving RMSI provide the UE with information about time and frequency resources of control resource set (CORESET) and monitoring parameters like periodicity and window duration for detecting PDCCH that schedules PDSCH carrying the actual RMSI data. The transmission of the PBCH is based on a single antenna port transmission using the same antenna port as PSS and SSS within an SS Block. While FDM precoder cycling is precluded, the gNB may use TDM precoder cycling by changing the precoder from one PBCH transmission to another.

DMRS of PBCH is mapped on every PBCH symbol with equal FDM density in all PRBs. In addition, PCID-based FDM shift is used to map DMRS on REs in each PRB as depicted in Figure 3.25.

DMRS sequence is based on PCID and N LSBs of the SS Block index. In case of carrier frequency range, the system is deployed for below 3 GHz is N = 2, otherwise N = 3.

3.3.1.4 SS Block Burst Set

To support flexible resource allocation and beamforming for the SS Block, multiple SS Blocks can be transmitted by the gNB within a certain period. The set of SS Blocks is called an SS Block burst set. Within one burst set the gNB transmits the SS Blocks throughout the whole sector thus enabling narrower transmit antenna radiation pattern with higher beamforming gain than with a sector wide radiation pattern. There are a

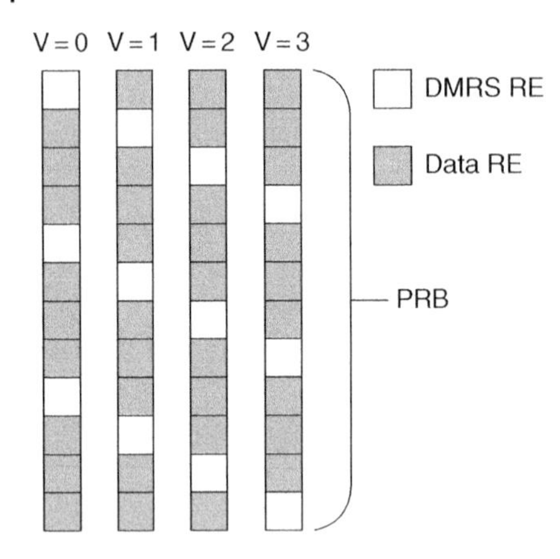

Figure 3.25 DMRS mapping on REs in an PRB based on Physical Cell ID.

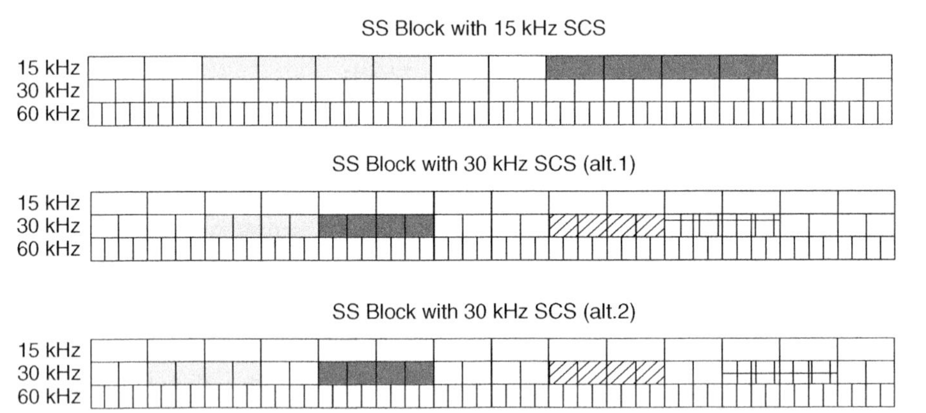

Figure 3.26 SS Block positions within a slot as a function of SS Block subcarrier spacing below 6 GHz.

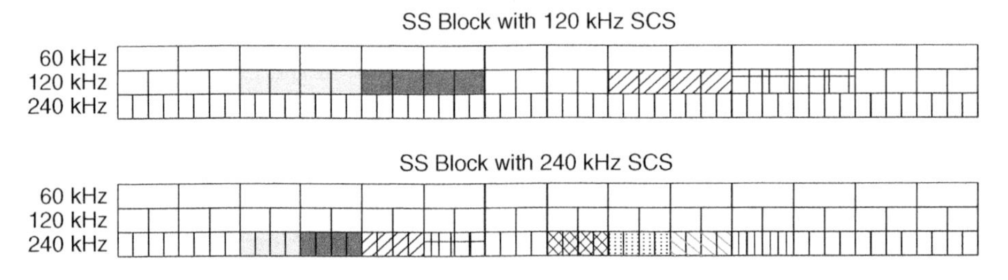

Figure 3.27 SS Block positions within a slot as a function of SS Block subcarrier spacing above 6 GHz.

number of fixed TDM locations within a 5 ms half frame defined where the SS Block can be transmitted. The number depends on the carrier frequency range. At below 3 GHz there are four locations available, between 3 and 6 GHz there are eight locations available as shown in Figure 3.26. At above 6 GHz there are up to 64 locations available for the SS Block transmissions per SS Block burst set as shown in Figure 3.27. That means, that at below 3 GHz the gNB may transmit SS Block throughout the sector using four different transmit beams while at above 6 GHz the gNB may use up to 64 transmit beams.

The fixed TDM locations of the SS Blocks of the SS Block burst set within a 5 ms half frame in the frame structure depend on the applied SCS for the SS Block. At below 6 GHz the block can be transmitted either using 15 or 30 kHz SCS, as shown in Table 3.6, and at above 6 GHz either using 120 or 240 kHz SCS.

With 15, 30, and 120 kHz SCS there are two SS Block positions within a slot of 14 symbols and with 240 kHz SCS there are four positions within a slot of 28 symbols as shown in Figures 3.26 and 3.27.

3.3.2 System Information Broadcast (SIB)

System information broadcast is divided into minimum system information (MSI) and other system information (OSI). MSI provides the UE information needed to access a cell. Most essential MSI parameters are conveyed in PBCH of the SS Block and the RMSI is carried in separate transmissions using PDSCH scheduled via PDCCH. For RMSI delivery, PBCH essentially provides CORESET configuration and monitoring parameters for the UE to be able to monitor PDCCH for RMSI scheduling. PBCH content, except the SS Block location index, is the same for all Synchronization Signal (SS) Blocks within an SS Block burst set for the same center frequency. Payload contents of the PBCH is illustrated in Table 3.7.

3.3.2.1 Remaining Minimum System Information (RMSI)

Three different multiplexing pattern options have been defined to multiplex PDCCH transmission of the RMSI CORESET and PDSCH for data within delivery of the SS Blocks. The first option is TDM multiplexing between SS Block and PDCCH RMSI CORESET, and PDSCH for RMSI delivery in separated slots. The time difference between SS Block transmission and RMSI transmission is not fixed and they can have different transmission periods. All signals are confined within a bandwidth defined by the RMSI CORESET, which is the initial active DL BWP, as shows in Figure 3.28. This option is supported both at below and above 6 GHz carrier frequency ranges.

Table 3.7 PBCH content.

Parameter	Number of bits	Comment
SFN	10	Indicates system frame number
Half-frame timing	1	Indicates first or second slot of the frame
SS block location index (3 MSBs at above 6 GHz)	3	Reserved at below 6 GHz
Reserved for higher layer signaling	3	
Offset between SS block frequency domain location and PRB grid in subcarrier level	5/4	5 bits for below 6 GHz and 4 bits above 6 GHz
DL numerology to be used for RMSI, Msg 2/4 for initial access and broadcasted OSI	1	15 or 30 kHz for below 6 GHz; 60 or 120 kHz above 6 GHz
Spare	0/1	
CRC	24	Polar code with 24-bit CRC
Total	56	

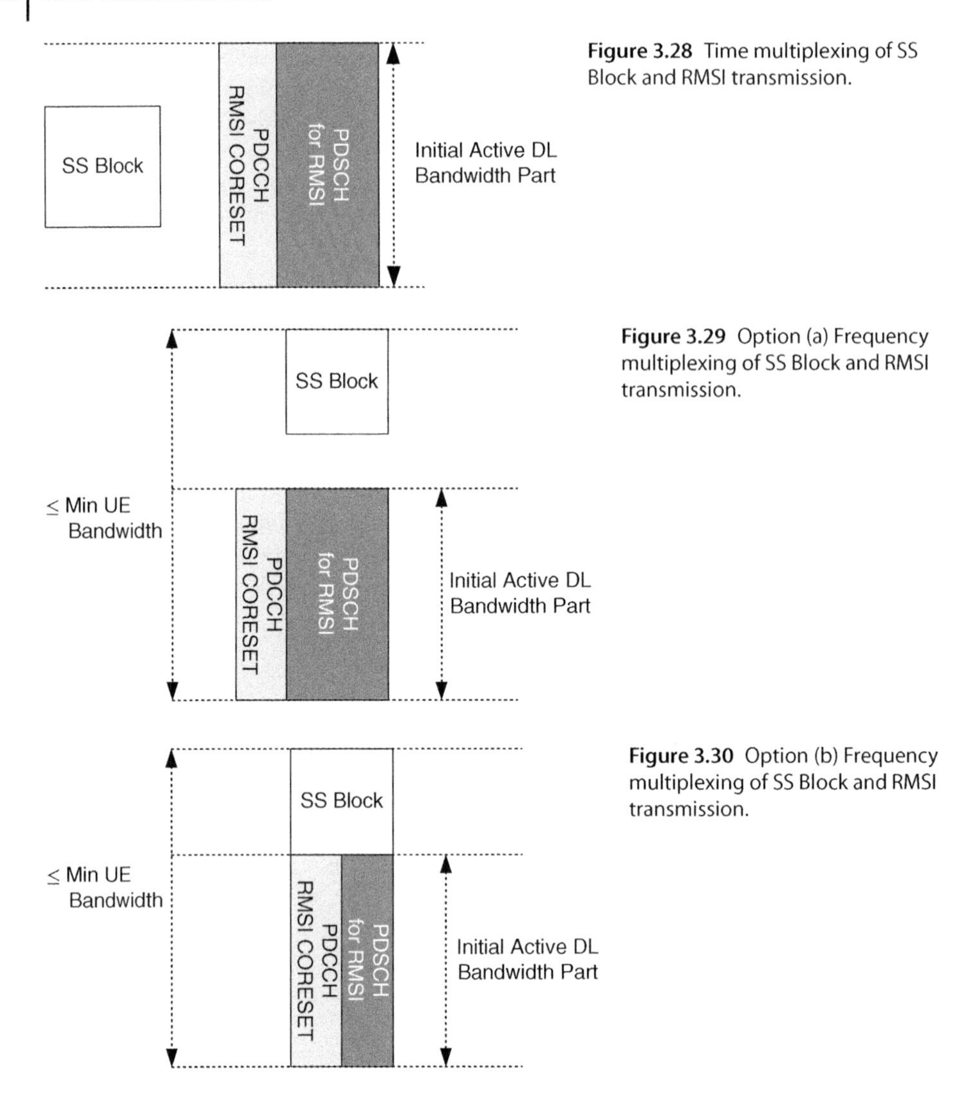

Figure 3.28 Time multiplexing of SS Block and RMSI transmission.

Figure 3.29 Option (a) Frequency multiplexing of SS Block and RMSI transmission.

Figure 3.30 Option (b) Frequency multiplexing of SS Block and RMSI transmission.

The second transmission option is the TDM multiplexing between SS Block and RMSI CORESET, and FDM multiplexing between SS Block and PDSCH for RMSI delivery where the signals are confined within a bandwidth supported by all UEs. There can be unused PRB(s) between SS Block PRBs and PRB used for PDSCH transmission, as shown in Figure 3.29, however this increases the needed minimum UE bandwidth. The SS Block and PDCCH RMSI CORESET are not overlapping in the frequency. This option is supported only at above 6 GHz.

The third option for the system is to utilize fully FDM multiplexing between SS Block and RMSI CORESET, and PDSCH for RMSI delivery where the signals are confined within a bandwidth supported by all NR UEs. The PDCCH carrying CORESET utilizes two OFDM symbols and PDSCH carrying RMSI utilizes another two OFDM symbols, having exactly the same time duration as the SS Block as shown in Figure 3.30. This option is supported only at above 6 GHz bands.

The numerology used for SS Block transmission is defined for each band. For CORESET and PDSCH carrying RMSI the used numerology is signaled in PBCH as shown in Table 3.7. For below 6 GHz, either 15 or 30 kHz SCS can be used, above 6 GHz either 60 or 120 kHz is used.

Multiplexing option 1 supports all different combinations whereas option 2 can be used with SS Block SCS of 120, or 240 kHz and RMSI numerologies of 60 or 120 kHz. For option 3, shown in Figure 3.30, the numerology of 120 kHz must be used for both SS Block and RMSI transmission.

3.3.2.2 Other System Information

The broadcast delivery of OSI is supported by PDSCH transmission scheduled via PDCCH (like RMSI delivery). The same DL numerology is used for broadcasted OSI as is used for RMSI and as informed in the PBCH payload. Both slot-based PDCCH and PDSCH, and non-slot-based PDSCH transmissions for broadcast OSI delivery are supported. For the non-slot-based transmission, 2, 4, and 7 OFDM symbol duration for the broadcast OSI PDSCH is supported. CORESET configuration will follow the one provided by PBCH for RMSI CORESET, but the TDM parameters (i.e. search space) are provided in RMSI.

3.3.3 Downlink Data Transmission

Section 3.2.3 discusses the basic frame structures and introduces the PDSCH/PUSCH resource allocation. The actual downlink data transmission and HARQ work within this setup and is not too different from the basic setup in LTE, but some critical differentiators exist. The PDCCH location can be freely placed within the slot, and the number of symbols allocated for PDSCH is very flexible, even though the typical mode of operation is to place the PDCCH in the beginning of the slot, followed by DMRS and PDSCH, not too alike from the LTE setup.

Unlike in LTE, the maximum allowed timing advance (TA) is not baked inside a fixed HARQ-ACK timeline, but the HARQ-ACK timing can be scheduled freely, if the UE has guaranteed sufficient time to process the received packet and prepare for the HARQ-ACK. This means that the gNB must factor in the TA budget on top of the UE processing time and delay the HARQ-ACK with large TA settings. This allows for supporting large cell radii without having to have the extra budget for the corresponding two-way propagation delay embedded in the HARQ loop latency in normal or small cell deployments. Unlike in LTE, where the PDSCH to HARQ-ACK transmission delay is defined by the standard, the NR specification just defines the minimum processing time the UE must be guaranteed before it can be expected to be able to reliably report HARQ-ACK.

In addition to the asynchronous HARQ-ACK timing, the downlink HARQ retransmissions are also asynchronous. The UE may be configured to maintain up to 16 HARQ processes, and the PDCCH scheduling the PDSCH carries a HARQ process number and a new data indicator. This enables asynchronous HARQ retransmissions with no pre-determined time relation between different transmission attempts and facilitates different gNB architectures, so that, e.g. a high latency fronthaul deployment can use a larger number of HARQ processes and longer retransmission delay and a low latency fronthaul deployment uses conversely a lesser number of HARQ processes and shorter retransmission delay.

The NR downlink also supports semi-persistent scheduling like LTE, where the first scheduling PDCCH triggers a periodically repeating transmission occasion without needing to schedule every packet separately. This solution is aimed at reducing the required control channel capacity when supporting many simultaneous voice links in a cell.

Multi-slot transmission (known also as slot repetition or slot aggregation) for improved coverage is possible both in downlink and uplink. As uplink is typically the coverage limiting link, the downlink multi-slot transmission may not be as useful though. RRC can configure the UE to expect that the gNB repeats each TB transmitted in the downlink over a specific number of consecutive slots. The redundancy version is cycled across the slots so that full IR combining gain can be achieved. If some of the slots in the span of the consecutive slots are configured to be uplink slots, then those slots are omitted, but still counted as part of the slot aggregate.

3.4 Uplink Physical Layer

3.4.1 Random Access

For contention-based random access (CBRA), NR supports a 4-step RACH procedure like LTE. A UE initiates the procedure by transmitting a Physical Random Access Channel (PRACH) preamble in Message 1 (Msg1). Upon detection of the preamble, the gNB responds with Message 2 (Msg2) containing Random Access Response (RAR). The gNB uses PDCCH for scheduling and PDSCH for transmitting Msg2 within a configured TDM RAR window. The RAR includes UL grant for the Message 3 (Msg3) that is transmitted by the UE using PUSCH. Finally, the gNB transmits a contention resolution message (Msg4) using PDCCH for scheduling and PDSCH for transmitting the message. Physical channels for Msg1 to Msg4 together with the main content of each message are illustrated in Table 3.8. For contention-free random access, only Msg1 and Msg2 are needed.

Numerology for PRACH preamble is provided in the RACH configuration (RMSI for standalone deployment). For Msg2 and Msg4 transmissions the SCS is the same as for RMSI (both PDCCH and PDSCH), for Msg3 the numerology is provided in the RACH configuration separately from the Msg1 SCS. For contention-free Random Access (RA)

Table 3.8 Physical channels for messages in NR RACH procedure.

Message	Physical layer channel	Content
Message 1	PRACH	RACH Preamble
Message 2	PDCCH, PDSCH	Detected RACH preamble ID, Timing Advance, UL grant, C-RNTI
Message 3	PUSCH	RRC Connection request, Scheduling request
Message 4	PDCCH, PDSCH	Contention resolution message

procedure for handover, the SCS for Msg1 and the SCS for Msg2 are provided in the handover command.

As a new component, corresponding to multiple SS Block locations within a half frame, NR supports receive beamforming in TDM for PRACH preamble reception by allocating multiple RACH occasions for which the gNB may use different receive beams. PRACH occasion and PRACH preamble selection by the UE also signals the preferred SS Block beam that will be used for Msg2 and Msg4 transmissions.

That is done by configuring an association between SS Block and a subset of RACH resources, one or multiple RACH occasions, and/or a subset of PRACH preamble indices, for determining Msg2 and Msg4 DL TX beam. Based on the DL measurement on SS Block(s) and the corresponding association, the UE selects the subset of RACH resources and/or the subset of RACH preamble indices for PRACH preamble selection. The association between one or multiple occasions for SS Block and a subset of RACH resources and/or subset of preamble indices is provided to the UE by RMSI (contention-based RACH) or known to the UE. The association between SS Blocks and RACH preamble indices and/or RACH resources is based on the transmitted SS blocks indicated in RMSI.

In handover case, a source cell can indicate in the handover command, the association between RACH resources and CSI-RS configuration(s) or association between RACH resources and SS blocks. In other words, for mobility the association may be defined between CSI-RS resource(s) and PRACH resources.

NR supports both long and short sequence-based PRACH preambles. Long sequences are targeted to macro, and extended coverage and long-range deployments. Short sequences are introduced to support efficient small cell deployments and efficient beamforming support. In addition, one design principle related to short sequences was the ability to configure the same SCS for PRACH preamble as for other uplink channels like NR-PUCCH and PUSCH.

3.4.1.1 Long Sequence

Sequence length for the long sequence-based PRACH preamble is 839 like in LTE. Also, Zadoff-Chu sequence family known from LTE is used. Long sequence preambles are supported for below 6 GHz deployments with two different SCS: 1.25 and 5 kHz. Three different formats have been defined for 1.25 kHz SCS targeting at LTE refarming, support of cells up to 100 km range and extended coverage. One format has been designed for the 5 kHz SCS option, especially for high speed cases. All long sequence-based formats support type A and type B restricted sets to support different mobility scenarios. Type A and B restricted sets consist of PRACH preambles that have cyclic shift distance allowing unambiguous preamble detection under doppler frequency up to ±1 SCS and up to ±2 SCS. Table 3.9 illustrates the long sequence-based PRACH preamble formats.

3.4.1.2 Short Sequence

Sequence length for short sequence-based PRACH preamble formats is 139 and the Zadoff-Chu sequence family is used. Short sequence preamble formats support different numerologies, namely 15 and 30 kHz at below 6 GHz, and 60 and 120 kHz at above 6 GHz. Starting symbol for the PRACH preamble format can be symbol 0 or 2 within a slot. The latter is to provide room for PDCCH in the beginning of the slot.

Table 3.9 Long sequence PRACH preambles.

Format	L_{RA}	Δf^{RA} (kHz)	N_u	NCPRA	T_{GP}	Use case	Support for restricted sets
0	839	1.25	24576κ	3168κ	2975	LTE reframing	Type A, Type B
1	839	1.25	2*24576κ	21024κ	21 904	Large cells, up to 100 km	Type A, Type B
2	839	1.25	4*24576κ	4688κ	4528	Coverage enhancement	Type A, Type B
3	839	5	4*6144κ	3168κ			Type A, Type B

Table 3.10 Base formats for short sequence-based PRACH preambles.

Format	L_{RA}	Δf^{RA} (kHz)	N_u	NCPRA
A1	139	$15 \times 2^{\mu}$	$2 \times 2048\kappa \times 2^{-\mu}$	$288\kappa \times 2^{-\mu}$
A2	139	$15 \times 2^{\mu}$	$4 \times 2048\kappa \times 2^{-\mu}$	$576\kappa \times 2^{-\mu}$
A3	139	$15 \times 2^{\mu}$	$6 \times 2048\kappa \times 2^{-\mu}$	$864\kappa \times 2^{-\mu}$
B1	139	$15 \times 2^{\mu}$	$2 \times 2048\kappa \times 2^{-\mu}$	$216\kappa \times 2^{-\mu}$
B2	139	$15 \times 2^{\mu}$	$4 \times 2048\kappa \times 2^{-\mu}$	$360\kappa \times 2^{-\mu}$
B3	139	$15 \times 2^{\mu}$	$6 \times 2048\kappa \times 2^{-\mu}$	$504\kappa \times 2^{-\mu}$
B4	139	$15 \times 2^{\mu}$	$12 \times 2048\kappa \times 2^{-\mu}$	$936\kappa \times 2^{-\mu}$
C0	139	$15 \times 2^{\mu}$	$2048\kappa \times 2^{-\mu}$	$1240\kappa \times 2^{-\mu}$
C2	139	$15 \times 2^{\mu}$	$4 \times 2048\kappa \times 2^{-\mu}$	$2048\kappa \times 2^{-\mu}$

A random access preamble format consists of one or multiple random access preamble(s). A random access preamble consists of one preamble sequence and (CP), and one preamble sequence consists of one or multiple RACH OFDM symbol(s). Furthermore, RACH occasion (RO) is defined as the time-frequency resources on which the PRACH preamble is transmitted using the configured PRACH preamble format with a single transmit beam at the UE. There are nine base formats as shown in Table 3.10.

Given the base formats, 10 different formats can be configured: A1, A2, A3, B1, B4, C0, C2, A1/B1, A2/B2, and A3/B3. In Ax/Bx all formats but last preamble use Ax format and the last preamble in a slot uses the format Bx. Correspondingly, there are the following number of occasions within a slot for different configurable formats as described in Table 3.11.

Table 3.11 Number of RACH occasions within a slot.

Format	Number of RACH occasions within a slot	Format	Number of RACH occasions within a slot
A1	6	C0	4
A2	3	C2	2
A3	2	A1/B1	6 or 7
B1	6 or 7	A2/B2	3
B4	1	A3/B3	2

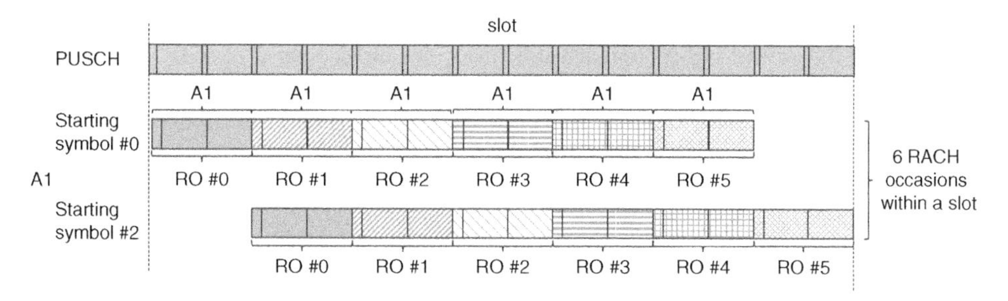

Figure 3.31 PRACH preamble formats A1 with two different starting symbols within a slot: #0 and #2.

Figure 3.31 exemplifies A1 format within a slot together with two different starting symbols: #0 and #2.

3.4.2 Uplink Data Transmission

Section 3.2.3 discusses the basic frame structures and introduces the PDSCH/PUSCH resource allocation. The uplink data transmission and HARQ work within this setup and are not too different from the basic setup in LTE, but some differentiators exist. The number of symbols allocated for PUSCH is fully flexible, and the timing relative to the scheduling PDCCH can be dynamically chosen by the gNB. As in the downlink case, the typical mode of operation is to place the PDCCH in the beginning of the slot, and the uplink DMRS and PUSCH will follow in the next slot (FDD) or in the next uplink slot (TDD).

As with the downlink HARQ-ACK timing, the maximum allowed TA is not baked inside the PDCCH-to-PUSCH delay, but the PUSCH timing can be scheduled freely with the scheduling PDCCH, if the UE is guaranteed sufficient time to process the PDCCH and encode the data packet for transmission on the PUSCH. This means that the gNB must factor in the TA budget on top the UE processing time defined in the standard. This allows for supporting large cell radii without having to have the extra budget for the corresponding two-way propagation delay embedded in the PUSCH preparation latency in normal or small cell deployments. Unlike in LTE, where the PDCCH to PUSCH transmission delay is defined by the standard to accommodate also the largest supported cell radius, the NR specification just defines the minimum processing time the UE must be guaranteed with before it can be expected to be ready to start the PUSCH.

In addition to the asynchronous PDCCH to PUSCH timing, there is no HARQ timing for PUSCH in the specifications. Each transmission is independently scheduled with PDCCH that carries the HARQ process number and new data indicator. This enables asynchronous HARQ retransmissions with no pre-determined time relation between different transmission attempts and facilitates different gNB architectures so that, e.g. a high latency fronthaul deployment can use a larger number of HARQ processes and longer retransmission delay and a low latency fronthaul deployment conversely uses a lesser number of HARQ processes and shorter retransmission delay. The UE always supports 16 HARQ processes, but the gNB only uses as many as it needs.

Multi-slot transmission (known also as slot repetition or slot aggregation) for improved coverage is possible both in downlink and in uplink. As uplink is typically the coverage limiting link, it is likely the more beneficial of the two in real deployments. RRC can configure the UE to repeat each transport block over a specific number of consecutive slots. The redundancy version is cycled across the slots so that full IR combining gain can be achieved. If some of the slots in the span of the consecutive slots are configured to be downlink slots, those slots are omitted, but still counted as part of the slot aggregate. The same symbol and frequency allocation is used across all the repeated slots.

Like in LTE, NR uplink supports frequency hopping for PUSCH to equalize the impact of frequency selective channels. Two types of frequency hopping are supported, but operation without frequency hopping is possible as well. When multi-slot transmission is not used, the frequency hop takes place once, in the middle of the allocated transmission duration, e.g. if 10 symbols were allocated for the PUSCH, then the frequency hop takes place in between the 5th and 6th symbol of the transmitted transport block (TB). The frequency offset to be applied is fully configurable. When multi-slot transmission is used, it is possible to configure the link to hop once per slot, rather than once within a slot.

3.4.3 Contention-Based Access

NR uplink supports two types of transmissions with configured grant (grant-free transmissions). One of the types, type 2, is often referred to as semi-persistent scheduling like for NR downlink and LTE as discussed in Section 3.3.3, where the first scheduling PDCCH triggers a periodically repeating transmission occasion without the need to schedule every packet separately. This solution is aimed at reducing the required control channel capacity when supporting many simultaneous voice links in a cell.

The other type of transmission with configured grant, type 1, is analogous to circuit switched radio, where the RRC configuration provides the UE with a set of time/frequency resources as transmission opportunities as well as the MCS to use, and the UE transmits on a given resource if it has any data. The gNB needs to detect the presence of the transmission, e.g. from the UE-specific DMRS, and to attempt to decode the transmission based on the provided configuration. Multi-slot transmission with RV cycling can be used in conjunction with the configured grant transmissions. Retransmissions after failed decoding attempts are scheduled using normal uplink scheduling procedure using PDCCH.

3.5 Radio Protocols

3.5.1 Overall Radio Protocol Architecture

In NR, a set of radio protocol layers are used to convey different IP-packet formats to the PHY as user plane data. In addition to IP-packets, NR supports transport of Ethernet MAC frames as user plane data. These data packets are transmitted over Data Radio Bearers (DRBs). NR radio protocols provide also reliable transport of the RRC protocol message via Signaling Radio Bearers (SRBs). The RRC provides an overall toolbox for

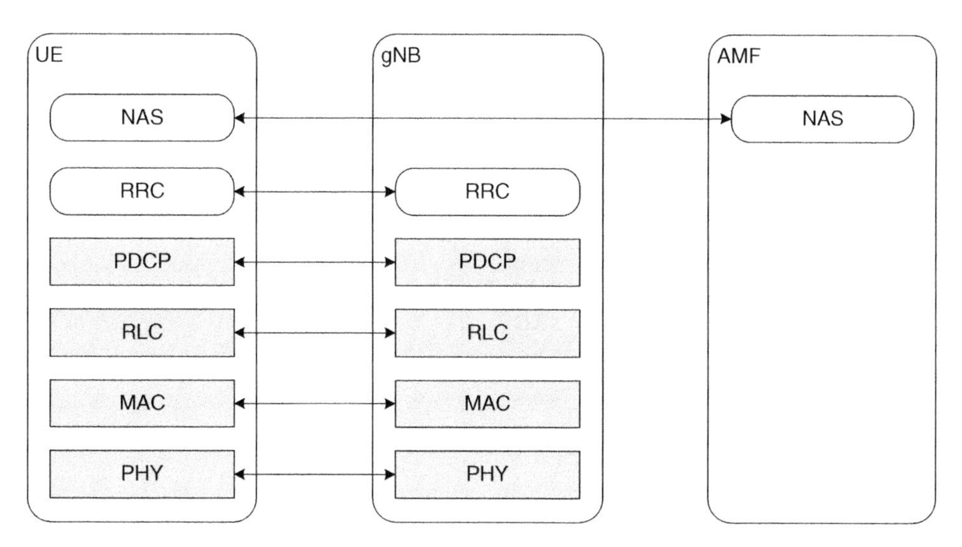

Figure 3.32 Control plane protocol stack.

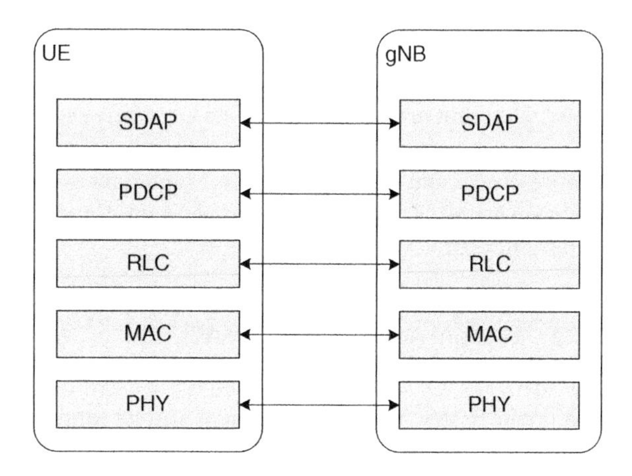

Figure 3.33 User plane protocol stack.

RRM and UE configuration as well as it conveys different Non-Access Stratum (NAS) signaling messages between core network (CN) and UE. To achieve this, and to support different network architectures, NR radio protocol layers have several functions for efficient radio operation. In the following sections the radio protocol is explained in detail.

The NR user plane and control plane radio protocol architecture is depicted in Figures 3.32 and 3.33 (see [36]) Compared to the previous generation, i.e. UMTS or LTE, the new Service Data Adaptation Protocol (SDAP) was introduced to the user plane. The main function of SDAP is to enable radio protocol support for the 5G Quality of Service (QoS) framework. The new QoS framework is discussed in Chapter 6, details of SDAP functions are presented in Section 3.5.5.

Most parts of user and control plane protocol layers are very similar, like in LTE. However, all protocols must be re-designed to meet 5G requirements and to support NR

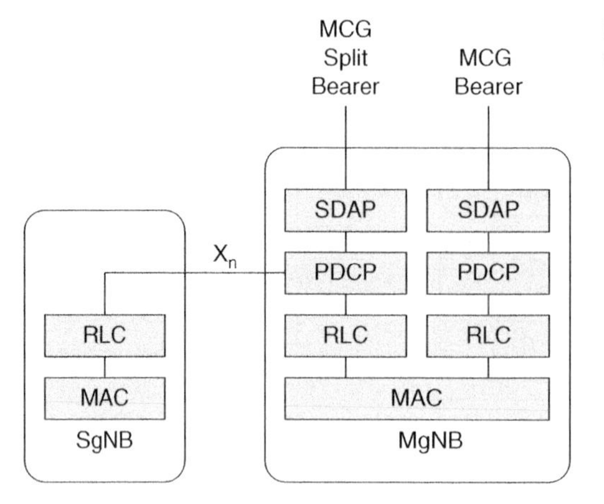

Figure 3.34 MgNB bearers for dual connectivity [36].

network architecture options. The options include different dual connectivity solutions essential for the UE having simultaneous connections on low operating frequencies and mmWave frequencies. The protocol architecture with master gNB bearer dual connectivity is depicted in Figure 3.34.

In addition, the radio protocol architecture is designed to support different radio network cloudification solutions, where parts of the protocols are implemented in the cloud. Different architecture options are discussed in Chapter 4.

Perhaps the most significant driver for re-design of the radio protocols, was to obtain a more processing friendly design for high data as well as low latency connections. This was seen essential for UE chipset implementations but gains for BS implementations are also significant.

3.5.2 Medium Access Control (MAC)

The MAC layer is in the center of all NR procedures. It is the interface between Layer 1 and upper layers. The MAC protocol and its functions are specified only for the UE in NR (see [37]):

- Mapping between logical channels (LCHs) and transport channels according to the transferred traffic type as presented in Section 3.5.2.1.
- Multiplexing/demultiplexing of MAC Service Data Units (SDUs) belonging to one or different LCHs into/from transport blocks (TB) delivered to/from the PHY on transport channels. This function includes concatenation of SDUs. Concatenation is solely performed by MAC in the NR radio protocol stack.
- Scheduling information/buffer status reporting (BSR) to the network for receiving UL grants by means of the Scheduling Request (SR) procedure, which may involve triggering of the Random Access procedure in the absence of dedicated SR resource configuration.
- Error correction through HARQ. Adaptive and asynchronous HARQ is used for both uplink and downlink directions, fully under control of the network
- Priority handling between LCHs by means of logical channel prioritization (LCP) procedure to support QoS enforcement among different services that run in parallel in the UE. The MAC entity may be configured to restrict mapping of a LCH to a grant

with certain L1 characteristics, which may include, e.g. the used numerology and/or the Transmission Time Interval (TTI) duration.

- Padding: MAC is the only radio protocol layer in charge of padding. Padding is performed in case not enough data can be multiplexed within one TB that is granted for transmission in L1. The MAC should always maximize the transmission of data, i.e. padding is usually only allowed in case there is not enough data in the buffer to fill the full grant.
- Beam failure management to support L1 procedures. MAC is responsible for triggering the UE-based beam management procedures like beam failure detection (BFD) and recovery.

In comparison with LTE, the MAC layer in NR holds the same set of functions added with the beam failure management procedures in support for systems operating at high frequencies. However, much thought has been put on the actual processing of data and control in the MAC layer as well as the design of the Protocol Data Unit (PDU) structure to be able to squeeze the UE's UL grant-to-data transmission time to an absolute minimum.

3.5.2.1 Logical Channels and Transport Channels

The MAC entity provides an interface to Radio Link Control (RLC) entities through LCHs, which are characterized either as control channels carrying different kind of control plane data or traffic channels carrying user plane data. The following LCHs are defined for control plane data:

- Broadcast Control Channel (BCCH): for the provisioning of system information messages to the UE which include either Master Information Block (MIB) or System Information Block(s) (SIB). Applicable only in DL direction.
- Paging Control Channel (PCCH): for the provisioning of paging messages to UEs in the cell which is used to page the UEs in IDLE or INACTIVE mode, to indicate change in system information, or to provide indication that a warning system message or public warning system (PWS) message is to be broadcasted. Applicable only in DL direction.
- Common Control Channel (CCCH): for the transmission of non-secured RRC messages of SRB0 between UE and network prior to establishment of SRB1 and ciphering. Applicable both in UL and DL.
- Dedicated Control Channel (DCCH): for the transmission of RRC messages of SRB1, SRB2, or SRB3 between UE and network to configure/maintain the RRC connection or to convey piggybacked NAS messages.

The following LCH is defined for user plane data:

- Dedicated Traffic Channel (DTCH): for the transmission of all user plane data. Multiple DTCHs can be established for different type of user plane data according to their QoS requirements.

Interface from L1 to serve MAC entity is handled through transport channels, whereas MAC is responsible for mapping the data from LCHs to proper transport channels. Transport channels are defined based on the L1 characteristics on how and when information is transmitted. Through transport channels L1 provides MAC with a transport block (TB) and its size, based on the mapping of PHY parameters like MCS and slot length. MAC fills the TB with a MAC PDU carrying the data.

The following transport channels are defined in downlink:

- Broadcast Channel (BCH): for the transmission of MIB from the BCCH LCH, i.e. it only conveys part of the system information carried in BCCH. The BCH uses a fixed sized container and has fixed position/periodicity in L1, so it does not need to be scheduled.
- Paging Channel (PCH): for the transmission of paging messages from the PCCH LCH. The PCH has configurable periodicity and has flexible message size, hence, it is scheduled each time using the Paging Radio Network Temporary Identifier (P-RNTI), which is common in the system. However, the used transport format is limited by the minimum supported UE category in the system.
- Downlink Shared Channel (DL-SCH): for the transmission of downlink control and user plane data from BCCH (excluding MIB), CCCH, DCCH, and DTCH LCHs. Each UE is allocated to a DL-SCH when scheduled, but the UE may be scheduled using multiple DL-SCHs within one time instant, e.g. through carrier aggregation. Furthermore, DL-SCH used for system information messages from BCCH may be transmitted simultaneously. DL-SCH is flexibly configurable on a per UE basis according to supported UE capabilities.

The following transport channels are defined in uplink:

- Uplink Shared Channel (UL-SCH): for the transmission of uplink control and user plane data from CCCH, DCCH and DTCH LCHs. It is the equivalent of D-SCH in uplink direction.
- RACH: for the transmission of random access preamble. No LCH maps to RACH, since RACH does not carry any data above MAC, the random access preamble and used resource selection is performed by the MAC entity.

Mapping between the LCHs and transport channels is presented in Figures 3.35 and 3.36 for downlink and uplink.

3.5.2.2 MAC PDU Structures for Efficient Processing

MAC PDU structure in NR used for data transmission, i.e. data transmitted over UL-SCH and DL-SCH, is designed to support very efficient processing both in the transmitter and receiver. This is motivated by the very high 5G data rates, need to enable concatenation at MAC layer as well as to remove the concatenation feature from RLC protocol. This necessitates much more MAC SDUs to be multiplexed within

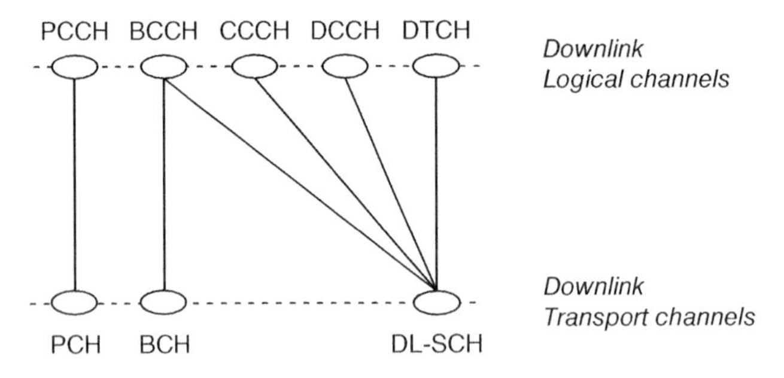

Figure 3.35 Downlink logical channel mapping to transport channels.

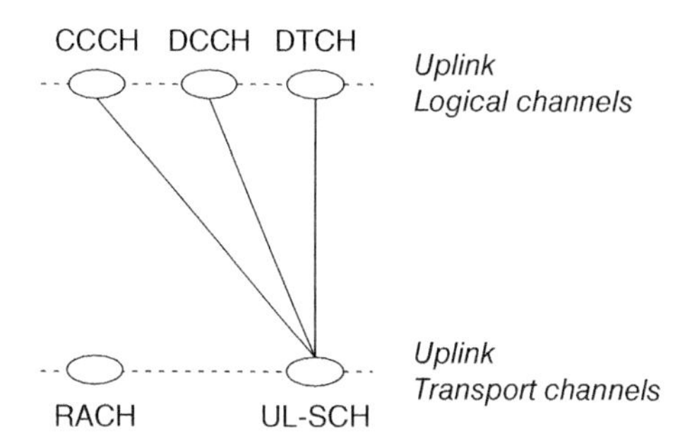

Figure 3.36 Uplink logical channel mapping to transport channels.

one MAC PDU. Furthermore, processing efficiency is mainly targeted in the UE to be able to support as short grant-to-transmission times and data reception-to-feedback transmission times as possible.

Unlike in conventional 3GPP systems where the MAC header with a number of sub-headers is packed in front of the MAC PDU before the data, SDUs or control elements (CEs), NR interlaces each MAC sub-header with the corresponding MAC SDU or CE. The combination of MAC sub-header and MAC SDU/CE is called "MAC subPDU," both of which are multiplexed one by one into the MAC PDU. Such a design enables pipeline processing to be used both in the transmitter as well as in the receiver. The transmitter can start feeding L1 with the MAC PDU in each MAC subPDU at a time before the whole MAC PDU has been constructed. Similarly, the receiver can pass a part of the MAC PDU for MAC processing even before the entire slot has been received. That is, parallel L2 and L1 processing can be performed even for one transmission/reception.

MAC PDU structures for downlink and uplink are illustrated in the following Figures 3.37 and 3.38.

As can be seen in Figures 3.37 and 3.38, for downlink MAC PDU the MAC CEs are multiplexed together in front of the PDU. For uplink MAC PDU they are still multiplexed at the end of the PDU but before padding. The downlink design allows the UE to process any MAC control received in downlink as soon as possible. For uplink, putting

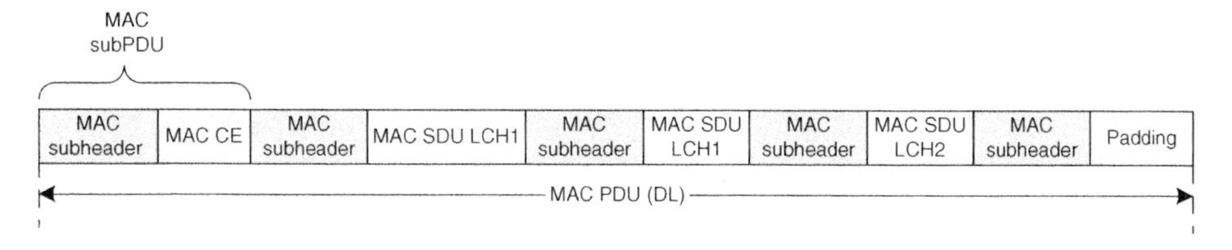

Figure 3.37 MAC PDU structure in DL.

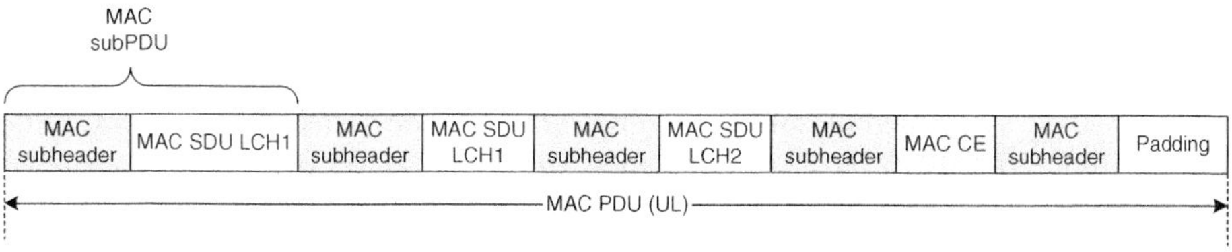

Figure 3.38 MAC PDU structure in UL.

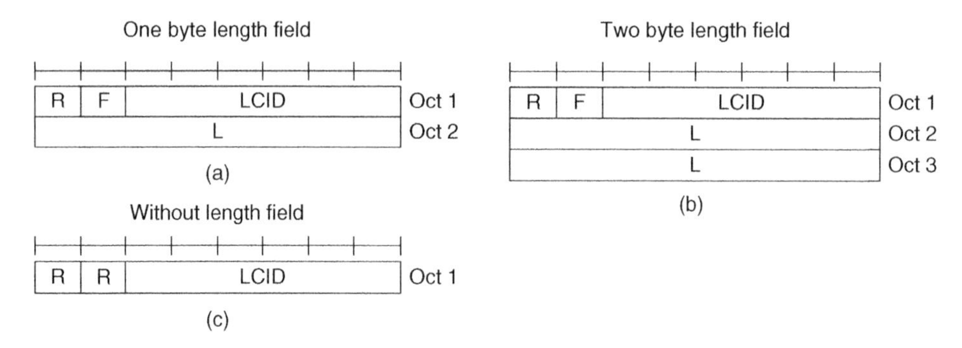

Figure 3.39 MAC sub-header structures.

the control information at the end of the MAC PDU allows the UE to process such control as late as possible in the transmitter, e.g. the buffer status report shall consider any data multiplexed in the PDU where the buffer is reported.

MAC sub-headers are also simplified compared to LTE. There is no explicit bit indicating whether another MAC SDU/CE will follow, this is implicitly determined from the MAC SDU/CE length, the remaining MAC PDU size, and the possible padding sub-header. MAC padding is also only multiplexed at the end of the MAC PDU by indicating one-byte MAC sub-header without length after which all remaining bits in the PDU are all padding. The used length field size is only indicated dynamically in the sub-header to support various SDU sizes, or otherwise the LCID (Logical Channel ID) field is used to determine whether the MAC sub-header comes with the L field or not (see Figure 3.39).

3.5.2.3 Procedures to Support UL Scheduling

The UE MAC layer implements a LCP function, which performs the actual scheduling of UL data from different LCHs to the available resources in the MAC PDU. Furthermore, several procedures are defined in the UE MAC layer to support the gNB with UL scheduling decisions. These functions are SR, BSR and power headroom reporting (PHR). These procedures are specified similarly to LTE, however, NR introduced an enhancement called LCP restrictions to support the various 5G services more efficiently also in UL direction.

In the LCP function of the UE the MAC entity uses a bucket size scheduler to decide how much data is multiplexed from each LCH to a certain MAC PDU. Each LCH is assigned a priority, prioritized bit rate (PBR), and bucket size duration (BSD) based on which the bucket size for each LCH is calculated. Furthermore, given the introduction of multiple sub-carrier spacings (SCS), non-slot-based scheduling, and in general the various services to be supported, the LCP function can be configured to apply LCP restrictions in the scheduling. With LCP restrictions, data from a certain LCH can be restricted to be mapped only to a certain type of grant categorized by its PHY characteristics. The grant is categorized by transmission duration, used SCS, used carrier, and whether it is a dynamically scheduled or configured grant. For instance, this allows URLLC data to be multiplexed only to "fast grants," which are usually short in terms of transmission duration and gNB feedback is fast. Furthermore, LCP restrictions are applied in case of Packet Data Convergence Protocol (PDCP) duplication to avoid multiplexing of duplicated data in one MAC PDU. PDCP functions are discussed in Section 3.5.4.4.

The SR procedure is used by the UE MAC entity to indicate to the gNB the available data in UL buffers. The MAC entity may be configured by the network with multiple SR configurations and each UL LCH can be associated with a certain SR configuration. This enables the serving gNB to determine the service for which the UE has UL data available through the LCH association already from the SR transmission and hence allows to provide the right grant type. This is especially useful when LCP restrictions are configured. In case UE has data in its buffers for different LCHs associated with different SR resources, it can have multiple SR procedures ongoing at any point in time.

A UE uses BSR to indicate to the serving gNB the amount of available data in the UL buffers. Buffer is reported on a per Logical Channel Group (LCG) basis which may consist of one or more LCHs. NR implements 8 LCGs compared to 4 in LTE which is justified by the support of twice more DRBs (29 as the maximum) and PDCP duplication (see Section 3.5.4.4). As the BSR is multiplexed to any type of grant in UL, enhancements are designed to avoid any additional latency for latency sensitive data. In case the BSR was triggered by latency sensitive data and the LCP restrictions do not allow multiplexing the data into the available grant, a SR can be triggered to indicate latency sensitive data presence regardless of the multiplexed BSR.

PHR is used to indicate to the serving gNB the available UL power the UE can still use. The power headroom (PH) is reported per each serving cell with configured UL carrier in case of carrier aggregation. By means of knowing the PH for each serving cell, the gNB can adjust its scheduling decisions such that the available UL maximum transmit power is not exceeded and hence the probability for errors is not increased.

3.5.2.4 Discontinuous Reception and Transmission

For UE power saving during active data transmission, NR supports discontinuous reception (DRX) and UL grant skipping, which are applicable when there are short moments without DL or UL data. The network configures DRX according to the traffic characteristics of the active service, however, as there may be multiple active services, the DRX concept also adapts to these by means of two different DRX cycle durations that can be used simultaneously.

DRX is applied to enable the UE Rx to sleep when there is no expected data transmission in DL. Basically, DRX controls the need for the UE to monitor the PDCCH scheduling decisions and is dictated by several timers including inactivity and HARQ Round Trip Time (RTT) timers and network commands through MAC CEs. Two different DRX cycles can be configured, a Short and a Long DRX cycle, from which the Short cycle is used for short sleeping periods when data comes to the buffer in short bursts. The Long cycle is applied when data is not expected for a longer time and a longer sleep duration can be applied. The gNB can also indicate with MAC CE to the UE to go directly to Long DRX, e.g. in case gNB determines it has no more DL data to be sent to the UE. The sleep times enabled by the Short and Long DRX vary from 2 milliseconds up to ~10 seconds, so various types of configurations and services can be supported. Compared to LTE, the main notable difference in the NR DRX concept is that the different cells configured for the UE may use different sub-carrier spacings which means that the slot boundaries are not aligned. Regardless of this, common DRX in the MAC entity is applied. As the HARQ RTT timers are cell specific, those are depending on the used sub-carrier spacing resulting in different timer values in the different cells.

Uplink grant skipping is used to avoid unnecessary transmissions in the UL whenever the UE has no data to transmit. This reduces UE energy consumption as well as UL interference introduced to neighboring cells. Such skipping is always used by the UE for configured grants (e.g. "grant-free scheduling") as the same grant may be configured for multiple UEs to be used, the collision probability is lowered as the UEs without data will skip the transmission in such a grant. For dynamic grants, the uplink grant skipping is configurable and enables the gNB to do blind scheduling without compromising the UE energy consumption. Given the fact that the UL grant skipping may be enforced because of configured LCP restrictions, in the case where there is data in the UL buffers and Periodic BSR is triggered, the BSR is included in a MAC PDU and the grant should not be skipped.

3.5.2.5 Random Access Procedure

Compared to LTE, the NR Random Access (RA) procedure is enhanced to support multi-beam operation and supplemental (supplementary) uplink (SUL) bands. Furthermore, Beam Failure Recovery (BFR) procedure (see Section 3.5.2.6) and on-demand system information request are added as new use cases for the RA procedure. The RA procedure is triggered by the MAC entity itself, e.g. for UL data arrival, SR failure, or BFR, by the RRC layer, e.g. for initial access or handover, or by the network through PDCCH order, e.g. for DL data arrival, or Secondary Cell (SCell) addition with different UL timing.

Both contention based random access (CBRA) and contention free random access (CFRA) procedures are supported. CBRA preamble is always associated to a certain SS Block beam which is configured by the network, whereas CFRA preamble can be associated to an SS Block beam or to a beam which is identified by using CSI-RS and is dedicatedly configured to the UE. Before each preamble transmission, the UE firstly selects an SS Block, in case of CBRA, or in case of CFRA an SS Block or CSI-RS based on which the preamble selection can be performed. The selection between SS Block and CSI-RS is based on network configured Reference Signal Received Power (RSRP) thresholds measured from the DL beams.

For CBRA, the preamble space from which the UE randomly selects a preamble may be allocated to two different preamble groups, Random Access preamble groups A or B. Group B is used to indicate that the UE has more data in its buffers than a configured threshold, being a rough indication about the UE buffer status. In such a case, the gNB can give the UE a bigger grant already in the RAR. Each SS Block includes preambles from both group A and B in case group B is configured in the cell.

CBRA is a four-step procedure which involves:

1. PRACH preamble transmission by the UE, where the UE selects randomly a preamble associated to the selected SSB and the selected RA preamble group (A or B).
2. RAR transmission by the gNB addressed to RA-RNTI (Random Access Radio Network Temporary Identifier), which may include responses to multiple preambles transmitted by multiple UEs in the first step. However, only one response is transmitted to one PRACH preamble regardless if the gNB can detect that multiple UEs transmitted the same preamble.
3. Msg3 transmission in UL by the UE, which is used by the UE to identify itself, i.e. the UE identity is multiplexed into the MAC PDU of Msg3 provided either by the RRC or MAC layer.

4. Contention resolution message transmission by the gNB, which identifies the UE by the UE ID provided in Msg3. In case multiple UEs transmitted the same preamble, only the UE with proper contention resolution ID can proceed after this point, other UEs must trigger a new preamble transmission.

CFRA is a two-step procedure which involves:

1. Pre-configured PRACH preamble transmission by the UE, where the UE transmits a gNB allocated preamble associated to the selected SSB or CSI-RS.
2. RAR transmission by the gNB addressed to either RA-RNTI or C-RNTI (Cell Radio Network Temporary Identifier). Only in the case of BFR, RAR transmission may be addressed directly to C-RNTI of the UE since UL timing alignment is available, otherwise the UE will decode the RA-RNTI associated to the PRACH where the preamble was transmitted.

Based on measurement results provided by UE, the gNB can only configure CFRA preambles to a subset of beams provided by the cell. This is to reduce signaling overhead as well as system flexibility, as the UE need not be allocated a CFRA preamble from beams where it is unlikely to be served. However, e.g. during the handover procedure when the UE is moving, the UE might find a beam which is not allocated with CFRA preamble as the one where it wants to get served. Hence, both CFRA and CBRA preamble transmissions may happen within the same RA procedure. The UE prioritizes the beams with allocated CFRA resources, however, if such a beam is unavailable (dictated by a RSRP threshold,) the UE will perform CBRA toward a selected SSB beam.

In support for SUL operation, when the RA procedure is triggered, the UE firstly selects the UL carrier (UL or SUL) to perform the RA. The selection is based on a DL RSRP threshold measured from the DL signal and in case the DL signal level is below the threshold, SUL is used. This means that the UE cannot switch to UL from SUL, or vice versa, during one RA procedure, but will run the procedure until the end through the initially selected carrier. Such a principle is adopted to simplify UE operation as no fresh DL measurements are needed for every preamble transmission attempt.

Furthermore, the RA procedure in NR includes a concept of prioritized random access where by network configuration, certain UEs may be prioritized in the RA procedure. The prioritized RA is only applicable to UEs performing BFR or handover and happens via different configuration of parameters used in the RA procedure. For instance, the UE can be configured to ramp up its UL transmission power more rapidly than other UEs or in case of overload in the PRACH, scale down the backoff time if this is indicated in response to the RA preamble transmission.

3.5.2.6 Beam Failure Management

Beam failure management is performed in the MAC layer by support of PHY measurements and involves two procedures: BFD and BFR. By means of the BFD and BFR procedures, the serving beams used for communication between UE and gNB can be recovered rapidly without involving any upper layer failure procedures, like radio link failure (RLF) procedure. These procedures complement the beam management procedures discussed in Chapter 5.

BFD is based on beam failure instance (BFI) indications from the PHY based on which MAC determines when the serving beams are in a failure condition. The UE is

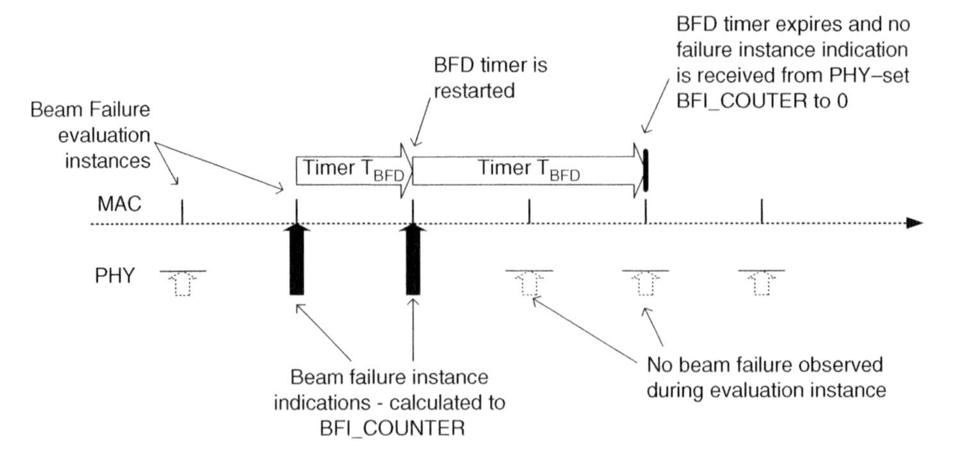

Figure 3.40 Beam failure detection principle.

configured to monitor certain BFD-RS(s) (Beam Failure Detection Reference Signal) to evaluate, if the serving beams are workable based on hypothetical PDCCH BLER (Block Error Rate). In case all the serving beams are determined to be in failure condition, PHY provides a BFI indication to the MAC layer. The BFD procedure in MAC is then dictated by a timer and a counter which calculates the number of BFI indications received. Every time the BFI indication is received, the timer is restarted and in case the timer expires, the counter is reset. Hence, the timer duration determines the sufficient period within which the beams are workable. However, in case the counter reaches a configured threshold value before the timer resets the counter, beam failure is detected which triggers the BFR procedure. The BFD operation principle is illustrated in Figure 3.40.

BFR is triggered because of detecting a beam failure, which is used by the UE to indicate to the serving gNB a new candidate beam to serve the UE. The procedure is performed by triggering the Random Access procedure (see Section 3.5.2.5). The network may configure the UE with a set of candidate beams based on SS Block or CSI-RS beams,

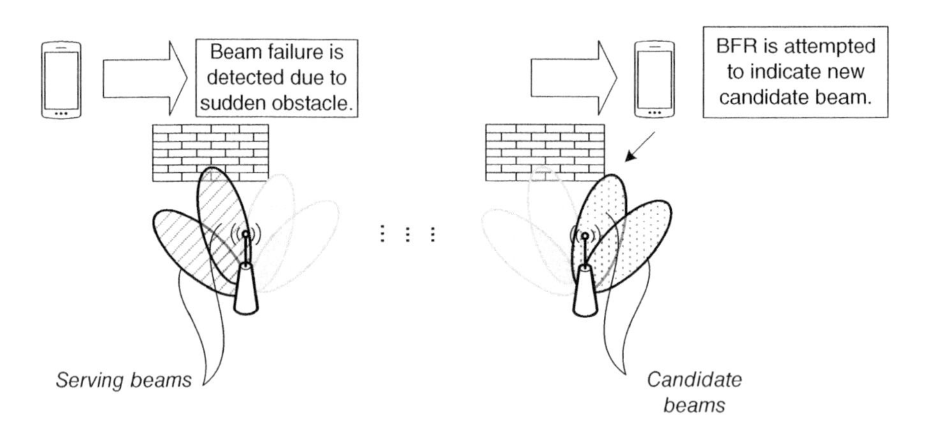

Figure 3.41 Beam failure recovery principle.

which are associated with CFRA preambles. For instance, the network may determine that such beams are the most likely candidates in case beam failure happens. By means of CFRA preamble, the new preferred DL beam is indicated from which the gNB can start scheduling the UE again and apply the beam management procedures in the PHY. As discussed in the previous RA procedure section, beams with allocated CFRA preamble resources are prioritized in the RA procedure and hence the UE prioritizes the candidate beams configured. However, in case no candidate beams are available (dictated by a configured RSRP threshold), fallback to CBRA is enforced and a workable SSB beam is indicated.

The BFR principle is illustrated in Figure 3.41.

The BFR procedure is successful in the case that the corresponding RA procedure is successfully completed. Hence, in the case where the RA procedure fails, the BFR procedure also fails, leading to declaration of RLF and enforcement of RRC level recovery.

3.5.3 Radio Link Control (RLC)

The RLC protocol is responsible for segmenting the PDCP PDUs to fit within the MAC PDU, minimizing the needed padding as well as ensuring lossless delivery of the data by means of the Automatic Repeat Request (ARQ) protocol, as defined in [38]. Compared to LTE, NR removed functions from RLC for processing efficiency moving them to PDCP and MAC, reordering/in-order delivery is done by PDCP and the concatenation function is performed along the MAC multiplexing.

Three data transfer modes can be configured for an RLC entity depending on the type of data it serves: Transparent Mode (TM), Unacknowledged Mode (UM) and Acknowledged Mode (AM).

The TM RLC entity is defined for the transmission of data in "one-shot" like system information broadcast or paging messages through BCCH or PCCH LCHs, respectively, or SRB0 data through the CCCH LCH. According to its name, the TM RLC entity does not modify the data nor does it include any headers when submitting it to the lower layer. TM RLC entity is uni-directional, i.e. it is configured either to transmit or receive data.

The UM RLC entity is defined for data services like voice or video streaming that do not require lossless delivery, i.e. packets are not acknowledged nor re-transmitted by the UM RLC entity. The UM RLC entity handles only the segmentation function. The UM RLC entity is uni-directional, i.e. it is configured either to transmit or receive data.

The AM RLC entity is defined for data services sensitive for any losses in the link like TCP protocol services. The AM RLC entity provides lossless delivery of upper layer data by means of the ARQ protocol through status reporting by the receiving entity and re-transmissions by the transmitting entity. The transmitting entity may also poll the receiving entity to transmit a status report called polling. The AM RLC entity is bi-directional, i.e. it can receive and transmit data and control information like status reports.

SRBs are always configured with AM RLC entity, while DRB may be configured with either AM RLC entity or one or two UM RLC entities. For instance, a video streaming service could only require an uni-directional link in downlink direction and hence only one UM RLC entity, while voice service requires sound to be both transmitted and received requiring a bi-directional link and hence two UM RLC entities for a DRB.

3.5.3.1 Segmentation

When MAC multiplexes data from one or multiple LCHs, the purpose of RLC segmentation is to ensure data can be multiplexed into the MAC PDU even though the grant size is insufficient to carry the whole PDCP PDU. Also, when multiple PDCP PDUs are multiplexed within the MAC PDU, for the remainder of the grant, one of the RLC entities segments a PDCP PDU to maximize transmission of data. This is also an enabler to support the wide range of data rates in the NR system. The receiving RLC entity is responsible for reassembling the RLC SDUs and PDCP PDUs from the received segments based on their sequence number (SN) and segmentation information carried in the RLC header.

Offset-based segmentation is always used by the RLC entity. That is, whenever a segment is created from an RLC SDU, offset information about the start position of the RLC SDU segment in bytes within the original RLC SDU is indicated to the receiving RLC entity in the RLC PDU header. Hence, whenever the first segment of the RLC SDU is transmitted, offset information is not included as it would naturally show only a field of zeros. This required the RLC PDU header being able to indicate whether the segment is the first segment, middle, or the last segment of the RLC SDU, however, allocating 2 bits for this purpose is justifiable given it reduces RLC header overhead by 16 bits for every segmented RLC SDU. Furthermore, this enables the RLC PDU header size to be known in advance in the transmitter without the knowledge of available grant size regardless of whether segmentation was to be performed or not providing better capabilities for pre-processing in the transmitter (see chapter 3.5.3.3).

The segmentation of RLC SDU into RLC PDUs is illustrated in Figure 3.42.

Since the RLC protocol does not support concatenation, in contrast to LTE, the exposed header overhead increases when many RLC SDUs are multiplexed within one MAC PDU as each one of those form an own RLC PDU and hence require its own header with SN, segmentation information, etc. For the UM RLC entity, however, as the SN serves no other purpose than differentiating segments of different RLC SDUs, the overhead is reduced by indicating the SN in the RLC PDU header only for segmented RLC SDUs. Consequently, for full RLC SDUs the RLC PDU header contains only the indication for the receiver that the RLC PDU consists of a complete RLC SDU. By means of such a principle, at most one new SN will be allocated per RLC entity for each MAC PDU transmission compared to one SN per RLC SDU. Thus, the SN lengths

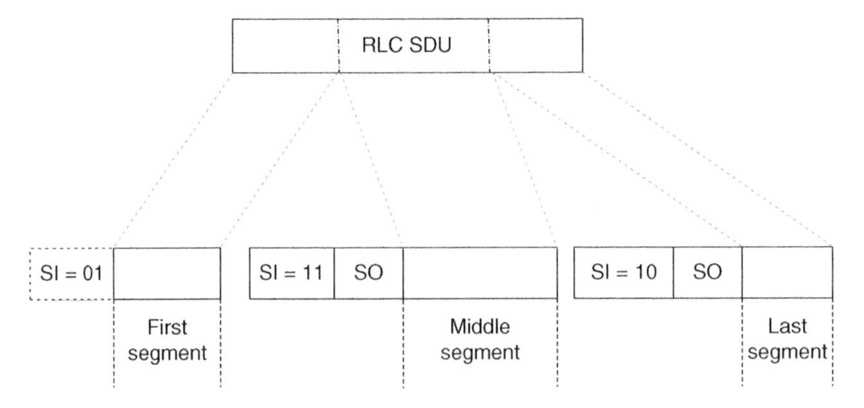

Figure 3.42 RLC SDU segmentation into RLC PDUs.

can be smaller for the UM RLC entity compared to the AM RLC entity and header overhead is further reduced.

3.5.3.2 Error Correction Through ARQ

The ARQ protocol is the second main function of RLC. It ensures lossless delivery of upper layer data by means of polling, status reporting and re-transmissions. Based on the SN and segmentation information in the RLC PDU header, the receiving RLC entity may track whether RLC SDUs or RLC SDU segments have been lost on the air interface.

Polling is performed by the transmitting RLC entity to trigger status reporting in the receiving RLC entity. By means of configured threshold values (number of transmitted RLC PDUs or bytes) or timer expiry, a poll bit is flagged in the RLC PDU header to ensure sufficient frequency in status reporting so that the transmit window can advance. It is notable that unless the transmitter gets feedback on a RLC SDU in the beginning of the transmit window and before the transmit window gets full (window size is half of the SN space), no new data can be transmitted so as not to confuse the receiver operation. Such window stalling limits the achievable throughput and hence polling is performed.

Status reporting is performed by the receiving RLC entity to inform the transmitting RLC entity about successfully received and missing RLC SDUs or RLC SDU segments. The missing SDUs or segments are indicated explicitly based on their SNs and offset information in the RLC Status PDU. At the same time, successfully received SDUs and segments are indicated implicitly by the absence of their SNs in the Status PDU by means of the highest status state variable. Only the highest SN is indicated explicitly in the Status PDU, which is neither indicated as missing nor successfully received. With such a structure all the SNs not explicitly indicated as missing below the highest status state variable are implicitly acknowledged to the transmitting RLC entity.

Re-transmissions are performed by the transmitting RLC entity based on the received status reports from the receiving RLC entity. Re-transmissions can be on a per full RLC SDU basis or per RLC SDU segment basis, i.e. the amount of re-transmitted data can be minimized. The number of maximum re-transmissions can be configured for the RLC entity, and in case such a threshold is reached, RLF is declared. Therefore, the AM RLC entity is also considered to be lossless. It attempts to transmit every byte until it is received successfully or RLF is detected. When RLF is detected, data loss happens also with the AM RLC entity as it is re-established during the RRC re-establishment procedure, hence, the error probability is not complete zero but still very small (10^{-8}–10^{-6} depending on the scenario of interest).

3.5.3.3 Reduced RLC Functions for Efficient Processing

As discussed at the beginning of the chapter, compared to LTE, NR removed concatenation and re-ordering/in-order delivery functions from the RLC entity. The removal of concatenation provides better processing efficiency in the transmitter and avoiding the reordering improves processing in the receiver side.

When concatenation is not performed at RLC, the higher layer functions are decoupled from real-time constraints in the transmitter. Each PDCP PDU submitted to the transmitting RLC entity can be pre-processed to an RLC PDU without any information about the available grant size from the MAC layer. Hence, the ARQ protocol in the RLC entity could be decoupled from the real-time processing constraints as the new RLC PDU creation as well as re-transmission decisions can be made in advance before the grant reception. Only the possible segmentation needs to be performed in real-time, however, this requires re-encoding of one-bit value in the RLC PDU header along with segmenting the data field, that is, the RLC PDU header size does never need to be changed because of segmentation leading to optimal pre-processing capabilities.

Avoiding reordering/in-order delivery at RLC allows immediate delivery of a received RLC SDU to the PDCP entity. As reordering support by the PDCP layer was required from the beginning to support dual-connectivity, the removal of reordering from RLC was backed up to avoid duplicate functions in the stack. Nevertheless, the main reason for such a design is to allow stable loading of PDCP deciphering function as PDCP PDUs can be deciphered immediately when fully received in the RLC entity before the reordering happens. In LTE, as the reordering is performed by RLC, the PDCP entity receives a burst of PDCP PDUs simultaneously when missing RLC PDU releases the reordering buffer to PDCP. Such behavior leads to very high processing load peaks in the receiver and reduced energy efficiency, ciphering and de-ciphering functions are one of the most energy hungry functions in the transmitter and receiver.

3.5.4 Packet Data Convergence Protocol (PDCP)

The number of functions at the PDCP layer, defined in [39], has increased in NR compared to LTE. Due to new architecture options in the radio network, namely split between distributed unit (DU) and centralized unit (CU), processing efficiency requirements in the UE, new services of 5G, and inherent support of dual connectivity, the new set of functions was justified. The PDCP is now solely in charge of reordering and in-order delivery as these functions were removed from RLC as discussed in the previous section. It should be noted that the in-order delivery can also be switched off by configuration, if the service in question does not require ordered delivery. Security is one of the main functions of PDCP, which includes ciphering and integrity protection of data. Integrity protection is always applied for SRBs like in LTE but can now be configured to be used with DRBs as well. Support of PDCP PDU duplication was introduced to provide more reliable and lower latency communication in support for URLLC services. Other PDCP functions inherited from LTE include: header compression and de-compression using Robust Header Compression (RoHC) protocol, routing for split bearers, timer-based SDU discard at the transmitter and duplicate discarding at the receiver side.

In dual connectivity operation the PDCP layer is in charge of routing data to different dual connectivity legs which each have their own UM or AM mode RLC entities. Thus, in the MAC BSR, also the PDCP buffer status is conveyed to the schedulers of both involved gNBs. PDCP data duplication can be used to improve reliability by using both dual connectivity links to transmit the same data. This can be especially useful for SRB signaling, e.g. in high mobility scenarios. As the same data is sent over the two links, the receiver may receive the same packet twice, hence, to avoid propagating this to upper layers, PDCP has the capability to discard any duplicates at the receiver.

3.5.4.1 Reordering

As discussed in the RLC chapter, the reordering and in-order delivery functions are now performed by the PDCP layer mainly due to the native support of dual-connectivity as well as to enable stable loading of the PDCP deciphering function. However, also the introduction of the NR network architecture option to split between DU and CU backed up such a decision, PDCP PDUs transferred over the F1-interface can be received out-of-order in the CU and DU. Reordering in the RLC layer would be just pointless as out-of-order delivery could happen in the network already in the transmitter side (PDCP PDU transfer from CU to DU in DL direction).

The reordering window in the receiving PDCP entity is a push-based window regardless of the RLC mode associated with the DRB, AM or UM. This means that each PDCP PDU received with SN out of the window is considered as an old PDU and is discarded by the receiving PDCP entity. The window is moved forward by receiving a PDCP PDU with the lowest SN still considered within the window (the next SN from the previously delivered SN of a PDCP SDU) or alternatively, in case the reordering timer expires. Thus, the window is kind of "pushed" forward from the low end. This is different to LTE, where the UM mode RLC entity is operating with pull-based reordering window where each SN received out of the window is considered as new and becomes the high edge of the window. The low edge is hence "pulled" by the high edge. The NR principle is a bit more complicated for UM mode DRBs as no more than half of the PDCP SN shall be in flight at any point in time and there is no RLC level feedback to determine the lowest SN delivered. However, given the big SN spaces specified for PDCP (12 and 18 bits) and the level of simplification the single window enables for the receiver, i.e. PDCP does not need to care about the associated RLC mode, such a principle makes sense.

In support for a wide range of 5G services, it is observed that not all services require the reordering and in-order delivery to operate. Consequently, NR specifies a possibility to configure out-of-order delivery for a DRB which is performed by the PDCP entity. That is, any PDCP data PDU received from RLC is delivered immediately to upper layers. However, it should be noted that the reordering window and receiver variables are similarly updated when in-order delivery is required enabling reconfiguration between these two modes if necessary.

Additionally, to support lossless mobility, both reordering and in-order delivery are needed in PDCP. This allows forwarding PDCP SDUs from the source gNB PDCP layer to the target gNB PDCP layer without knowing exactly which PDCP PDUs the UE has received correctly, i.e. they have not been acknowledged by the RLC layer, at the same time as new upper layer data is transmitted from core network to the target gNB. The target gNB may obtain the status of correctly received PDCP SDUs by the UE and continue the transmission immediately from any available PDCP SDU that is not yet received by the UE. This is done to avoid transmission of unnecessary duplicates which the UE has already received. The receiver in the UE is then able to reorder and deliver all PDCP SDUs to the application layer in-order.

For QoS support, to avoid transmitting outdated data, timer based SDU discard is used in the transmitter to discard data in the transmission buffer. Given that any missing SN in the sequence will stall the reordering window in the receiver, using the timer-based SDU discard should mainly be applied to SDUs without an assigned SN. It is left to UE and network implementations how to achieve such behavior enabling the best possible throughput.

3.5.4.2 Security

The ciphering at the PDCP layer is performed for all user plane traffic flows provided from SDAP layer as well as for RRC signaling. However, the integrity protection is applied always for SRBs carrying RRC signaling but can be configured for DRB in case UE supports it. The UE will indicate its capability for the aggregated data rate of user plane integrity protected data over all DRBs configured with integrity protection, lowest possible value for the data rate is as low as 64 kbps. Additionally, as all the signaling between UE and core network is transmitted inside RRC containers in the RRC signaling, the PDCP layer provides ciphering and integrity protection also for those messages.

The design of PDCP for security and integrity protection is such that different ciphering and integrity protection algorithms, defined in [40], can be adopted without modifications to the actual PDCP protocol if supported input parameters for ciphering and integrity protection are sufficient. Currently supported input parameters are:

- COUNT: A value which is updated by the PDCP SN that are the least significant bits of the COUNT.
- DIRECTION: Defining the direction of the transmission (uplink or downlink).
- BEARER: The radio bearer identifier in [40]. The RRC provides this value and uses the value RB identity −1.
- KEY: The ciphering and integrity keys for the control plane and for the user plane are K_{RRCenc}, K_{UPenc}, K_{RRCint}, and K_{UPint}.

For both, ciphering and integrity protection, the algorithm and the key that the PDCP entity shall use are configured by RRC and are PDCP entity specific. This allows that different bearers utilize different keys and algorithms. This enables Radio Access Network (RAN) architectures where PDCP entities are processed independently in different independent locations without utilizing the same key (see Chapter 4 on different architecture options).

The integrity protection is based on a standard method by including MAC-I field into the PDCP PDU header field of data packets. The transmitter computes the value of the MAC-I field before ciphering. At the receiver the integrity of a PDCP PDU is verified by calculating the X-MAC based on the input parameters as specified above. If the calculated X-MAC corresponds to the received MAC-I, integrity protection is verified successfully.

The ciphering is applied only for the data part of a PDCP data PDU, however, SDAP header or SDAP control PDU included in PDCP SDU are not ciphered. This was done mainly for two reasons: the ROHC protocol running in the PDCP layer requires to jump over the SDAP header to be able to find the IP header of the PDCP SDU. This also allows implementations where the Quality of Service Flow Identifier (QFI) information is used for improved scheduling decisions also in the DU side of the network. The former is also required to stick with the fixed size of the SDAP header (1 byte) to allow PDCP just to blindly remove the SDAP header part of the given SDU.

The PDCP architecture on ciphering and integrity protection allows processing of the PDCP SDU independently of the actual data transmission time in the radio interface. At the UE side, this allows PDCP PDUs to be pre-processed in memory waiting for uplink transmission opportunities. In the network this allows PDCP to be in a CU and a cloud-based implementation.

3.5.4.3 Header Compression

The PDCP performs IP header compression and decompression using the ROHC protocol. Different ROHC profiles are supported by defining different profile identifiers for

Table 3.12 Supported header compression protocols and profiles.

Profile Identifier	Usage	Reference
0x0000	No compression	RFC 5795
0x0001	RTP/UDP/IP	RFC 3095, RFC 4815
0x0002	UDP/IP	RFC 3095, RFC 4815
0x0003	ESP/IP	RFC 3095, RFC 4815
0x0004	IP	RFC 3843, RFC 4815
0x0006	TCP/IP	RFC 6846
0x0101	RTP/UDP/IP	RFC 5225
0x0102	UDP/IP	RFC 5225
0x0103	ESP/IP	RFC 5225
0x0104	IP	RFC 5225

each supported ROHC profile as shown in Table 3.12. The actual profile definition, how to map a data flow to each profile is not defined by 3GPP but RFC 5795 [41] is referenced. As the ROHC profile may send independent control packets, called interspersed ROHC feedback, between ROHC decompressor and ROHC compressor, PDCP supports special PDCP control PDU types for such packets.

3.5.4.4 Duplication

The PDCP duplication function allows PDCP data PDUs to be duplicated at the transmitting PDCP entity and sent over two associated RLC entities. Such a feature was introduced mainly to support URLLC-based services by enabling the duplication for DRBs but also enhances, mobility robustness in high mobility scenarios by duplicating the SRB data. Thus, the duplication is configurable on a per DRB or SRB basis by RRC. Additionally, duplication for DRBs is controlled in MAC by means of a MAC CE, which can activate or deactivate duplication for a DRB configured with duplication.

The RLC entities associated with such a PDCP entity are called primary and secondary RLC entities. From the naming one can already determine that the primary RLC entity is always active regardless of the duplication activation status, i.e. the MAC control applies to the activation status of the secondary RLC entity. Hence, PDCP control PDUs, like the PDCP status report, that are not duplicated are always transmitted over the primary RLC entity. The RLC entities can either belong to the same cell group with Carrier Aggregation (CA) based duplication or to different cell groups with dual-connectivity-based duplication, hence, PDCP duplication with only one carrier is not possible. Whenever the same cell group is used, restrictions in MAC are put in place to guarantee that the two duplicates never end up on the same carrier. Otherwise, they might fail at the same time if multiplexed to the same MAC PDU, destroying the benefits of duplication.

To avoid unnecessary duplication overhead, when receiving ACK from either of the RLC entities for a certain PDCP PDU, the PDCP discard mechanism is used to discard the duplicated PDU from the buffer of the other RLC entity. This also enables the slower RLC entity to advance its transmission window in case the transmission of the old packets is not meaningful anymore.

3.5.5 Service Data Adaptation Protocol (SDAP)

SDAP serves as an interface between the core network (CN) and RAN and provides the key part of the QoS framework in NR. SDAP handles the mapping between QoS flows from a PDU session and DRBs. Hence, there is one SDAP entity configured for each PDU session and each DRB is only serving one SDAP entity, i.e. data from multiple PDU sessions are not mixed into the same DRB. Another key function of SDAP is the marking of each transmitted packet in both DL and UL direction with a QFI. The marking is used in the DL direction to enforce an implicit mapping of QoS flow to DRB by means of reflective QoS whereas it is used in the UL direction to enable the gNB to replicate the marking on the RAN-CN interface for QoS enforcement in the CN (see Chapter 6).

The SDAP entity can be configured either with or without the SDAP header, the latter being known as "transparent mode."(TM). The TM allows the removal of any SDAP introduced overhead, e.g. in case of DRBs for which the gNB does not plan to apply reflective QoS. Otherwise, the 1-byte SDAP header is used, which consists of a 6-bit QFI field and reflective QoS indicators in DL direction (see Section 3.5.5.1) as well as a 6-bit QFI field and control PDU indicator in UL direction (see Section 3.5.5.2). The 1-byte fixed header size is designed to support efficient processing and ease the PDCP ciphering function, which does not cipher a SDAP header as discussed before.

3.5.5.1 Mapping of QoS Flows to Data Radio Bearer

The mapping between QoS flows and DRBs is controlled by the gNB for both DL and UL directions, whereas the CN configures the UE with IP flow to QoS flow mapping rules as discussed in Chapter 6. Each QoS flow is assigned a QFI and each QFI is assigned a QoS profile by the CN, which is provided to the gNB upon PDU session establishment. The QoS profile defines the required QoS characteristics for a certain QoS flow. Based on the QoS profile, the gNB knows how to treat each individual QoS flow and determines whether and which kind of different DRBs are needed for the PDU session. Finally, the gNB provides the mapping rules to the UE for the QoS flow to DRB mapping in UL direction via explicit RRC signaling or through implicit mapping via reflective QoS.

As not all QoS flows are active at the same time and as there is the possibility of having up to 64 QFIs for a given PDU session, configuring immediately all QFI to DRB mapping rules would generate too much signaling overhead and would slow down the PDU session setup. For this purpose, the concept of a default DRB is introduced in UL direction where the UE maps all QFIs for which no mapping rule has been provided by the gNB. Thus, each PDU session comes with at least the default DRB. Dedicated DRBs can be established on need basis when new QFIs emerge in the communication link. Nevertheless, it should be noted that it is an option for the gNB to configure all the mapping rules immediately during the PDU session setup in which case no default DRB needs to be established.

Reflective Quality of Service (RQoS) allows implicit update of the QFI to DRB mapping rule in the user plane. Whenever a new QFI appears in the DL (or the first DL packet for a given QFI is buffered), the gNB maps the packet into the desired DRB (existing one or newly added) and transmits this to the UE with Reflective QoS flow to DRB mapping Indication (RDI) bit flagged. By means of the flagged RDI bit, the UE determines that the QFI to DRB mapping rule need to be updated by the given packet and reads the QFI field in the SDAP header. The QFI is then implicitly mapped to the corresponding UL part of the DRB whenever a packet with such QFI is transmitted by the UE. The SDAP header carries also the NAS level RQI (Reflective Quality of Service Indication) bit in DL packets as the CN does not apply any headers to the packet that will be carried over the air. The RQoS is especially efficient in mapping new QFIs to existing DRBs as no explicit RRC signaling is required to configure the UE with new mapping rules.

3.5.5.2 QoS Flow Remapping between Data Radio Bearer

Due to UE mobility or any other reason determined in the network, the gNB may trigger QoS flow remapping. One common scenario to remap a QoS flow is also to move it from the default DRB to a dedicated DRB. As for any other mobility case, lossless and ordered delivery should be able to be enforced for the given QoS flow, and since such remapping involves two DRBs and PDCP entities, this is performed by the SDAP entity. Due to the additional buffering, the in-order delivery necessitates (data in the new DRB needs to be buffered if data remains in the old DRB) the QoS flow remapping design in a way that the additional buffering requirement only applies to the gNB. Hence, in the DL direction it is left for network implementation to ensure the old DRB is "emptied" from the packets of the remapped QoS flow before transmitting them in the new DRB. In the UL, the UE transmits an SDAP control PDU with a control bit flagged and the QFI after the last packet of the remapped QoS flow via the old DRB which serves as an end marker. Based on the end marker, the gNB can release buffered packets of the remapped QoS flow to upper layers, maintaining the in-order delivery.

3.5.6 Radio Resource Control (RRC)

For an efficient radio communication system, controlling available radio resources, maintaining radio connections and allocating available resources to connections based on the end user QoS requirements are vital operations. In NR, the RRC protocol, defined in [42], is providing necessary configuration signaling and procedures for these operations. RRC has following functions:

- Broadcast of system information,
- RRC connection control,
- Measurement configuration and reporting,
- Inter-radio access technology (RAT) mobility,
- Set of other functions including transfer of dedicated NAS and UE radio access capability information.

The RRC protocol defines these functions and corresponding signaling procedures including UE actions. However, the protocol defines only a limited set of procedures for the network, i.e. how the gNB shall operate. The majority of the gNB operations, how and when network utilizes different signaling procedures, is left to the RRM function of the radio network, which is implementation dependent.

Through the system information broadcast, RRM controls the transmission and transmitted parameters in MIB, SIB type 1 (SIB1) and other SIB types. The MIB is transmitted in PBCH of the SS block, and SIB1 as part of the RMSI as discussed in Section 3.3.2. The RRC protocol defines the UE actions and procedures for acquiring MIB, SIB1, and other SIBs from downlink broadcast messages. In addition to acquiring system information from broadcast messages, NR supports also the UE initiated system information request procedure for other SIB types.

This system information acquisition process is part of the cell selection and cell reselection process needed for Idle mode mobility, see Section 5.5. Based on the received MIB, the UE can continue reception of OSI for detecting parameters or deciding whether a detected cell is suitable. The UE determines cell quality based on RSRP measurement and parameters provided in SIB1 and other SIBs. In case UE has detected multiple cells fulfilling cell quality requirements, the UE ranks all detected cells based on cell ranking criteria. During cell ranking the UE orders all suitable cell parameters provided in SIBs and selects the highest ranked cell. The RRM of the network can control both cell quality criteria as well as cell ranking by a cell specific offset and priorities, so that the UE reselects the most appropriate cell for Idle mode camping. Therefore, the RRM has sophisticated means to control Idle mode mobility of the UE by appropriate parameter settings in SIBs. In addition, by system information broadcast the network can configure a cell to be barred, control or forbid the access for a set of subscribers or a set of service types.

The RRC connection control defines UE actions when the network is controlling the radio connection between UE and gNB. It defines how the network can contact the UE in Idle mode or RRC Inactive state by paging, see Section 5.5.2, and controls the RRC connection establishment procedure and initial security activation.

The RRC connection reconfiguration procedure is used to configure the PHY functions, and all radio protocols for active signaling and/or data transfer between UE and gNB. In addition, it is used for RRC state management between RRC CONNECTED and RRC INACTIVE state. The RRC state management principles are discussed in Section 5.5.2.

In RRC CONNECTED state, the UE has SRBs established for dedicated RRC signaling, utilizing error free delivery provided by AM-RLC. Additionally, DRB are configured based on service needs to transfer user plane data. For both SRBs and DRBs the used configuration is signaled by using the RRC protocol. In RRC CONNECTED state the radio link exists between UE and gNB or multiple gNBs in case of dual-connectivity, and therefore PHY features supporting efficient data transmission on PDSCH and PUSCH are enabled. The network can configure the UE using RRC signaling toward a preferred PHY configuration, including channel state information (CSI) measurement configuration and reporting, as well as preferred transmission and reception configuration.

The configuration of PHY features and transmission modes needs to be according to UE capabilities as the network shall avoid configurations that are not supported by the UE. The network learns the UE capabilities from the UE capability transfer procedure supported by RRC. In addition, the RRC protocol supports storing UE capabilities in the network avoiding UE capability transfer procedure at every RRC IDLE to RRC CONNECTED state transition.

In RRC CONNECTED state the connection control maintains the radio link by using network-controlled intra system handovers and beam management functions performed by MAC. To perform these functions a set of measurement reports is needed in RRM. The RRC measurement configuration and reporting function is used to configure UE measurement objects, measurement quantities as well as reporting conditions in both MAC and RRC layer of the UE. The RRC connection control defines the criteria when the UE considers the radio link failed and UE actions for radio link recovery.

The RRC protocol is also used for control mobility to other RAT such as LTE. The radio network's RRM can configure appropriate measurements and reporting criteria to the UE regarding another RAT, e.g. LTE or UMTS. The UE shall perform measurement on the other RAT and reporting based on reporting criteria. Based on received measurement reports RRM can initiate handover toward the target RAT using RRC. The inter-RAT handover message can include the physical and radio protocol layers configuration of the target RAT to the UE. Similarly, when the gNB receives a request on incoming handover from another RAT, RRC signaling is used to provide radio configuration to be used in NR at handover from the other RAT. The other RAT forwards received radio configuration parameters as transparent container to the UE in the handover command.

Finally, the RRC protocol contains a set of other functions including transfer of dedicated NAS information. For dedicated NAS information messages, the RRC signaling defines transparent containers, so that NAS messages are not processed by the RRC protocol. This provides robust means to transfer NAS messages in sequence and avoids race conditions between NAS and RRC state machines in mobility and state functions.

The RRC signaling is implemented by using Abstract Syntax Notation Number 1 (ASN.1) encoding, supporting flexible extendable message definitions. It can be expected that the RRC protocol will be extended by many new parameters and procedures in coming NR releases. These new parameters and procedures are needed to allow configuration of new PHY or radio protocol features, as well as to extend the capabilities of the NR radio system and RRM of the radio network.

3.6 Mobile Broadband

3.6.1 Introduction

Broadband data communication was originally available via wired connections by utilizing Ethernet and ADSL during the 1990s and even the early years of the twenty-first century. The first steps of wireless broadband communication were taken with WLANs radio generations, which were simply meant to adopt LAN communication to wireless connections. The most dominant solutions were based on IEEE 802.11b and IEEE 802.11g, also known was Wi-Fi, used to detach computers from the cable connection allowing limited mobility to the user. At the same time people started to utilize their first smartphones to read emails and synchronize calendars over cellular connections such as GPRS and later UMTS. Wi-Fi provided better broadband type data connections in terms of bitrate and latency and allowed to use normal web-browsers, but was hardly mobile as WLAN connectivity was only available in limited areas like at home, in offices or airports (still today most of the data traffic worldwide goes via WLAN). However, several manual steps are needed when entering a new Wi-Fi network.

Smartphones had limited processing capabilities, and their screens were small with low resolution. The available cellular data connections provided modest data rates and long latencies, thus those connections where there were hardly broadband connections even though providing support for wide area coverage and mobile users. The user experience and device (laptops and smartphones) usage were quite different compared to wired connections. Different service providers developed web-pages optimized for mobile phones, or services where provided over i-mode or WAP protocol.

During that time, it was apparent that these different environments will merge, however, it was unclear when this will happen and which technology components will enable this change. From a communication technology point of view, the following evolution steps can be identified:

- Introduction of HSPA, i.e. 3GPP Release 5 in 2003 (see [43]) and Release 6 in 2006 (see [44]) and compatible cellular networks and mobile devices.
- Introduction of IEEE 802.11n in 2009 (see [45]) and Wi-Fi Alliance certified devices and access points (APs).
- Introduction of LTE Release 8 in 2008 (see [46])and LTE Release 10 for ITU-R 4G submission in 2011 (see [47]) and compatible products.

From the devices perspective, significant changes were:

- Introduction of the first iPhone in 2007 and iPhone 3G in 2008. The iPhone 3G supported 3G HSPA and the App Store for downloading third party applications.
- Induction of the Android 1.5 (Cupcake) operating system for mobile devices in 2009.
- Introduction of HTML-5 during 2014.

From the devices perspective, one should not forget other significant technology improvements for processor and camera technologies for example. This development has introduced current MBB and services we know so well today, where main data traffic is due to video downloads and picture uploads to social media such as Facebook and Instagram. The popularity of these services is highly dependent on the availability of affordable mobile devices with sufficient memory, display, camera and processing capability, and the ability of and being always connected with high-bandwidth and low latency cellular connection.

For the future, it can be anticipated that average traffic consumption is increasing, and it is even less depending on the user's location. Currently, the majority of video is consumed in buildings, which may seem self-evident. However, one can anticipate and already see that habits will change toward video consumption being more dependent on available time than location of the user. Therefore, data consumption has and will increase when people are traveling in cars, buses, trains, and even planes requiring more efficient solutions to avoid high number of high-speed UEs each individually connecting to cellular base stations.

3.6.2 Indoor Solutions

During the past decade, the de-facto indoor wireless solution has been IEEE 802.11n-based WLAN known as Wi-Fi (see [18]). The previous generations of 802.11 are also used by legacy devices but the industry has now changed the focus on IEEE 802.11ac with even higher data rates. Development of IEEE 802.11 technologies is ongoing with IEEE 802.11ax and other standard amendments. During the 5G era, the IEEE 802.11 family is expected to remain a vital component of indoor solutions. The 802.11 technology is supported by an extremely high number of different device types, such as TV equipment, game consoles, printers, etc. operational at 2.4 GHz Industrial Scientific and Medical (ISM) band and increasingly also on the 5 GHz band as discussed in Chapter 2. These bands are very attractive for deploying privately owned indoor hot spots and networks.

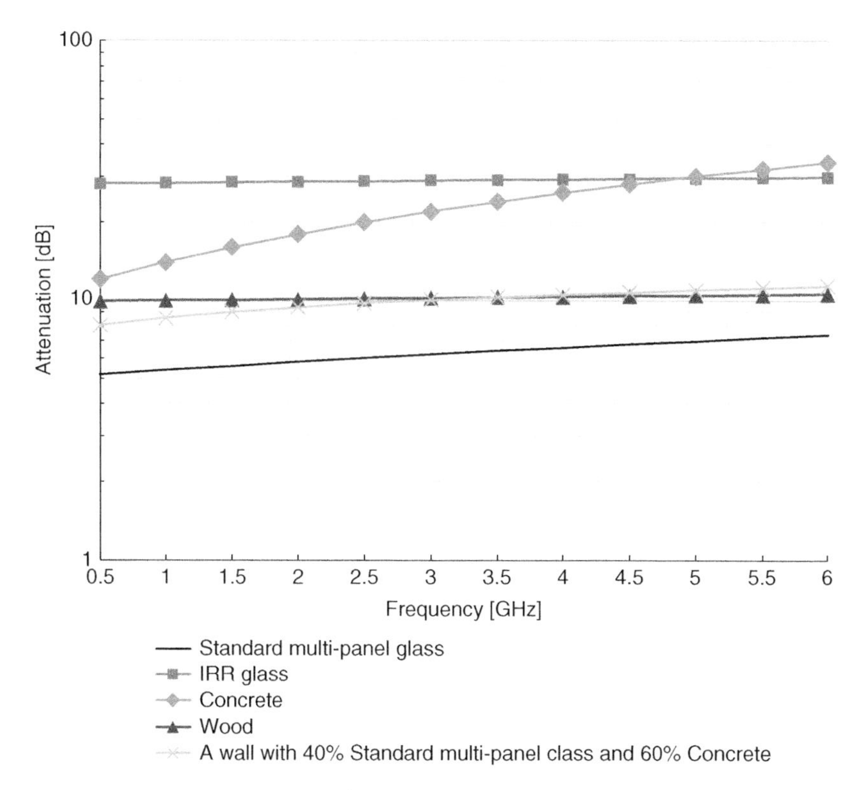

Figure 3.43 Wall penetration loss for O2I below 6GHz.

For future solutions enabling cost-efficient NR BS indoor deployments would provide significant benefits. This can be easily understood when looking at Figure 3.43 that illustrates wall penetration loss caused by different materials as a function of carrier frequency. The result is based on 3GPP channel model studies and agreed for simulations modeling, without taking any attenuation modeling inaccuracy terms into account (see [48]). The inaccuracy is modeled as normal distribution with zero mean and 4–5 dB standard deviation.

Figure 3.43 shows that modern infrared reflective, i.e. selective windows introduce very high 25–29 dB attenuation for all radio signals independently of the frequency. The concrete wall has linear increase from 12 to 34 dB when frequency is changed from ~500 MHz to 6 GHz. Standard multi-panel glass and wood have sufficiently flat attenuation in terms of frequency, but still significant 5–7.5 dB attenuation for glass and 9–10.5 dB for wood.

Even though buildings are not constructed by a single material, but are rather a combination of glass, concrete, and wooden parts, it is apparent that providing high quality signal from outside to inside buildings using concrete and infrared reflective glass is difficult even with relatively low frequencies, i.e. frequencies around 2 GHz. Naturally, such attenuation reduces the maximum available coverage and reliability of the network even with low data rate services such as voice. Operators will continue to provide indoor coverage as today by utilizing frequencies below 1 GHz and frequencies around 2 GHz and

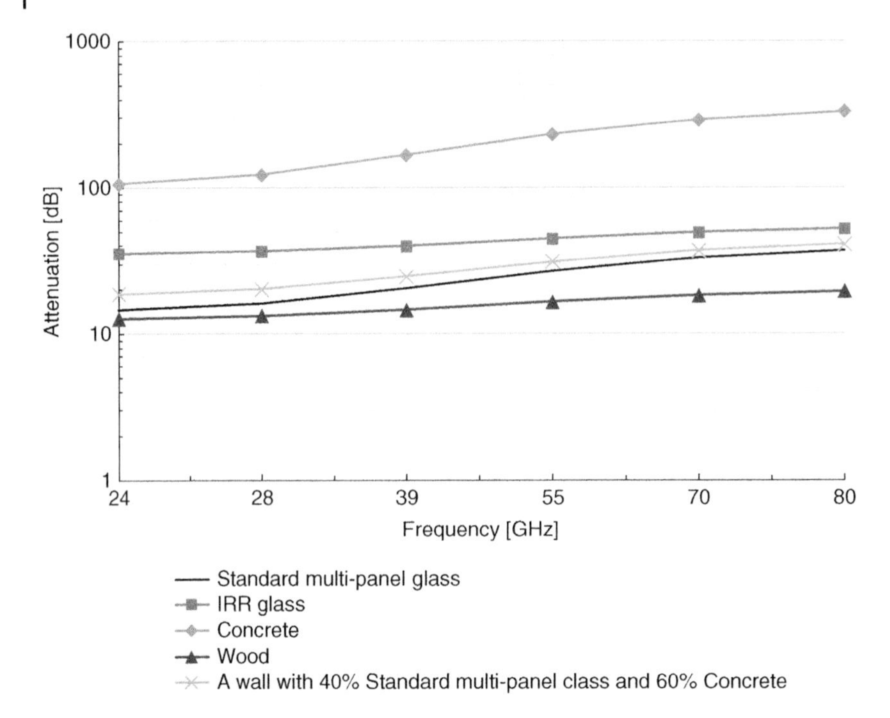

Figure 3.44 Wall penetration loss for O2I at mmWave frequencies.

with high frequencies when possible, but available bandwidth on the low frequencies is limited to 5–20 MHz per band and operator. In some limited cases a single operator may have up to 40 MHz of low band available. Only at 3.5 GHz frequency operators will have the possibility for 80–100 MHz bandwidth or more on a single band, but there the penetration loss can be an additional 10 dB or more as show in Figure 3.43.

When considering mmWave frequencies, the situation is even more problematic as shown in Figure 3.44, presenting the wall attenuation in logarithmic scale (see [48]). It is apparent that for mmWave frequencies serving indoor users from outside is not really feasible. Even if the 5G base station would be able to utilize high transmission powers, the uplink would be very limited due to the end user's device maximum TX power. Additionally, even if the uplink connection could be maintained, the device power consumption would be significantly higher when communicating to the outside compared to communicating with 5G base stations or Wi-Fi access points located inside the building.

The wall attenuation results in other factors that need to be noted. The first one being that when base stations are located inside, the interference caused from inside to the outside system or vice versa is greatly reduced, making cells more isolated and thus suffering less from interference of other cells. Secondly, different parts of the indoor networks can also experience higher isolation, even if the interior walls might not be concrete, interior floors and bearing walls will still attenuate signals significantly. Therefore, when utilizing higher frequencies, more 5G base stations are required to cover single and separated floors, while the interference of base stations with each other is less than in lower frequencies.

To be able to install indoor base stations, one needs to consider at least following points:

- What is the available wired backhauling to the building?
- How can the base-stations be installed?
- How to provide access and authorization to the network?

When the building has an existing fiber connection and internal Ethernet cabling, the first requirement is solved easily as now multiple 5G base stations or Wi-Fi access points can be installed in different locations where Ethernet is available, or different base stations can share the fiber connections. However, this can be a rare case in many cities with older houses and even historical city centers. Existing twisted pair copper lines originally installed for PSTN telephony will easily become a limiting factor and replacing cabling to the building with fiber becomes too expensive. In such cases, fixed wireless type solutions as discussed in Section 3.1.2 as the use case of 5G-TF, becomes of interest.

Notably, the solution can utilize 5G technology, including both below 6 GHz and mmWave frequencies to provide connection from 5G base stations to separate CPE. CPEs can have external antennas mounted on outside walls of the building preferable with LOS conditions to the base station(s). The motivation for using both below 6 GHz and mmWave frequencies allows utilization of mmWave frequencies to those CPEs in LOS condition, and e.g. bands n77, n78, or n79 (see Chapter 2) for those CPEs that are in NLOS or momentarily blocked by moving obstacles. Such combination of different frequencies has been shown to provide high system capacity as mmWave frequency is used opportunistically for high quality LOS links, leaving more bandwidth on lower frequencies (n77, n78, or n79) bands available for devices in NLOS conditions, while avoiding the need of extensive mmWave frequency resources allocation to these users.

A short cable through the wall or window is used to connect the external unit to the indoor CPE, which can then operate as Wi-Fi AP or 5G gNB providing connection to the building's internal Ethernet network. In cases where internal cabling is not available, repeaters or perhaps self-backhauling solutions are needed to have full coverage in any large home or office building with concrete walls or floors.

The second question becomes relevant when considering, how installation and maintenance can be done in individual houses and small offices. Today telecom operators are owning the base station infrastructure, including antennas cabling and the base station physical cabinet. Additionally, a preferred location of the base station is a public location where operator has own secured access. Alternatively, base stations are installed in basements or other technical spaces of a block house with cabling to the roof of the house. In such cases, an operator needs to have leasing deals with property owners.

However, it can be anticipated that such base station installation models are not economically feasible to address small offices and individual houses and new complementing business models are needed to address these installations. Additionally, ease of installation, configuration and operation needs to be designed for consumers rather than telecommunication operators with trained installation and maintenance personnel. Naturally, a consumer product aimed to serve a small number of users must have a significantly lower equipment price, energy consumption, cost and complexity of installation than the operator-installed macro base station serving hundreds or even thousands of users.

The third question is related to who can control the access to the BS installed in an individual home or office. Preferably, the access and authorization to the network in operator provided networks would be based on existing cellular network authentication solutions utilizing the universal subscriber identity module (USIM). Similarly, for operator hosted WLAN the USIM can be used but when the network provider is the property owner this becomes typically much more cumbersome. There might be users with different operators' subscriptions, thus solutions should support multiple operators. Additionally, home or property owners will have their own preferences on how to control network access and granting access to the local area network for accessing local services, such as printers, media servers, etc.

Today cellular standard-based home or office networks hardly exist, and it must be seen if those will become available in the early 5G time frame. In the Wi-Fi solution space, the most common method is to provide network ID and password manually for, e.g. hotel customers at check-in, or to configure the Wi-Fi access profile at the device for office access. Providing passwords manually is still very cumbersome and not really providing any additional security, as passwords are not individual and updates are infrequent. In many public places such as shopping malls and airports, the solution is simply to allow limited access to the network without authentication by accepting the terms of use.

It is clear, that commonly used Wi-Fi authentication solutions are not going to be sufficient in the future. The requirement for easy installation, configuration and use is not only a technical, but also an economical imperative and the related legal responsibilities need to be addressed in the market place.

However, when these issues are addressed, 5G, with extreme wide channel bandwidths, high data rates and low radio interface latency, can clearly be revolutionary in terms of providing MBB connectivity indoors, nicely coupled to mobile outdoor solutions, so that MBB service is not dependent on place or mobility.

3.6.3 Outdoor-Urban Areas

For urban and dense urban areas, it is evident that initial 5G deployments will reuse existing cell sites currently utilized for LTE and 3G as well as the remaining 2G deployments. The utilization of these sites can be quite straightforward; however, limitations may arise in terms of practical installation of antennas, sufficient backhaul capacity, etc.

For initial deployments in urban and dense urban areas operators address two alternative routes, depending on their overall technology strategy and investment timeline. The first option is to utilize existing LTE networks as the coverage and connectivity layer for 5G connections and couple new higher frequency bands, mainly n77, n78, or n79, with the LTE coverage layer. This solution relays on the dual-connectivity (DC) architecture between LTE and NR where LTE operates as the master of the connection and NR is used as a capacity booster when available. The benefit of this is that no full coverage is needed for an initial NR deployment in the first place, and typically there is no need to initially refarm any spectrum resources from earlier systems to NR. This is enabled by the DC solution, allowing data connections to be continued without interruption at LTE side when NR coverage is lost without the need of inter system handovers. In addition, the operator may also utilize high mmWave frequencies, bands n257, n258, n260, or n261, in dense urban deployments or for fixed wireless services together with LTE.

The obvious drawback of this approach is that when connection to 5G is lost, the service is downgraded to LTE service quality (smaller data rates and higher latency). Additionally, the new core network architecture and features are not available and therefore system operation is based on the existing LTE network. Therefore, this option is seen as a transition solution toward a standalone NR operation with NR radio network and new core network infrastructure.

The second approach for 5G deployments, is to deploy a standalone NR solution. Such deployments in urban and dense urban areas would benefit from initial usage of 1 GHz frequencies to obtain full coverage, including indoor coverage by using existing cell sites. As in the initial phase, the number of NR capable devices will be small compared to LTE devices, operators need to carefully balance 1 GHz spectrum usage between LTE and 5G. In addition to sub 1 GHz network deployments, additional deployments on n77, n78, or n79, similarly to the first approach are needed to achieve data rates that differentiate 5G from LTE and other earlier generations. The benefit of this approach is the possibility to utilize the new core network architecture and services immediately during 5G launch. In addition, the utilization of low frequencies in the first place, will provide more unique NR coverage for the operator in urban areas, and finally no transition to standalone 5G operation is needed. The drawback of this approach is the needed spectrum balancing between LTE and NR as well as the number of sites to be upgraded to NR needs in initial phase, as unnecessary frequent mobility between different RATs will have negative impacts to the end user service experience.

Notably, in both strategies the bands n77, n78, and n79 are vital for delivering the promise of eMBB in NR. Simply these bands each have a total spectrum of 500 MHz or more, and can provide continuous system bandwidth of 100 MHz in an operator's network. Such amount of spectrum is a significant boost to any operator's frequency assets, and it also matches nicely to the first NR UE capabilities supporting 100 MHz RX bandwidth for eMBB on bands below 6 GHz.

Especially in a dense urban environment, the existing sites may already provide full coverage by utilizing n77, n78, and n79 5G bands alone, without sub 1 GHz coverage layer. This can be achieved, especially outdoor and less attenuated indoor locations, in cities with small ISDs. The NR standard provides several tools for improving NR coverage compared to LTE. These solutions, such as downlink control channel beamforming by utilizing mMIMO, are band agnostic and can play an interesting role on bands n77, n78, and n79. Additionally, the concept of SUL was introduced by utilizing a carrier aggregation framework. In this case, the connection can use bands n80 – n84 or n86, located on sub 1 GHz or below 2 GHz for uplink transmission but maintain downlink transmission on higher wideband carriers such as n77. This solution can relief the uplink from coverage problems, which is often the limiting factor due to lower transmission power. However, as uplink bandwidth on these SUL bands is limited, it is expected that utilization of SUL is a short-term solution and not widely used. A more straightforward solution would be to aggregate the higher frequency band carriers with an available lower frequency carrier and get additional benefits of the two carriers also for the downlink.

References

All 3GPP specifications can be found under http://www.3gpp.org/ftp/Specs/latest. The acronym "TS" stands for Technical Specification, "TR" for Technical Report.

1 3GPP TS 25.306: "UE Radio Access capabilities".
2 3GPP TS 25.993: "Typical examples of Radio Access Bearers (RABs) and Radio Bearers (RBs) supported by Universal Terrestrial Radio Access (UTRA)"
3 Mogensen, P., Pajukoski, K., Tiirola, E., et al.: "Centimeter-wave concept for 5G ultra-dense small cells", IEEE VTC2014.
4 Levanen, T., Pirskanen, J., Koskela, T., et al.: "Low Latency Radio Interface For 5G Flexible TDD Local Area Communications", IEEE ICC2014.
5 5G Test Network Finland (5GTNF): http://5gtnf.fi/overview.
6 Parkvall, S., Furuskog, J., Kishiyama, Y., et al.: "A trial system for 5G wireless Access", VTC2015.
7 5G-SIG (Special Interest Group), KT will provide world's first 5G Experience www-page at https://corp.kt.com/eng/html/biz/services/sig.html.
8 Verizon 5G Technical Forum Web page at http://www.5gtf.net.
9 KT PyeongChang 5G Special Interest Group, TS 5G.300: KT 5th Generation Radio Access; Overall Description.
10 Vakilian, V., Wild, T., Schaich, F, et al.: "Universal-filtered multi-carrier technique for wireless systems beyond LTE," 2013 IEEE Globecom Workshops (GC Wkshps), Atlanta, GA, 2013, pp. 223–228.
11 Ahmed, R., Wild, T., Schaich, F., et al.: "Coexistence of UF-OFDM and CPOFDM," in IEEE 83rd Vehicular Technology Conference (VTC Spring), 2016, pp. 1–5.
12 Zhang, X., Jia, M., Chen, L., et al.: "Filtered-OFDM – Enabler for flexible waveform in the 5th generation cellular networks," in 2015 IEEE Global Communications Conference (GLOBECOM), 2015, pp. 1–6.
13 3GPP TR 38.802: "Study on New Radio Access Technology Physical Layer Aspects".
14 Levanen, T., Pirskanen, J., Pajukoski, K., et al.: "Transparent Tx and Rx Waveform Processing for 5G New Radio Mobile Communications", accepted to IEEE Wireless Communications Magazine.
15 3GPP R1-1609568 "Out of band emissions, in-band emissions and EVM requirement considerations for NR", Nokia, Alcatel-Lucent Shanghai Bell, 3GPP TSG-RAN WG1 #86bis Lisbon, Portugal, October 10–14, 2016.
16 3GPP R1-1610113: "Coverage analysis of DFT-s-OFDM and OFDM with low PAPR", Qualcomm Incorporated, October 10–14, 2016 October 10–14, 2016.
17 3GPP TS 38.104: "NR; Base Station (BS) radio transmission and reception".
18 IEEE Standard for Information technology: "IEEE802.11–2016, Part 11: Wireless LAN Medium Access Control (MAC) and Physical Layer (PHY) Specifications".
19 3GPP TS 38.211: "NR; Physical channels and modulation".
20 Mohsenin, T., Truong, D.N., Baas, B.M. (2010). A low-complexity message-passing algorithm for reduced routing congestion in LDPC decoders. *IEEE Transactions on Circuits and Systems I* 57 (5): 1048–1061.
21 Zhang, Z., Anantharam, V., Wainwright, M.J. et al. (2010). An efficient 10GBASE-T Ethernet LDPC decoder design with low error floors. *IEEE Journal of Solid-State Circuits* 45 (4): 843–855.

22 Mohsenin, T., Shirani-mehr, H., Baas, B.M. (2013). LDPC decoder with an adaptive word width data path for energy and BER co-optimization. *VLSI Design* 2013 (1): 1–14.
23 Li, M., Naessens, F., Debacker, P., et al.: "An area and energy efficient half row-paralleled layer LDPC decoder for the 802.11ad standard", in Proc. IEEE Workshop on Signal Processing Systems (SiPS'13), Taipei City, pp. 112–117, Oct. 2013.
24 Yin, B., Wu, M., Wang, G., et al.: "A 3.8 Gb/s large-scale MIMO detector for 3GPP LTE-advanced", in Proc. IEEE ICASSP, Florence, Italy, May 2014.
25 Li, A., Xiang, L., Chen, T. et al. (2016). VLSI implementation of fully-parallel LTE turbo decoders. *IEEE Access* 4: 323–346.
26 Dizdar, O. and Arıkan, E. (2014). A high-throughput energy-efficient implementation of successive-cancellation decoder for polar codes using combinational logic. *IEEE Transactions on Circuits and Systems I: Regular Papers*, vol. abs/1412.3829.
27 Park, Y.S., Tao, Y., Sun, S., et al.: "A 4.68Gb/s belief propagation polar decoder with bit-splitting register file", in Symp. on VLSI Circuits Digest of Technical Papers, pp. 1–2 Jun. 2014.
28 Giard, P., Sarkis, G., Thibeault, C., et al.: "Unrolled polar decoders, Part I: Hardware Architectures", May 2015 [online]. Available at http://arxiv.org/pdf/1505.01459.pdf.
29 Lee, X.R., Chen, C.L., Chang, H.C., et al.: "A 7.92 Gb/s 437.2 mW stochastic LDPC decoder chip for IEEE 802.15. 3c applications", IEEE Transactions on Circuits and Systems I: Regular Papers 62.2 (2015): 507–516.
30 Weiner, M., Blagojevic, M., Skotnikov S., et al.: "27.7 A scalable 1.5-to-6Gb/s 6.2-to-38.1 mW LDPC decoder for 60GHz wireless networks in 28nm UTBB FDSOI", 2014 IEEE International Solid-State Circuits Conference Digest of Technical Papers (ISSCC), 2014.
31 Ajaz, S. and Lee, H.: "Multi-Gb/s multi-mode LDPC decoder architecture for IEEE 802.11ad standard", IEEE Asia Pacific Conference on in Circuits and Systems (APC-CAS), pp. 153–156, Nov. 2014.
32 Zhang, K., Huang, X., Wang, Z.: "A high-throughput LDPC decoder architecture with rate compatibility", IEEE Transactions on Circuits and Systems I: Regular Papers 58.4 (2011): 839–847.
33 Belfanti, S., Roth, C., Gautschi, M., et al.: "A 1Gbps LTE-advanced turbo-decoder ASIC in 65nm CMOS", IEEE Symposium on VLSI Circuits (VLSIC), 2013.
34 Roth, C., Belfanti, S., Benkeser, C. et al. (Jun. 2014). Efficient parallel turbo-decoding for high-throughput wireless systems. *IEEE Transactions on Circuits and Systems I: Regular Papers* 61 (6): 1824–1835.
35 3GPP TS 38.212: "NR; Multiplexing and channel coding".
36 3GPP TS 38.300: "NR; NR and NG-RAN Overall Description; Stage 2".
37 3GPP TS 38.321: "NR; Medium Access Control (MAC) protocol specification".
38 3GPP TS 38.322: "NR; Radio Link Control (RLC) protocol specification".
39 3GPP TS 38.323: "NR; Packet Data Convergence Protocol (PDCP) specification".
40 3GPP TS 33.501: "Security architecture and procedures for 5G system".
41 IETF RFC 5795: "The RObust Header Compression (ROHC) Framework".
42 3GPP TS 38.331: "NR; Radio Resource Control (RRC) protocol specification".
43 *Overview of 3GPP Release 5, Summary of all Release 5 Features*. ETSI Mobile Competence Centre, Version 9th September 2003.

44 (2006). *Overview of 3GPP Release 6, Summary of all Release 6 Features, Version TSG #33*. ETSI Mobile Competence Centre.
45 IEEE 802.11-09/0991r0, TGn Closing Report, Bruce Kraemer Marvell, 2009.
46 (2008). *Overview of 3GPP Release 8*. ETSI Mobile Competence Centre.
47 (2011). *Overview of 3GPP Release 10*. ETSI Mobile Competence Centre.
48 3GPP TR 38.901: "Study on channel model for frequencies from 0.5 to 100 GHz".

4

Next Generation Network Architecture

Devaki Chandramouli[1], *Subramanya Chandrashekar*[2], *Andreas Maeder*[3], *Tuomas Niemela*[4], *Thomas Theimer*[3] *and Laurent Thiebaut*[5]

[1] *Nokia, Irving, TX, USA*
[2] *Nokia, Bangalore, India*
[3] *Nokia, Munich, Germany*
[4] *Nokia, Espoo, Finland*
[5] *Nokia, Paris-Saclay, France*

4.1 Drivers and Motivation for a New Architecture

4.1.1 New Services Emerging

Consumer demand continues to be insatiable with ever growing appetite for bandwidth that is needed for 4K or even 8K video streaming, augmented and virtual reality and multiple devices, among other use cases. On the same token, operators want the network to be "better, faster and cheaper" without compromising any of these three elements. Furthermore, operator demands are not homogeneous in terms of requirements or in terms of timing for deployment of new services.

In recent years, the emergence of Internet of Things (IoTs) has provided us with abundance of different use cases, each of them with their own demands for the network. This creates a business landscape for network operators where many new business partners, verticals, come and go, each with a need for a service level agreement (SLA) of their own.

As opposed to previous generations that targeted one type of device (a mobile phone for 2G, a smartphone for 3G/4G) and one type of user, the 5G network needs to serve a diverse set of devices and applications, e.g. from super secured and expensive phones for security forces to cheap and "dumb" IoT devices. Furthermore, the same device may host or relay very different applications ranging from resilient voice services to simple IoT services.

Consumer needs for more bandwidth continue to increase, with virtual and augmented reality leading the way toward connections with ultra-high bandwidths and very low latency. The challenge for the network operator is to deploy a network with these characteristics at reasonable costs. The most challenging part is to provide it cheaper than previous generation of networks.

Beyond providing better service for the several billion humans already connected, there is a need to also connect the currently non-connected ones (humans and things) and in some cases, offer 5G as a replacement for fixed access (e.g. Industries, enterprise,

5G for the Connected World, First Edition.
Edited by Devaki Chandramouli, Rainer Liebhart and Juho Pirskanen.

etc.). This requires cost efficient solutions. As 5G will be a mature technology some day, it is beneficial to have a system that enables providing essential functions at minimal complexity and costs.

As a summary, following are key drivers for the new architecture:

1. Enabling support for new business cases.
2. Enabling operational agility for rapid deployment of new services.
3. Enabling extreme automation to reduce Total Cost of Ownership (TCO).

In short, the new network is expected to be more "flexible, faster and cheaper."

4.1.2 Targets for the New Architecture

Translating service expectations to objectives for the new architecture has led to the following solution categories:

1. Service flexibility
2. Modular and extendable
3. Unified policy control
4. Access agnostic
5. Automated network deployments

Service flexibility means that the architecture needs to support different lifecycles for different network services. It should be easy to add a new service to the network, control which users are allowed using it, parametrize the service per group of users and control network resource usage per group of users. Equally important is the capability to easily remove services from the network.

Modular and extendable architecture allows to add new network functions (NFs) and modify existing network functions with limited or no impact to existing functions. Extending functionality while maintaining key paradigms of the architecture is essential for enabling the architecture to evolve without increasing overall complexity.

An architecture based on unified policy control allows to build consistent policies covering the entire network behavior and to some extent also the user equipment (UE) behavior. This way diverse use cases can co-exist, and operators keep full control over use cases and offered services. Assignment of functions like policy control to single functional elements of the architecture allows for consistent and less complex introduction and offering of new network services.

An access agnostic architecture allows to build a system utilizing multiple different access technologies. One device may simultaneously be served by multiple accesses, e.g. 3rd Generation Partnership Project (3GPP) access and non-3GPP access. This applies not only to 3GPP and non-3GPP accesses but also to different variants of 3GPP access. Thus, the 5G system allows access via "NR" (New Radio) as well as long term evolution (LTE) radio. This also implies that mobility of a UE (device) from one access technology to another should be seamless for the end user (i.e. without service interruption).

Automated network deployments are crucial for being able to take advantage of the architecture flexibility. This implies that the new 5G Core should be designed to enable operational agility. Without full automation, operational costs will prevent customized support for verticals and IoT service providers.

4.1.3 Shortcomings of the Current Architecture

One of the most frequently asked question was: "Can the existing Evolved Packet Core (EPC) efficiently support the new services of the 5G era?" Considering the expectations on the network, device and application evolution, the answer was clearly "No." The EPC architecture was designed mainly for the mobile broadband use case. Furthermore, the functional split defined for EPC was "box-driven," i.e. assuming dedicated hardware (bar-metal) deployments. For the 5G architecture, virtualization was considered as a basic design principle. Furthermore, the new 5G system is required to support ultra-low latency and high reliability use cases simultaneously at reasonable costs.

EPC was developed for Internet and operator voice services in mind. Therefore, it is lacking support for the diverse service landscape of the 5G era and is not being inherently cloud native.

Service flexibility is supported in evolved packet system (EPS) only in terms of providing network differentiation on a per UE basis. There is no method available to introduce multiple services per UE with different network parametrization and a different set and topology of network functions.

Modularity and extendibility is lacking in EPS because of three main reasons:

1) All interfaces are point-to-point interfaces, allowing only two predefined functions to interconnect, with each function being part of a solution for a predefined problem statement. Extending the functionality of the system requires new functions and/or new reference points.
2) Network functions are not orthogonal in their scope. As an example, session management (SM) functionality is distributed between mobility management entity (MME), serving gateway (S-GW), packet data network gateway (P-GW) and Policy and Charging Rules Function (PCRF). This means adding further capabilities to session or quality of service (QoS) management requires updating multiple network functions and interfaces between them, which hinders service flexibility.
3) Each network function hides all the session context it maintains per UE. There is no easy method for network functions to expose and share their session data. In addition, session context is partly overlapping in different network functions like MME, S-GW and P-GW.

Policy control is limited to QoS and charging aspects. There is no unified method for providing other kinds of policies, e.g. mobility policies and network selection policies.

The Core Network architecture is not access agnostic; it requires dedicated interworking functions (enhanced packet data gateway [ePDG]/trusted wireless LAN access gateway [TWAG]) for non-3GPP access each replicating a lot of the functionalities supported by MME, and S-GW. Thus, as an example, when introducing features like emergency sessions almost the same functionality needs to be defined, tested (inter-operability tests) and deployed in multiple network functions (MME, S-GW, ePDG, TWAG).

There is no unified core network architecture with common control and user plane (UP) functions supporting 3GPP and non-3GPP accesses.

Automation is not supported natively in the EPS network architecture beyond European Telecommunications Standards Institute (ETSI) defined Network Function Virtualization (NFV) related implementation and instantiation of virtual network functions.

Automation is built only as vendor specific functionality in the management plane, there is no structural support for automation in the network architecture.

4.2 Architecture Requirements and Principles

4.2.1 Overview

Based on the drivers and motivations for the new architecture, the following requirements can be identified for this new architecture:

1. The architecture shall consist of independent domains that enable the independent evolution of each domain and the aggregation of capacity and coverage of multiple access technologies with a single architecture.
2. Network functions are orthogonal and self-contained, i.e. each main function of the network needs to be assigned to a single Network Function entity, e.g. access control and mobility management (MM) functionality is supported by a single Network Function and not distributed across multiple Network Functions such as MME, ePDG, TWAG in EPC.
3. The architecture shall enable a UE to connect to multiple services simultaneously in a way that provides optimum user plane data paths for each service.
4. The architecture shall use a flexible connectivity framework between network functions without relying on pre-determined interconnections. This allows new usage scenarios for the services provided by each network function and enables a "plug and play" model for newly introduced network functions. Thus, operators shouldn't have to ask 3GPP to define a new reference point every time they wish to deploy a new function and interface with 3GPP defined functions. In other words, a service based architecture (SBA) is required. Service based interface allows multi-point connectivity for authorized network functions.
5. The architecture shall support the versatility of network usage by different services requiring different QoS but also different Control Plane (CP) resiliency, scalability and security. For example, the (internet protocol, IP) connectivity management may have different resiliency, scalability and security requirements depending on whether it applies for public safety devices, for a baseline smartphone or for a non-critical cheap IoT devices.
6. The architecture shall apply a policy driven model where policies can be flexibly defined and applied for any function including the UE.
7. The architecture shall define programmable network functions enabling (external) applications to influence the services of the network functions and enabling automation of all aspects of network operation.
8. The architecture shall support cloud natively designed network functions that are instantiated on a cloud infrastructure and are able to accommodate cloud optimized resiliency models.
9. The architecture shall be able to unify 3GPP and non-3GPP accesses in a single control and user plane in the core network.
10. The architecture shall enable services to be deployed either locally (geographically close to the UE) and/or centrally in a way that is transparent to the UE.

11. The architecture shall enable connectivity services to support new vertical markets including for example industrial control applications requiring very low latency (possibly coupled with high reliability) or demanding non-IP connectivity services (e.g. Ethernet like services).
12. The architecture shall provide a simple mechanism to expose network events, e.g. UE registration to the network and establishment of connectivity on behalf of a UE.

Besides all the expectations and requirements for the new architecture, another non-technical challenge with the new architecture development is the "timing aspect." Market needs are not homogeneous around the globe.

Some operators wish to deploy 5G radio early as a capacity boost leveraging existing LTE and EPC deployments, mainly to offer enhanced broadband services.

Some operators wish to deploy both 5G NR and new core network early to offer enhanced broadband services including network slicing.

Some operators wish to deploy a revolutionary new core network with many new features but had no urgency regarding the deployment timeline.

To ensure that we have a global ecosystem with the right balance on schedule and features supported, 3GPP agreed to a phased approach.

1. *Phase 1 (3GPP Release 15).* An early availability of a basic system that fulfills the goals and expectations set for commercial deployments for the Next Generation System including the Core Network.
2. *Phase 2 (3GPP Release 16).* Building a complete and feature rich system using the basic system defined as a foundation thereby ensuring backwards compatibility.

To ensure phase 1 is completed in a timely manner, 3GPP agreed to prioritize essential requirements for foundational network.

4.2.2 Architecture Domains

Architecture domains can be defined in many ways, depending on the viewpoint and targeted characteristics. There are practical constrains on the number of domains. Too few domains lead to monolithic structure that prevents "divide and conquer" approach for managing complexity, whereas too many domains lead too overly complex interactions between domains and unclear responsibilities for each domain.

The first principle for splitting the network is difference between distributed radio access nodes and central processing functions. Antenna and radio frequency (RF) units need to be distributed due to laws of physics while processing should be centralized due to laws of economics. Adding computing power to highly distributed sites will never be as cost efficient as adding it to central sites, but it might be necessary for some use cases and services.

Second principle is the split of management, control and user plane processing functions. Management and control plane are a natural split based on the very different tasks these planes have, management dealing with infrequent needs of operating personnel while control plane dealing with the very frequent needs of end user devices and their users. This layering allows to keep the control plane more centralized while enabling the user plane to be more distributed for improving end user experience and to some degree reducing transport costs, in alignment with the locations where the services are

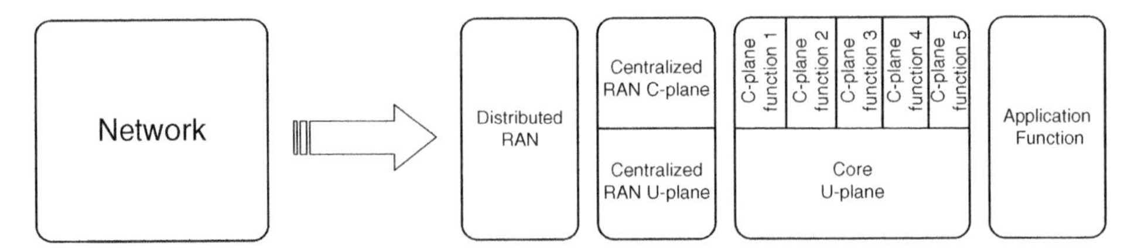

Figure 4.1 Architecture domains.

provided. This also acknowledges the different implementation approaches needed for control plane (focusing on how to achieve an easily maintainable and extendable solution) and user plane (focusing on how to achieve high throughput and low latency cost effectively).

Third principle is the separation of access and core related processing. This also creates a natural split between managing the radio resources and managing the user connections (the latter predominantly a core function).

Fourth principle is the breakdown of core control plane into several Network Functions. This is necessary to manage the high complexity of this domain.

Fifth principle is the separation of the core functions from the applications. Figure 4.1 shows the architecture domains resulting from the described principles.

4.2.3 Flexible Connectivity Models

Connectivity services are required from various locations. Internet connectivity (i.e. inter-connection between mobile network and Internet service provider) can be highly centralized especially from the perspective of a mobile network user. But it is also possible that some services, such as video streaming, are provided from a local cache to optimize transmission resources of the operator and to reduce the delay experienced by the end-user. Connections to enterprise networks, be that to an office premises or to a factory can be highly local in nature. Cities may offer its residents and visitors services that are hosted within the city boundaries.

For these and many more scenarios it is beneficial to optimize the data path from the mobile network user to the service's point of presence. This optimization is relevant for latency as well as for bandwidth purposes. Latency depends both on the physical distance as well on how many processing hops the connection needs to traverse, thereby localizing the service provides a direct benefit by reducing latency. Bandwidth optimization is a slightly more complex topic but having a high bandwidth connection across a wide network area puts more strain to the transport network than being able to provide the service locally. This is especially true for very popular videos where local caching of data content may provide fair savings in the operator backbone.

Since not all services and content can be localized, the architecture must support flexible connectivity models that allows a single UE to connect to multiple data networks (DNs) and to multiple access points simultaneously, enabling optimizing the topology per service. This requires that the UE can connect to multiple interfaces to a DN simultaneously, each interface to a DN being in various places in the topology.

Figure 4.2 shows this kind of a flexible connectivity model with radio access network (RAN) connecting to multiple user plane (UP) entities which connect to different data

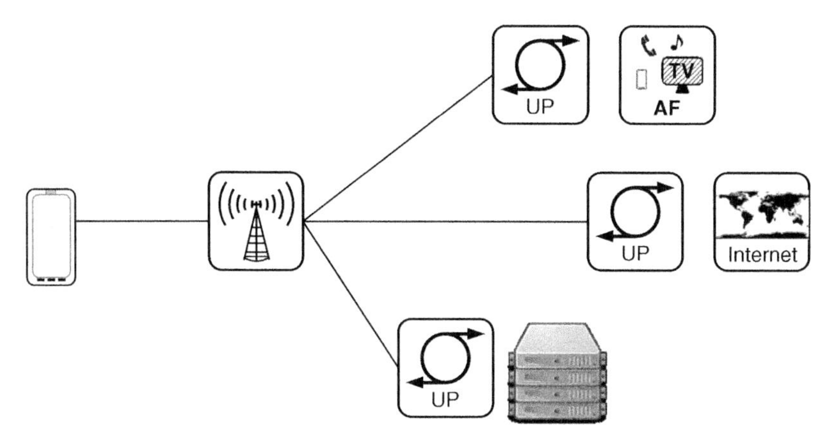

Figure 4.2 Flexible connectivity model.

networks. The architecture defined to support these principles is further described in Section 4.3.

4.2.4 Service Based Architecture

Traditional telco architectures are based on point-to-point interfaces between two network elements. Often those interfaces are designed to be different, not only in the structure of the application layer but on the transport stack as well. This tradition has led to architectures that are unnecessarily complex, require a lot of effort in extending or modifying the functionality and don't facilitate reuse functions across the system.

To design a system that is flexible and future proof, the following requirements need to be supported:

- Interconnections between functional entities of the network are flexible and dynamic, without any predetermined point-to-point structure.
- This interconnection must be based on a common protocol framework for all services.
- Each network function provides its capabilities as services any other network function can use if operator policy allows.
- Multiple versions of the same service must be able to co-exist simultaneously.
- Each network function should only be concerned about the services it offers and the services it uses, no other services should have any impact to this network function.
- Each service must be self-contained in a sense that there should not be any assumption, implicit or explicit, about a network function using the services in certain order.
- All operations that may modify the same contextual data are grouped in the same service.

To fulfill these requirements, the basic paradigm is for each network function to provide its' services via application programming interfaces (APIs), for each network function instance to register itself as a provider of these services to a central repository function and for all network functions intending to use those services to query the central repository on the instances providing the services (Figure 4.3).

This paradigm enables each service to have its' own lifecycle, to easily add new services without impact to any other service and to decouple the service provider and service

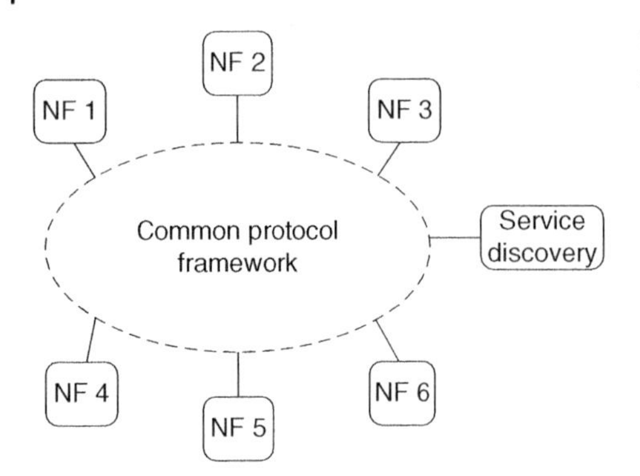

Figure 4.3 Service based architecture – use of service discovery.

user in a sense that a new version of a service, or a completely new service, doesn't impact the user until the service user is ready to use this new version or the new service (with the old version of the service being simultaneously available for as long needed).

4.2.5 Unified Policy Framework

An operator can define multiple policies on how the network should behave, or how it should not behave, and what behavior is allowed or not allowed for a device. In a diverse service landscape, different policies are needed to control how the network interprets the required characteristics of a service for a user. Policy control ensures a systematic and unified behavior on those aspects where this is needed. Policy control is the way how new functions can be introduced into the network and gain awareness of the required behavior. Therefore, the requirements for a policy control architecture are as follows:

- Policies must be able to cover all aspects of the network behavior; the architecture must not limit the policies to specific use cases only.
- Policies may be of different scope in terms of service, user, time of day, network conditions, etc.
- It must be easy to add new policies.
- Any network function can request a certain policy.
- All available policies are capable of being exposed to all the network functions as needed.

4.2.6 Programmable Network

A programmable network is a network where the network functions, or rather their services, and the data they store, can be easily tailored for various applications to fit their needs.

On one hand this requires the capability to orchestrate network functions as use case specific resources with their own parameter sets, isolated from those functions serving a different use case. This is called network slicing, and is described in Section 4.7.

On the other hand, the architecture needs to enable external or internal applications to influence the behavior of network functions and enable network functions to expose network events to applications.

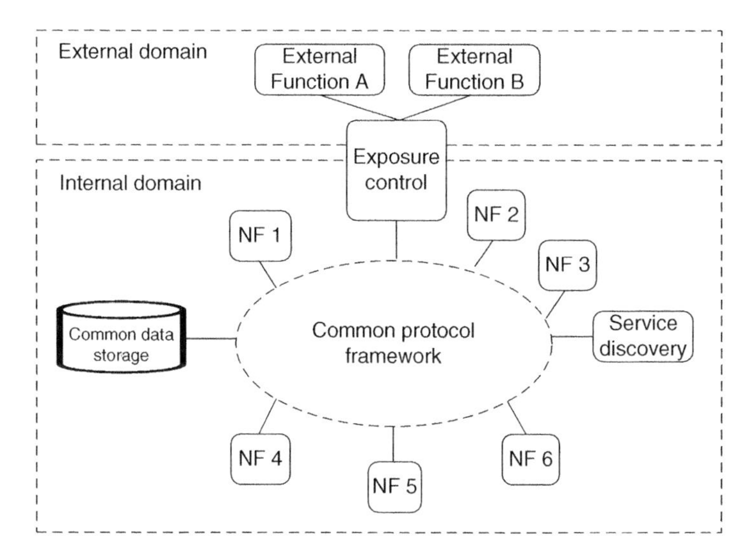

Figure 4.4 Exposure control.

To control what is exposed and to whom, network functions should not expose information directly to the external network but via a well-defined (central) exposure function. This function should authenticate and authorize access to services and data and hide the network internal architecture from the external domain. Within the core control plane domain, the network functions can use each other's services directly. The architecture for controlling the exposure of information is shown in Figure 4.4.

4.2.7 Cloud-Native Network Functions

Cloud native refers to network functions being inherently designed to operate in a cloud environment. This means not only the ability for the network function to be efficiently executed on top of a virtualized infrastructure but also to fulfill the expectations on automation and business agility.

Cloud native network functions should be:

- Designed for failure, i.e. tolerating any type of failures that may occur frequently without impacting the end user service.
- Stateless to the degree possible, taking advantage of common solutions for resiliency and enabling use of common solutions for session data storage, as well as to simplify and improve scalability.
- Programmable, or software defined, to enable orchestration and automated configuration, performance monitoring, fault monitoring and troubleshooting. This includes providing all related data to analytics functions to process and to feed into the automation life cycle.
- Modular to enable frequent and independent additions, updates and removal of parts of the network function; in practice this means a micro-services architecture supporting a continuous delivery model in accordance with DevOps paradigm.

In general, cloud native means building the network function as a flexible software product that can be easily tailored, managed and deployed.

4.2.8 Architectures for Different Spectrum Options

Traditionally new 3GPP access technology generations have been defined as standalone (SA) overlay networks where the new RAN is independent from the old RAN with core network interworking capabilities enabling inter-system mobility. This kind of deployment model is still valid and needed for 5G as well. However, considering the diverse spectrum options available for 5G, the need for a more tightly LTE integrated deployment model, called Non-Standalone (NSA) is needed. This is primarily resulting from the following issues:

- High frequency (e.g. 28 GHz) radio propagation conditions can lead to rapid signal loss for due to an obstacle between sender and receiver, making it more difficult to provide a robust connection and handovers triggered and executed fast enough to maintain that connection while user is moving.
- Deployments using high frequency bands also lead to smaller coverage areas (inter-site distances of 100–200 m), making initial 5G coverage very spotty, leading to high load for inter-system handover.
- Low frequency band deployments provide excellent coverage but lack of bandwidth to provide the throughput expected from 5G.
- High costs associated to rapid build-out of wide area (WA) 5G coverage.

These issues can be solved by the NSA model, where 5G NR is a secondary cell to the LTE master cell (or vice versa), using aggregation via dual connectivity to achieve:

- Robust connection via the LTE or 5G NR layer of the master cell, depending on whether LTE or 5G NR is the master. This requires wide 5G coverage.
- Aggregated throughput via LTE and 5G NR access.
- Ability to keep selected services such as Voice over Long Term Evolution (VoLTE) calls on the LTE layer.

When 5G coverage becomes good in selected areas it may be beneficial to reverse the roles between LTE and 5G by making 5G the master and LTE the secondary access. This enables faster setup times compared to a NSA solution with signaling via LTE.

4.2.9 RAN Architecture Principles

RAN architecture is mainly driven by cost efficiency, providing high throughput per user and cell, low latency for services and multi-access aggregation of different RAT(s) (Radio Access Technology). Thus, the 5G RAN architecture follows to below mentioned key principles:

- Centralization of baseband processing, leading to a RAN architecture split into centralized and distributed units for cost efficient pooling of RAN control plane and higher-layer baseband processing while keeping RF related functions close to the cell site. Such deployments require proper planning of site configurations and the transport network connecting distributed and central units.
- Multi-access aggregation covering 5G and LTE using the centralized RAN unit as an anchor and aggregation point for increased capacity, coverage and connection reliability (interference control).

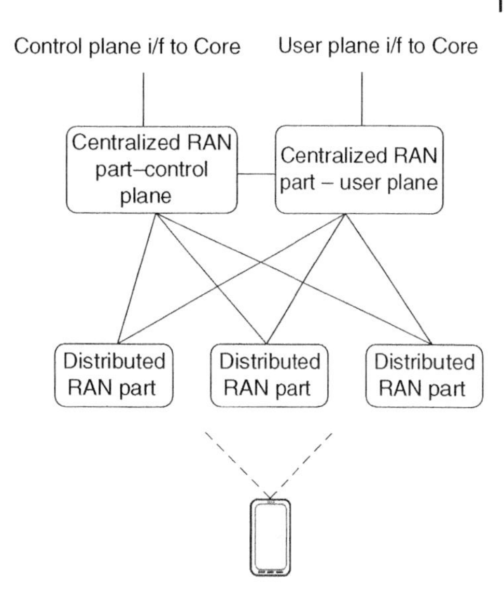

Figure 4.5 RAN architecture principles.

- Enabling integrated deployments with core functions, especially user plane functions (UPFs), as well as applications on top of the centralized RAN unit for efficient delivery of high throughput and low latency content (MEC).

The centralized RAN unit can be realized as a virtual network function running in an edge or radio cloud. Figure 4.5 shows the high-level RAN architecture.

4.2.10 Interworking Principles

Interworking can be divided into two levels:

- Access interworking, i.e. how to enable the use of multiple access networks for a single service and device together with the same core network, while preserving radio connectivity to the UE. In this case, access technologies can either be 3GPP or non-3GPP access (e.g. Wi-Fi) based.
- Core level interworking, i.e. how to enable mobility of an UE from one access network to another access network served by different core networks. This allows seamless mobility with IP address preservation when the user moves between access networks (e.g. due to loss of coverage).

The capability to aggregate multiple RATs into a single access connection is a very desirable property as it provides increased capacity, coverage and reliability for a connection. This kind of aggregation is relevant for LTE and 5G and requires a common anchor point in the RAN as well as common procedures over the radio interface to flexibly and dynamically use any of these access technologies based on whichever is best suited considering the service requirements, radio resource availability and coverage situation.

Even though the 5G System allows integrating LTE radio access, this requires upgrades to the LTE eNB as the eNB need to support the (new or updated) procedures and interfaces defined for the 5G Core. To enable an independent deployment and evolution of

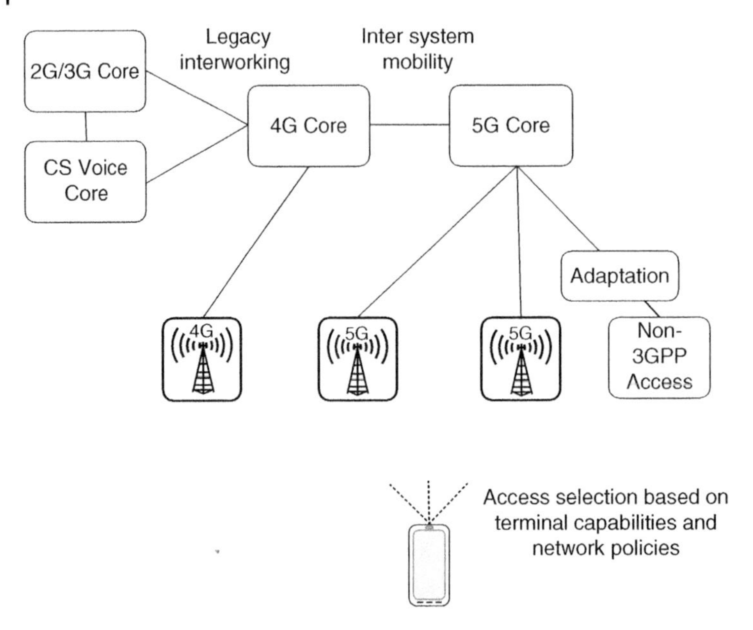

Figure 4.6 Inter system, multi-access interworking view.

the different access technologies, a core network based interworking solution between legacy core (EPC) and new core (5GC) is needed (Figure 4.6).

For applications such as internet protocol multimedia subsystem (IMS) it should be transparent whether a user is currently served by the legacy core (EPC) or by the new 5G Core (5GC). Thus, core entities facing the applications (API exposure, subscriber database) shall hide from applications which core network (EPC/5GC) is currently serving the user.

The requirement to support interworking for all existing legacy mobile access systems has created considerable complexity to EPC with the definition of complex mechanisms for circuit switched (CS) voice interworking such as Circuit Switched Fall Back (CSFB) and Single Radio Voice Call Continuity (SRVCC). Since in the advent of 5G the role and coverage of legacy systems like 2G and 3G will be further reduced and since LTE provides already good coverage in many countries, defining direct interworking between 2G/3G and 5G is no longer justified. The same logic applies to the voice service interworking, making IP based (IMS and over-the-top (OTT)) voice services the only supported voice services in 5G. In cases when there is need for interworking with 2G/3G, including CS voice interworking, it can be done with the help of LTE/EPC. However, some operators expressed strong demand to enable voice call continuity between 5G and 3G (which is being defined in 3GPP Rel-16). They used the following two scenarios to justify their requirement:

1. Operators with both Voice over New Radio (VoNR) and LTE enabled, but no VoLTE: Some operators deploy VoNR and LTE, but do not launch VoLTE in their LTE network, which means VoNR will be dropped, if the UE moves from 5G coverage to 4G coverage.

2. Operators with no or poor LTE coverage (nor VoLTE): In some countries, operators only have 2G/3G deployments. Deploying a 5G System and VoNR directly would not be feasible for them since voice continuity cannot be guaranteed.

4.3 5G System Architecture

4.3.1 5G System Architecture Reference Model

The 5GS architecture is built around NF (Network Functions) that are the smallest set of functionalities deployable in a multi-vendor environment. Internal interfaces, structures and contexts within a NF are not subject to standardization.

Figure 4.7 shows the non-roaming architecture model, while Figure 4.8 shows the roaming architecture model.

Some key principles that were adopted for the new architecture are the following:

1) Converged core to support multiple access technologies. The 5G Core network supports NR, E-UTRA and non-3GPP access (Wi-Fi, Fixed). Untrusted wireless local area network (WLAN) access was the only non-3GPP access connected to 5GC as part of 3GPP Rel-15 while trusted and Wireline access were deferred to future releases (Rel-16).
2) Separate User and Control Plane functions following software defined networking (SDN) principles, allowing independent scalability and evolution of CP and UP. This allows for a flexible deployment of CP functions (geographically) separated from

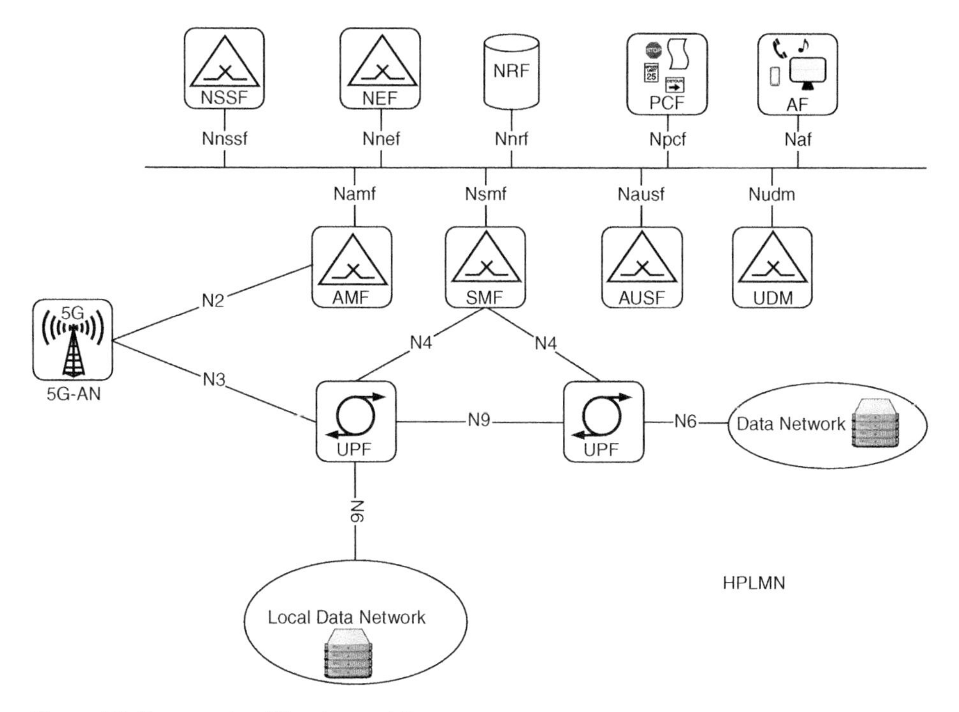

Figure 4.7 Non-roaming 5G system architecture.

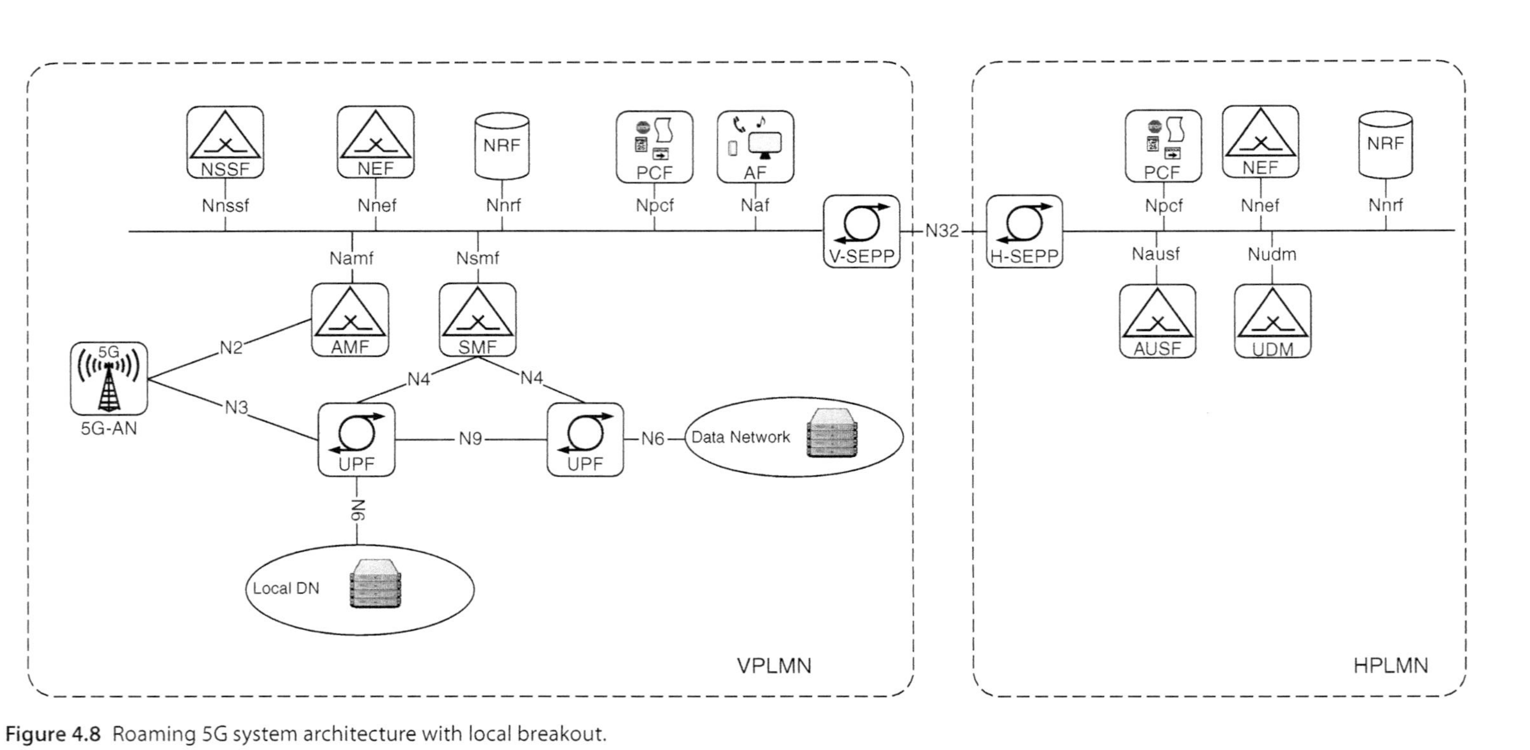

Figure 4.8 Roaming 5G system architecture with local breakout.

UP functions. Each of them being possibly deployed in centralized or distributed locations (data centers).

3) As opposed to EPC which defined three types of UPFs (SGW-u, PGW-u, traffic detection function [TDF]) with statically defined capabilities, the 5G System contains a generic UPF with capabilities that can be programmed by the CP allowing much more flexibility for the deployment of 5G user plane features.
4) Compute and storage separation: capability of a NF to store UE and protocol data unit (PDU) session context in a database (unstructured data storage function, UDSF), allowing the context to be shared across multiple instances of this network function. This supports multiple features such as scaling of network functions and 1 : N-resiliency models. In both cases, when a new instance of the NF is selected to serve an UE/PDU session, the new NF instance can recover the UE/PDU session context from the database.
5) A SBA (Service Based Architecture) supporting the principles stated in Section 4.3.2.
6) Modularization of the functional and interface design: ensure a proper design where specification and deployment updates due to the addition of new features are minimized.

Compared to previous generations, the 3GPP 5G System architecture is service-based. That means wherever suitable the architecture elements are defined as network functions that offer their services via interfaces of a common framework to other network functions that are authorized to make use of the provided services (either using request/response or subscribe/notify transactions). The Network Repository Function (NRF) allows every network function to discover the services offered by other network functions. This architecture model, which further adopts principles like modularity, reusability and self-containment of network functions, was chosen to enable deployments considering the latest virtualization and software technologies. The related SBA figures depict those service-based principles by showing the network functions, primarily Core Network functions, with a single interconnect to the rest of the system. Reference point-based architecture figures are also provided by 3GPP in TS 23.501 [1], which represent the interactions between network functions for providing system level functionality and to show inter-public land mobile network (PLMN) interconnections across various network functions.

The 5G System architecture comprises of the following Network Functions:

- Authentication Server Function (AUSF)
- Access and Mobility Management Function (AMF)
- Data Network (DN), e.g. operator services, Internet access or 3rd party services
- Unstructured Data Storage Function (UDSF)
- Network Exposure Function (NEF)
- NF Repository Function (NRF)
- Network Slice Selection Function (NSSF)
- Policy Control Function (PCF)
- Session Management Function (SMF)
- Unified Data Management (UDM)
- Unified Data Repository (UDR)
- User plane Function (UPF)
- Application Function (AF)

- User Equipment (UE)
- (Radio) Access Network ((R)AN)
- 5G-Equipment Identity Register (5G-EIR)
- Security Edge Protection Proxy (SEPP)

For the reader's reference, the outcome of 3GPP normative 5G System architecture work is documented in:

- 3GPP TS 23.501 [1]: "5G System Architecture,"
- 3GPP TS 23.502 [2]: "5G System Procedure and Flows,"
- 3GPP TS 23.503 [3]: "5G System Policy Control and Charging Framework."

Stage 3 Technical realization of Service Based architecture is documented in 3GPP TS 29.500 [14]. In case of roaming with local breakout, the roaming UE connects to the Data Network (DN) in the visited public land mobile network (VPLMN). NSSF, AMF, SMF, UPF and AF are in the VPLMN. The SEPP protects and simplifies the interactions between PLMNs. The home public land mobile network (HPLMN) provides UDM, AUSF and PCF. Other scenarios and corresponding architecture figures are all covered in 3GPP TS 23.501 [1].

Figure 4.9 shows the network functions (NF) that make up the 5G core control plane. The following further describe the role and functions of each of them.

In accordance with SBA principles, each network function offers services to other network functions. Interfaces to other domains are still point-to-point in nature and are thus not directly connected to the SBA framework.

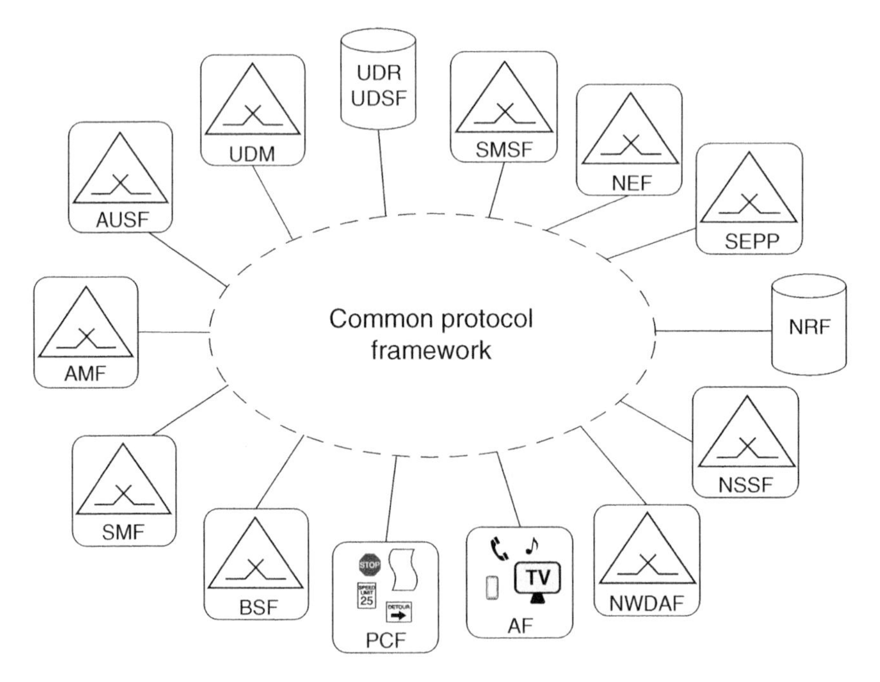

Figure 4.9 5G core with SBA framework.

The SBA protocol framework enables flexible definition and re-use of services, easy modifications and easy integration to external domains via the NEF.

4.3.2 Functional Description

4.3.2.1 Access and Mobility Management Function (AMF)

The Access and Mobility Management (network) function (AMF) terminates the Access Network to Core Network control plane interface (N2) as well as the UE to Core Network signaling (non-access stratum, "NAS") interface (N1). AMF features can be divided into following categories:

- UE access control
- UE access security management
- UE mobility management
- NAS message routing
- Location Services management
- Transport for Location Services messages between UE and location management function (LMF) as well as between RAN and LMF
- Lawful interception (LI) support (for AMF related LI events and interface to the LI system)
- Public Warning System (PWS) message delivery

UE access control refers to the control of UE registration in the 5G System, including:

- triggering UE authentication (carried out by the AUSF)
- authorizing UE access to the network (e.g. check of roaming restrictions)
- managing the registration area of the UE (an UE shall issue a Registration Area update when it goes out of the Registration Area provided by the AMF).

UE access security management consists of two functions:

- Security Anchor Function (SEA) for receiving the intermediate key resulting from the UE access authentication and the
- Security Context Management (SCM) for deriving access specific security keys from the key provided by the SEA.

UE mobility management includes:

- Idle mode mobility management procedures for registration updates within 5G access as well as between EPC and 5G Core, i.e. inter system idle mode mobility;
- connected mode mobility management procedures, i.e. handovers within 5G access and N26 based inter-system handovers between EPC and 5G Core.

NAS message routing provides a generic capability for other network functions to communicate with the UE using the NAS protocol. The initial use cases for this capability are SM signaling from/to SMF and short message service (SMS) delivery from/to short message service function (SMSF). This feature includes managing the UE signaling connection via the access network and paging the UE for transitioning from idle to connected mode when needed.

LI support includes delivery of AMF events to LI system.

AMF provides these functions for both 5G access (3GPP access) and non-3GPP access such as untrusted non-3GPP Access (e.g. access to 5G core via WLAN). Separate UE context is maintained for 3GPP and non-3GPP accesses. Some functions (e.g. paging, handover procedures) may not be applicable to non-3GPP access.

The AMF serving an UE is in the PLMN serving the UE: in the Visited PLMN when the UE is roaming, in the Home PLMN or an equivalent PLMN when the UE is not roaming.

4.3.2.2 Session Management Function (SMF)

Features supported by the SMF can be divided into following categories:

- PDU Session control;
- Termination of the SM part of the NAS interface;
- UE IP address allocation;
- UPF selection and control;
- Control of policy enforcement; and
- Charging and LI support.

The "PDU session" is the (new) abstraction for 5G user plane services replacing the concept of a packet data network (PDN) connection in EPC. A PDU session provides connectivity between applications on an UE and a Data Network (DN) such as the "Internet" or private corporate networks in case of IP.

PDU session control supports establishing, modifying and releasing PDU sessions, including related tunnels between access network and UPF as well as between UPFs, when relevant. All characteristics of the PDU session, such as session and service continuity (SSC) mode are determined by the SMF based on UE request (NAS signaling) and on information received from the subscription data base UDM/UDR. Optional authentication and authorization of the PDU session by the external Data Network is also supported. Session control includes User Plane path establishment between UE and 5G Core when triggered by downlink data reception.

UE IP address allocation includes assigning an IP address, either from SMF's own internally managed IP address pool or via external entity like a dynamic host configuration protocol (DHCP) server or an authentication, authorization and accounting (AAA) server.

UPF selection and control includes:

- The SMF selects one, or in the case of local and central UPF, multiple UPF instances used to host the session.
- The SMF controls the functions of the UPF: traffic detection, traffic forwarding, QoS enforcement, traffic monitoring (e.g. to support charging), replication to the LI system.
- In case the selected UPF becomes unavailable or no more suitable due to UE mobility, the SMF selects another UPF and incorporates this UPF into the data path.

Control of policy enforcement includes interacting with the PCF for obtaining policy rules and decisions related to the QoS and charging properties of the PDU session. These are propagated by the SMF to the UPF for enforcing policies on the data flow. The SMF supports interfaces to the charging system for usage reporting and credit control-based charging related to PDU sessions. LI support includes delivery of SMF events to the LI system. Depending on the configuration, when the UE is roaming, a PDU session may be controlled by a SMF in the Visited PLMN and by a SMF in the Home PLMN. SMF features and PDU session related functionalities are further detailed in Chapter 6.

4.3.2.3 Policy Control Function (PCF)

The PCF provides a single framework for defining any type of policies in the network and for delivering related policy rules to the other control plane network functions (NF), as relevant per each function.

In case the target NF is an SMF the policies sent by the PCF to the SMF may correspond to:

- Charging properties of the PDU session;
- The different QoS and/or charging and/or traffic forwarding rules to apply to different service flows within the PDU session;
- Policies related to local traffic switching which may influence the SMF selection of the UPF.

In the case where the target NF is an AMF the policies sent by the PCF are of two distinct categories:

- Policy information sent from the PCF to the AMF may correspond to service area restrictions for the UE and the priorities of various access types (e.g. LTE and NR) the UE may use;
- Policy information sent from the PCF to the UE via the AMF:
  - Access Network Discovery and Selection Policy (ANDSP) used by the UE for selecting non-3GPP accesses.
  - User Equipment Route Selection Policy (URSP) used by the UE to determine how to route outgoing traffic. Traffic can be routed to an established PDU session, can be offloaded to non-3GPP access outside a PDU session or can trigger the establishment of a new PDU session.

Such kind of policy information was handled by the Access Network Discovery and Selection Function (ANDSF) in EPC. 5GC allows the PCF to co-ordinate between connectivity related policies sent to the UE and policies sent to the network (SMF).

The PCF features can be divided into following categories:

- Retrieving subscription information or specific application requirements.
- Retrieving network conditions (e.g. information on the roaming status or on the access type the UE is currently using).
- Policy rule determination.
- Policy rule delivery to a NF.

Policy rule determination contains methods how policies governing different aspects of the network behavior can be defined and managed, considering:

- User subscription data or application requirements;
- Network conditions;
- Local operator policies.

How policy rules are determined and how they are influenced by local operator policies are not subject to 3GPP standardization.

Policy rule delivery contains methods for the PCF to:

- Receive network conditions from a NF;
- Instruct the NF when to request new policy rules (Policy Rule Request Triggers);
- Provide policies upon NF request (Policy Rule Request Triggers have been met) or push policies to the NF.

The PCF can get subscription information and global application requirements from the UDR. The information being retrieved from the UDR can target an individual subscription, a group of users, all users of a roaming partner or any user.

The PCF may get application requirements targeting an individual PDU session via its service authorization exposure. Via this service authorization exposure an AF can provide requests targeting an individual PDU session identified by the UE address (IP or Ethernet address).

4.3.2.4 Unified Data Management (UDM)

The UDM is a control function in the Home PLMN that supports access to the data storage. It allows for:

- Subscription data management, access and service authorization;
- user identification storage and management;
- user authentication; and
- support of SMS service.

Subscription data management includes accessing the subscription data in the UDR, delivering relevant data to the NFs serving the UE (AMF, SMF) and updating the UE location and serving nodes (e.g. AMF) in the UDR. The UDM is always located in the Home PLMN of the UE.

4.3.2.5 Authentication Server Function (AUSF)

The AUSF is responsible of the primary authentication and key agreement procedures carried out to enable mutual authentication between UE and network. It provides keying material that can be used between the UE and network in subsequent security procedures. The AUSF is closely linked to the UDM. The AUSF is always located in the Home PLMN of the subscriber.

4.3.2.6 Unified Data Repository (UDR)

The UDR provides storage and retrieval, of subscription data (by the UDM), policy data (by the PCF) and structured data for exposure (by the NEF).

In the roaming case, UDR in Home PLMN and UDR in Visited PLMN may be used to serve an UE. The UDR is in the same PLMN as the Network Functions storing data in and retrieving data from it using Nudr. Nudr is an intra-PLMN interface. Subscription data, are stored in an UDR located in the HPLMN, while policy data or data for exposure can be stored in an UDR located in the VPLMN.

4.3.2.7 Unstructured Data Storage Function (UDSF)

The UDSF provides storage and retrieval of unstructured data, such as session related data, for any NF (Figure 4.10). Deployment of the UDSF is optional, depending on the implementations of the NFs and operator desires. UDSF may be co-located with the UDR to allow for a single data layer solution.

In the case of roaming, UDSF in Home PLMN and UDSF in Visited PLMN can be used to context for a given UE.

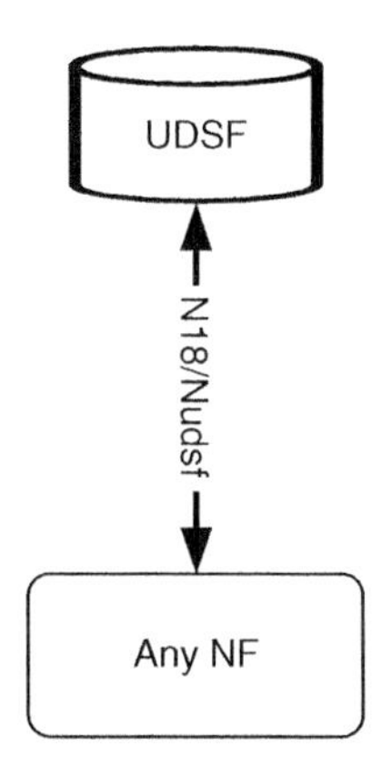

Figure 4.10 Compute and storage separation – data storage architecture.

4.3.2.8 Network Repository Function (NRF) and Network Slice Selection Function (NSSF)

The NSSF and NRF are auxiliary functions deployed between the Operation, Administration and Maintenance (OAM) of the network and the Network Functions providing services for UEs. The NRF provides a service discovery function for the other Network Functions to use. NRF features include NF profile and NF instance management and NF service discovery.

NF profile and NF instance management includes NRF maintaining the list of NF instances and a profile for each instance. This profile includes, e.g. the address of the instance, the services supported and key attributes defining the domain in which these services are authorized to be used, such as network slice related identifiers and the PLMN ID.

NF service discovery contains the methods for NFs to query the NRF for discovery of NF instances providing a requested service or set of services and for the NRF to notify when the set of such instances has been modified.

Each PLMN runs its own NRF. Each part of the network that is separated from the rest of the network due to slicing may have its own NRF.

The NSSF selects the network slice(s) to be used for an UE. Basically, this consists of mapping Single Network Slice Selection Assistance Information (S-NSSAI) to the actual Network Slice Instance Identifier (NSI). Additionally, the NSSF determines the set of AMF instances that can be used to serve the UE.

Each PLMN runs its own NSSF. The NSSF has an overview of all slices of the network.

4.3.2.9 Network Exposure Function (NEF)

The NEF provides a means for external domains to access data and capabilities of the 5G core control plane and to expose structured data within the core control plane domain.

The clients of the NEF may be external AFs possibly deployed by 3rd party operators or internal 5GC Control Plane NF(s); the NEF provides the capability to (re-)expose APIs and data provided by other NFs. The Multi-Access Edge Computing (MEC) platform is an example of a NEF client.

NEF features can be categorized in the following way:

- Secure exposure;
- structured data exposure management; and
- mapping of identifiers and concepts.

Secure exposure provides means to authenticate and authorize (possibly throttle) the external domain entity to which data or capabilities are exposed. It may also create usage data that can be used to charge the external domain for the services rendered by the PLMN. The capabilities exposed by the NEF are described in Section 4.2.6.

Structured data exposure management consists of a service provided by the NEF to other NFs such that the NEF stores NF data as structured data in the UDR. These data can afterwards be provided to other NFs.

Mapping of identifiers and concepts consists of translating between identities used in the AF domain and identities used in the 5G core domain.

4.3.2.10 Security Edge Protection Proxy (SEPP)

The SEPP is a non-transparent proxy. It provides message filtering and policing on inter-PLMN control plane interfaces, and topology hiding.

Detailed functionality of the SEPP is specified in 3GPP TS 33.501 [6].

Since the SEPP is a proxy, i.e. it transparently forwards messages from a NF in one PLMN to a NF in another PLMN without parsing the message, no service-based interface is needed for the SEPP.

4.3.2.11 User Plane Function (UPF)

The 5G System relies on a generic UPF to handle the user plane traffic.

UPF(s) offer capabilities that can be programmed over the N4 interface by the SMF allowing flexibility for the deployment of 5G user plane features. UPFs can be geographically distributed or centralized. The 3GPP specifications place no restriction on the number of UPFs (cascaded UPFs) being used to serve a PDU session.

The following functions are offered by a UPF instance:

- Traffic detection including application detection;
- traffic forwarding based on rules defined by the SMF;
- data buffering and forwarding to ensure data integrity during handover;
- QoS enforcement;
- counting of resource usage. UPF reports the usage data toward the SMF via the N4 interface;
- event reporting to the SMF based on various trigger conditions; and
- replication of user plane traffic for monitoring and LI purposes.

UPF features are further detailed in Section 6.7.

4.3.2.12 Application Function (AF)

AF is not part of the 5G core but it can interact with the 5G core for various purposes, such as accessing 5G core data and capabilities via NEF, influencing how the traffic related to applications should be routed (e.g. via local UPFs) or informing the 5G core about the QoS needs of applications.

While external to the 5G core, some AFs may be considered by the operator to be in a trusted domain and thus allowed to access 5G core functions directly without using the NEF.

4.4 NG RAN Architecture

4.4.1 Principles and Objectives

The motivation for a new system architecture for 5G as described in Section 4.1 applies in general to the RAN as well. However, there are some unique challenges which need to be considered. The design of the NG (5G) RAN architecture needs to have a healthy balance between evolution and revolution, to fulfill the novel requirements and operation paradigms on the one hand, and cost-efficiency and best utilization of already existing deployments (e.g. LTE) on the other hand. NG-RAN is defined as in Section 4.22.

In summary, the main drivers of the 5G RAN architecture design are the following:

- *New services and business models.* Efficient support of new and enhanced services, including mobile broadband with very high data rates, massive machine-type communication and ultra-reliable and ultra-low latency communication. Here, network slicing is widely recognized as a key technology enabler to tailor network operations to new services and business in an efficient way.
- *Cloud RAN* will drive a paradigm shift in operation and deployment of mobile networks. Based on on-demand provisioning of resources, centralization of network functions and NFV will impact functional components and interfaces of the logical network architecture.
- *Integration of multiple Radio Access Technologies (RATs)* to achieve an improved user experience and cost efficiency, 5G should provide a framework for tight integration of new 5G and existing radio interfaces in the RAN, including Wi-Fi and LTE.

The impact of the new 5G service requirements on RAN are vastly different in terms of network performance indicators, but also in the categories of cost efficiency and functional requirements. As an example, for massive IoT the cost factor for chipsets and modules is the most important due to the expected number of devices, low average revenue per device (ARPD), low peak rates and restricted periods of service and mobility.

- For enhanced mobile broadband (eMBB), a peak cell rate of 20 Gbps and an experienced user data rate of 1 Gbps with at least 100 Mbps at cell edges is expected. The end-to-end latency should not be above 10 ms. Vehicular speeds up to 500 km h^{-1} shall be supported for high-speed train use cases.
- For mIoT, while experiencing usually low data rates not above 100 kbps, a density of up to 10^6 devices/km^2 can be expected.
- For ultra-reliable low-latency communication (URLLC), less than 0.5 ms over-the-air latency is expected for certain use cases.

The main challenge for the 5G RAN architecture is to support all requirements within one flexible framework, without compromising interoperability and manageability.

Figure 4.11 illustrates how the 5G main use cases integrate into a common RAN perspective. Important aspects are the tight integration of LTE with support for Dual Connectivity between LTE and NR, Fronthaul split architecture for different

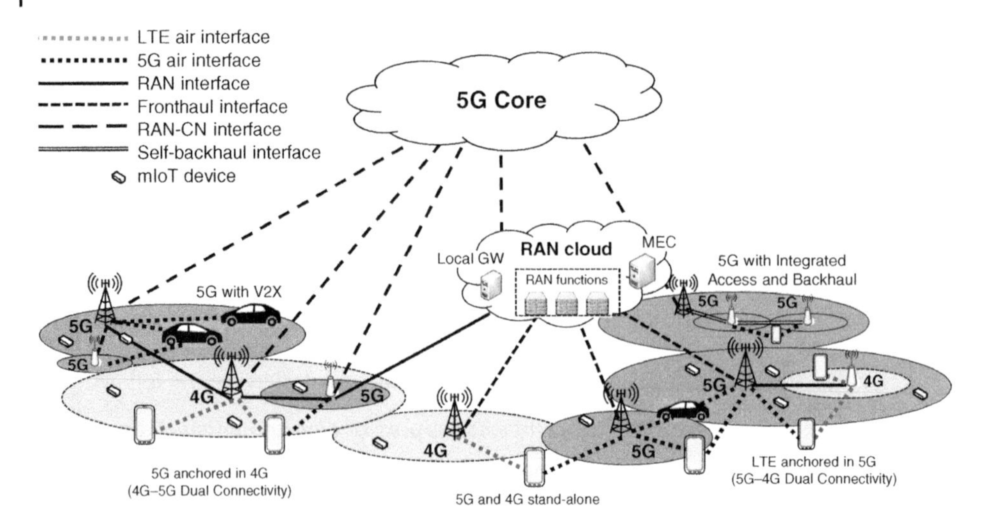

Figure 4.11 5G RAN scenarios.

infrastructure and transport capabilities, multi-hop self-backhauling, and support for MEC (see Section 4.8), enabling for support local services operating on the same virtualized or physical infrastructure as RAN functions.

The 5G RAN functional architecture aims to achieve a unified framework which minimizes the need for specialized functions for different RATs (like 5G, LTE, Wi-Fi) or radio interfaces (like wide area [WA], cmWave, and mmWave air interfaces).

Figure 4.12 illustrates these general principles. The 5G core network is access-agnostic as far as possible, aiming to utilize common functionality for different RATs as far as possible. This facilitates independent evolution of core and radio, e.g. if a new RAT is to be deployed, and the use of a single RAN-CN interface for both 3GPP and non-3GPP access [30].

In RAN, common functions for different RATs are more challenging to achieve. However, there are strong functional similarities for NR and LTE, which is reflected in the NG-RAN architecture. This re-use of parts of the higher-layer user-plane protocol as anchor for multiple RAT and multi-connectivity (MC) features, including routing, traffic handling and bearer mapping, and QoS/QoE enforcement. The control plane includes radio resource and mobility management, as well as radio resource control (RRC) protocol for different RATs which can have common and access-specific aspects.

Furthermore, integrated radio resource management and control of multiple RATs for efficient coordination of radio resources Cloud RAN enables gains for system and cell edge capacity, as well as improved user experience with seamless and transparent inter-RAT access.

RAN may be deployed in a radio cloud with a functional split between virtualized and physical network functions which reflects the infrastructure and service requirements of the operator. This is illustrated by the real-time (RT) and non-real time (NRT) parts of the radio protocols, which can be in a radio cloud or at the cell site.

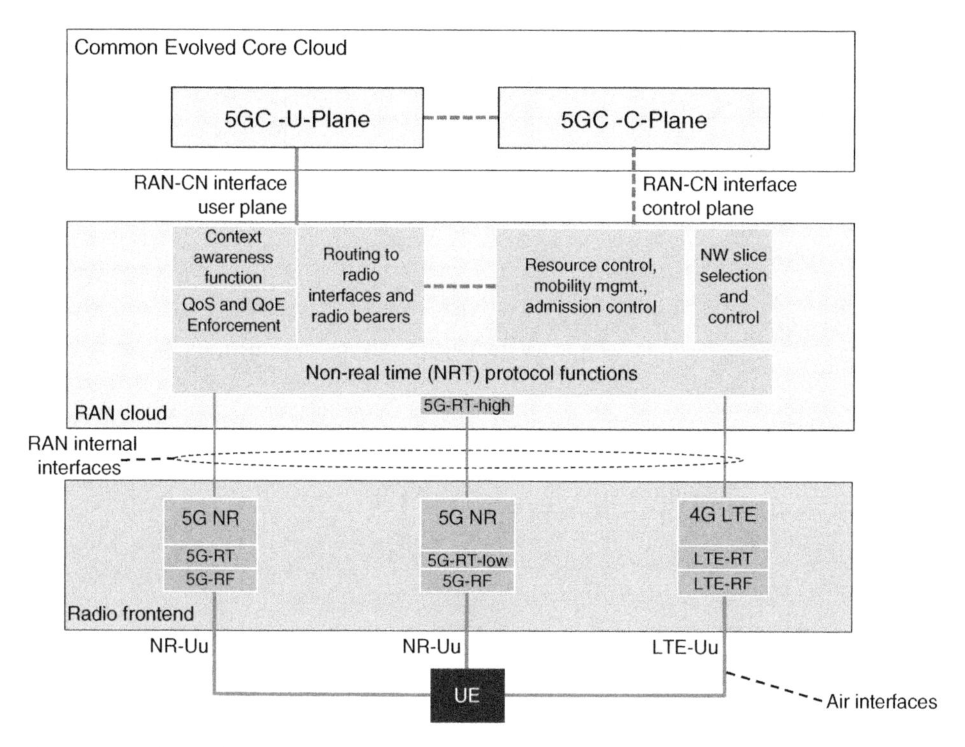

Figure 4.12 5G RAN functional architecture.

4.4.2 Overall NG-RAN Architecture

The overall NG-RAN architecture is reflected in Figure 4.13.

As described in the previous section, the overall 5G RAN architecture addresses the following main requirements:

- Tight integration of RAN nodes with both LTE (E-UTRA) and NR air interface;
- common RAN-CN interface for all nodes; and
- flat and simple architecture with clear functional split between RAN and core.

These requirements are reflected in the overall NG-RAN architecture as illustrated in Figure 4.13. It comprises the following elements:

- gNB radio nodes, hosting NR air interface;
- ng-eNB radio nodes, hosting LTE (E-UTRA) air interface;
- NG interface consisting of a control (NG-C, corresponding to the N2 reference point) and user plane (NG-U, corresponding to the N3 reference point) part connecting NG RAN radio nodes to the 5G core; and
- Xn interface, consisting of a control plane (Xn-C) and user-plane part (Xn-U) interconnecting both gNB and ng-eNB radio nodes [31].

The functional split between RAN and Core follows the principle that access specific functions which are concerned with radio resources are hosted in RAN nodes. This includes radio resource management in general, connection setup, scheduling, as well as functions for AMF selection and user-plane routing. For more details, refer to [4].

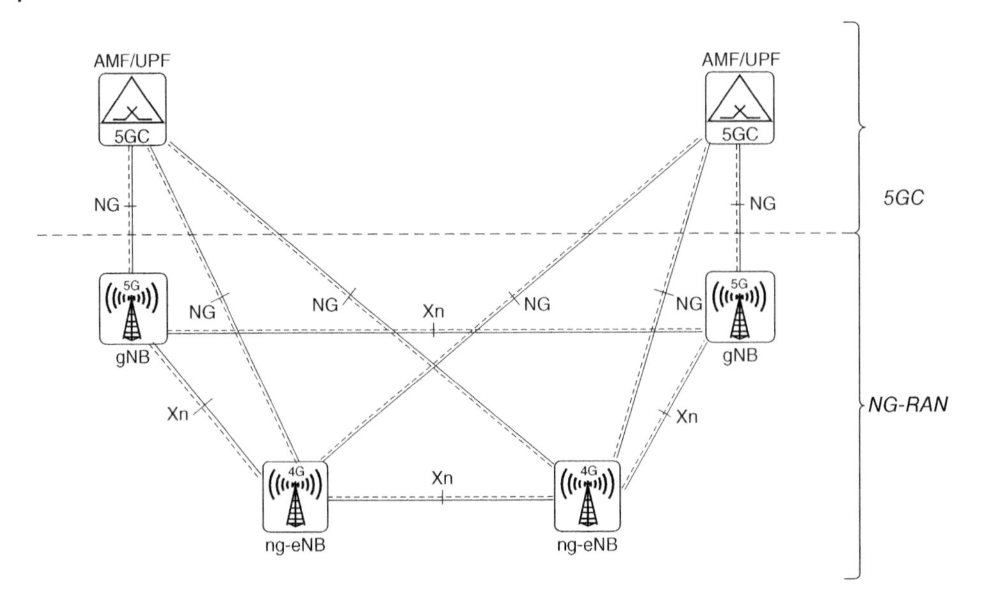

Figure 4.13 NG-RAN architecture.

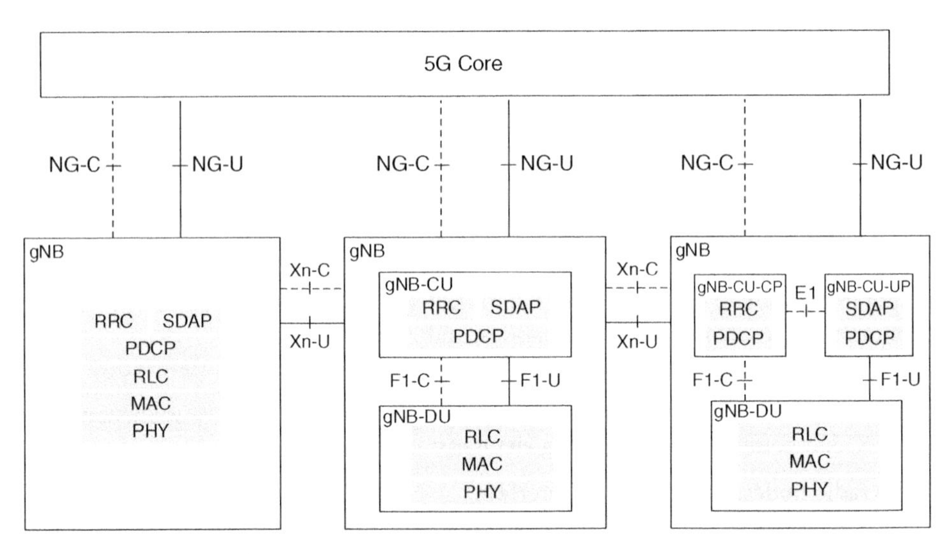

Figure 4.14 Logical NG-RAN architecture with split options.

4.4.3 Logical NG-RAN Split

The logical NG-RAN architecture [25] supports interfaces such that a flexible placement and independent scaling of logical entities is enabled. Both centralized and distributed network deployments are supported by the architecture design. It depends on the operator's infrastructure and offered services which functions will be centralized and where to be placed. This is illustrated in Figure 4.14.

Figure 4.14 shows three possible configurations of a logical gNB, possibly deployed in the same network:

- On the left-hand side, a "classical" monolithic base station.
- In the center, a gNB with a higher-layer Fronthaul split interface, which enables deployment of RRC, service data adaptation protocol (SDAP), and packet data convergence protocol (PDCP) in edge cloud facilities. The gNB is logically separated in a central unit (gNB-CU), and one or more distributed units (gNB-DU), which are interconnected with the F1 (F1-C, F1-U) interface [26].
- On the right-hand side, a gNB with control/user plane separation for the higher-layer split, where the gNB-CU is further separated in gNB-CU-CP, hosting RRC and PDCP protocols. And gNB-CU-UP, hosting SDAP and PDCP protocols. The E1 interface connects gNB-CU-CP and gNB-CU-UP with a strict 1 : N relation [27].

Depending on the fronthaul capabilities, both C-plane and U-plane RAN functionalities can be placed in centralized nodes. In this configuration, the smaller signaling effort between functions in the cloud enables an efficient coordination between them.

4.4.4 Lower-Layer Split

The very high bandwidth requirements on 5G make classical centralized radio access network (C-RAN) deployments impractical, since those rely on the transmission of quantized radio symbols (raw data) between central and remote unit. This approach results in very high bandwidth requirements on the Fronthaul interface coming with high costs. The solution in 5G is to develop a new lower-layer split (LLS) which requires less bandwidth, but still enables the benefits of C-RAN such as gains from coordination and joint processing.

In case of lower layer split, the Fronthaul interface interconnects a central unit and many remote units where the latter hosts functions of the lower physical layer (PHY) of the logical radio node (see Figure 4.15 left-hand side). Note that lower layer split and higher layer split options can be in principle deployed in parallel, i.e. either in a central unit which implements both higher and lower layer split interfaces, or in a multi-tier deployment where both higher and lower layer split implementations are present for the same logical node.

Due to the complexity of the PHY, many different implementation options are possible, and it is less clear (compared to higher layers) how to group and split functions. This makes it more difficult to converge to a single split point, which is illustrated in the right-hand side of Figure 4.15. In this example, four different split points are defined, each with its implementation-specific advantages and disadvantages. Nevertheless, there are some common principles in all split options which need to be supported:

- Support for different beamforming techniques and algorithms for digital, analog and hybrid beamforming;
- interoperability with preferably small number of user and implementation specific parameters;
- support for advanced receiver and coordination features (such as Cooperative Multi Point); and
- future proof for upgrades of both central and remote units.

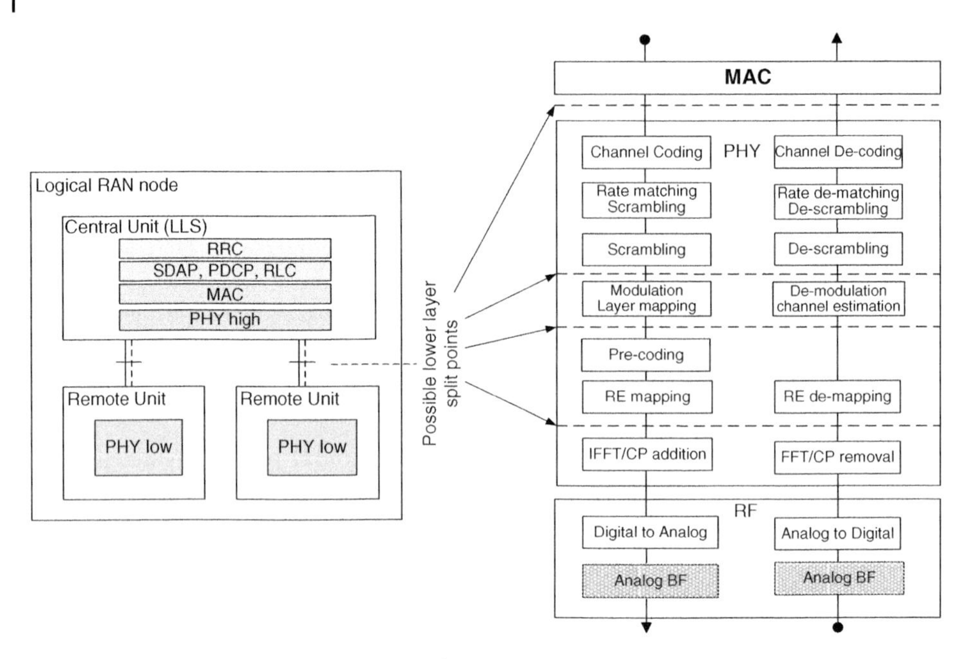

Figure 4.15 Lower-layer split architecture and options.

LLS is addressed in various industry fora and standardization bodies, including the Common Public Radio Interface (CPRI) industry cooperation, IEEE 1914.1 and in 3GPP.

4.4.5 Service-Aware Function Placement

The flexible functional split is a key feature for enabling (ultra-)low latency services. Due to the strong requirements of below 1 ms end-to-end latency for certain services (see Chapter 1), service-related functions need to be placed close to the access. How close depends on the specific service requirements. This necessity is adversarial to the objective of centralization and cloudification, which benefits from aggregation of processing-intensive functions due to scaling effects. The solution to resolve this contradiction is the NG-RAN logical split architecture with service-aware placement of RAN functions as illustrated in Figure 4.16.

- A common c-plane function hosted in the gNB-CU-CP in an edge cloud controls several dislocated user-plane components.
- MEC platform functions enable hosting of URLLC services in local radio clouds, as well as low latency or mIoT services in edge clouds.
- 5G core control plane is co-located with fully centralized services.

This example illustrates how the flexible functional split architecture can be used for service-aware placement of functions. It also enables many other deployment scenarios which are not necessarily driven by service requirements, but by infrastructure capabilities, cost, or legacy deployments.

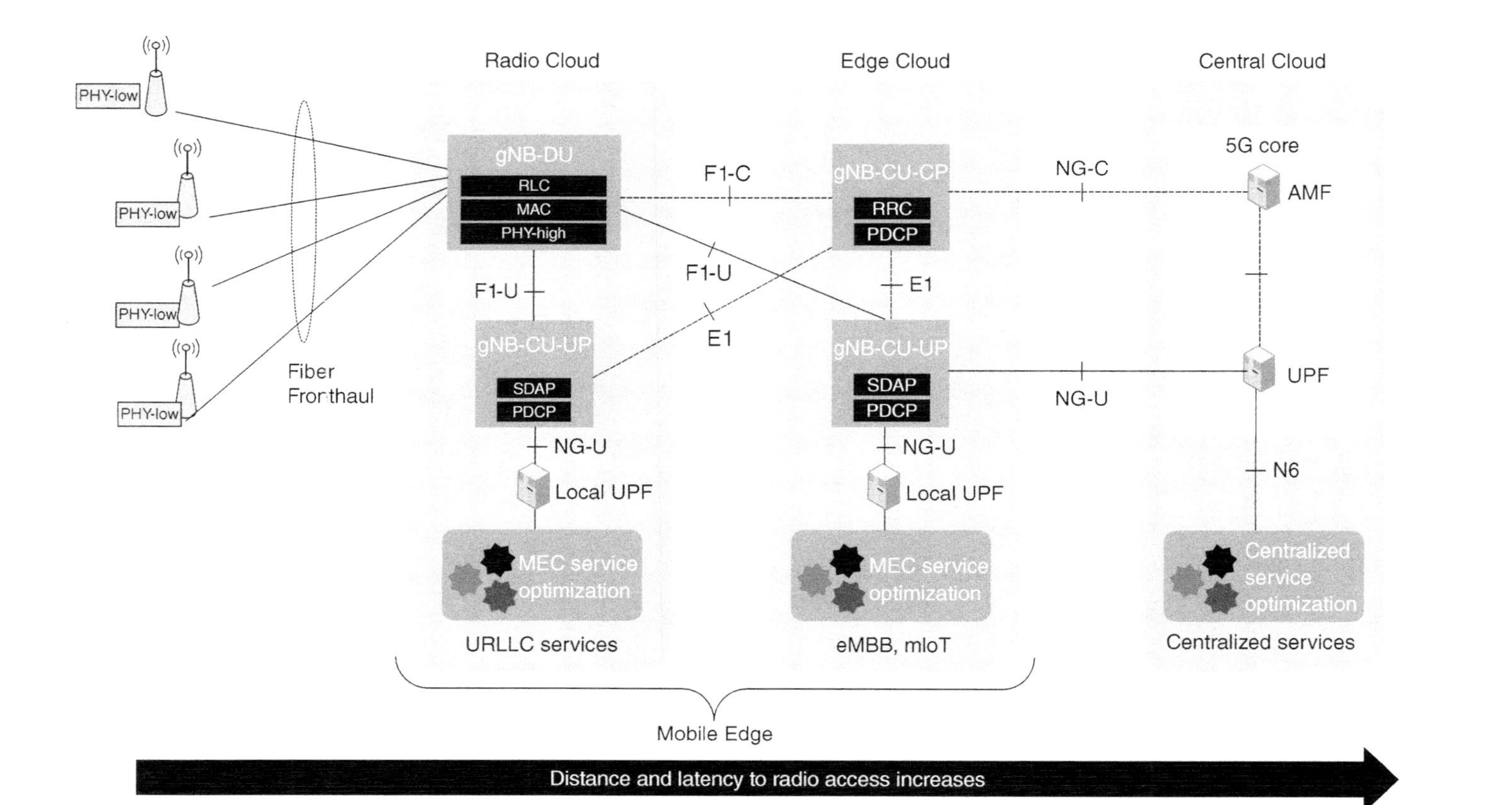

Figure 4.16 Service-aware placement of RAN functions.

4.4.6 Connectivity to Multiple RAT

- One of the key features of 5G is the tight integration of multiple RATs in the radio access, specifically by using Multi-Connectivity (MC) techniques. This refers, in general, to the case if a UE is served by more than one transmission/reception points (TRPs) at the same time, potentially using different RATs. This definition includes a fairly wide set of techniques ranging from lower layer and higher layer radio, and even core network or OTT methods such as Multi-Path TCP (MP-TCP). Here, we concentrate on higher-layer radio access methods enabled by NR and NG-RAN to achieve Multi-Connectivity to enhance user throughput, packet reliability and connection robustness. In the first release, 5G focuses on a specific subset of Multi-Connectivity called Dual Connectivity which involves exactly two RAN nodes where one is the Master Node (MN) which takes most of the radio resource decisions, and the other a Secondary Node (SN) in a supplementary role. In future releases, other RATs such as WiFi will be integrated as well. The following terms are introduced [26]:
- Multi-Radio Access Technology Dual Connectivity (MR-DC) as a generic term for Dual Connectivity which involves NR and LTE nodes;
- NR-DC for Dual Connectivity between gNBs;
- NE-DC for NR-E-UTRA Dual Connectivity where the gNB is the master node and the ng-eNB is the secondary node
- NGEN-DC for NG-RAN E-UTRA Dual Connectivity where an ng-eNB is the master node and gNB is the secondary node;
- EN-DC for E-UTRA-NR Dual Connectivity in option 3 deployments (see Section 4.5) where an eNB has the master role and a gNB the secondary role. Note that this variant is strictly interpreted as being not part of the NG-RAN architecture, but of evolved universal mobile telecommunications system terrestrial radio access network (E-UTRAN) Rel-15.

Figure 4.17 illustrates the generic control-plane architecture of MR-DC and EN-DC, where the involved RAN nodes can host NR or LTE radio. The Master Node (MN) or Master eNB (MeNB) for EN-DC takes RRC decisions and initiates procedures for SN (SeNB) management. In general, the UE follows the RRC state of the master to avoid state confusion. The MN is generally also the node which provides c-plane connectivity to the 5GC (or EPC in case of EN-DC). This is illustrated in Figure 4.17, which shows the involved RAN and CN interface depending on the DC type. Apart from some differences in terminology, both 5GC and EPC-connected DC types employ connectivity from master and from secondary node to the core network, depending whether higher layers (i.e. at least PDCP) of the radio user plane is anchored in the master or secondary node, respectively (Figure 4.18).

Irrespective of the DC type, the functions for secondary node management (addition/modification/release/change), bearer setup, QoS mapping and others are very similar. For enhanced connection robustness, split transmission of RRC messages over the two involved radios is supported (via a so-called split signaling radio bearer [SRB]). This reduces the probability of radio link failures, where the UE/RAN node ignores duplicate reception of messages, if both transmissions were successful. Furthermore, an additional SRB3 can be established to the SN to enable more autonomy of SN decisions and reduced signaling between SN and MN.

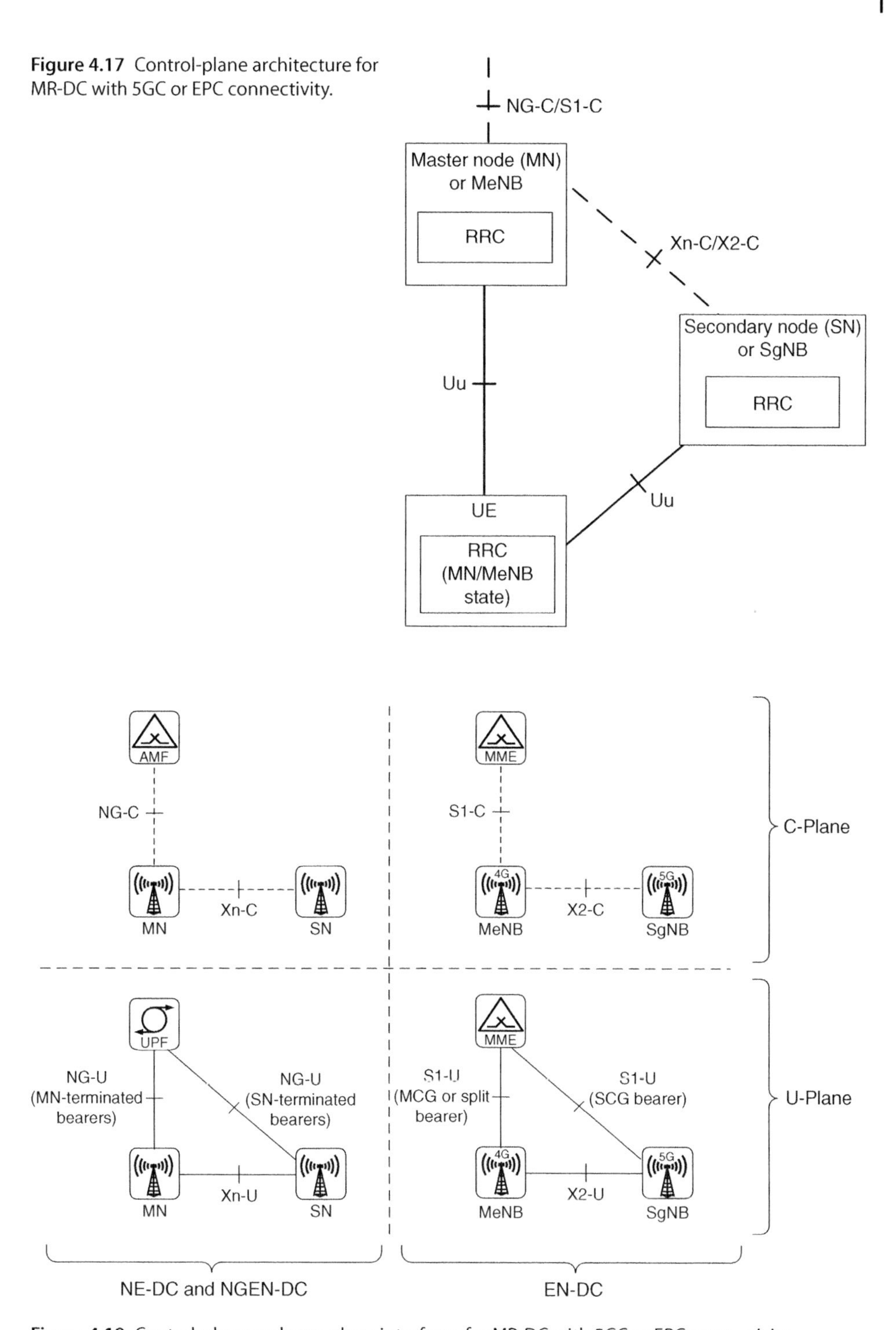

Figure 4.17 Control-plane architecture for MR-DC with 5GC or EPC connectivity.

Figure 4.18 Control-plane and user-plane interfaces for MR-DC with 5GC or EPC connectivity.

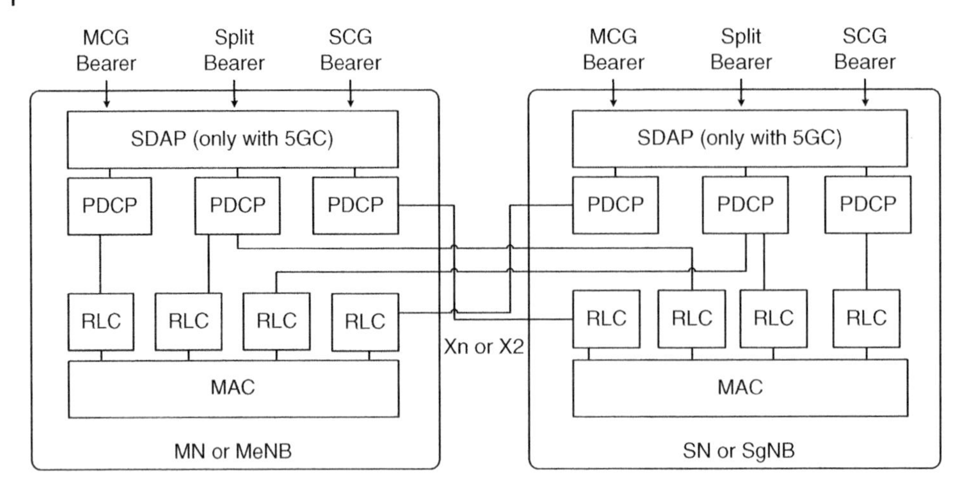

Figure 4.19 User-plane architecture for MR-DC and EN-DC.

Different configurations for data radio bearers (DRB) are supported as illustrated in Figure 4.19. The PDCP layer is responsible for splitting and aggregating, or duplicating packets depending on the service requirements. A flow control mechanism (the same as employed by the F1-U interface) between MN and SN avoids data starvation between RAN nodes. For EN-DC, the X2 interface with flow control is used.

4.5 Non-Standalone and Standalone Deployment Options

4.5.1 Architecture Options

Since NR could be deployed over a wide range of spectrum (e.g. 600 MHz to 100 GHz) bands and the fact that operators have different needs, and the timeline to deploy NR, 3GPP specifications allow various NSA and standalone deployment options. Traditionally, 2G, 3G, and 4G were always introduced as standalone system in the first release of the specifications. NSA options were considered later, e.g. in LTE with the introduction of small cells in 3GPP Rel-12. However, with NR, both NSA and standalone deployments options will be possible based already with the first 5G 3GPP release 15.

Figures 4.20–4.23 show the possible NR deployment options with a 5G Core.

Spectrum availability and radio coverage will be one of the key considerations for operators to determine the right deployment option. Following are the key considerations for operators to determine the right option:

1) Spectrum availability:
   a. NR capacity bands: 3.5 GHz, mmWave
   b. NR coverage bands: <2 GHz, best under 1 GHz
   c. LTE/NR spectrum sharing
   d. 3G refarming strategy
   e. Unlicensed band

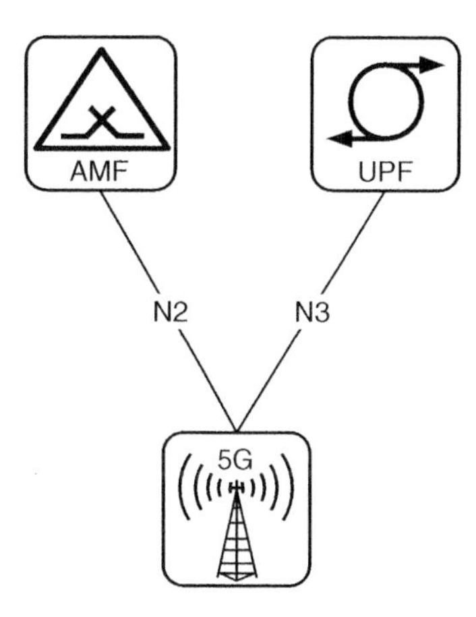

Figure 4.20 3GPP Option 2 NR standalone architecture with 5G Core.

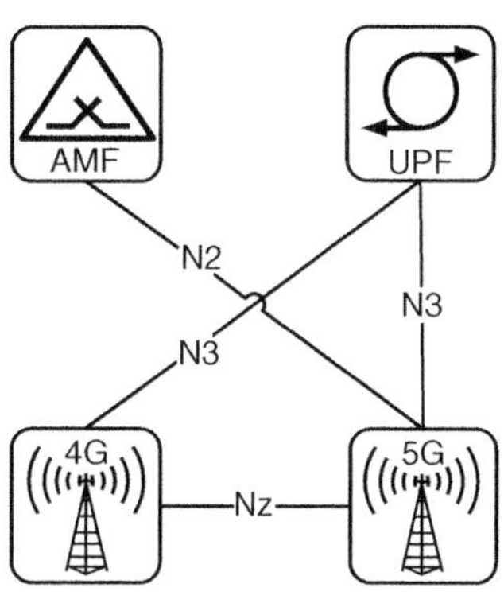

Figure 4.21 3GPP Option 4 NR non-standalone architecture with 5G Core.

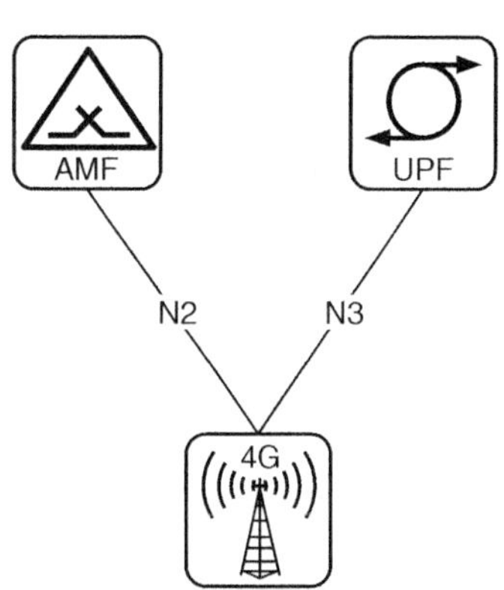

Figure 4.22 3GPP Option 5 E-UTRA standalone architecture with 5G Core.

2) Device availability:
   a. Most of the early devices will support option 3 NSA (3GPP Rel-15 early drop)
   b. Standalone and 5G System based deployment options (2/4/7) will come in a later step
   c. Support of FDD bands is expected after TDD band support (in higher frequencies enough bandwidth available to go for TDD)

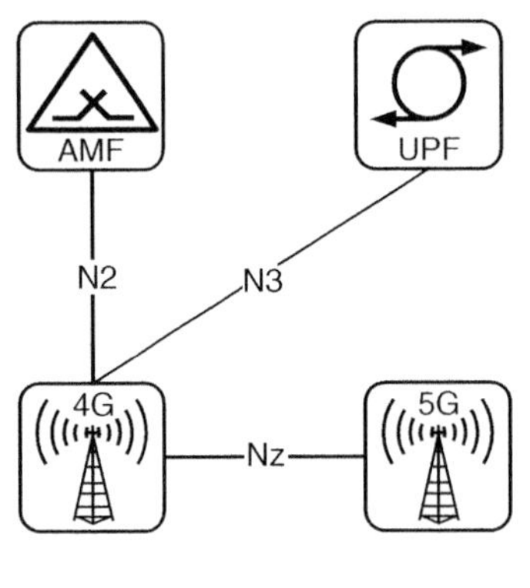

Figure 4.23 3GPP Option 7 E-UTRA non-standalone architecture with 5G Core.

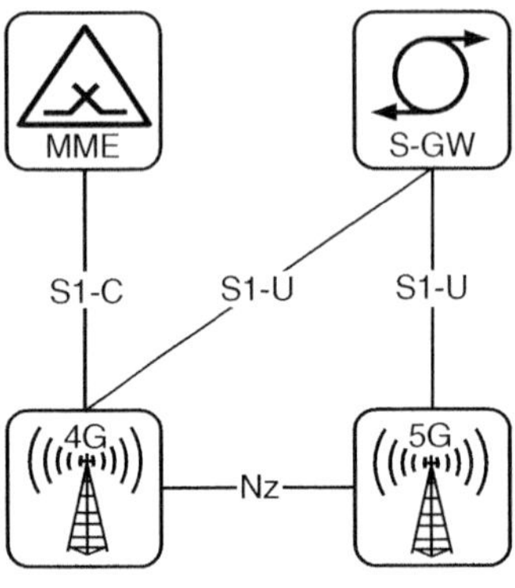

Figure 4.24 NR non-standalone architecture with EPS (also referred to as Option 3 in 3GPP).

3) Cell density:
   a. NR 3.5 GHz massive multiple-input-multiple-output (mMIMO) matches LTE 1800 MHz grid in high density areas
   b. Network wide coverage will need sub 1 GHz spectrum or rely on LTE coverage
4) Services and business Strategy:
   a. eMBB only or wide set of 5G use cases exploiting low latency and high reliability
   b. Vertical opportunities
   c. Cost reduction versus new revenue
   d. Long term LTE and EPC investment plans

Based on spectrum availability and radio coverage, most of the operators publicly expressed that they will start with a NR NSA architecture re-using existing EPS (also referred to as Option 3 in 3GPP) before moving toward one or several of the options 2, 7, and 4 deployment options stated above. Due to the importance of option 3, we summarize the rationale, objective, and technical impacts of this deployment option 3 below (Figure 4.24).

4.5.2 Non-Standalone Architecture with EPS

For operators who want to quickly deploy NR to boost their radio capacity and user plane bandwidth in places where LTE starts to get saturated, 3GPP has defined an early drop of Rel-15 specifications (completed in March 2018) where a gNB (RAN node that supports NR) is used as a secondary RAN node of a LTE eNB connected to EPC (Figure 4.25).

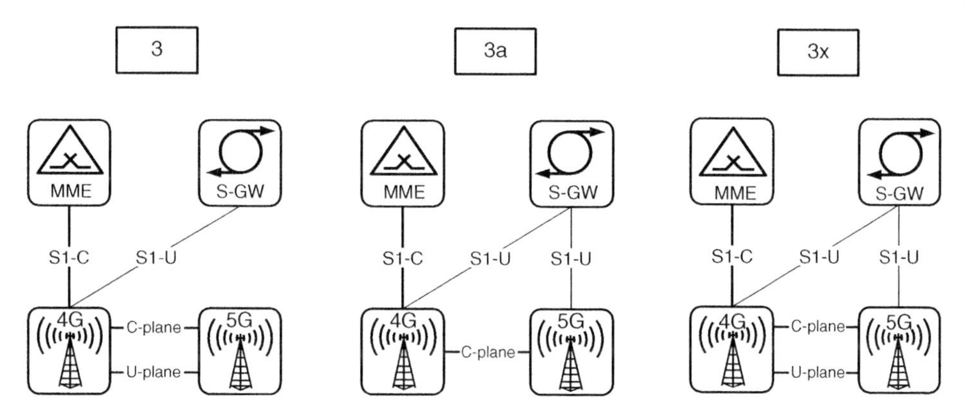

Figure 4.25 User plane architecture options.

However, deployments using this option can't benefit of all the functionalities and features offered by the 5G System Architecture but can just provide a higher user plane throughput (where an EPS service can be simultaneously supported by both LTE and NR radio). Enhancements to EPS were introduced for connectivity of devices that support NR NSA with EPS NAS. It mainly encompasses the following enhancements:

- *Subscription based access control.* Mechanism for the HPLMN and VPLMN to control UE's access to 5G NR based on subscription data (when 5G NR is used as a secondary RAT).
- *Gateway selection based on NR capability.* UE "5G Radio Access Capability" handling in MME and enabling selection of S-GW and P-GW optimized for NR capable devices (e.g. allowing for higher throughput).
- *Extended throughput/bit rate.* Extend the maximum range of maximum bit rate values to cater with 5G demands.
- *Secondary RAT type for charging.* Mechanism to provide "secondary RAT type" information from RAN to CN (e.g. for S-GW and P-GW charging records, for PCC and possibly IMS).
- Optional support for new security algorithms.

It should be noted that the enhancements to EPS are purely optional. This implies that NR can be deployed in NSA mode with EPS. the UE can be offered basic connectivity with NR anchored to E-UTRAN, without upgrading the existing EPC network elements (i.e. MME, S-GW, P-GW, home subscriber server [HSS]) in deployed networks.

4.6 Identifiers

4.6.1 Overview

Each subscriber and/or subscription to the 5G System is allocated one 5G Subscription Permanent Identifier (SUPI) for use within the 3GPP system. The 5G System supports identification of subscriptions independent of the UE. Each UE accessing the 5G System is assigned a Permanent Equipment Identifier (PEI). It is also required by the 5G System to support allocation of a temporary identifier 5G Globally Unique Temporary Identifier (5G-GUTI) to support user confidentiality protection.

4.6.2 Subscription Permanent Identifier

A globally unique 5G SUPI must be allocated to each subscriber in the 5G System and provisioned in the UDM/UDR. The SUPI is used only inside the network. The SUPI may either be based on international mobile subscriber identity (IMSI) as defined in 3GPP TS 23.003 [5], or it can be a network-specific identifier, used for a private network.

To enable roaming scenarios, the SUPI must contain the PLMN ID (mobile country code (MCC) plus mobile network code [MNC]), especially for a IMSI based SUPI. SUPI is not expected to be used over the air (i.e. in clear) in the paging, access stratum (AS) or NAS messages exchanged with the UE.

4.6.3 Subscription Concealed Identifier

The Subscription Concealed Identifier (SUCI) is a privacy preserving identifier containing the concealed SUPI. It is expected to be used in the NAS messages.

4.6.4 Temporary Identifier

The AMF assigns a Temporary Identifier 5G-GUTI to the UE for supporting subscriber confidentiality. The AMF leverages the same identifier to include routing information that the 5G-AN node can use to determine how to route the NAS message. The UE is required to derive the 5G-S-TMSI (5G S-Temporary Mobile Subscription Identifier), which includes the routing information, from the identifier (5G-GUTI) assigned by the AMF and to provide that in the RRC message toward the 5G-AN. The 5G-AN uses the routing information present in the 5G-S-TMSI and routes the NAS message accordingly. Structure of the 5G-GUTI and 5G-S-TMSI are as follows:

$$<5G-GUTI>:=<GUAMI><5G-TMSI>$$

where

<GUAMI> = <MCC> <MNC> <AMF Region ID> <AMF Set ID> <AMF Pointer>
<5G-S-TMSI> = <AMF Set ID> <AMF Pointer> <5G-TMSI>
GUAMI = identifies one or more AMFs within an AMF Set.
MCC = Mobile Country Code of the PLMN.
MNC = Mobile Network Code of the PLMN.
AMF Set ID = identifies an AMF Set. An AMF Set comprises of AMFs that support the same set of Network Slices.
AMF Pointer = identifies one or more AMFs within an AMF Set.
AMF Region ID = identifies AMF Region. An AMF Region comprises of multiple AMF Sets.
5G-TMSI = identifies the UE within the AMF.

When an AMF assigns a 5G-GUTI to the UE with a GUAMI (comprising an AMF Pointer) value that identifies only one AMF, the 5G-TMSI identifies the UE uniquely within the corresponding AMF. However, when AMF assigns a 5G-GUTI to the UE with a GUAMI (comprising an AMF Pointer) value used by more than one AMF, the AMF shall ensure that the 5G-TMSI value used within the assigned 5G-GUTI is not already in use by the other AMF(s) sharing that GUAMI value. The reason for allowing the GUAMI (comprising an AMF Pointer value) identify one or more AMFs is to allow one or AMF process a given UE transaction, enable stateless or state-efficient AMF behavior, explained further in Section 4.9.4.

4.7 Network Slicing

4.7.1 Introduction and Definition

The emerging wide variety of services with distinct characteristics and requirements have led to the question how these requirements can be cost effectively met by the network. Different services together with personalization of tenants' networks create a long tail of requirements that become impractical to meet with a single network since the complexity of each network function within that network grows along with the requirements. Network slicing uses a divide-and-conquer approach where each network function instance handles only requirements relevant for some users, services, or tenants.

In its most basic form network slicing refers to an implementation model where particular sets of end users, services, or more generally tenants are assigned to dedicated sets of network function instances (Figure 4.26). Contrasted with the non-sliced implementation where all of them would be sharing common instances, slicing provides a universal method for user, service, or tenant specific customization without directly requiring this customization capability from the network functions themselves. Slicing requires full automated orchestration capability as otherwise deploying various custom behaviors would be prohibitively costly.

While this arrangement is possible with physical network functions as well, for cost efficiency reasons virtual network functions are the default approach for slicing. This enables each tenant to have a virtual network, or a part of a network, on their own.

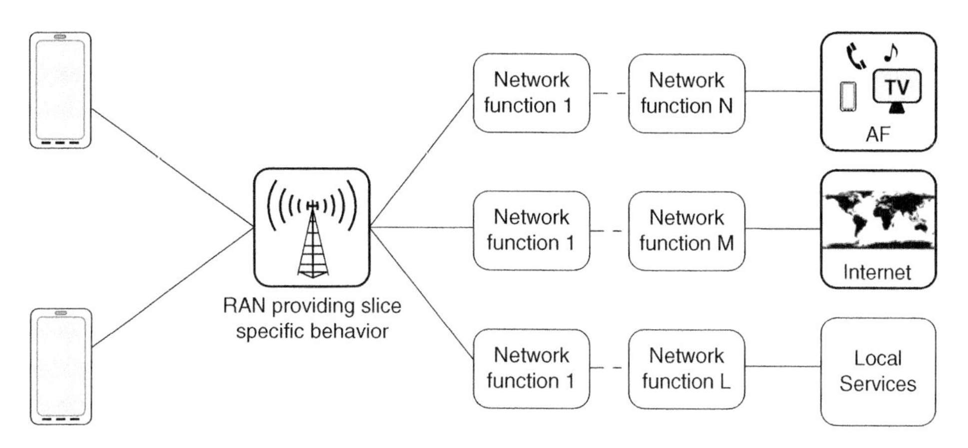

Figure 4.26 Network slices.

This basic form of slicing is however inadequate when considering the whole network. Virtualized network functions need to be provided with computing, networking, and storage resources, and functions at different sites must connect to the transport network. Some network functions, such as antennas or power amplifiers, can't be virtualized. For these physical entities a different approach is required. Physical resources can be connected to slices by having policies controlling the use of these resources, reflecting the characteristics of a slice. Thus, network slicing can be defined as the ability to create isolated behavior across the entire network in an automated way.

Slice specific physical instances are possible, thereby enabling the entire network including the antenna to be sliced, but such approaches are usually impractical from a cost and network deployment perspective, except perhaps for most specialized use cases.

Network slicing assumes each network function instance can be easily tailored for meeting specific requirements, creating a need for programmable network functions to allow for automated slice creation and maintenance.

With the growing number of network slices, automation of operations becomes of paramount importance. Operating multiple network function instances and configurations can easily increase operational costs, if an adequate level of automation is not available. Further, different functions and resources used by the slices must be configured and monitored coherently, as any misalignment breaks the behavior of the slice. This requires network level mechanisms, which can be referred to as a "network operating system."

Network slicing and automation together create a programmable network, capable of adapting to the needs of any service or user by providing isolated network behaviors for each of them.

4.7.2 Isolation Properties

Slicing provides isolation in many ways and for various properties of the network. The most relevant ones are:

- *Isolation of functions.* Slices don't require the same network functions, some of the functions can be omitted completely and some functions may have a very different set of services that are relevant in a slice.
- *Isolation of configurations.* Even for the same functions and services, different configurations can be used, such as parameters impacting the behavior of a function (e.g. level of high availability) or which neighboring functions to use.
- *Isolation of resource usage policies.* Different slices may have different policies on how much resources are provided by, e.g. the virtualization infrastructure or by the physical network functions such as throughput in switches and routers. This means the load domains are isolated, enabling the prevention of an overload in a manner consistent with the criticality of the traffic in the slice.
- *Isolation of lifecycles.* Timelines for when a slice is introduced, updated, and finally removed from the network can vary greatly from slice to slice. Functions can also be tested in one slice before extending the use to other slices.
- *Isolation of failure domains.* Slicing enables

- the impact of a failure to be kept within the slice. In this case the failed function is not present in other slices. The failure is completely isolated to a single slice.
- *Isolation of security domains.* Since slices should operate on isolated virtual networks, attackers should not be able to access other slices even when one slice is compromised.

These isolation properties are the real benefits of slicing. These properties enable to have use case tailored networks, to move away from "one size fits all" approaches. For these properties to materialize, the 5G System must accommodate the slicing paradigm in the network architecture and definition of the network functions.

4.7.3 Slicing Architecture

Network slicing requires certain specific capabilities from the network, particularly the slice selection function, but equally important is how the network architecture supports meaningful ways of isolating the functions and the configuration of services within those functions. The key targeted properties are:

- Network functions capable of being either: slice specific, common to a group of slices or common to all slices.
- Orthogonal architecture where each functional domain is fully confined in a single network function.
- Service oriented architecture where each service a network function provides can be individually enabled, tailored or disabled.

In the 5G network architecture these properties are supported in multiple ways. User plane and control plane are separated, with user plane being confined into a single versatile function, the UPF. Since a UE can connect to multiple UPFs simultaneously, dedicated UPFs can be assigned to a service and thus to a slice. This way entire user plane traffic of, e.g. a certain data network can be fully confined within that slice only.

Access and mobility management (AMF) is separated from session management (SMF). This enables the SMF to be fully confined into a slice while all UE specific functions can be common to multiple slices. Naturally different UEs can be assigned to different AMFs and in this way two level slicing can be achieved: common functions creating the first level and service specific functions the second level.

Subscriber data in the UDM are separated into access, mobility related and session related parts, thereby enabling to slice the subscription data as well. This matches the functional split between AMF and SMF and enables to assign corresponding parts of the UDM functions into same slices as their AMF and SMF counterparts, if so desired.

PCF can be assigned as a common entity, slice specific entity or both, i.e. some policies being common to all slices while others being slice specific. This likely needs to follow the assignment of the network functions that are the consumers of those policies. For example, a PCF should be assigned to the same slice to which the SMF using the QoS and charging policies provided by the PCF is assigned to, while a PCF instance providing access and mobility related policies can reside in the common functions together with the AMF.

While all the above network functions can be assigned to a slice, none of them is aware of that. This is the key architectural property of slicing, it allows for multi-tenancy without making the network functions multi-tenant aware.

The only functions aware of slicing are the NEF, NRF and the NSSF. What these functions have in common are that their focus is controlling which network function and instance of it should serve a request rather than providing the needed service by themselves. Thus, they can be said to be the "control plane of the control plane." NEF can be assigned as a common entity, slice specific entity, or both. Again, this structure should be defined according to the network functions whose capabilities are being exposed. Additionally, since the external domain is likely not aware of the slicing architecture of the network, a common NEF is needed to provide a single point of access toward the external domain and determining the correct slice where the slice specific NEF resides. This is where NEF needs to be aware of slices. The NRF is slice aware, and thus able to select the right network function instances for each slice. This is where the SBA can be effectively used by providing different services or different versions of services in different slices. The NRF acts as the single control point of this exposure. The NSSF can support the AMF in selecting the allowed network slice selection assistance information (NSSAI) and enables the network to control which slices are used for UE's PDU session(s).

Figure 4.27 shows an example of how network functions can be assigned to slices or applied as common functions.

In the example, there are three slices proving services. Slices #1 and #2 share a common AMF while slice #3 has a dedicated AMF. This means that on UE level the network is sliced into two, each UE assigned to only one of the AMFs. Those UEs assigned to the top AMF can access services from slices #1 and #2, while those assigned to the bottom AMF

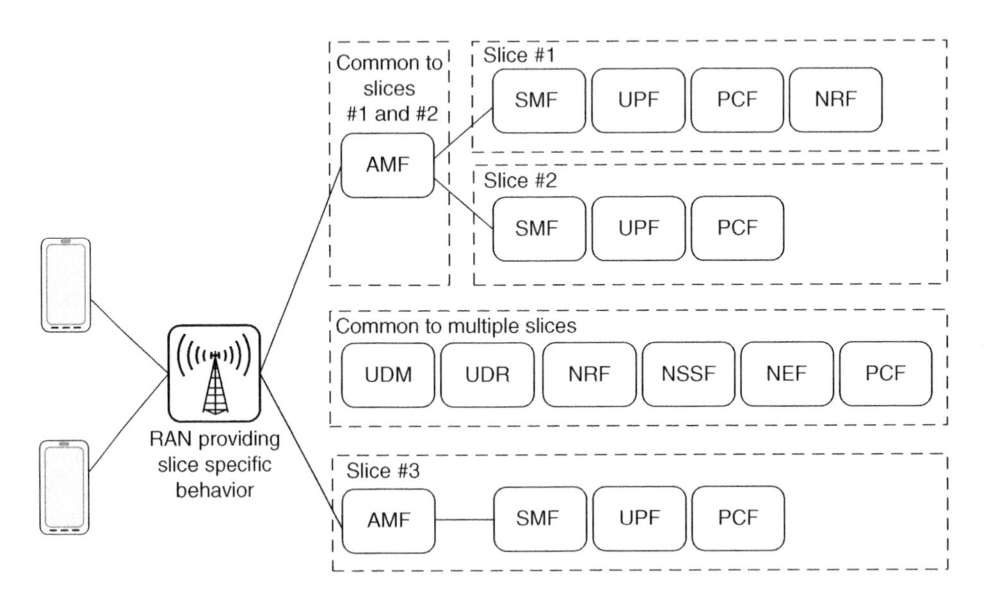

Figure 4.27 Network slice with 5G system.

can access services from slice #3 only. Each slice contains SMF, UPF and PCF instances. SMF related policies are thus controlled by the PCF within the slice while AMF and UE related policies are controlled by the PCF that is common to all slices. Slice #1 has its own NRF; this can be relevant, for example, in cases where the slice belongs to a tenant that manages slice specific instances. In this example, UDM, UDR, NSSF and NEF are common to all slices, with NRF being common to slices #2 and #3.

In 3GPP Rel-15, 5G RAN needs to support AMF selection based on slice identifiers provided by the UE. In addition, AMF and gNB exchange information which slices are supported per Registration Area via the N2 interface. Although Rel-15 enables RAN to be slice aware, slice specific behavior of 5G RAN was not specified in Rel-15 and is still open at the time of writing this book (planned for 3GPP Rel-16). However, the following RAN specific slicing features can be envisioned (some of them are implementation and deployment specific):

- In case of CU-DU split RAN architecture, CU control and user plane (CU-CP, CU-UP) resources can be dedicated to certain slices, or common to several slices. This allows the implementation of a high-throughput CU user plane instance just for an eMBB slice.
- Flexible deployment of CU-CP and CU-UP resources based on the needs of a given slice. This allows CU user plane location geographically close to the DU for URLLC slices (although one CU-UP can still serve many DU), while CU control plane is centralized.
- Slice aware and slice optimized scheduler in the DU. This allows the allocation of resources to specific slices based on the given SLA and dynamically shift resources from one slice to another, if there is the need and possibility (i.e. one slice does not consume the allocated resources that were promised to it) to do so.
- Support of virtual local area network (VLAN), other virtual private network techniques or DiffServ Code Point (DSCP) marking to separate slices on the backhaul interface between gNB and core network.

5G is targeted to be a natively virtualized system, enabling use of cloud paradigm for maximizing flexibility and automation. This way slicing becomes a cost-effective solution and proper tailoring of functions per each slice can be achieved.

4.7.4 Slice Selection

Slices to be used for an UE and it's PDU sessions need to be decided and controlled by the operator and need to correspond to the user's subscription. The 5GC provides UE(s) with rules associating applications on the UE with corresponding Data Network Names (DNN) and slicing information.

The selection of the AMF serving an UE should consider the various slices the UE may need to use. Since the initial AMF selection is done by the 5G-AN (e.g. gNB in NG-RAN) which is not aware of the subscription data, 3GPP has decided to use a UE-assisted slice selection paradigm where the network provides the UE with NSSAI. The UE relays this information to the 5G-AN, enabling the 5G-AN to select the correct AMF. The NSSAI is provided to the UE during initial registration.

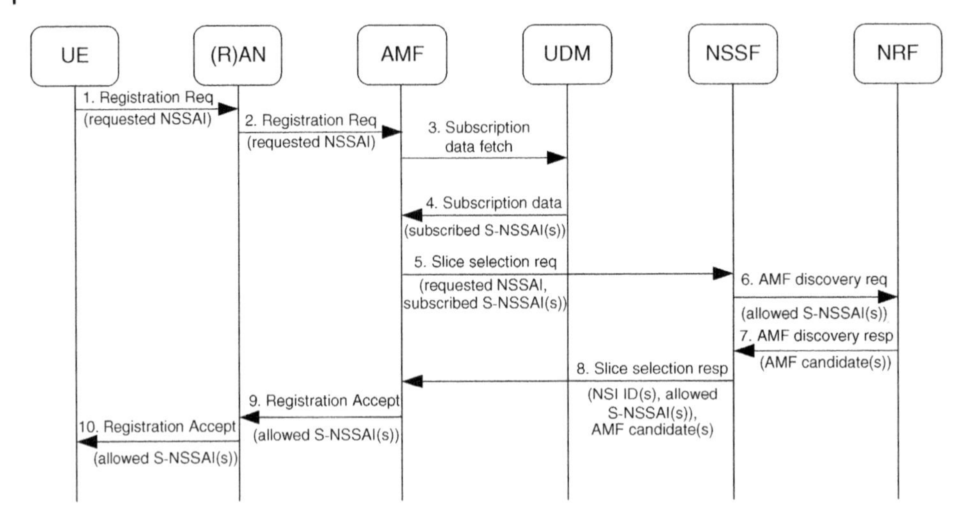

Figure 4.28 Network slice selection call flow.

A NSSAI is made up of one or more S-NSSAI each of which comprises the following:

- Slice Service Type (SST) indicating the slice characteristics.
- Slice Differentiator (SD) indicating further differentiation in case multiple slice instances comply with the expectations set by the SST. SD can be used to differentiate between slices of different tenants.

S-NSSAI can contain standard values or PLMN specific values; in the latter case the UE will not use the S-NSSAI in other PLMNs except the one to which the S-NSSAI is associated. Standard SST values can contain service indicators such as eMBB, URLLC or MIoT. SD values are not standardized.

While S-NSSAI indicates the characteristics expected, it does not necessarily uniquely identify a single slice as the network may also choose to use other additional information when selecting the slice. Furthermore, multiple S-NSSAI values can lead to select a single slice. This is consistent with the assisting nature of the NSSAI.

When requiring PDU session related resources in the 5G-AN, the 5GC provides the S-NSSAI corresponding to the PDU session to the 5G-AN. Thus, the 5G-AN can select resources and policies consistent with the S-NSSAI.

A slice selection message flow example when the UE is registering in the network is shown in Figure 4.28. Please note that the flow is simplified in terms of message names and contents details.

1. UE indicates requested NSSAI as part of Registration Request. The requested NSSAI is a configured NSSAI for the PLMN (i.e. provisioned to the UE) when accessing the PLMN for the first time.
2. RAN selects an AMF based on requested NSSAI and sends the request to the selected AMF.
3. AMF fetches subscription data from UDM, providing also the requested NSSAI

4. UDM returns the subscription data, including subscribed S-NSSAI(s) for those requested S-NSSAI(s) for which a subscribed S-NSSAI exists.
5. AMF requests NSSF to select a Network Slice Instance(s) (NSI) matching the subscribed S-NSSAI(s) and to provide information on AMF(s) capable of serving the allowed NSI(s).
6. NSSF discovers the suitable AMF(s) by providing allowed S-NSSAI(s) to NRF.
7. NRF provides AMF candidate(s) to NSSF.
8. NSSF provides NSI identifier(s), allowed S-NSSAI(s) and AMF candidate(s) to AMF.
9. AMF sends Registration Accept to RAN, including allowed S-NSSAI(s). In case a different AMF needs to be selected, the AMF processing the Registration Request redirects the request to an AMF that belongs to the AMF candidate set.
10. RAN sends the Registration Accept to UE. UE uses the provided allowed S-NSSAI(s) as requested NSSAI is subsequent procedures.

4.7.5 Interworking with EPS (e)DECOR

While actual network slicing is not supported in EPS, there is a similar concept for enabling a dedicated core network (DECOR) for a set of subscribers, as well as an enhancement of it (enhanced dedicated core network [eDECOR]). Service specific slicing, with UE simultaneously being able to access multiple slices, is not possible in LTE. RAN slicing is also not supported by LTE but some basic support for this may emerge, if suitable UE type or Core Network Identity (CNID) information is provided by the core to the RAN.

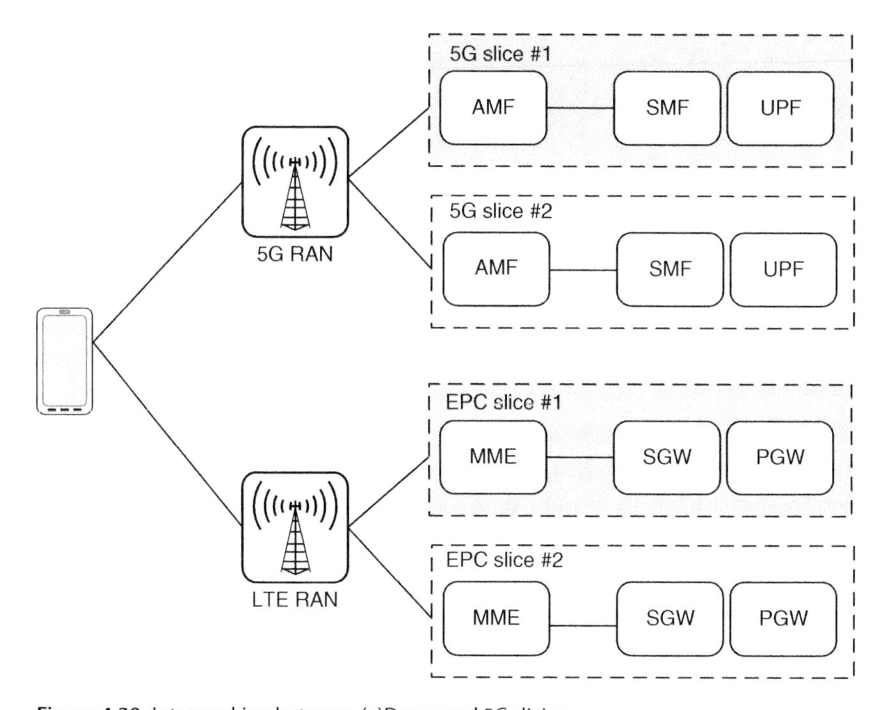

Figure 4.29 Interworking between (e)Decor and 5G slicing.

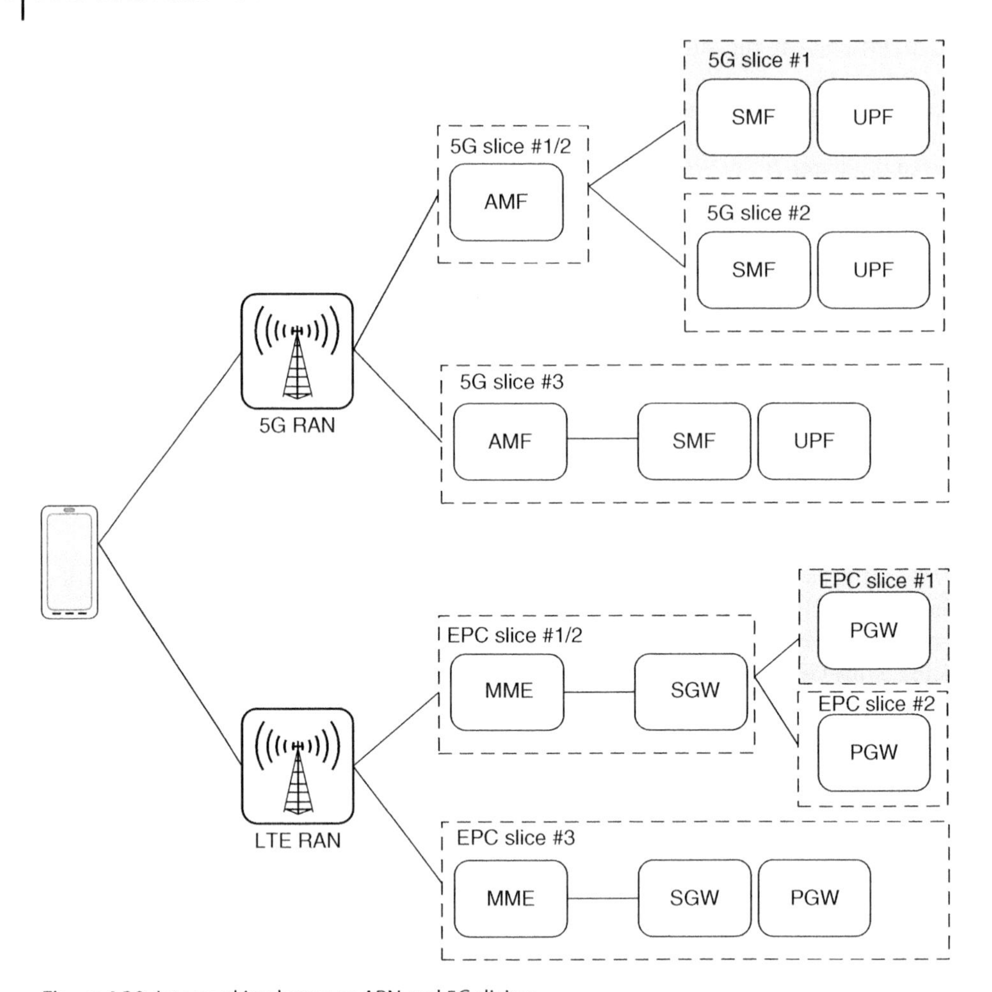

Figure 4.30 Interworking between APN and 5G slicing.

Given the different capabilities, 3GPP specified mechanisms to interwork with EPS networks that support or do not support (e)DECOR. A detailed overview of EPS interworking principles is provided in Section 4.11. These mechanisms can be used to provide inter-system handover with IP address preservation between EPC and 5GC in the following configurations:

- EPC consisting of a dedicated core network containing MME, S-GW and P-GW, and 5G slice consisting of AMF, SMF and UPF.
- EPC consisting of P-GW and 5G slice consisting of SMF and UPF (Figure 4.29)

In Figure 4.29, the UE IP address is preserved when UE moves from EPC dedicated core/slice #1 to 5G slice #1 or vice versa.

In Figure 4.30, the UE IP address is preserved when UE moves from EPC dedicated core/slice #1 (specific P-GW based on access point name [APN]) to 5G slice #1 or vice versa.

4.8 Multi-Access Edge Computing

Multi-Access Edge Computing (MEC) is an ETSI ISG (industry specification group), focusing on an architecture framework and environment for hosting performance-critical applications at the network edge (refer to [8]). Use cases such as video analytics, location services, augmented reality, optimized local content hosting and caching can benefit from very low latency. A set of MEC APIs are defined to expose network capabilities and services such as radio network information.

Figure 4.31 shows a high-level structure of the MEC framework, consisting of an umbrella MEC system-level management, a local host (edge cloud) level and the network infrastructure domain including 3GPP and other networks. The MEC host level includes virtualization infrastructure, which is hosting MEC applications and a common MEC platform. The platform offers application-enabling functions such as service registration and discovery, MEC APIs for information exposure and it interacts with the network layer including the 5G core.

The MEC local hosts (MEC platform and MEC enabled applications) are deployed in data centers at the network edge, i.e. close to the network access, possibly in the same data centers than the RAN functions.

MEC was created independently of 5G and many use cases can also be deployed in a 4G network environment. However, 5G is the first network generation specifically designed for MEC applications requiring mobility, lowest latency and service continuity at the same time. EPC networks can satisfy each of those needs individually but hardly their combination.

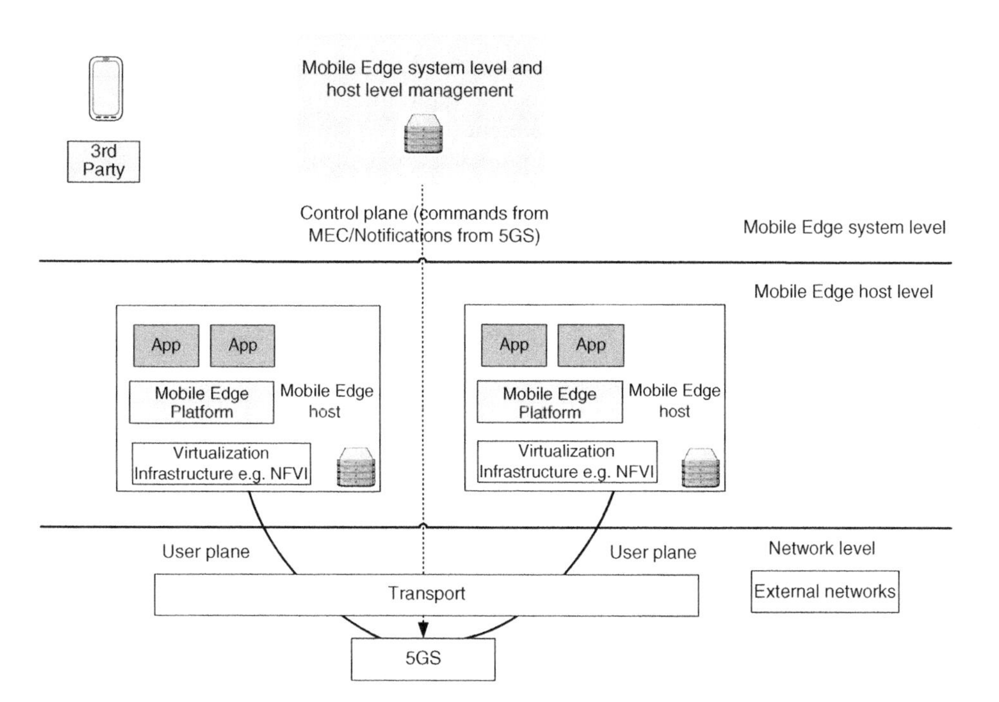

Figure 4.31 Multi-access edge computing framework.

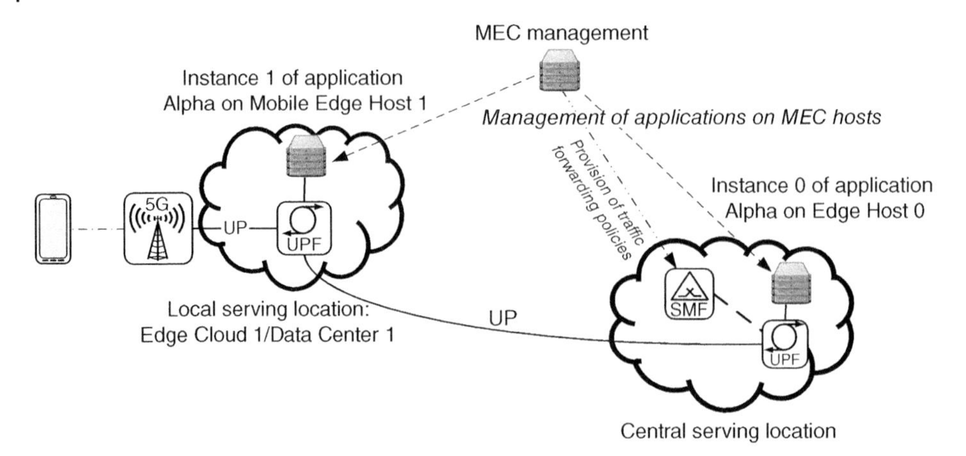

Figure 4.32 Session establishment and initial UPF selection.

MEC enabled applications (like any other application) are accessed by a UE via "PDU Sessions" providing user plane connectivity. "PDU Sessions" are further defined in Chapter 6.

As MEC is hosting applications at the network edge, it requires the capability to deliver user plane traffic of PDU Sessions to "local" application instances. In the network example shown in Figure 4.32, 5GC CP (Control Plane) functions are deployed in a centralized core cloud, whereas 5GC UPF are available in both the edge clouds (also hosting RAN higher layer functions) and in the core cloud. One or more UPF instances, e.g. a local and a central UPF, can be selected for a PDU Session depending on network policies and interactions with the MEC system-level management platform.

The MEC system-level management deploys application instances. It provides information to the 5GC PDU Session management (SMF) where application instances are deployed. This is to be used when selecting suitable UPFs so that these UPF can forward traffic related with an application to the closest deployed instance of that application.

The information on where instances of an application are deployed is provided by MEC as:

- Information to identify the application traffic (traffic filters).
- A set of "DNAI(s)" (Data Network Access Identifier) identifying where instances of the application are locally deployed. A DNAI can, for example, refer to a data center where local hosts and local applications, e.g. a MEC platform and local applications hosted by MEC, are deployed (possibly co-located with 5GC UPF(s) and RAN functions).

In the example of Figure 4.32, based on requirements received from the MEC system-level management and on UE location (5G-AN location) the SMF

- selects a local UPF-1 in an edge cloud close to the UE. The local UPF is responsible of forwarding selected uplink (UL) traffic to applications on the local MEC host. Such a local UPF may support an uplink classifier (UL CL) as defined in Section 6.3.1;
- configures traffic handling rules of the local UPF based on requirements received from MEC system-level management.

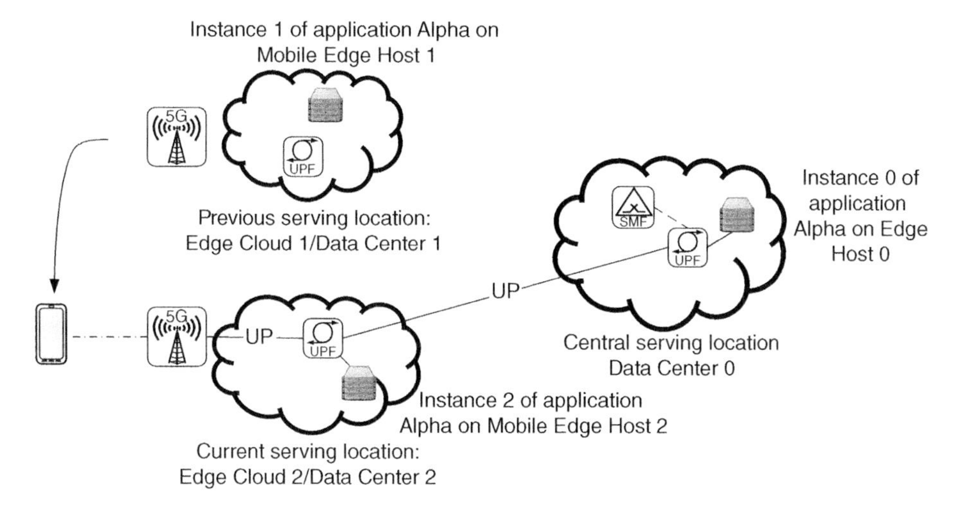

Figure 4.33 Reselection of UPF and application following UE mobility.

At this point the UE has an established connectivity to local services in the Edge Cloud (via UPF-1) as well as to the public Internet through the central Telco Cloud (via UPF-0).

Mobility may lead to a change in the (R)AN attachment, anchoring the UE in a different Edge Cloud (e.g. Edge Cloud #2 shown in Figure 4.33). The 5GC CP determines that another Edge Cloud 2 has become the new primary serving location for the UE, and hence selects UPF-2 as the new local UPF.

Further steps depend on the specific needs and characteristics of the application:

- Some applications are stateless, i.e. they do not require a specific UE-based context which allows to simply redirect the UE to a new application instance (App2 in this example) without any further interactions with the MEC platform.
- Stateful applications supporting application mobility need to transfer the UE context to the new application instance. Hence, the 5GC needs to notify the MEC environment about the UE mobility (transfer from source to target DNAI) so that the MEC environment can organize the transfer of the UE context from the application instance in the source DNAI to the application instance in the target DNAI.
- For stateful applications not supporting application mobility, the SMF keeps the connectivity to the old instance of the application until the usage of the application in the UE is terminated.

4.9 Data Storage Architecture

4.9.1 Introduction

In a 5G System, the network is expected to manage different kinds of information:

1) Subscription data based on user's subscription contract which are expected to be semi-static.
2) Policy data based on operator's policies, subscription and dynamic events occurring in the network (e.g. current session, location, session status, etc.).

3) Information targeted for exposure toward external AFs for different purposes: application behavior adaption, application's ability to influence the network, dimensioning and statistics purposes.
4) Transient UE context that is created in each NF based on UE's registration and session in the respective NF.

Traditionally (e.g. in EPS), the above four different types of data were stored and processed independently. Also, compute and storage layer was already split in the case of subscription data, optional UDC architecture for the HSS with the Ud interface toward UDR. In the 5G System, one objective of Data Storage architecture was to specify a unified framework, especially for the storage layer. Also, the objective was to enable deployment of stateless functions for all NFs/NF services within 5GC to improve resiliency, reliability and availability of the system.

4.9.2 Compute-Storage Split

In the 5G System Architecture, decoupling of Compute from Storage enables support for improved resiliency of a NF, almost "near zero downtime" in terms of service interruption for the UE. It also saves NF processing resources (e.g. for mIoT devices). It allows unified data storage for a network function (i.e. multiple instances of a NF can share the same storage resource), thus enables independent dimensioning of the NF and storage resource. Therefore, the storage resource also serves as a repository (like an "one stop shop") to support network capability exposure and data analytics. Some benefits of having a common data storage layer are:

- Ability to support high reliability with less hardware footprint. Allows support for newer redundancy models: N + m Geo-Redundancy, rather than 1 : 1 mated-pair Geo-Redundancy.
- Introduces the ability to support dynamic run time load (re-)balancing with no impact to user's services.
- Opportunity to leverage contexts/state information for data analytics and internal and external capability exposure, e.g. toward edge applications.

Ability to support higher reliability is achieved by introduction of the UDSF in the 5G System Architecture with the caveat that the data stored in the UDSF are "opaque." This implies that the data stored in the UDSF can only be understood and interpreted by the NF (and the NF instances from the same vendor) storing it. The structure of the data stored in UDSF is not specified by 3GPP.

Opportunities to use the context data for data analytics, internal and external capability exposure is enabled by introduction of the UDR in the 5G System Architecture. Data stored in the UDR are standardized allowing NFs (NF instance from multiple vendors) to store and retrieve {subscription, policy, application, exposure related} data from UDR.

It is an implementation decision whether UDSF and UDR are independent functions or collocated.

4.9.3 What is "Stateless"? How "Stateless" is "Stateless"?

In simple words, the term "stateless" is used to refer to a state of a compute function (in the context of the 5G System) that does not store UE context within its cache (i.e.

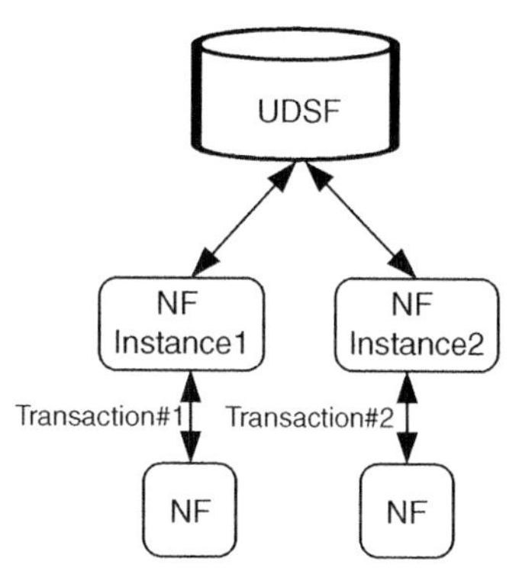

Figure 4.34 No state NF.

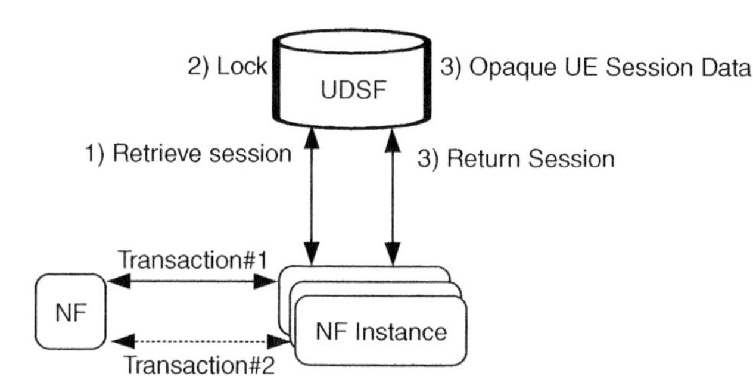

Figure 4.35 Stateless NFs.

for a certain duration when it remains stateless). How long this duration is, depends on the type of Network Function, ranging from a NF that requires no state to a NF that is completely stateless and a NF that is always stateful. Broadly, there are four possible implementation choices for a Network Function in the 5G System:

1) *No state* (Figure 4.34). Such NFs do not store UE state information (UDM frontend connected to one database). This allows NF consumers to select any NF instance for processing a UE transaction.
2) *Stateless NF* (Figure 4.35). Such NFs hold UE state only for the duration of a transaction, and any NF instance can be selected to process the UE transaction. This allows any NF instance to process UE transactions, thus enabling high scalability and resiliency. However, the risk of race conditions due to two consumers (frequently) selecting a different NF instance for processing the same UE transaction increases (possibly have an impact on system performance).
3) *State efficient NF* (Figure 4.36). The NF holds UE state during periods of frequent activity (e.g. for some seconds or minutes while the UE is in connected state) in the middle of a certain procedures, and it stores UE state in the Storage resource when the UE is inactive (e.g. no control plane procedure ongoing). This implies all transactions within a certain procedure for a given UE are expected to be processed by the same NF instance. It stores stable state in the UDSF at the end of a certain procedure. When the Producer NF does not hold state, it has the freedom to release the long-lived association with a Consumer NF instance for a given UE. Thus, after the release of an association, the Consumer NF can select any NF instance for

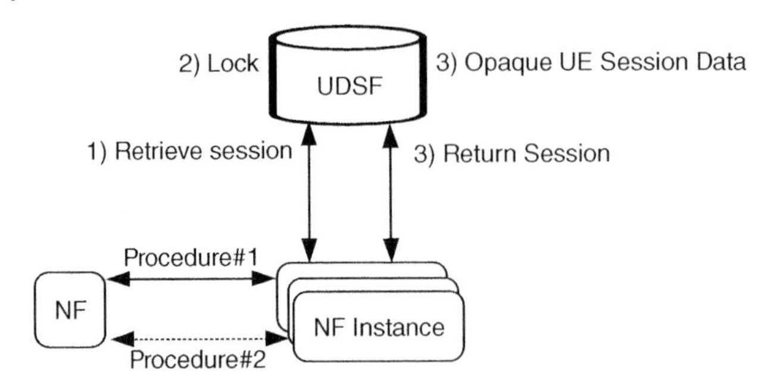

Figure 4.36 State-efficient NFs.

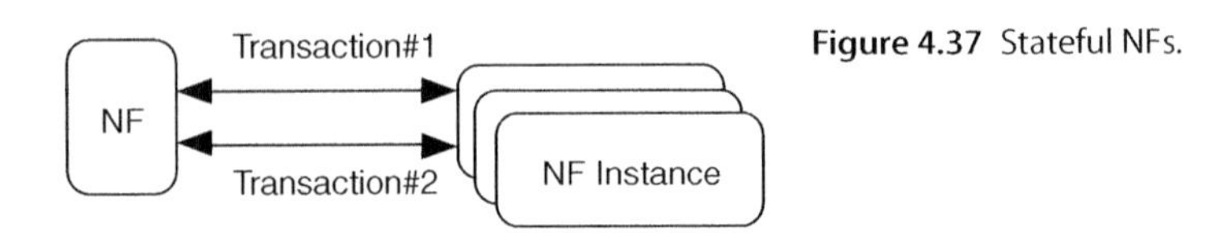

Figure 4.37 Stateful NFs.

processing UE transaction, however since the frequency of NF instance selection is limited compared to purely stateless NF configuration, the probability of race conditions (i.e. two different NF consumers selecting a different NF instance for processing a certain UE transaction) is minimized. This configuration allows change of NF instances for resource scaling up and down, NF planned maintenance, NF failure handling, etc. Thus, it enables scalability up to "n" NF instances processing UE transactions.

4) *Stateful NF* (Figure 4.37). The NF holds UE state information permanently.

Depending on the type of NF and deployment needs, a wide range of implementation configurations are possible. Table 4.1 summarizes the four different implementation options described above.

Compute-Storage split alone is not sufficient to enable a scalable NF (i.e. N + 1 scalability), especially for NFs like AMF, SMF. As illustrated above, it is also the behavior of the NF that it exhibits toward its peer NFs that matters, e.g. whether the peer NF can invoke any NF instance or invoke the same NF instance for processing a certain transaction, to determine whether the NF instance is stateless with N + 1 scalability.

Generally, for NFs like AMF, SMF, state-efficient (#3 described above) NF implementation options might be reasonable. This implies that the NF instances are transaction-stateful (i.e. efficiently access context/states stored within the cache in the middle of a certain procedure) but stateless outside the procedure (i.e. stateless after the completion of a certain procedure with the need to retrieve externally stored context/states only once in the beginning of the procedure). For support of stateless NF or state efficient NF, some standard enablers are required as this also impacts interoperable interfaces between producer and consumer NF, but it also allows scalability up to "n" NF instances. In 3GPP Release 15, this was specified mainly for the AMF. For NFs like UDM and AUSF, stateless NF or no-state NF kind of implementation options might be reasonable.

Table 4.1 NF types in terms of states.

	No-state NF	Stateless NF	State-efficient NF	Stateful NF
Select NFs at transaction start	Yes	Yes	Yes	No
Interactions with UDSF	None	Highest, context stored at the end of every transaction	Medium, stable state stored at the end of a procedure, e.g. Registration, Service Request, PDU Session	None
Supports failure	Yes	Yes	Yes, but the transient UE session may be lost	No, but needs specialized recovery procedures
Scale-in (NF removal from service)	Yes	Yes	Yes	No – sessions must be moved
Scale-out with load balancing (LB)	Yes	Yes – LB for all sessions	Yes – LB for new UE activity	No
Candidate NFs	UDM FE	PCF	AMF, SMF	

4.9.4 AMF Resiliency and State-Efficient AMF

In the first 3GPP release of the 5G System, significant effort was undertaken to specify standard enablers for state-efficient AMF and AMF resiliency. The reason for starting with the AMF was that the AMF maintains associations to the UE, RAN and other NFs. Thus, some of the enablers, e.g. temporary identifier encoding for the UE and its semantics, presence of certain fields in AS/NAS messages to allow for a state-efficient AMF, had to be introduced in the very first release of the new system to ensure that this is working with all 5GS UE(s) in the field from the very beginning of 5G introduction. Such identifiers, (fields within AS, NAS messages) that are critical to basic connectivity and functioning of the system cannot be changed in subsequent releases easily without adverse impact. Thus, it was important and critical to introduce the necessary standard enablers for support of state-efficient AMF in the very first 3GPP release as part of clean slate 5G System architecture to ensure forward compatibility.

The 5G System standards specify enablers for implementations to support configuration #4 (see above) for state-efficient AMF. The concept of an AMF Set was introduced to allow load sharing and scalability up to n AMF instances (where n = number of AMF instances within an AMF Set) for processing UE transactions. An AMF is referred to by the AMF Pointer. This allows load balancing of AMF instances sharing the same AMF pointer value. Also, an AMF pointer can point to one or more AMFs. An AMF set comprises of AMFs that can store and retrieve data from the same data base (UDSF). The AMFs within the same AMF Set are assumed to support the same network slice(s). An AMF Set is identified by the AMF Set ID. Figure 4.38 illustrates the AMF structure.

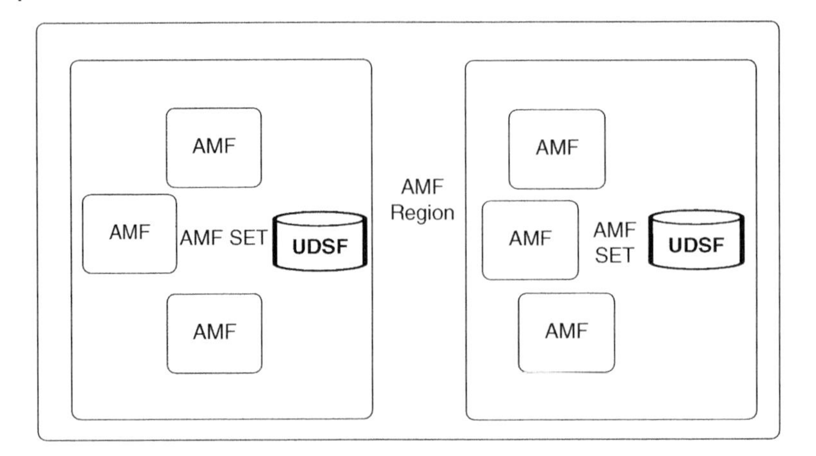

Figure 4.38 AMF structure.

The AMF assigns a Temporary Identifier (5G-GUTI) to the UE for supporting subscriber confidentiality. The AMF leverages the same identifier to include routing information that the 5G-AN node can use to determine the target NF instance(s) for NAS messages. The UE derives the 5G-S-TMSI including the routing information (AMF Set ID and AMF Pointer value) from the 5G-GUTI assigned by the AMF, and provides it in the RRC message toward the 5G-AN. The 5G-AN uses the routing information present in the 5G-S-TMSI and routes the NAS message accordingly.

An AMF can advertise a distinct and/or common address toward the 5G-AN: The standard specified the ability for an AMF to provide a distinct AMF Pointer value (that identifies a specific AMF) and a common AMF Pointer value (that is shared by multiple AMFs within an AMF Set) toward the 5G-AN over the N2 interface using next generation application protocol (NGAP).

AMF Set ID and AMF Pointer value can be included within the 5G-S-TMSI provided by the UE. The AMF assigns the 5G-GUTI including the GUAMI which itself includes AMF Set ID and AMF Pointer value. If the AMF wishes to remain stateless, it can allocate a 5G-GUTI with a common AMF Pointer value.

When the UE sends an RRC establishment request, it includes the 5G-S-TMSI including AMF Set ID and AMF Pointer value. The 5G-AN can use the AMF Set ID and AMF Pointer value in the 5G-S-TMSI to determine how to route the encapsulated NAS message toward the AMF. If the AMF Pointer value provides a distinct address for a specific AMF, then the 5G-AN will forward it to this specific AMF. If the AMF Pointer value maps to multiple AMFs, then the 5G-AN can forward it to any of these AMFs. It can use the AMF Set ID to select any AMF within the AMF Set, if the specific AMF has failed or reported out-of-service. This allows for scalability up to an AMF Set.

Figure 4.39 illustrates the principles above.

Ability to release signaling (per UE TNL – Transport Network Layer) association was introduced to avoid long lived associations between 5G-AN and a specific AMF instance for a given UE in connected (RRC_CONNECTED/CM-CONNECTED) state. An AMF

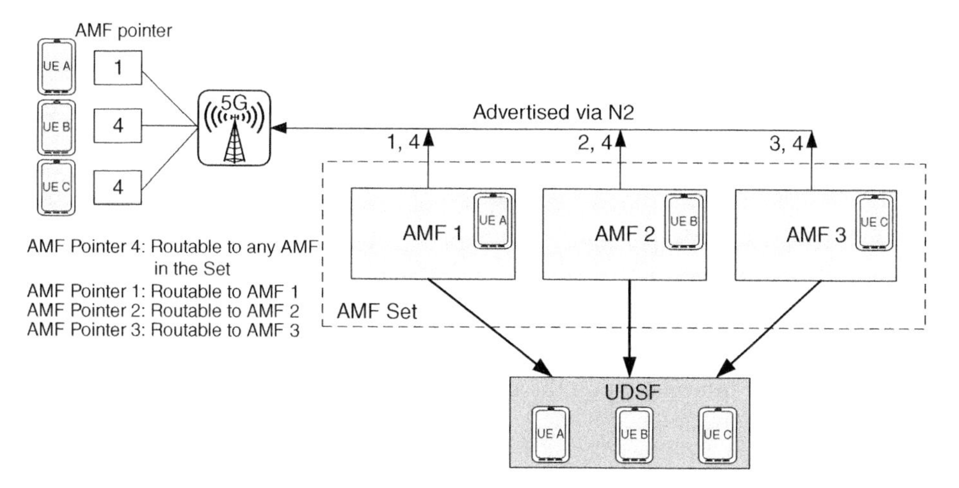

Figure 4.39 Routing N1/N2 messages to any AMF.

can release "per UE" TNL association between AMF and 5G-AN without releasing UP connectivity and/or RRC connectivity for the given UE. This allows the AMF to become stateless or state efficient enabling the AN to select any AMF from the AMF Set (even for UE(s) that are in RRC_CONNECTED state) for subsequent N2 messages targeted for a given UE.

Ability for other CP NFs (e.g. SMF, PCF, UDM) to select a new AMF instance when the AMF releases the signaling association (or it has failed or reported out-of-service) was also introduced. Control plane NFs can also select another AMF instance for processing the UE transaction.

If there is a race condition between mobile originating (MO) and mobile terminating (MT) transactions, e.g. an AMF instance is already processing a UE's MT transaction while another AMF has been selected for a MO transaction, then the wrong (newly selected) AMF is expected to forward the transaction to the right (currently serving) AMF for processing the UE transaction and respond accordingly to the requester.

In the past generations (i.e. EPS, general packet radio service [GPRS]), these enablers were not present thus it was not possible to support "zero downtime" for MME without impacting the UE's service and/or incurring excessive signaling. To avoid service disruption, fully redundant NFs (1 : 1 mated-pair) had to be deployed. With the 5GS enablers for stateless/state-efficient AMF, operators can avoid deploying fully redundant NFs at the same time avoid service disruption and support scalability up to n AMF instance, where n is the number of AMFs within an AMF Set. More importantly, this introduces the ability to support "dynamic run time load balancing" between AMFs with no service disruption and it also avoids heavy signaling traffic toward the UE. This avoids also concurrent signaling toward millions of UE(s) for AMF load re-balancing.

Detailed descriptions of the technical solutions and procedures for support of state-efficient AMF, AMF resiliency, N2 management, AMF planned removal and AMF load re-balancing can be found in 3GPP TS 23.501 [1] and TS 23.502 [2].

4.10 Network Capability Exposure

4.10.1 Introduction

With network capability exposure via APIs provided by the NEF, operators can give applications (either operator's own or third-party applications) access to network capabilities affecting the services provided to the users. This is intended to provide additional value to the application by optimizing the application and/or network behavior and to help operators to better monetize their network. Such interactions may be limited by a SLA between operator and third-party service provider.

Network capability exposure allows support for following features:

- Operator can get information on the UE behavior (e.g. traffic patterns) to tune the network accordingly.
- An application can monitor UEs by getting, e.g. information on the location of the UE, on the change of association between UE and serving location (e.g. an edge data center) or on the reachability of the UE.
- To support application instances deployed in the edge influence the flow of user plane traffic and user plane connectivity, the network exposes requested information (e.g. regarding mobility events) toward the application. This allows proper selection and configuration of UPFs in the network.
- An application can request an application trigger to be sent to an application instance running on a UE or in the network; this can trigger the UE to establish communication with the proper application instance in the network.
- An application can "sponsor" some data (like advertisement) ensuring that the corresponding data transmission is not charged to the end-user.
- An application can provide descriptors, e.g. Packet Flow Descriptors (PFDs), and forwarding rules corresponding to an application identifier.
- A third party can negotiate cheaper usage of the network but only in some conditions, e.g. based on time of day or based on the location where the service is delivered.

4.10.2 Bulk Subscription

The NEF is burdened with significant overhead resulting from per UE events:

1) Subscription events need to be supported for every UE.
2) Serving NF needs to be determined for every UE.
3) Mobility events need to be managed as the serving NFs (e.g. AMF) are modified for every UE.
4) Identifiers must be maintained on a per UE basis.

To reduce this overhead, the 5G System supports capability exposure using bulk subscriptions to reduce signaling and to reduce the overhead incurred due to "per UE" subscription events (i.e. there is no need to keep track of serving NF for a given UE using bulk subscription). This allows the NEF to subscribe with registered NF instances for a set of events targeting a group of UEs or any UE. The two main principles are:

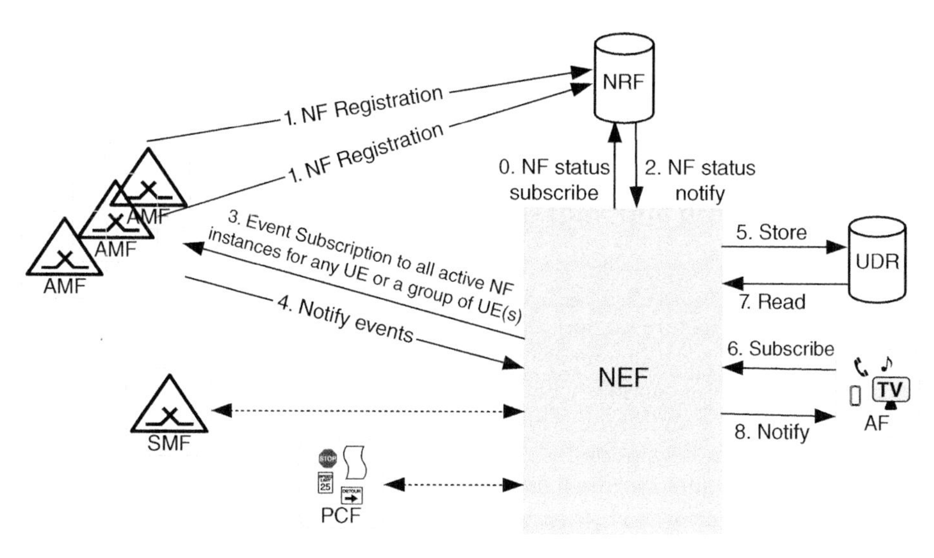

Figure 4.40 Network capability exposure with bulk subscription.

1) Exposure subscription for one UE, group of UE(s), any UE with the respective NFs, e.g. for all UE(s) served by an AMF.
2) Exposure Subscription from the time of NF instantiation, e.g. immediately after AMF instantiation. This is to ensure that no event is missed by the NEF.

Figure 4.40 illustrates the exposure procedure with bulk subscription:

4.10.3 NEF Capabilities

Network capability and event exposure are key enablers to support network programmability as defined in Section 4.2.6.

In case an external application is seeking to use Network capability exposure and Event exposure services, the NEF acts as the network interface to the external application and is responsible for:

- Authenticating the application.
- Authorizing the application; this may include checking the SLA for the application. This corresponds to a global authorization for the application to invoke network APIs; per-UE authorization is performed by the UDM upon request of the NEF.
- Issuing usage data collection to keep track of the application requests and to potentially charge the provider of the external application.
- Support for bulk subscription with all NFs or individual subscription with the targeted NF for exposure events related to a single UE, group of UE(s) or any UE.
- Translation of parameters, e.g.:
  - o Translating between internal location information (Tracking Area and/or cell ID) and geographical information e.g. latitude and longitude.
  - o Using knowledge on the application to determine network parameters such as the slice or the DNN that is the target of the application request.
  - o Translating the external identification of the UE into the PLMN identification of the UE (SUPI); for this purpose, the NEF may need to invoke the UDM.

In summary, the NEF has the role of an API gateway toward the 5G network, controlling access to network capabilities, and hiding internal network details such as serving network functions and internal identifiers from external applications.

4.11 Interworking and Migration

4.11.1 Background

As stated in the architecture section, the 5G System natively integrates various access technologies, thus the focus of interworking and migration was mainly to ensure seamless mobility and service continuity between existing deployed EPS networks and the new 5G System. Note seamless interworking with IP address preservation between 5G System and 2G/3G is not supported.

As with any new technology, initial deployments are expected in small regions within a country (urban scenarios) and coverage will be sparse. Besides that, following are the additional challenges faced with the development of an interworking and migration solution for the 5G System:

1) There are many variants expected for 5G network deployments.
2) The 5G Radio is expected to be deployed in high and low bands. In high bands, the UE will lose 5G coverage quite often. Thus, support for an efficient interworking solution is essential.

To ensure that various forms of mobility and coverage scenarios are considered, two levels of interworking are supported:

1. *System level interworking with EPS.* Interworking supported between NG-RAN/5GC and LTE/EPC with or without a signaling reference point. This is explained in Section 4.9.2.
2. *Access Network interworking.* Interworking supported between ePDG/EPC and Non-3GPP Interworking Function (N3IWF)/5GC for UE(s) connected via untrusted non-3GPP access. This is explained in Section 4.11.3.

4.11.2 Migration from EPS Toward 5GS

Deployments based on different 3GPP architecture options, EPC based or 5GC based, and UEs with different capabilities (supporting EPC NAS and 5GC NAS) may coexist at the same time within one PLMN.

The ability to deploy two radio technologies in the same spectrum simplifies migration from EPS toward 5GS. How this is done is explained in Chapter 2.

Figure 4.41 illustrates the migration support from EPS to 5GS.

To support smooth migration, following are some assumptions that have been made regarding UE and network support:

1) EPC and 5GC have access to a common subscriber database, that is HSS in case of EPC and UDM in case of 5GC, acting as the master data base for a given user.
2) A UE that is capable of supporting 5GC NAS procedures may also be capable of supporting EPC NAS to operate in legacy networks, i.e. in case of roaming.

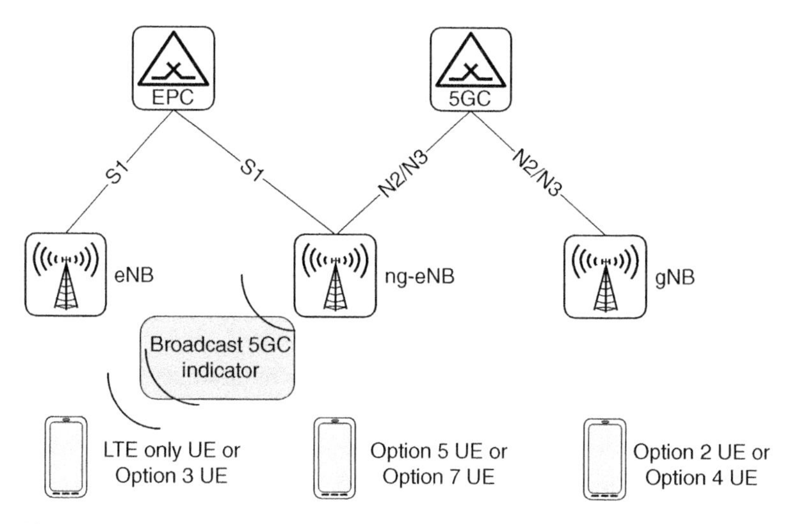

Figure 4.41 EPS to 5GS migration.

3) The UE can use EPC NAS or 5GC NAS procedures depending on the serving core network it obtains services from. The procedure for the UE to select EPC versus 5GC NAS is specified in 3GPP TS 24.501 [12].
4) A UE that supports only EPC based Dual Connectivity with secondary RAT NR
   - always performs initial access through E-UTRA (LTE-Uu) but never through 5G NR;
   - performs EPC NAS procedures over E-UTRA (i.e. Mobility Management, Session Management, etc.) as defined in 3GPP TS 24.301 [13].
5) A UE that supports camping on the 5G System with 5GC NAS
   - can perform initial access either through E-UTRA or 5G NR connected to 5GC; or
   - can perform initial access through E-UTRAN toward EPC, if supported and needed.

Supporting different UEs with different capabilities in the same network, UEs that are capable of only EPC NAS (possibly with EPC based Dual Connectivity with secondary NR) and UEs that support both EPC NAS and 5GC NAS procedures in the same network, is illustrated in Figure 4.42. The figure shows the principles that apply for ng-eNB and UE to enable appropriate system selection and routing.

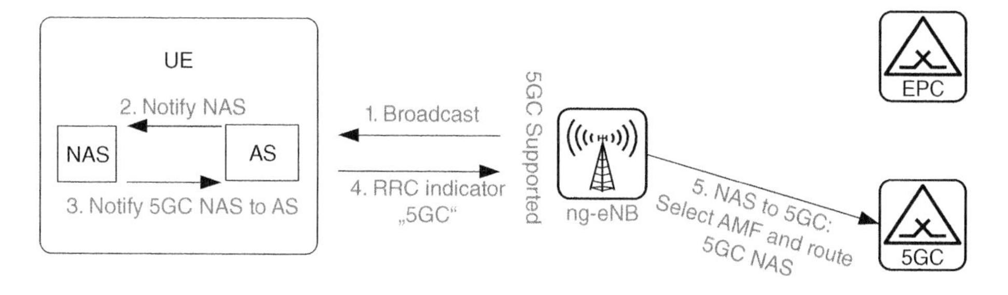

Figure 4.42 EPS to 5GS migration – system selection and routing.

The following steps are performed:

1) The eNB connected to 5GC will broadcast this information via the air interface. Based on that, the UE AS layer indicates "E-UTRA connected to 5GC" capability to the UE NAS layer.
2) The UE NAS layer selects 5GC NAS or EPC NAS based on network support and operator policies as specified in 3GPP TS 24.501 [12].
3) Upon selection of EPC versus 5GC NAS, the UE AS layer is informed by the NAS layer whether a NAS signaling connection must be initiated toward 5GC.
4) Based on that information, UE AS layer indicates to the RAN whether it is requesting 5GC access (sends "5GC requested" indication).
5) The RAN uses this indication to determine whether a UE is requesting 5GC or EPC access. The RAN routes NAS signaling to a MME or AMF accordingly.

4.11.3 System Level Interworking with EPS

To support seamless service continuity between 5GS and EPS, an interworking architecture requires

- a common UPF that serves as an IP anchor so that the UE IP address can be preserved;
- a combo SMF/P-GW-C function. UPF and SMF are regarded as PGW-U and PGW-C, respectively, from the viewpoint of a S-GW;
- a common subscription database HSS/UDM; and
- a signaling interface N26 between MME and AMF for context transfer. N26 is expected to be based on S10 interface to allow interworking with a MME that will not be upgraded to support interworking with the 5GC.

Figure 4.43 shows the interworking architecture:

The interworking solution was specified to cater for different deployment needs:

1) Interworking between EPC and 5GC using N26 interface between MME and AMF.
2) Interworking between EPC and 5GC without any direct interface between EPC and 5GC network functions.

Furthermore, the solution supports two different kinds of UEs:

1) *Single Registration Mode UE* (Figure 4.44). This UE has only one active NAS state (either RM state in 5GC or evolved packet system mobility management [EMM] state in EPC) and it is either working in 5GC NAS mode or in EPC NAS mode depending on whether it is connected to 5GC or EPC. The UE maintains a single coordinated registration state for 5GC and EPC. To enable re-use of a previously established 5G security context when returning to 5GC, the UE also keeps the native 5G-GUTI and the native 5G security context when moving from 5GC to EPC.
2) *Dual Registration Mode UE* (Figure 4.45). The UE can handle independent registrations in 5GC and EPC. In this mode, the UE maintains two NAS states independently. It may be registered to 5GC only, EPC only, or to both 5GC and EPC at the same time.

While Dual Registration mode is optional for the UE, Single Registration mode is mandatory for the UE in 3GPP Rel-15. Networks that support interworking with EPC, may support interworking procedures with or without N26 interface. Interworking

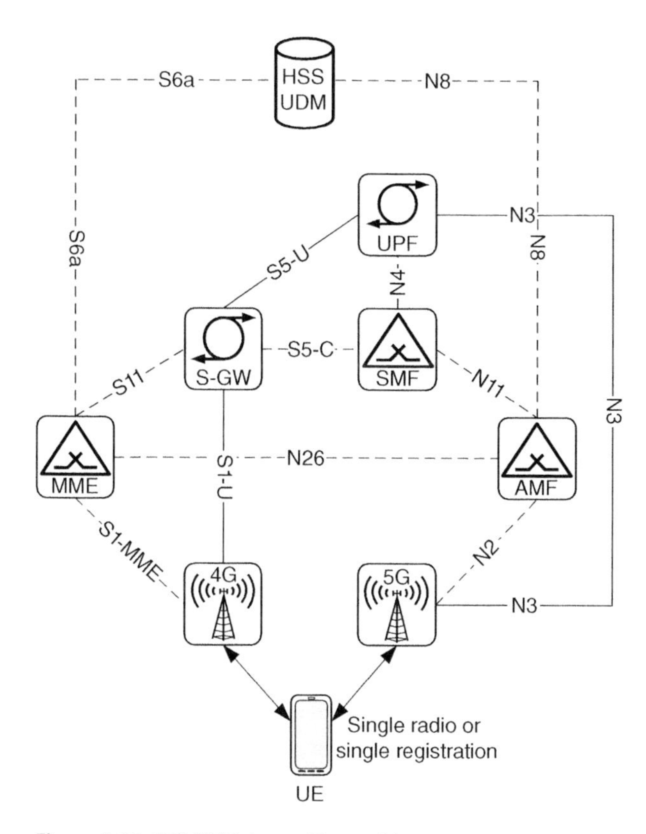

Figure 4.43 5GS-EPS interworking architecture.

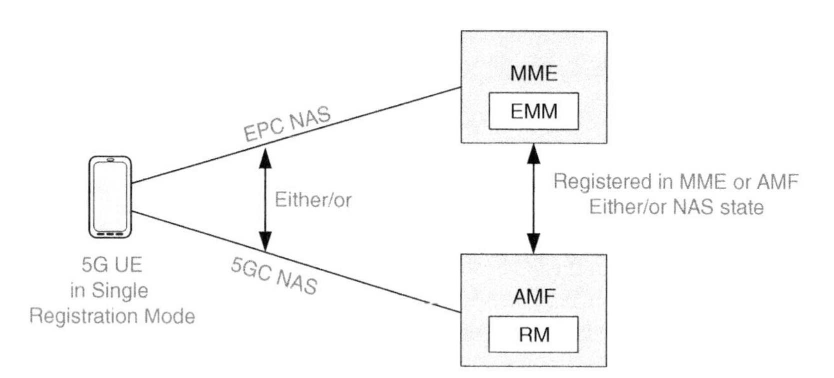

Figure 4.44 Single registration mode UE.

procedures using N26 are providing IP address continuity on inter-system mobility for UEs that support 5GC NAS and EPC NAS. Networks that support interworking procedures without using the N26 interface, also support procedures to provide IP address continuity on inter-system mobility for UEs operating in single-registration and dual-registration mode.

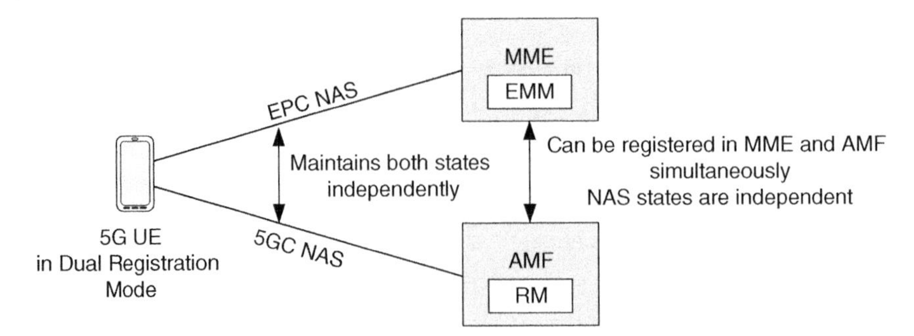

Figure 4.45 Dual registration mode UE.

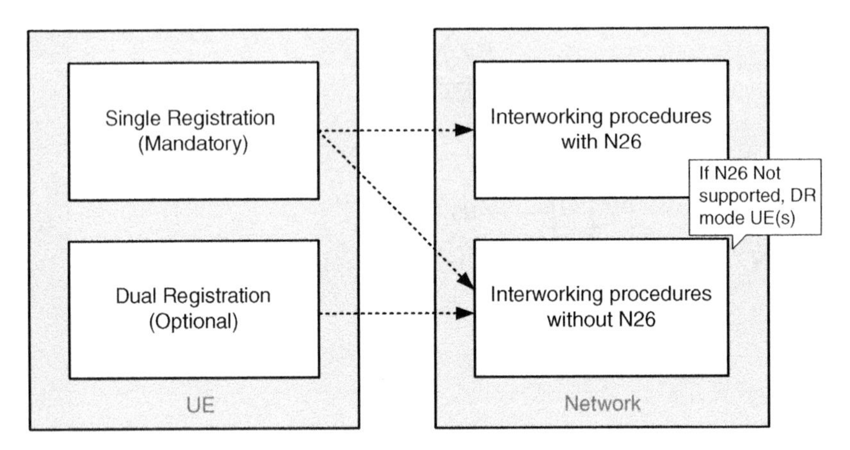

Figure 4.46 UE and network support.

Figure 4.46 summarizes the (mandatory, optional) capabilities supported by a UE and the default mode of operation that can be expected from the network for EPS interworking.

The following list provides an overview of some high-level principles adopted for interworking with EPS:

1) Interworking with MME is based on the S10 interface used within EPC at inter MME mobility. An MME (that is not upgraded) assumes that UE is moving from another MME when the UE moves from the 5G System to EPS. With this approach, the MME treats the AMF as if it was an MME. The AMF acts as an MME toward the MME. This avoids mandatory and/or non-backward compatible enhancements for an MME at the protocol level.
2) AMF needs to ensure that MME does not assume the UE was previously served by a serving general packet radio service support node (SGSN). After it moves from 5GC to EPC and if the MME would assume the UE was served by a SGSN, security is assumed to be based on 3G – 4G interworking which is less inferior. Also, this would lead to an incorrect context retrieval request toward SGSN.
3) Retain MM/security contexts in the AMF for a configured period to allow support for interworking using the native security context and at the same time support security separation.

4) Minimize signaling to and from HSS/UDM in case of frequent ping-pong of the UE between 4G and 5G coverage.
5) Minimize impacts to the MME but allow essential (optional) backward compatible enhancements and (e.g. for simplification in the context transfer and improved security) to be specified for an optimal interworking with an MME that can support these optional enhancements. One such justified reason for enhancements – improved security, security separation between EPS and 5GS (i.e. to avoid attack in one system causing an attack also in the other system). Necessary and well justified evolution (e.g. for security reasons) should be possible and not prohibited.

PDU Session types "Ethernet" and "Unstructured" are transferred to EPC as "non-IP" PDN type (when supported by UE and network). The UE sets the PDN type to non-IP when it moves from 5GS to EPS and after the transfer to EPS, the UE and the SMF shall maintain information about the PDU Session type used in 5GS, i.e. information indicating that the PDN Connection with "non-IP" PDN type corresponds to PDU Session type Ethernet or Unstructured respectively. This ensures that the appropriate PDU Session type will be used when the UE moves and transfers PDU Sessions to 5GS.

4.11.3.1 Understanding Terminology

Since there are many related concepts discussed such as dual connectivity, dual registration, dual radio etc., Table 4.2 provides a comparison and summary of these concepts to put things in the right perspective.

4.11.4 Interworking Between EPC and 5GC Using N26

Interworking procedures using the N26 interface enable the exchange of MM and SM states and the security context between source and target network. The UE operates in single-registration mode.

The support for N26 interface between AMF in 5GC and MME in EPC is required to enable seamless session continuity (e.g. for voice services) for inter-system change. It allows for fast handover and minimized service interruption. When the UE moves from 5GS to EPS, the SMF determines which PDU sessions can be relocated to EPS, e.g. based on capability of the deployed EPS, operator policies for which PDU sessions, seamless session continuity should be supported etc. The SMF can release the PDU sessions that cannot be transferred as part of the handover process to EPS. However, whether the PDU Session is successfully moved to the target network is determined by the target EPS.

When the UE supports single-registration mode and the network supports interworking with the N26 interface, the following principles apply:

- In case of idle-mode mobility from 5GC to EPC, the UE performs TAU (Tracking Area Updating) procedure with 4G-GUTI mapped from 5G-GUTI. The MME retrieves the UE's MM and SM context from 5GC, if the UE has a PDU session established or if the UE or the EPC support "Attach without PDN connectivity." The UE performs an Attach procedure, if it is registered without PDU session in 5GC and the UE or the EPC do not support Attach without PDN connectivity. For connected-mode mobility from 5GC to EPC, inter-system handover is performed.
- In case of idle-mode mobility from EPC to 5GC, the UE performs Registration procedure with native 5G-GUTI and mapped 5G-GUTI (mapped from 4G-GUTI). If the

Table 4.2 Used terminology.

Concept	Dual connectivity	Dual registration	Dual radio
Core registration	Single IMSI, single registration	Single IMSI, dual registration; UE can register in both networks independently and the network is required to maintain the two registrations without canceling them as part of HO	Dual registration (single or dual credential)
RAN-Core interface	Single	One at a time	One per credential
RRC interface	RRC mainly anchored using the Master RAT (including containers over Master RRC); Light RRC for Secondary RAT	Single RRC	Dual RRC
Services	Allows services to be supported in different RATs	Allows services to be supported in different systems	Allows services to be supported in different RATs
Paging	Listens to paging in the Master RAT only	Depends on UE's radio capability; 3GPP solution does not require the UE to listen to paging in both RATs	Listens to paging in both RATs
NAS states	One NAS state only	One NAS state per registration! UE maintains the NAS states independently	One NAS state per registration! UE maintains the NAS states independently

AMF can retrieve the stored security context (e.g. from its own cache, from UDSF or from another AMF) using the native 5G-GUTI, the AMF uses this security context as opposed to the security context retrieved from EPC. In addition, AMF and SMF retrieve the UE's MM and SM context from EPC. For connected-mode mobility from EPC to 5GC, inter-system handover is performed.

4.11.5 Interworking Between EPC and 5GC without N26

For interworking without the N26 interface, IP address continuity is provided for the UE on inter-system mobility by storing and fetching the combined PGW-C + SMF address and corresponding APN/DDN information via the combined HSS + UDM. Such networks also provide an indication that "interworking without N26" is supported to UEs during initial Registration in 5GC. This indication is valid for the entire PLMN. UEs that support dual-registration mode may use this indication to decide whether to register "early" in the target system, i.e. before the mobility occurs.

When the UE supports single-registration mode and the network supports interworking procedure without N26 interface, the following principles apply:

- In case of mobility from 5GC to EPC, the UE may perform TAU procedure with 4G-GUTI mapped from 5G-GUTI. The MME determines that the old (source) node is an AMF, rejects the TAU with a "Handover PDN Connection Setup Support" indication to the UE. Based on this indication, the UE may perform Attach in EPC with "handover" indication in PDN Connection Request message and it subsequently moves all its other PDU sessions from 5GC to EPC using the UE initiated PDN connection establishment procedure with "handover" flag.
- In case of mobility from EPC to 5GC, the UE performs Registration of type "mobility registration update" in 5GC with 5G-GUTI mapped from 4G-GUTI. The AMF determines that old (source) node is an MME, but proceeds as if the Registration is of type "initial registration." The Registration Accept message includes "Handover PDU Session Setup support" indication to the UE. Based on this indication, the UE may subsequently move all its PDN connections from EPC to 5GC using the UE initiated PDU session establishment procedure with "Existing PDU Sessions."

It should be noted that the network does not automatically cancel the UE registration when the UE moves from one system to another.

4.12 Non-3GPP Access

4.12.1 Introduction

The 5G System is built to support multiple kinds of accesses in a generic way. This means the 5GC provides a single control plane and user plane interface toward the UE and/or the access network to allow for UE reachability over

- 3GPP accesses (the procedures and protocols on the "last mile access" are defined by 3GPP); and
- Non-3GPP accesses (the procedures and protocols on the "last mile access" are not defined by 3GPP but possibly by another standards forum).

Regardless of the access the UE is using, the same 5GC NF instances are used to serve the UE in most of the cases. This allows the UE security context that was established on one access (e.g. 3GPP access) to be re-used on different access systems (e.g. non-3GPP access). The same master key can be used to derive access related ciphering and integrity keys, without the need to re-authenticate the UE.

The same protocol ("NAS") and procedures are used between the UE and the 5GC. This corresponds to the N1 interface of the reference architecture (Section 4.3.1). This is an improvement compared to EPC where three different protocols, NAS, Internet Key Exchange (IKE) extensions, Wireless Local Area Network Control Plane Protocol WLCP, and three different Network Functions (MME/SGW, ePDG, TWAN) for three different types of accesses (3GPP, untrusted Non-3GPP and trusted Non-3GPP access) were necessary.

The same protocol and procedures are used between the different kinds of access networks and the 5GC. This corresponds to the N2 (Control Plane) and N3 (User Plane) interface of the reference architecture (Section 4.3.1). The same UPFs can serve a PDU session regardless of the (3GPP or Non-3GPP) access used to carry the UP flows of this PDU session. This allows for an efficient PDU session mobility between 3GPP and Non-3GPP access as only two NFs (a SMF and a UPF) may need to be involved due to mobility. Which access type (3GPP or Non-3GPP) a PDU session may use is controlled by policies on the UE.

The 3GPP Rel-15 specifications do not support multi-access PDU sessions, i.e. PDU sessions concurrently served by both 3GPP and Non-3GPP access.

However, there are still some access specific aspects in 5GC, such as the following:

- Some control information exchanged between 5GC and the access network depend on the access type, e.g. the User Location Information (ULI) or the information about Discontinuous Reception (DRX) (discontinuous UE reception, which is only defined for 3GPP access). The 3GPP cell ID can be used as ULI when the UE is camping over WLAN or over Wireline.
- Some procedures do not apply on some access system, e.g. paging and periodic registration procedures do not apply to untrusted non-3GPP access.
- The mechanisms used for mobility between 3GPP and Non-3GPP access are different from the handover procedures within 3GPP access. This is because handover within 3GPP access is controlled by the network (RAN) while mobility between 3GPP and Non-3GPP access is under control of the UE.

Apart from the differences listed above (that correspond to optional information and procedures for some accesses), the same 5GC protocol and state machines apply for the different accesses. The specific protocol and procedural adaptations related to each access technology are confined in the access network itself, i.e. the RAN for 3GPP access and an interworking function called N3IWF for non 3GPP accesses (see Figure 4.47).

3GPP has defined "untrusted non-3GPP access" as the only type of non-3GPP access connected to 5GC in Release 15. The support of other kinds of accesses like wireline access is deferred to 3GPP Release 16.

Establishment of the signaling connectivity between a UE and the 5GC over un-trusted Non-3GPP access is depicted in Figure 4.48 and follows these steps:

1) The UE connects to an untrusted Non-3GPP access network (e.g. WLAN as defined in IEEE 802.11 [10]) with procedures outside the scope of 3GPP. Any non-3GPP

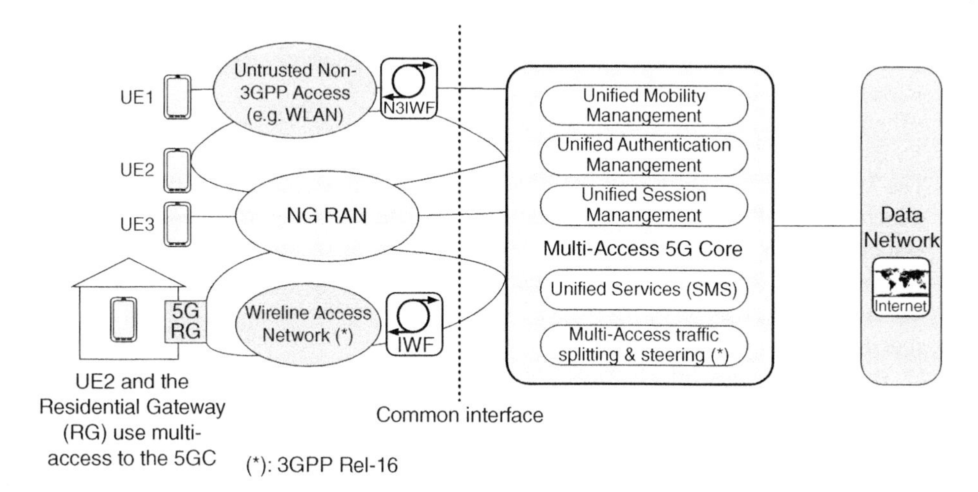

Figure 4.47 5G common (multi-access) core.

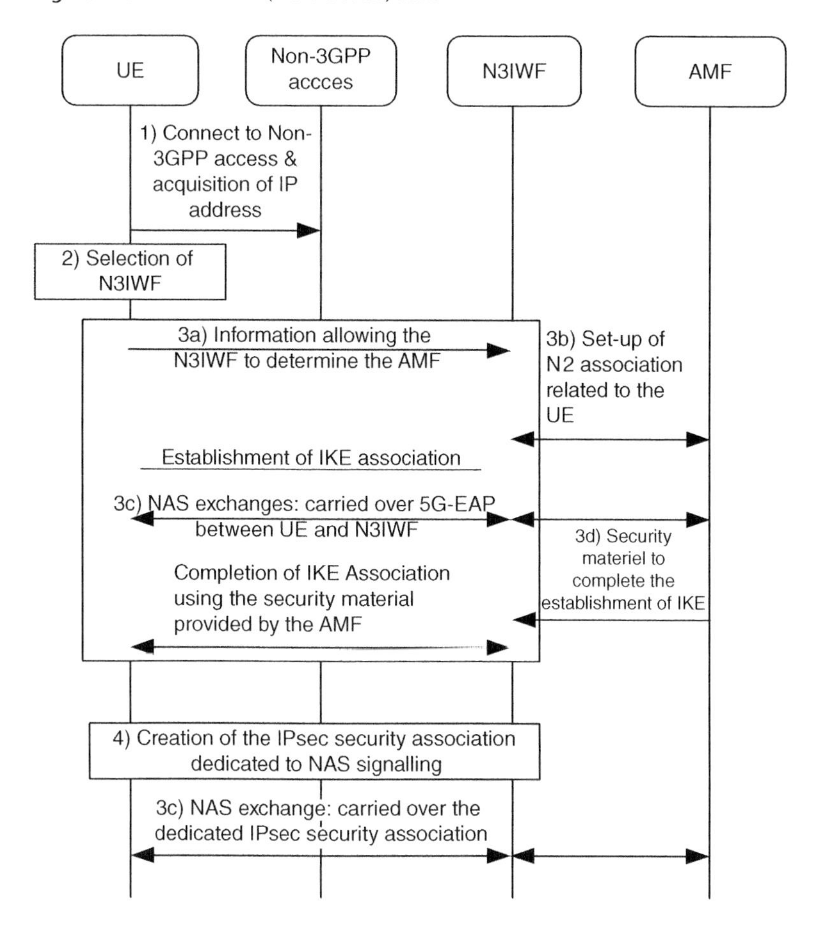

Figure 4.48 Establishment of the signaling connectivity between a UE and the 5GC over un-trusted non 3GPP access.

authentication method can be used, e.g. no authentication (in case of a free WLAN), pre-shared key, username and password, etc. The UE receives an IP address from the access network that will be used to contact the N3IWF.
2) When the UE decides to attach to 5GC, it selects an N3IWF as described in 3GPP TS 23.501 [1] Section 6.3.6.
3) The UE proceeds with the establishment of an IKE (see [22]) association with the selected N3IWF possibly reachable over other un-trusted access networks. The IKE association protects both UE and N3IWF against security attacks coming from any of these un-trusted networks. The establishment of the IKE association follows internet engineering task force (IETF) RFC 7296 [22] complemented by following 3GPP specific procedures:
   a. The UE provides information (temporary UE identifier, target set of slices…) allowing the N3IWF to select an AMF.
   b. The N3IWF establishes an NGAP association with the AMF for a give UE.
   c. The UE exchanges NAS signaling with the 5G Core (Registration, Service Request and possibly authentication related signaling). As the IKE association is still in authentication phase, NAS signaling is carried over EAP-5G (a specific flavor of the Extensible Authentication Protocol IETF RFC 3748 [17]) on the interface between the UE and the N3IWF and over the N2 association related to the UE between N3IWF and AMF. At the end of this exchange the AMF sends a NAS Security Mode Command to the UE.
   d. Access is now granted to the UE. The AMF provides the N3IWF with security material to complete the establishment of the IKE association.
4) An IPsec security association dedicated to NAS signaling is created.
5) From now on NAS signaling is carried over the IPsec security association dedicated to NAS signaling between UE and N3IWF and over the N2 association related to the UE between N3IWF and AMF.

The IKE protocol between UE and N3IWF plays an equivalent role to the RRC protocol between UE and NG-RAN: both IKE and RRC provide a secured link for the UE to exchange following information with the network:

- NAS signaling relayed to or from the N2 interface with the AMF.
- Add-on information like 5G UE identity (5G-S-TMSI or GUAMI), slice information (NSSAI) allowing the AN to select the correct AMF.
- Access related signaling used to negotiate PDU Session (and QoS flow) related resources on the link between UE and 5G AN (radio link for the RRC protocol, IPSec link in case of N3IWF/Untrusted Access to 5GS).

4.12.2 Interworking with EPC in Case of Non-3GPP Access

To support seamless service continuity between 5GS and EPS in case of Non-3GPP access, the interworking architecture requires support for the following requirements (which are common requirements like for 3GPP access as exposed in Section 4.11.1):

- Common UPF/PGW-u that serves as an IP anchor so that the UE IP address can be preserved.
- Combo SMF/P-GW-c function. UPF and SMF are regarded as PGW-U and PGW-C, respectively, from the viewpoint of the S-GW.

- Common subscription database HSS/UDM.
- Seamless service continuity between 5GS and EPS in case of Non-3GPP access does not rely on N26 interface between MME and AMF or between ePDG and AMF. This is because no network-based handover can take place in case of Non-3GPP access (the mobility is UE driven) and because the PDG does not support any equivalent interface to the MME to MME S10 interface in EPC.

Once a UE loses Non-3GPP access connectivity to the N3IWF, the UE may try to move PDU Sessions to 3GPP access, if this is allowed by local policies:

- When the 3GPP access serving the UE provides connectivity to the 5GC, the UE issues 5G NAS signaling to the SMF (via AMF) to move the corresponding PDU Sessions. The AMF that is reachable over 3GPP access is the same AMF reachable over N3IWF/Non-3GPP access, at least in non-roaming cases. This AMF is aware of the SMF address where to forward UE requests.
- When the 3GPP access serving the UE does not provide connectivity to the 5GC, the UE needs to ensure it is attached to a MME and to issue a 4G PDN Connection establishment request with a handover request to trigger handover of the PDU Session. This process is equivalent to the mobility in Dual-Registration mode as described in Section 4.11.3.

For a UE desiring to access the 3GPP Core network over non 3GPP access, operator policies guide the UE choice of whether to use a ePDG in EPC or a N3IWF in 5GC.

4.12.3 Multi-Access PDU Sessions

3GPP Rel-15 specifications do not support multi-access PDU sessions, i.e. PDU sessions concurrently served by both 3GPP and Non-3GPP access. This feature is nevertheless being studied for further 3GPP releases as discussed in Section 4.21.

Multi-access PDU sessions imply the support of traffic splitting between accesses enforced in the UE for UL traffic and in an UPF of the 5GC for DL traffic.

Depending on specific use cases and application requirements, traffic flows can be placed on one or both access networks under policy control. Changes in access network performance can lead to dynamic re-allocation of traffic flows, especially when mobile or WLAN access networks are involved. Another challenge is making the steering function in the 5GC aware that downstream network conditions have changed at the UE side.

MPTCP (IETF RFC 6824 [21]) is one of the solutions which addresses the traffic steering problem in an elegant way, as it adapts itself in real-time to changes in network performance. It also has the capability to support TCP "elephant flows" that exceed the capacity of a single access network connection, by splitting those into multiple sub-flows. While this helps to improve the end user perception of high peak data rates, it comes at the price of increased latency due to buffering and synchronization of flows in the receiver.

4.13 Fixed Mobile Convergence

Fixed-Mobile Convergence (FMC) has been on the agenda of network operators for several decades. In the context of FMC, we use the terms fixed as well as wireline networks interchangeably. In the past, the biggest obstacle to convergence has been the internal structure of large incumbent operators where different organizations are typically responsible for fixed and mobile networks. Hence, the opportunities for convergence were limited to infrastructure sharing, e.g. transport, aggregation and IP core networks, also more recently virtualized cloud infrastructure, and service convergence (mainly converged voice and broadband data services).

As capabilities and data rates, reliability of the connectivity, and services offered by fixed and mobile networks are becoming similar (especially with 5G), most of applications can work over any access network. Commercial service offers, and end user devices are also increasingly bundling fixed and mobile connectivity, which has led to a growing interest in converged network architectures. A converged 5G Core creates a unique control point to leverage intelligent traffic steering and multi-access integration and to ensure a seamless end user experience through selecting the best available access technology.

The virtualization of (wireline) access networks is another potential enabler to consider FMC now as both FMC and virtualization imply the need to re-think network architectures and deployments.

Hybrid access is one of the main convergence use cases, where a residential gateway has both a wireline and a cellular (5G) radio connection to the same core network (see Figure 4.49). To optimize both user experience and access network usage, the residential gateway and the 5G Core require intelligent traffic combining, splitting and steering functions. Hybrid access enables both higher peak data rates as well as better resiliency by simultaneous use of both access networks.

Figure 4.49 shows the full range of convergence use cases that are of interest to operators driving FMC in the Broadband Forum (BBF). The BBF is responsible for DSL (Digital Subscriber Line) as well as PON (Passive Optical Network) access networks which are typically owned by incumbent network operators; many of those also operate cellular access networks. Competition on peak data rates is driving converged operators to bundle both access technologies. A variation of this is fixed wireless access where a wireless access network is used to provide fixed broadband services including Internet access and IPTV (Internet Protocol Television).

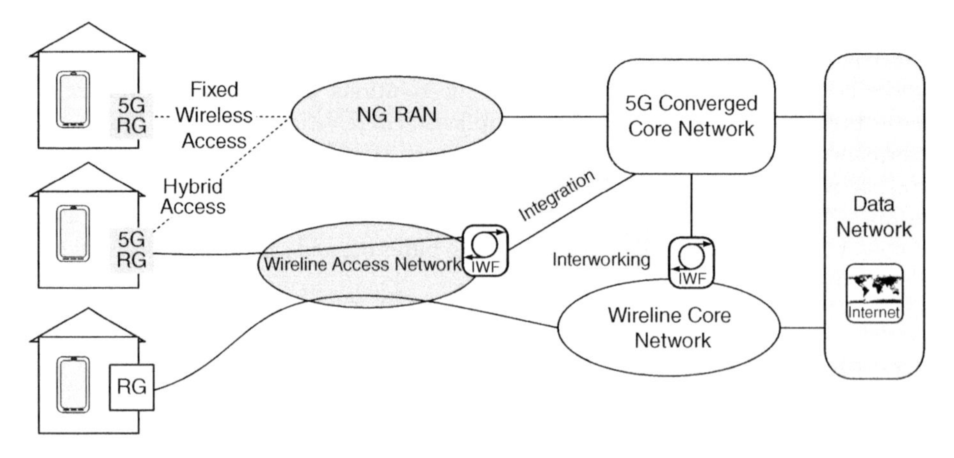

Figure 4.49 FMC use cases.

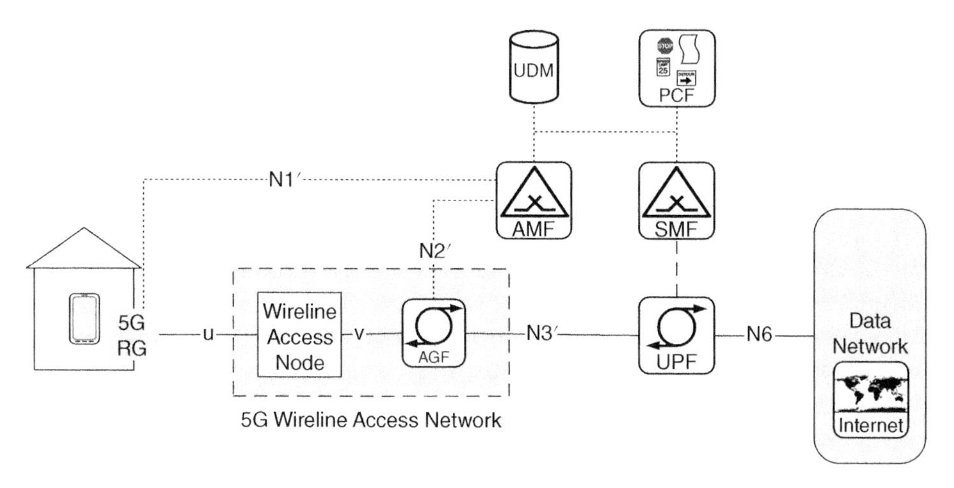

Figure 4.50 Integration of wireline access into the 5G Core.

A second class of use cases are related to the use of multi-access devices like smartphones which can connect simultaneously to two different access networks, e.g. 5G cellular and WLAN operating over wireline access network. Traffic combining, splitting and steering functions required for this scenario are like that for hybrid access, but operator policies and user preferences may be slightly different as the use case aims at improving coverage and reducing the impact of handovers between access networks.

Wireline access integration will be based on interworking functions that adapt between the wireline access network and the converged 5G Core, which has the advantage of minimizing impacts to wireline access nodes as well as 5G Core functions. As illustrated in Figure 4.50, a 5G Residential Gateway (5G RG) can support N1 (NAS) signaling toward the AMF for registration and SM. A new encapsulation for NAS signaling needs to be defined for wireline access, transporting NAS frames between the 5G RG and a new interworking function that connects to the core network via N2 and N3 interfaces.

Existing interfaces such as N1, N2 and N3 will be reused with minimal changes, ideally there would be no impacts to the protocols defined for 5GS in 3GPP Rel-15. Broadband subscriber management is taken over by the 5G SMF/UPF combination, which will need to support various flavors of residential gateways and offer the same data connectivity services than a Wireline BNG (Broadband Network Gateway). The full details of wireline access integration will be defined in 3GPP Rel-16, hence the corresponding interfaces have been denoted as N1′, N2′ and N3′ to indicate the possibility of small deviations from Rel-15 specifications.

At the time of writing this book, BBF and 3GPP are collaborating to standardize wireline access integration into the 5G Core. Studies conducted by BBF are focusing on requirements and gap analysis (see BBF SD-407 [9]), allowing 3GPP to study solutions, which will most likely be based on a trusted Non-3GPP interworking model. Normative specifications for FMC are targeted for completion in 3GPP release 16 (mid 2019).

4.14 Network Function Service Framework

4.14.1 Principles of a Service Framework

The Network Function Service Framework is mainly about using design principles that enable "simple, modular and flexible" design. Following are some of the key principles:

1) NF services offer a common interface framework to authorized consumers.
2) NF services offer capabilities to authorized consumers. Authorized consumers are identified based on operator policies.
3) NF services shall be self-contained (operate on own context), reusable and with an independent life cycle management.
4) NF services across Network Functions interact with each other either using "Request-Response" or "Subscribe-Notify" service operation types. Request-Response transactions are generally for 1-1 communication and guarded by a timer.
5) System procedures should be described by a sequence of service invocations.
6) NF services shall be accessible by means of an interface. An interface may consist of one or several operations.

Furthermore, to ensure that the NF services defined in 3GPP are bound by certain criteria, following principles applied in Rel-15:

1) Standardized NF services can be defined only when it is needed for 5G System Procedures.
2) Interaction between NF services within the NF is not subject to standards.

4.14.2 What is an NF Service?

An NF service is a capability of a certain Network Function (e.g. AMF or SMF) exposed by the NF (NF service producer) to authorized NFs (NF service consumers) through a service-based interface. An NF can expose one or more NF services. Similarly, an NF can consume one or more NF services.

Network Functions may offer different capabilities and thus, different NF services to distinct consumers. An NF can be composed of NF services as shown in Figure 4.51.

Each NF service shall be accessible by means of an interface. An interface may consist of one or several operations (Figure 4.52).

5G System procedures can be built by invocation of multiple NF services as shown in Figure 4.53. Figure 4.53 shows an illustrative example on how a procedure can be built; it is not expected that system procedures depict the details of the NF services within each Network Function.

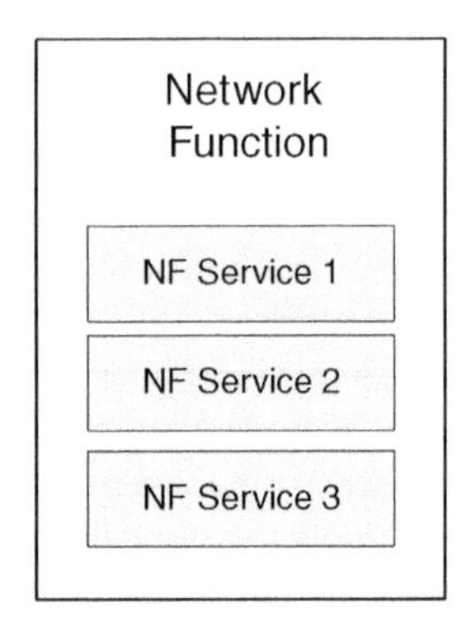

Figure 4.51 NF and NF service.

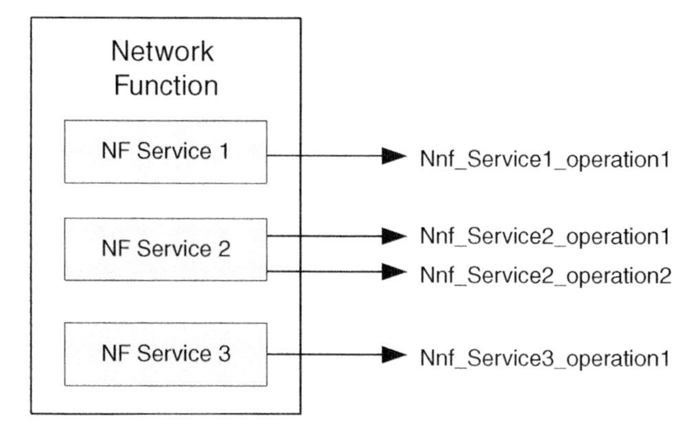

Figure 4.52 NF, NF service and NF service operation.

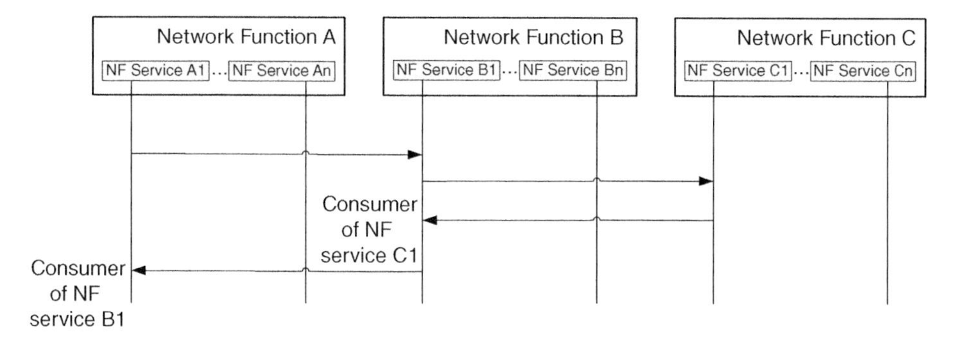

Figure 4.53 System procedures and NF services.

The following are the design criteria for specifying NF services of a certain NF within the 5G System:

- Each NF service operates on its own set of context data. A context refers to a state or a software resource or an internal data storage. The NF service operations can create, read, update or delete the context data.
- Any direct access of a context owned by a NF service is to be made by the service operations of that NF service. Services provided by the same NF can communicate internally within the NF.

Figures 4.54 and 4.55 illustrate the above criteria.

The use of NF services in operator deployed network is not restricted to the identified consumers. But it has also been agreed in 3GPP that no specific effort will be made to develop the description of NF services beyond what is necessary to support their use as part of the5G System procedures.

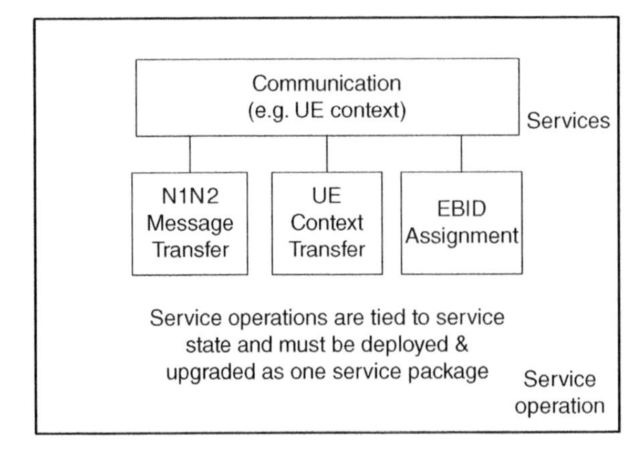

Figure 4.54 Self-contained service.

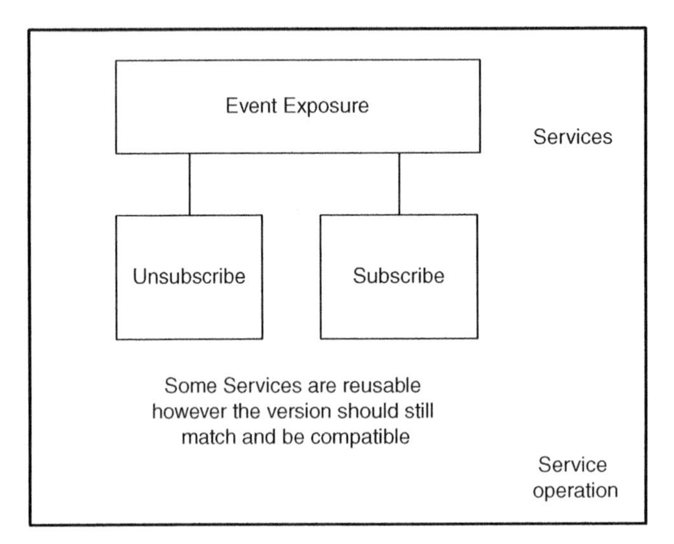

Figure 4.55 Re-usable services.

4.14.3 Consumer/Producer Interactions

The Network Function that is offering a service is called NF Service Producer. The Network Function that is consuming a service is called NF Service Consumer.

The interaction between consumer and producer within the NF service framework follows two mechanisms:

1) *"Request-Response"* (Figure 4.56). A control plane NF_B (NF Service producer) is requested by another control plane NF_A (NF Service consumer) to provide a certain NF service, which either performs an action or provides information or both. NF_B provides NF service based on the request by NF_A. To fulfill the request, NF_B may in turn consume NF services from other NFs. For the Request-Response mechanism, communication is one-to-one between consumer and producer. A one-time response from the producer to a request from the consumer is expected within a certain timeframe.

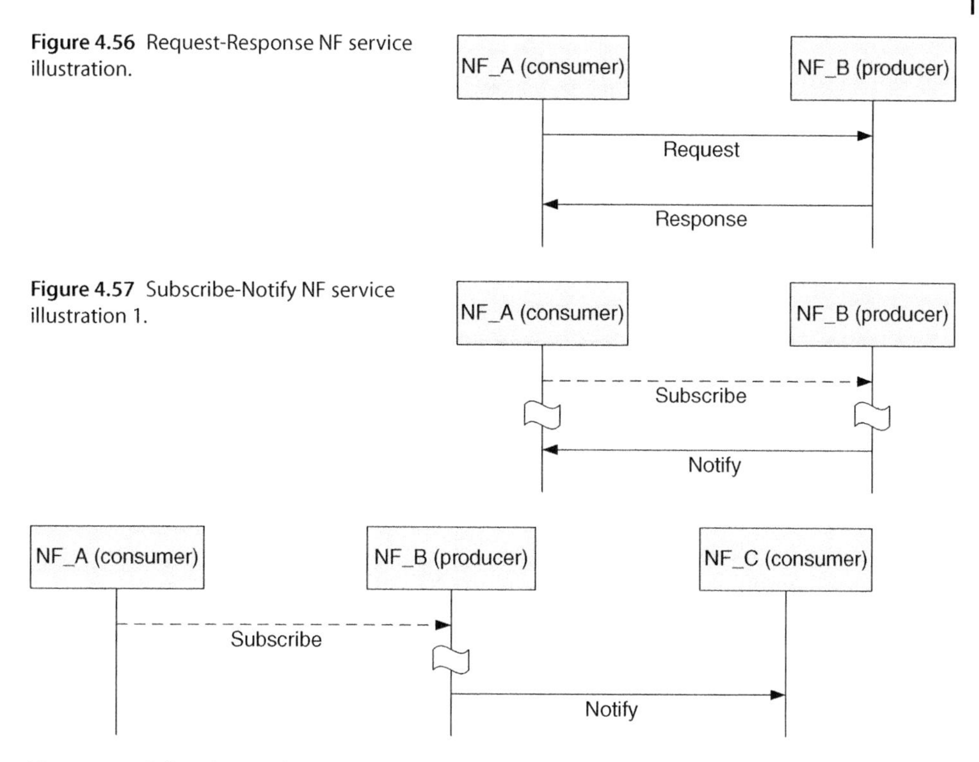

Figure 4.56 Request-Response NF service illustration.

Figure 4.57 Subscribe-Notify NF service illustration 1.

Figure 4.58 Subscribe-Notify NF service illustration 2.

2) *"Subscribe-Notify"* (Figure 4.57). A control plane NF_A (NF Service consumer) subscribes to a NF service offered by another control plane NF_B (NF Service producer). One or more control plane NFs may subscribe to the same control plane NF Service. NF_B notifies all subscribed NFs about the results of this NF service. The subscription request normally includes:
   1) The notification endpoint (e.g. the notification URL) of the NF Service consumer.
   2) The need receiving periodic updates.
   3) Events that are of interest for notification, e.g. the content of information requested has changed, reaches certain thresholds, etc.

A control plane NF_A may also subscribe to an NF Service offered by control plane NF_B on behalf of another control plane NF_C, i.e. it requests the NF Service producer to send the event notification to other consumers (Figure 4.58). In this case, NF_A includes the notification endpoint of the NF_C in the subscription request.

4.14.4 Network Function Service Authorization

The NF service authorization framework ensures that the NF Service consumer is authorized to access the NF service provided by the NF Service provider, according to the policies of the NF, the policies of the serving operator and inter-operator agreements. Service authorization information is configured as part of the NF profile of the NF Service producer. It includes the NF type and NF realms/origins allowed to consume NF Service(s) of the producer. The service authorization entails two steps:

- Check whether the NF Service consumer is permitted to discover the requested NF Service producer instance during the NF service discovery procedure. This is performed on a per NF granularity by the NRF.
- Check whether the NF Service consumer is permitted to access the requested NF Service producer for consuming the NF service on a per request type, per UE, per subscription and roaming agreement granularity. This type of NF Service authorization is embedded in the related NF service logic of the NF.

4.14.5 Network Function Registration and De-registration

The NRF maintains the information of available NF instances and their supported services. Each NF instance registers at the NRF the list of NF services it supports.

The NF instance may register this information with the NRF when the NF instance becomes operative for the first time (registration operation) or upon individual NF service instance activation/de-activation within the NF instance (update operation), e.g. triggered after a scaling operation. During the registration operation, the NF instance provides an NF profile that is maintained in the NRF. The NF profile includes:

- NF instance ID;
- NF type;
- PLMN ID;
- Network Slice related identifier(s), e.g. S-NSSAI, NSI ID;
- Fully qualified domain name (FQDN) or IP address of the NF;
- NF capacity information;
- NF specific service authorization information;
- Names of supported services;
- Endpoint information of instance(s) of each supported service; and
- Identification of stored data/information.

The NF instance can register with the NRF directly using an NRF service or via operations, administration & maintenance (OA&M).

The NF instance should also de-register from the NRF when it is about to (gracefully) shut down or disconnect from the network in a controlled way. If an NF instance become unavailable or unreachable due to unplanned errors (e.g. NF crashes), an authorized entity should deregister the NF instance from the NRF.

4.14.6 Network Function Discovery

The NF discovery and NF service discovery framework (Figure 4.59) enables one NF to discover a set of NF instances with a specific NF service or target NF type.

Unless the expected NF and NF service information is locally configured on the requester NF, e.g. the expected NF service or NF is in the same PLMN as the requester NF, the NF and NF service discovery is implemented in the NRF. The NRF is the logical function that is used to support the functionality of NF and NF service discovery.

To enable access to a requested NF type or NF service, the requester NF initiates the NF or NF service discovery by providing information such as the following:

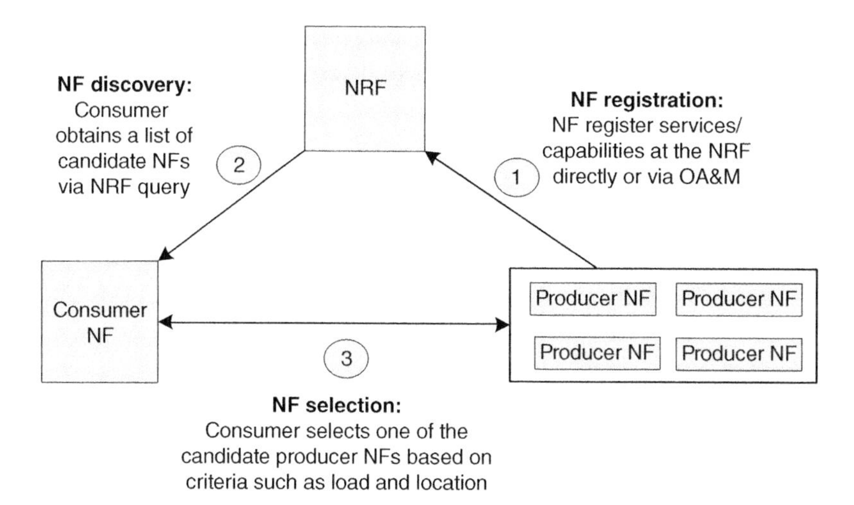

Figure 4.59 Network function discovery framework.

1) The type of the NF or the specific service it is attempting to discover, e.g. SMF, PCF, UE Location Reporting.
2) The NF service related parameters, e.g. slicing related information to discover the target NF.
3) The detailed service parameters used for specific NF discovery (refer to the related NF discovery and selection clause).

In the discovery response, the NRF provides information such as IP address, FQDN, or the identifier of relevant services and/or NF instances to the requester.

Based on that information, the requester NF can select one specific NF instance or a NF instance that is able to provide a NF Service, e.g. an instance of the PCF providing Policy Authorization.

For NF discovery across PLMNs, the requester NF provides to the NRF the PLMN ID of the target NF. The NRF in the local/visited PLMN interacts with the appropriate NRF in the target/home PLMN. The NRF in the local PLMN reaches the NRF in the target PLMN by forming a target PLMN specific query using the PLMN ID provided by the requester NF. The NRF in the local PLMN interacts with the NRF in the target PLMN to retrieve the IP address, FQDN or the identifier of relevant services of the target NF instances.

To enable network topology hiding, it is also possible that the IP address or the FQDN of proxy function(s) (SEPP) instead of the target NF instances are provided to the requester NF. The proxy functions are transparent to the requester NF.

4.14.7 Network Function Services

This section summarizes the NF services defined in 3GPP for the control plane Network Functions within the 5G Core. As discussed in the previous sections, the NF services can be consumed by any authorized consumers, thus the consumers are not fixed rather the consumers are meant only as examples. The examples are listed mainly to allow deployments getting an idea on the reason for specifying each NF service and service operation. Having an idea of the NF consumer also helps with trouble shooting with a given NF service in the field, furthermore it helps to determine the capabilities that should be offered by the given NF service.

NF services for 5GC NFs should be defined at a reasonable granularity (avoiding explosion of services) along with detailed NF service operations for each NF service. Here we summarize the NF services agreed for different 5GC NFs, taking AMF, SMF, NRF and UDM as examples. For details about 5GC NF services and NF service operations, please refer to 3GPP TS 23.502 [2].

4.14.7.1 AMF Services

The following NF services are specified for the AMF (Table 4.3).

4.14.7.2 SMF Services

The following NF services are specified for the SMF (Table 4.4).

4.14.7.3 UDM Services

The following NF services are specified for the UDM (Table 4.5).

4.14.7.4 NRF Services

The following NF services are specified for the NRF (Table 4.6).

Table 4.3 NF services provided by AMF.

Service name	Description	Example consumers
Namf_Communication	This service enables an NF to communicate with the UE and/or the AN through the AMF. This service enables SMF to request EBI allocation to support interworking with EPS.	Peer AMF, SMF, PCF, NEF, SMSF, UDM, NEF, LMF
Namf_EventExposure	This service enables other NFs to subscribe or get notified of the mobility related events and statistics.	NEF, SMF, PCF, UDM
Namf_MT	This service enables an NF to make sure UE is reachable.	GMLC

Table 4.4 NF services provided by SMF.

Service name	Description	Example consumers
Nsmf_PDU Session	This service manages the PDU sessions and uses the policy and charging rules received from the PCF. The service operations exposed by this NF service allows the consumer NFs to handle the PDU sessions.	Peer V/H-SMF, AMF
Nsmf_Event Exposure	This service exposes the events happening on the PDU sessions to the consumer NFs.	PCF, NEF, AMF

Table 4.5 NF services provided by UDM.

Service name	Description	Example consumers
Nudm_UE Context Management	1. Provide the NF consumer of the information related to UE's transaction information, e.g. UE's serving NF identifier, UE status, etc. 2. Allow the NF consumer to register and deregister its information for the serving UE in the UDM. 3. Allow the NF consumer to update some UE context information in the UDM.	AMF, SMF, SMSF, NEF, SMSF, GMLC
Nudm_Subscriber Data Management	1. Allow NF consumer to retrieve user subscription data when necessary 2. Provide updated user subscriber data to the subscribed NF consumer;	AMF, SMF, SMSF
Nudm_UE Authentication	1. Provide updated authentication related subscriber data to the subscribed NF consumer.	AUSF
Nudm_Event Exposure	1. Allow the NF consumer to subscribe to receive an event. 2. Provide monitoring indication of the event to the subscribed NF consumer.	NEF

Table 4.6 NF services provided by NRF.

Service name	Description	Example consumer(s)
Nnrf_NF Management	Provides support for register, deregister and update service to NF and NF services. Provide consumers with notifications of newly registered NF along with its NF services.	AMF, SMF, UDM, AUSF, NEF, PCF, SMSF, NSSF
Nnrf_NF Discovery	Enables one NF service consumer to discover a set of NF instances with specific NF service or a target NF type. Also enables one NF service to discover a specific NF service.	AMF, SMF, PCF, NEF, NSSF, SMSF, AUSF

4.15 IMS Services

4.15.1 Overview

The 5G System is a pure packet-based system without an in-built CS voice component like global system for mobile communications (GSM) and universal mobile telecommunications system (UMTS). Thus, the preferred way to provide voice and other multimedia services is via the IMS. It can be assumed that the same or a similar voice profile defined by global system for mobile communications association (GSMA) for VoLTE becomes also applicable for VoNR. For example, GSMA IR.92 has specified a well-known APN, the so called "IMS APN," to enable IMS roaming.

The 5G Core Network indicates to the UE that Voice over Internet Protocol (VoIP) in 5GS is supported; more precisely that the tracking area(s) the UE is currently camping on provide sufficient QoS and coverage for VoIP support. Therefore, a UE that has been provisioned as IMS VoIP capable can start an IMS voice session at its current location based on the received network indication.

When IMS services are supported for UEs registered in the 5G Core, the services can be provided to the UEs either via 3GPP access or Non-3GPP access.

However, during the initial stages of 5G System roll out, actual NR coverage is expected to be spotty. If a UE starts an IMS voice session in 5GS, it may need to continue the voice call when it moves out of NR coverage. To allow continuation of voice calls in LTE support for interworking with E-UTRA/EPS was introduced. It extends the voice service coverage area and provides a good voice experience in the early days of NR.

Some 5G System based deployed networks may not support IMS based voice service natively either because NR roll-out is in the high frequency band (i.e. rapid signaling loss can occur) or because early NR deployments are not supporting the required QoS. For such scenarios, support for voice fallback toward E-UTRA in combination with 5GC (RAT Fallback) and E-UTRAN in combination with EPC (EPS Fallback or System Fallback) was introduced. Even when fallback for voice is supported, the UE can remain registered in the 5G System for IMS signaling over NR and 5GC as fallback toward E-UTRA/5GC happens only when a MT or MO voice call is initiated.

4.15.2 Support for System/EPS Fallback for Voice

When a MO or MT voice call is initiated and the IMS requests the 5GS to establish the desired QoS flow for voice, NG-RAN detects the need for voice call setup, e.g. based on the respective 5QI = 1 (5G QoS indicator 1). Based on this trigger, NG-RAN can initiate the fallback toward EPS (Figure 4.60) and reject the QoS flow setup with "fallback toward EPS." This can either result in a network assisted handover from 5GS toward EPS using the interworking principles mentioned in Section 4.9.2 or using RRC release with redirection causing the UE to re-select E-UTRA/EPC (network assisted redirection), and the voice call to be successfully setup upon mobility toward EPS. Support for EPS Fallback does not assume the existence of signaling reference point N26 between MME and AMF.

However, if there is no N26 reference point, there is a risk that IMS signaling messages are lost due to the delay incurred during the fallback, which will require support for

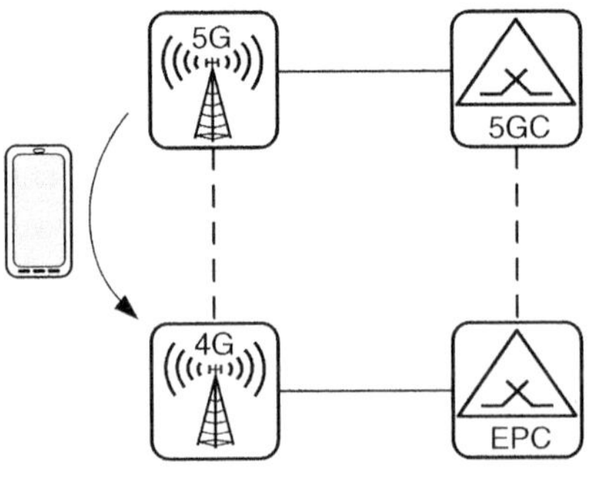

Figure 4.60 System fallback toward E-UTRAN/EPC (EPS fallback).

Figure 4.61 RAT fallback toward E-UTRA/5GC.

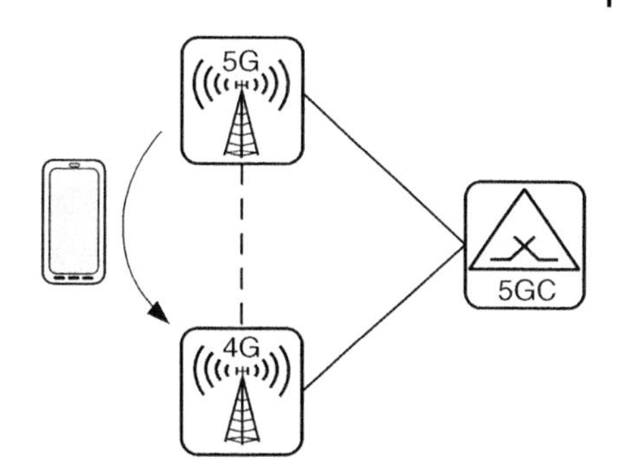

retransmissions at the protocol layer (e.g. TCP in case of session initiation protocol [SIP] signaling) to avoid several seconds of voice call interruption. With the N26 reference point in place the delay is reduced as no re-authentication is needed at the target side.

4.15.3 Support for RAT/E-UTRA Fallback for Voice

When a MO or MT voice call is initiated and the IMS requests the 5GS to establish the desired QoS flow for voice, NG-RAN detects the need for voice call setup, e.g. based on 5QI 1. Based on this trigger, NG-RAN can initiate the fallback toward E-UTRA/5GC (Figure 4.61) and reject the QoS flow setup with "fallback toward EPS." This can either result in network assisted handover from NR/5GC toward E-UTRA/5GC using N2 handover as explained in Chapter 5 or using RRC release with redirection causing the UE to re-select E-UTRA/5GC (network assisted redirection) and the voice call to be successfully setup upon mobility toward E-UTRA/5GC.

4.16 Emergency Services

4.16.1 Overview

Emergency services are available in the 5G System in 3GPP Rel-15 based on IMS (IMS Emergency Calls).

Emergency calls will be routed to a call center called PSAP (Public Safety Answering Point). An emergency call is routed to a PSAP that is best suited for this call, e.g. regarding caller location load situation. Since IMS is mostly access agnostic (with very few access specific adaptations) and supports emergency services since Release 9, it doesn't require any enhancements to support emergency calls over 5G-NR. However, the newly introduced 5G System needs to support some specific functionalities to provide support for emergency services.

1) The network must be capable of supporting IMS services as explained in the previous Section 4.14.
2) The network must consider emergency services with high priority both for control plane signaling related to emergency services establishment and user plane traffic related to emergency services communication. This implies that even under congestion situation, requests for emergency services and the actual emergency session shall be treated with high priority by RAN and each Network Function within the 5GC.

3) The network may need to support emergency services for unauthenticated UEs depending on regulatory requirements. In some countries, emergency services must be supported also for UEs without universal subscriber identity module (USIM) (e.g. North America) in addition to supporting emergency services for UEs with USIM but without active subscription. In some countries (e.g. Germany), emergency services are not provided to UEs without USIM, but the network may still need to support emergency services for unauthenticated UEs.
4) The network must be able to detect UE's location (Location Services) to route the UE to the right PSAP.
5) The network must indicate support for Emergency Services on a per RAT basis to the UE. The Emergency Services support indication is valid within the current Registration Area. Based on this indication, the UE determines whether a PDU session for emergency services can be initiated. If the network indicates support for Emergency Services in the RAT and in the Registration Area the UE is camping on, the UE is required to initiate a PDU session for Emergency Services.
6) The radio network needs to broadcast support for Emergency Services. This is needed for UEs that are in limited service state (UE not registered on the network) to determine which network natively supports emergency services for unauthenticated UEs.
7) When a PLMN supports IMS Emergency Services, all AMFs in that PLMN and at least one SMF shall have the capability to support Emergency Services.

The UE shall not initiate Emergency Services over untrusted Non-3GPP access to 5GC, if the emergency session can be established via 3GPP access. To use 5GC for emergency services in case of untrusted Non-3GPP access, the UE may select any N3IWF. For further details on native support for Emergency Services, please refer to 3GPP TS 23.501 [1] Section 5.16.4.

In addition to native support for emergency services, support for Emergency Services using fallback to another technology has also been introduced to cater to varying deployment needs. As mentioned above for voice services, some 5G deployments may not support IMS-based Emergency Services natively either because NR roll-out is in the high frequency band (i.e. radio propagation conditions are such that rapid loss of signal may occur) or because initial stages of NR deployments are unable to support required QoS and/or Location services. For such scenarios, support for Emergency Services fallback via RAT Fallback or EPS Fallback was defined.

4.16.2 Support for Emergency Services Fallback

The network can initiate fallback toward EPS and/or E-UTRA just as it does fallback for regular voice calls as outlined in Section 4.12. In addition, Emergency Services fallback triggered based on the UE initiated Service Request procedure was introduced.

The UE and 5GC may support the mechanism to direct or redirect the UE either toward E-UTRA connected to 5GC (RAT fallback) when only NR does not support Emergency Services but 5GC supports it or toward EPS (E-UTRAN connected to EPC System fallback) when the 5GC does not support Emergency Services.

If the AMF indicates support for Emergency Services using fallback (for a given RAT) in the Registration Accept message, a normally registered UE supporting Emergency

Services fallback shall initiate a Service Request with Service Type set to "Emergency." The AMF uses the Service Type Indication within the Service Request to initiate redirection of the UE toward the appropriate RAT/System. The 5GS can trigger redirection to EPS or redirection to E-UTRA connected to 5GC. After receiving the Service Request for Emergency Fallback, the AMF triggers an appropriate N2 procedure (e.g. paging) resulting in CONNECTED state mobility (handover procedure) or IDLE state mobility (redirection) to either E-UTRA/5GC or to E-UTRAN/EPC depending on factors such as N26 availability, network configuration and radio conditions. The AMF indicates to the RAN whether it should trigger RAT fallback or System fallback by including the Core Network type. RAN determines whether to trigger N2 (intra-NG-RAN) handover/redirection toward E-UTRA/5GC or N26 inter-system handover/redirection toward E-UTRAN/EPC. If the RAN decides to perform redirection, it indicates the Core Network type to the UE as part of the RRC Redirection message for the UE to determine whether it should initiate an EPC NAS or 5GC NAS procedure upon redirection to E-UTRA.

4.17 Location Services

Support for Location Services is optional for 5G System deployments. Also, the support for Location Services is restricted to provide regulatory services in 3GPP Release 15, e.g. determine location of the user in case of emergency calls. Figure 4.62 shows support for location services for non-roaming scenarios using the service-based interface representation. Only entities directly relevant to location services are shown.

The functional split between AMF and LMF is almost the same as the functional split between MME and serving mobile location center (SMLC). The main addition for the

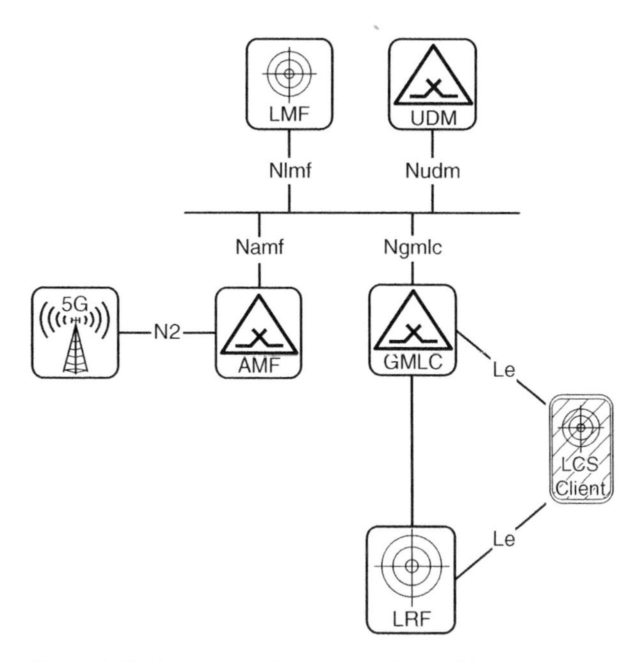

Figure 4.62 Non-roaming location services architecture.

5G System Location Services architecture is that the LMF is integrated into the SBA. Support for Location Services was included within 3GPP Release 15, but only RAT independent mechanisms are used in NR. This implies that Location Services support from E-UTRA and other mechanisms that do not require support for it directly in the gNB are used. At the time of writing this book, it is expected that the following principles will be employed for Location Services in NG-RAN Rel-15:

- NR will introduce basic positioning methods to meet regulatory requirements based on E-UTRA but defer other new high accuracy positioning methods to Rel-16.
- Aim is to introduce RAT independent positioning methods, E-UTRA RAT dependent positioning methods and network-based NR Cell ID and cell portion positioning methods.
- RAT independent schemes include e.g. A-GNSS, WLAN, Bluetooth, Sensor, MBS and E-UTRA RAT dependent schemes includes e.g. observed time difference of arrival (OTDOA).
- NR-PP positioning protocols will be the baseline for Rel-15 NR work.
- Transport of Location Services messages between 5GC and UE through the gNB.
- Transport of Location Services messages between 5GC and NG-RAN hosting E-UTRA (eNB).
- Support of measurement gaps and idle periods for location related inter-RAT measurements.

The LMF supports the following functionality:

- Location determination for a UE.
- Obtains downlink location measurements or a location estimate from the UE.
- Obtains UL location measurements from the NG-RAN.
- Obtains non-UE associated assistance data from the NG-RAN.

4.18 Short Message Service

4.18.1 Overview

With the advent of many popular over the top messaging services using smart phone apps, SMS traffic is decreasing but continues to still exist for many reasons. One classical use for SMS is support for "human to human" text messaging service and this is what is being replaced by over the top applications. In addition, SMS was used for device triggering purposes, especially in case of IoT devices. Recently, SMS has also been used for "two-factor" authentication by many companies, e.g. when using corporate Office applications outside the Intranet. In fact, SMS traffic has increased recently mainly due to use of "two-factor" authentication and IoT device management. Thus, classic SMS service must be supported in the 5G System.

SMS can be provided in LTE natively over IP (called "SMS over IP") via IMS (see 3GPP TS 23.204 [23]). There is no special requirement being placed on the 5G System for this feature; 5G is just used to provide IP connectivity. The precondition is that the UE must have successfully performed an IMS registration, is supporting the "SMS over IP" feature and has been configured to use the feature. If this is the case, a SMS is encoded by the network or UE into a special SIP message and sent to the peer. This requires support for IMS (SIP) client in the UE.

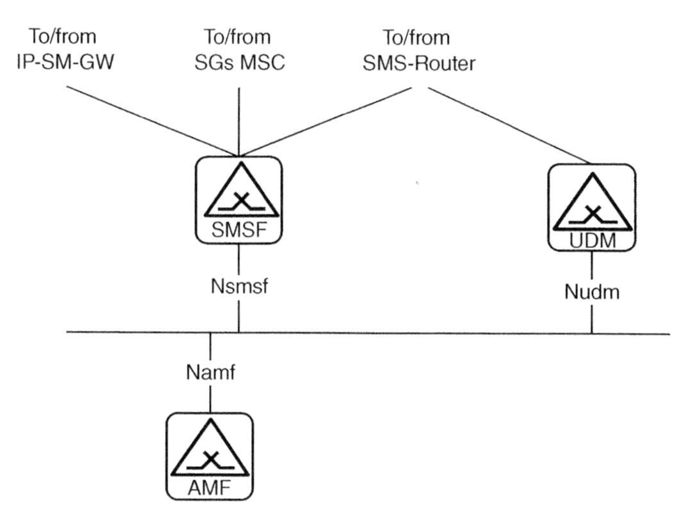

Figure 4.63 Non-roaming architecture for SMS over NAS.

For the scenario where there is no deployment of IMS in the network or no IMS (SIP) client support in the UE, a solution was specified to allow native support for SMS in the 5G System. Since the architecture leverages NAS messages to transmit and receive SMS, this architecture is called "SMS over NAS architecture."

4.18.2 SMS over NAS

Figure 4.63 shows the non-roaming architecture to support SMS over NAS using Service-based interfaces within the 5G Core control plane.

The System Architecture for SMS over NAS follows mostly the SMS over SGs solution approach specified for LTE/EPC with some subtle differences:

1) The SMSF need not assign any temporary identifier such as TMSI for the UE as it is only the AMF that is assigning temporary identifiers for a UE.
2) The SMSF does not have to assign any location area for the UE as it is only the AMF that is assigning a Registration Area for a UE.

Thus, the SMSF existence is completely transparent to the UE. Similarly, the transmission of SMS (and SMS payload) within a NAS message is completely transparent to the AMF. SMS relocation is not supported in Rel-15. Each UE is associated with only one SMSF at any time. Support for SMS incurs the following functionality:

- Support for SMS over NAS transport between UE and AMF. This applies to both 3GPP and Non-3GPP accesses.
- The AMF determines the SMSF for a given UE.
- Support for subscription checking and actual transfer of MO-SMS and MT-SMS by the SMSF.
- Support for MO-SMS and MT-SMS transmission for both roaming and non-roaming scenarios.
- Support for selecting proper domains for MT-SMS message delivery (e.g. over 5G, LTE or CS) including initial delivery and re-attempting in other domains.

The SMSF connects the legacy SMS architecture to the 5G Core, mainly to the AMF. The SMSF is responsible for SMS authorization, SMS relay and protocols, SMS charging and LI.

In the SMSF, SMS authorization checks the UE requests regarding SMS related subscription data retrieved from the UDM. SMS relay and protocols consist in relaying SMS, via AMF, from UE to SMS-GMSC/IW-MSC/SMS-Router or vice versa, and supporting SM-RP and SM-CP protocols toward the UE. The UE availability notifications are also included. SMS charging creates SMS related Charging Data Records (CDR). LI support includes delivery of SMS events to the LI system.

During the registration procedure, a UE that wants to use SMS provides an "SMS supported" indication over NAS signaling indicating the UE's capability. "SMS supported" indicates whether UE can support SMS delivery over NAS via 3GPP access or via both 3GPP and Non-3GPP access. If the core network supports SMS, the AMF includes "SMS supported" indication to the UE, and whether SMS delivery over NAS via 3GPP access or via both the 3GPP and Non-3GPP access is accepted by the network. SMS is transported over NAS without the need to establish DRBs, using NAS transport message carrying the SMS as payload.

4.19 Public Warning System

PWS is a service offered to local, regional, or national authorities to reach out to a public mass warning them of an impending emergency. Such warning messages may be necessary for reasons such as:

- weather related emergency, tornados, ice storms, hurricane;
- geological disasters, earthquake, tsunami;
- industrial disasters, release of toxic gas or contamination;
- radiological disasters, nuclear plant disaster;
- medical emergencies, outbreak of a fast-moving infectious disease; and
- warfare or acts of terrorism.

Some cities use emergency warnings to advise people of prison escapes, abducted children, emergency telephone number outages and other events.

The 5G System supports warning message delivery as specified in 3GPP TS 23.041 [24] but permits a number of unacknowledged warning messages to be broadcasted to UEs within a service area. The term "unacknowledged" implies that there exists no retry mechanism for warning messages.

There are different regional variants requiring support for PWS:

1) The Earthquake and Tsunami Warning System (ETWS)) feature that is based on Japanese requirements.
2) The Commercial Mobile Alert System (CMAS)) feature based on US requirements.
3) The Korean flavor called Korean Public Alert System (KPAS)).
4) The European flavor called European PWS (EU-ALERT).

The Cell Broadcast Center (CBC) belongs to the serving operator's domain. It is under the responsibility of public authorities and network operators in each country to agree on a national standard for the interface between Cell Broadcast Entity (CBE) and CBC (Figure 4.64).

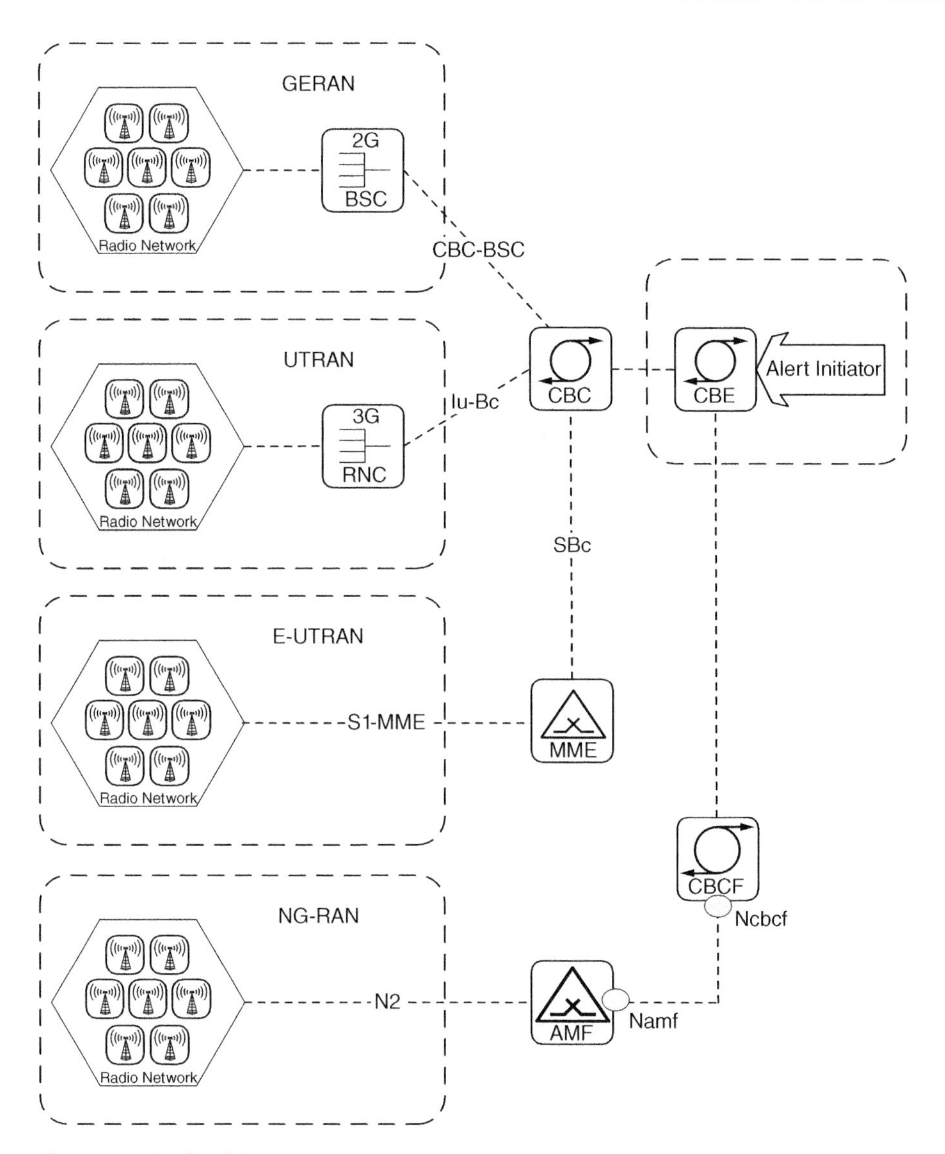

Figure 4.64 PWS architecture.

A simplified description of a warning message delivery is as follows:

1. The CBC identifies which AMFs need to be contacted and sends a request to the AMFs for the given region. Depending on the deployments, request to the AMF may go via CBCF. The request contains the warning message to be broadcasted, the warning area information and the delivery attributes.
2. The AMF forwards the warning message to NG-RAN nodes and uses the Tracking Area ID list to determine the NG-RAN nodes in the delivery area. If this list is empty, the message is forwarded to all NG-RAN nodes that are connected to the AMF.

3. The NG-RAN nodes use the warning area information to determine the cells where the message must be broadcasted. If the cell ID information is not present, the NG-RAN nodes broadcast messages in all cells. The NG-RAN nodes are responsible for scheduling the broadcast of messages and message repetitions in each cell.

4.20 Protocol Stacks

4.20.1 Control Plane Protocol Stacks

4.20.1.1 Control Plane Protocol Stacks Between the 5G-AN and the 5G Core

The Control Plane interface (N2/NG-C) between the 5G-AN and the 5G Core is defined according to following principles:

- Different types of 5G-AN (e.g. 3GPP RAN, N3IWF for Untrusted Non-3GPP access to 5GC) connect to the 5CG via a unique control plane protocol, the NGAP.
- There is a unique N2 termination point in the AMF per access for a given UE regardless of the number (possibly zero) of PDU sessions of the UE and regardless of the number of 5GC NF needing to send signaling to the 5G-AN (e.g. SMF(s) and SMSF serving the UE, Cell Broadcast Service).
- From the 5G-AN perspective, there is a single N2 termination point, the AMF.
- NGAP supports a decoupling between AMF and other functions such as SMF that may need to control the services supported by 5G-AN(s) (e.g. control of the UP resources in the 5G-AN for a PDU session).

Following procedures and information exchange are defined over N2:

- Procedures related to N2 interface management and not related to an individual UE, such as configuration or reset of the N2 interface. Via these procedures the 5G-AN and the AMF can exchange configuration information, e.g. about:
  - Tracking Areas, PLMN ID and slices (S-NSSAI) the 5G-AN support.
  - AMF identifier, slices (S-NSSAI) and transport connections the AMF supports.
  - AMF overload status.
- Procedures related with an individual UE:
  - Procedures related to NAS transport between the UE and the AMF.
  - Procedures related with UE context management, e.g. to provide the 5G-AN with relevant security material to secure the AN-related procedures (RRC or interface between the N3IWF and the UE).
  - Procedures related to resources for PDU sessions. They may correspond to messages that carry information, e.g. related to N3 addressing and QoS requirements, that is transparently forwarded by the AMF between the 5G-AN and the SMF.
  - Procedures related to handover management. These procedures are intended for 3GPP access only.

NGAP protocol stack is depicted in Figures 4.65 and 4.66. NG Application Protocol (NG-AP): 3GPP defined Application Layer Protocol between the 5G-AN node and the AMF (Figure 4.65). NG-AP is defined in TS 38.413 [16].

Stream Control Transmission Protocol (SCTP): This protocol guarantees delivery of signaling messages between AMF and 5G-AN node (N2). SCTP is defined in RFC 4960 [18].

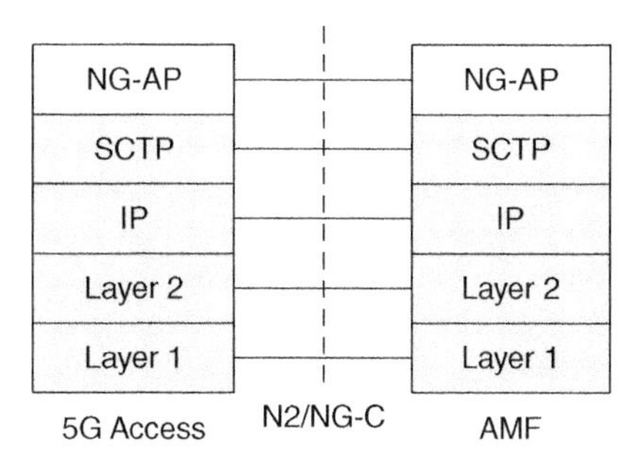

Figure 4.65 NGAP protocol stack.

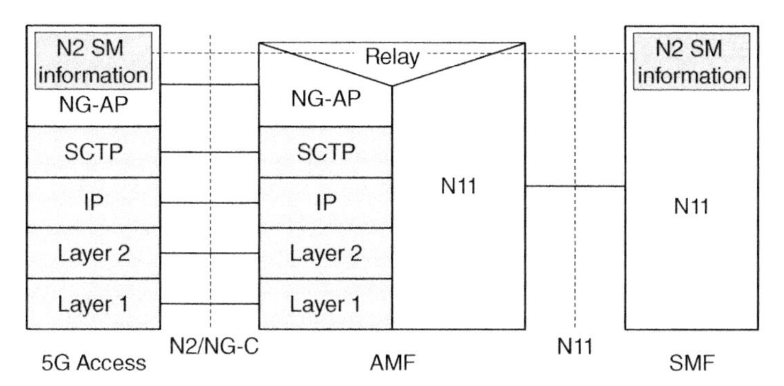

Figure 4.66 Control plane between AN and SMF.

N2 SM information: This is the subset of NG-AP information that the AMF transparently relays between the 5G-AN and the SMF (Figure 4.66).

4.20.1.2 Control Plane Protocol Stacks Between the UE and the 5GC

A single NAS protocol applies on any (3GPP and Non-3GPP) access for signaling between the UE and the 5G Core (Figure 4.67). The 5G NAS protocol is defined in 3GPP TS 24.501 [12].

A N1 NAS connection is used for each access to which the UE is connected. The N1 termination point is the AMF. The single N1 NAS connection is used for multiples purposes and 5GC Network Functions such as Registration Management and Connection Management (RM/CM controlled by the AMF itself and defined in Chapter 5), SM-related messages and procedures for a UE (supported by the SMF and defined in Chapter 6), SMS delivery (supported by the SMSF and defined in Section 4.14).

The NAS protocol on N1 comprises a NAS-MM and a NAS-SM component. The NAS-MM supports:

- Registration Management and Connection Management state machines and procedures between the AMF and the UE (see Chapter 5).
- Transmission of other types of NAS information (e.g. NAS SM) on behalf of other NFs.
- Providing a secure NAS connection (integrity protection, ciphering) between the UE and the 5GC, including for the transport of payload targeting other Network Functions (e.g. SMS, SMSF).
- Provide access control if applicable.

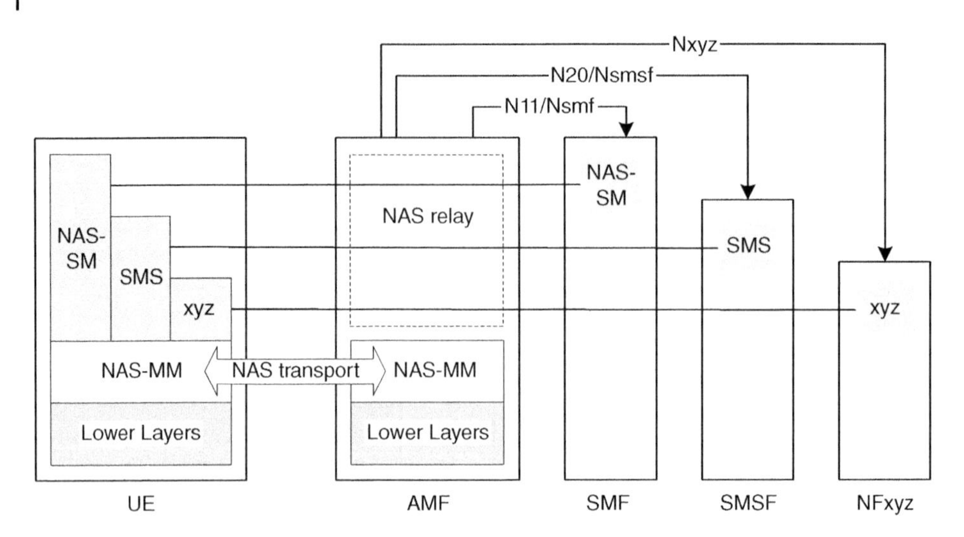

Figure 4.67 NAS transport for SM, SMS and other services.

Security of NAS messages is provided based on the security context established between the UE and the AMF after UE authentication. For this purpose, the AMF receives a Master Key from the AUSF from which it derives the security material to be used to secure NAS messages.

The NAS protocol for SM (NAS-SM) functionality supports user plane PDU session establishment, modification and release [12].

SM signaling messages are handled, i.e. created and processed, in the NAS-SM layer of the UE and in the SMF. The content of the SM signaling messages is not interpreted by the AMF. The NAS-MM layer transfers the SM signaling (NAS-SM) as follows:

- *For transmission of SM signaling*. The NAS-MM layer creates a NAS-MM message, including SM signaling and the corresponding PDU session ID, an indication that SM signaling is carried and finally a NAS security header.
- *For reception of SM signaling*. The receiving NAS-MM processes the NAS-MM part of the message, i.e. performs integrity check based on the NAS security header, and based on the indication that SM signaling is carried and on the PDU session ID, forwards the SM message to the relevant SM entity.

4.20.1.3 Control Plane Protocol Stacks Between the Network Functions in 5GC

The common protocol framework for SBA consists of:

- HTTP/2 as described in IETF RFC 7540 [19].
- TCP as transport layer protocol.
- The JavaScript Object Notation (JSON) format described in IETF RFC 8259 [20] as serialization protocol.
- Applying a RESTful (Representational State Transfer) framework for the API design whenever possible.
- OpenAPI 3.0.0 as the Interface Definition Language.

For transmitting large parts of opaque binary data along with the JSON format, multipart messaging is supported using:

- A multipart/related media type.
- A 3GPP vendor specific content subtype.
- A cross-referencing from the JSON payload using the Content-ID field.

This protocol suite enables a flexible definition of services and supports re-use, easy modifications and easy integration to external domains via the NEF.

4.20.1.4 Control Plane Protocol Stack for the N4 Interface Between SMF and UPF

The control plane protocol between SMF and UPF (N4 reference point) is described in Chapter 6 and specified in 3GPP TS 29.244 [15].

4.20.1.5 Control Plane Protocol Stack for Untrusted Non-3GPP Access

While the IKE security association between the UE and the N3IWF is not established (see Section 4.12), the UE exchanges signaling with the network using EAP-5G carried over IKE (Figure 4.68): EAP-5G allows the exchange of NAS signaling and auxiliary information such as the UE temporary identifier (5G GUTI) and the slices requested by the UE (requested NSSAI). The N3IWF uses the auxiliary information to select the proper AMF for this UE. EAP-5G is defined in 3GPP TS 24.502 [13].

When the IKE association between the UE and the N3IWF is established (see Section 4.12), the UE exchanges NAS signaling with the network using a dedicated IPsec security association (Figure 4.69).

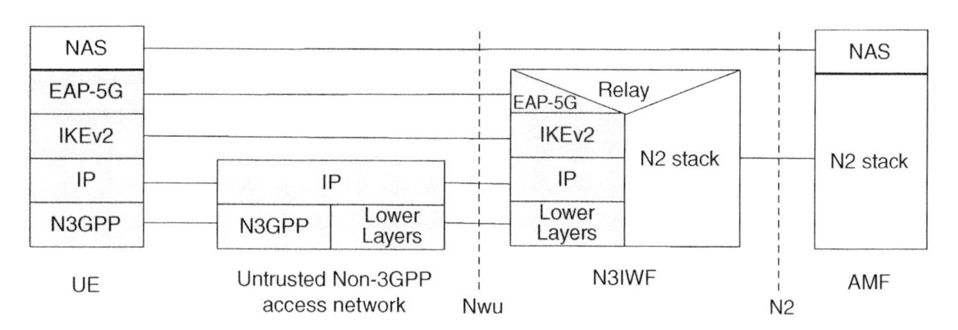

Figure 4.68 Control plane before the signaling IPsec SA is established between UE and N3IWF.

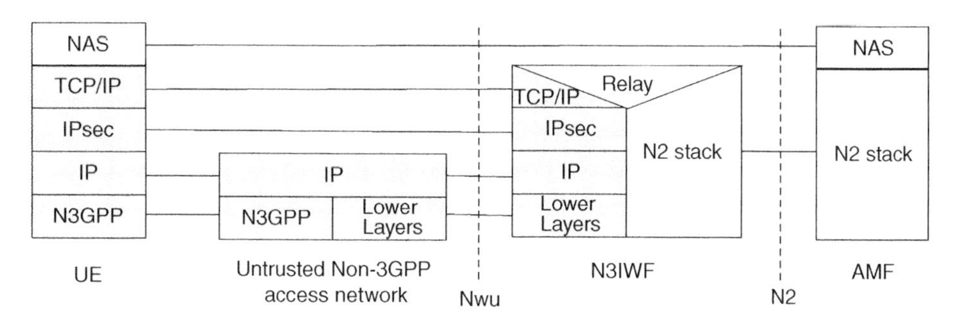

Figure 4.69 Control plane after the signaling IPsec SA is established between UE and N3IWF.

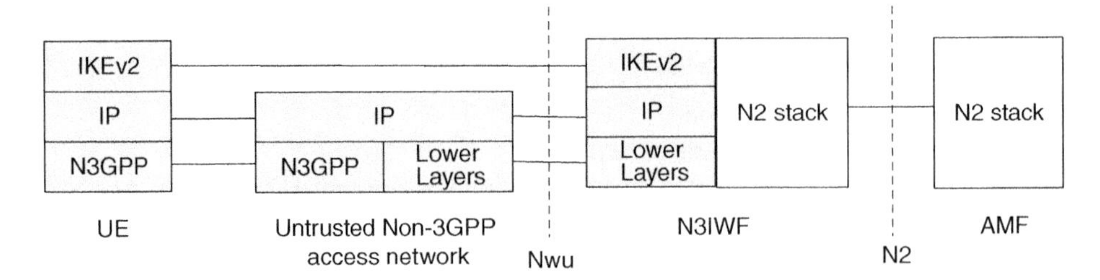

Figure 4.70 Control plane for establishment of user-plane via N3IWF.

When the N3IWF receives requests from the SMF related to user plane resources for a PDU session, it exchanges related signaling, creating, modifying, releasing such user plane resources on SWu for the PDU session, with the UE using dedicated signaling (as defined in 3GPP TS 24.502 [13]) carried over IKE (Figure 4.70).

In the three figures, the UDP protocol may be used between the UE and N3IWF to enable network address translation (NAT) traversal for IKEv2 and IPsec traffic.

4.20.2 User Plane Protocol Stacks

This clause illustrates the protocol stack for the user plane transport related with a PDU session (Figure 4.71).

PDU layer: This layer corresponds to the PDUs carried between the UE and the DN over the PDU session. When the PDU session type is IPv6, it corresponds to IPv6 packets. When the PDU Session Type is Ethernet, it corresponds to Ethernet frames, etc.

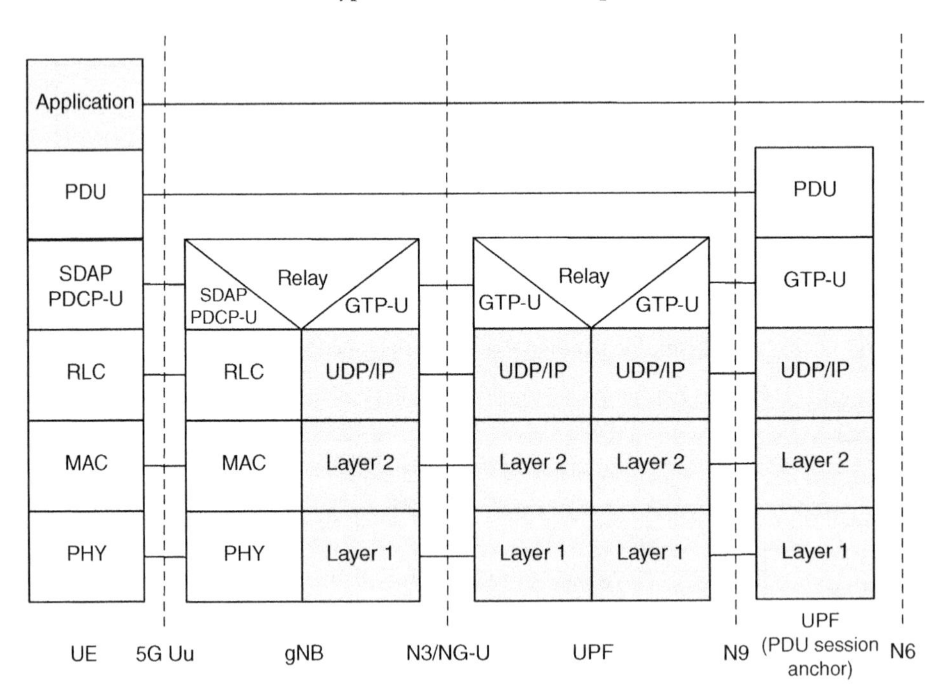

Figure 4.71 User plane protocol stack.

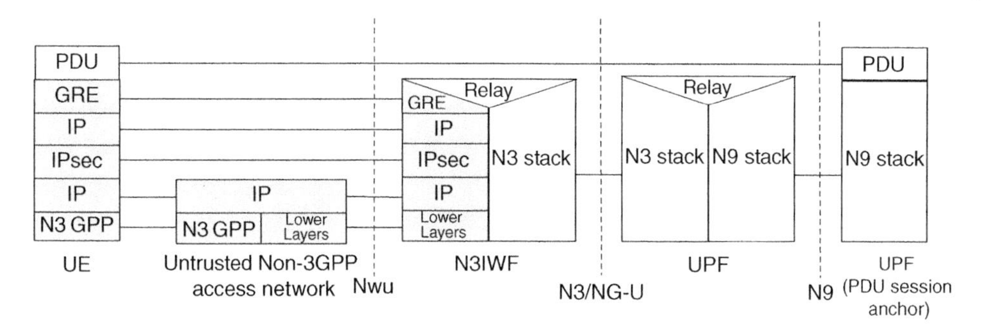

Figure 4.72 User plane protocol stack for untrusted non-3GPP access.

General Packet Radio Service Tunneling Protocol for the user plane (GTP-U): This protocol (defined in [11]) supports multiplexing traffic of different PDU sessions by providing encapsulation on a per PDU session level. It also carries the marking associated with a QoS flow as described in Section 6.4.

5G-AN protocol stack: This set of protocols is Access Network specific. When the 5G-AN is a 5G-NR access, these protocols are defined in Section 4.4.3 and in TS 38.300 [4]. When the 5G-AN is an Untrusted Non-3GPP access these protocols are defined in Figure 4.72.

IPsec is used in tunnel mode for the following reasons:

- Support MOBIKE [7] preserving the IKE association even though the UE has changed the local IP address, e.g. the UE is camping on a new WLAN access that has allocated a new IP address,
- Support of IP packet fragmentation.

UDP can be used below the IPsec layer to enable NAT traversal.

The NG-RAN user plane protocol stack for the gNB is illustrated in Figure 4.73. The figure shows the case for a gNB with F1 Fronthaul interface [25] between gNB-CU and gNB-DU. In case of monolithic deployment, F1 interface collapses. The F1-U interface like Xn-U and NG-U utilizes GTP-U on top of UDP and IP as transport protocol. The NR User Plane protocol [28] (denoted as "NR UP" in Figure 4.73) is introduced on top of F1-U and Xn-U for flow control and frame retransmission. On the NG-U interface (corresponding to N3 reference point in the 5G System), the PDU Session User Plane Protocol [29] is introduced which encapsulates user data and encodes additional information such as the Quality of Service Flow Identifier (QFI).

The user-plane protocol stack is similar like for the gNB with fronthaul interface, if master and secondary nodes are in Dual Connectivity mode involving the Xn interface as illustrated in Figure 4.74.

This figure applies to the cases of NR-E-UTRA-DC (NE-DC) and NG-RAN-E-UTRA-NR-DC (NGEN-DC) [26], i.e. for Dual Connectivity between 4G and 5G nodes irrespective of the master node role, but with all involved nodes connected to the 5G Core. If the F1 interface is present, Xn-U terminates at the gNB-CU. The gNB-DU relays the PDCP and higher layers between gNB-CU and UE as illustrated in Figure 4.73.

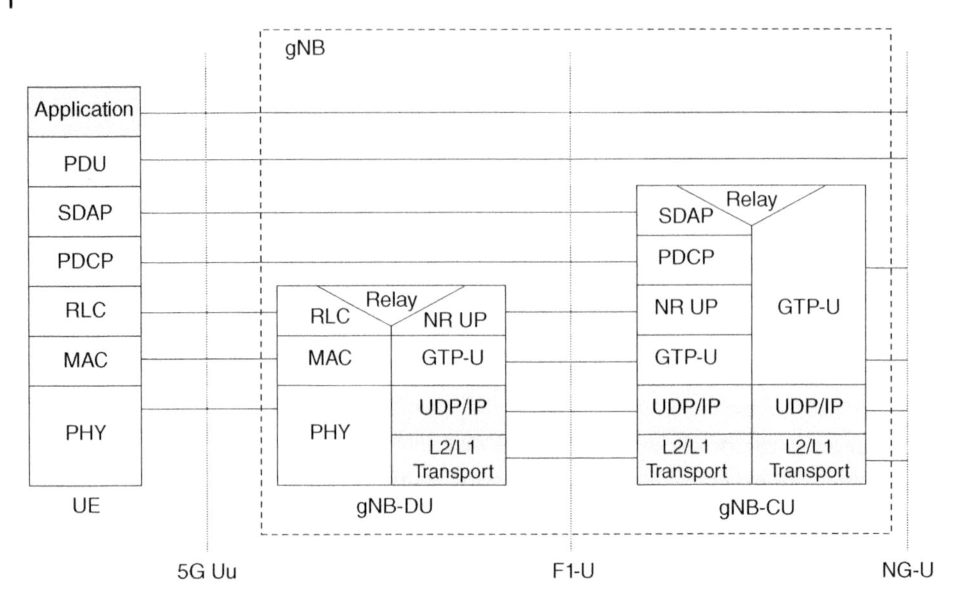

Figure 4.73 NG-RAN user plane protocol stack for gNB with F1 interface.

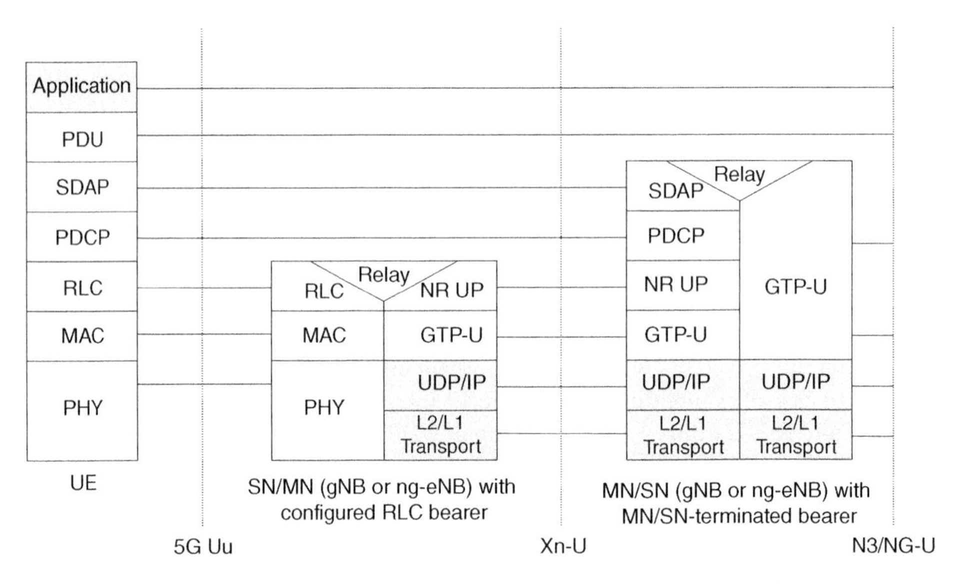

Figure 4.74 NG-RAN user plane protocol stack for MN/SN in dual connectivity mode.

4.21 Charging

Usage of the 5G System is subject to usage data collection. Usage data collection is a prerequisite for charging but may also be used to gather statistics and to monitor network usage and UE behavior. Usage data collection allows end-user off-line charging where the amount to be charged to the end-user is determined on a regular basis, e.g. every month; it allows end-user on-line charging where the amount to be charged to the end-user is determined as the telecommunication service is being delivered. This allows the network to cancel the service delivery when the end-user has not previously filled the account; it also allows inter-operator charging, where the visited PLMN collects usage data related with the telecommunication service it delivers to subscribers of roaming partners.

The SMF is responsible to collect usage monitoring data from the UPF and to transfer these data to the Charging Function (CHF). The CHF is responsible for creating CDR and transferring them to the billing system; it is responsible for checking whether the user account allows the delivery of the telecommunication service and controlling the network behavior, if this is not the case; for example, the CHF may instruct the SMF to have the user traffic redirected to an operator Web Page where the user is invited to refill the account; it is responsible for indicating to the PCF when the user has crossed some spending limits; this allows the PCF considering this information in the policies related with a user session and e.g. restrict the QoS for a given PDU session.

The SMF controls the UPFs to report traffic usage on a PDU session possibly considering the usage of traffic diversion at an UL classifier or a branching point as defined in Chapter 6. Charging information collected by the SMF includes the user ID (SUPI), the target DNN, the slice being used (S-NSSAI); the amount of data exchanged per access type (e.g. 3GPP or Non-3GPP access), rating group (a traffic identification that the PCF can associate with a traffic filter), it this allows to monitor separately traffic of different applications or traffic targeting different IP domains, time of the day.

4.22 Summary and Outlook of 5G System Features

Features specified for the 5G System as part of 3GPP Rel-15 include, but are not limited to (see [32]):

- Service-based architecture with service-based interfaces within 5GC Control Plane; definition for NF services.
- End-to-End Network Slicing.
  - UEs using multiple Network Slices simultaneously including Slice Selection policies in the UE linking applications to Network Slices.
  - Standardized and Operator-defined slice types.
  - Operator defined differentiation among slices with same slice type.
  - RAN and CN capability and business-level-rules-based availability of Network Slices per Tracking area.
- Roaming support for Network Slicing.
- Network Slicing interworking with EPS (with or without (e)DECOR).
- Data Storage architecture enabling Compute and Storage separation using UDSF, along with support for UDR.
- Architectural enablers for virtualized deployment.
- AMF Resiliency supported for all UE(s) returning from IDLE mode (ability for any AMF within an AMF SET to process a given UE's request).
- Support for resiliency of AMF, ability to handle AMF planned maintenance, AMF auto-recovery with no service disruption and/or adverse impact to the UE.
- Converged core network architecture with common interfaces N1, N2 and N3 for 3GPP and untrusted Non-3GPP access.

- Support of customized Mobility Management.
- Support of various PDU session types including IPv4, IPv6, Ethernet and Unstructured.
- Support for Edge Computing, including concurrent (e.g. local and central) access to a data network, an architectural enabler for low-latency services.
- Application influence on traffic routing.
- Support of URLLC services.
- Support for different SSCs.
- Separation of Control Plane and User Plane.
- Support for Interworking with E-UTRAN connected to EPC (with or without a signaling reference point between EPC and 5GC).
- Support for Interworking between ePDG connected to EPC and the 5G System.
- Policy framework for Session, Access and Mobility control, QoS and charging enforcement, policy provisioning in the UE. Introduction of NWDAF for data analytics support.
- Support of services: SMS over NAS (including over Non-3GPP access), IMS services, PWS.
- Flow-based QoS framework, including reflective QoS.
- Support for the new RRC Inactive state.
- Support of IMS services (including support for voice and for network HO based RAT fallback and EPS fallback when IMS services are not supported natively in 5GS deployments).
- Support of IMS Emergency Services over 3GPP and Non-3GPP access including support for Emergency Services using RAT fallback and EPS fallback when these are not supported in 5GS deployments.
- Location Services for regulatory use.
- SEPP to secure and hide the topology for inter-PLMN interconnections.
- Support for 5G MOCN (Mobile Operator Core Network) Network Sharing solutions.
- Control plane load control, congestion and overload control.
  - AMF load balancing, AMF load-rebalancing, TNL (Transport Network Layer between 5GC and 5G-AN) load (re-)balancing.
  - AMF and SMF overload control.
  - NAS level, DNN based and S-NSSAI (slicing) based congestion control.

Future 3GPP releases will be built on top of 3GPP 5G System Rel-15 features and capabilities. It is expected that the following features are covered in 3GPP Release 16 for the 5G System:

The new features (not exhaustive list) targeted for the overall 5G System Architecture in Release 16 could be grouped under the following themes:

- Verticals:
  - Cellular IoT support in 5G
  - Enhancements for URLLC
  - Vertical and LAN Services including Private Networks and Time Sensitive Networking (TSN)
  - Location Services
  - Vehicle-to-X (V2X) Services

- General System Architecture enhancements:
  - Service Based 5G System Architecture enhancements
  - Network Slicing Enhancements
  - Enhancements to SMF and UPF topology
  - User data migration
- Enablers for Network Automation
- Support for additional access
  - Wireless and Wireline Convergence
  - Access Traffic Steering, Switching and Splitting
  - Architecture aspects for using Satellite access in 5G

4.23 Terminology and Definitions

4.23.1 NG-Radio Access Network (NG-RAN)

A RAN that supports one or more of the following options with the common characteristics that it connects to 5GC:

1) Standalone NR.
2) NR is the anchor with E-UTRA extension.
3) Standalone E-UTRA.
4) E-UTRA is the anchor with NR extension.

4.23.2 5G-Access Network (5G-AN)

An access network comprising of a NG-RAN and/or Non-3GPP AN connecting to a 5G Core Network.

4.23.3 5G Core (5GC)

The core network specified in 3GPP TS 23.501 [1]. It connects to a 5G Access Network.

4.23.4 5G System (5GS)

The 3GPP system consisting of 5G Access Network (AN), 5G Core Network (5GC) and UE.

4.23.5 Access Stratum (AS) and Non-Access Stratum (NAS)

The Latin word "Stratum" means layer and was chosen to avoid confusion with other layers such as the Open Systems Interconnection (OSI) layers.

The Access Stratum (AS) layer consists of all functions that are directly related to the RAN and the control of connections between end user device and radio network. Protocols on AS layer run between the device and the radio base station to establish and maintain radio channels.

The NAS layer on the other hand is on top of the AS layer and consists of functions that are related to connection control, access control, mobility and SM. Protocols on

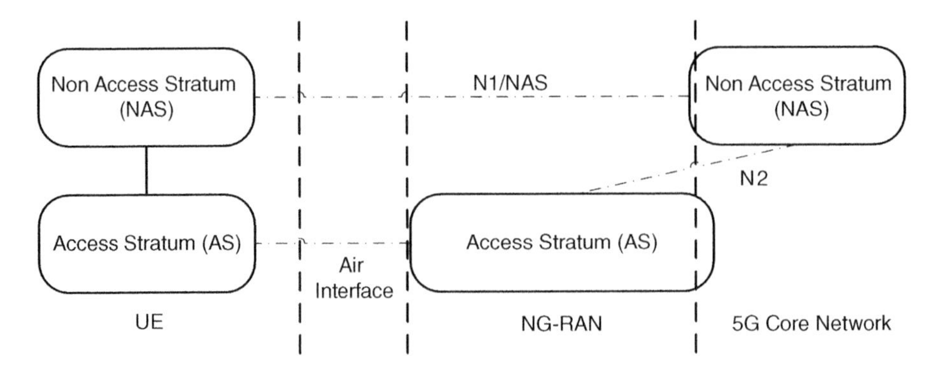

Figure 4.75 AS and NAS layer.

the NAS layer are exchanged between the device and the core network. The NAS and AS layer in the device and core network can communicate with each other. Figure 4.75 provides a simplified overview of the two layers.

References

All 3GPP specifications can be found under http://www.3gpp.org/ftp/Specs/latest. The acronym "TS" stands for Technical Specification and "TR" stands for Technical Report (document a priori not maintained by 3GPP).

1 3GPP TS 23.501: "System Architecture for the 5G System".
2 3GPP TS 23.502: "Procedures for the 5G System".
3 3GPP TS 23.503: "Policy and charging control architecture".
4 3GPP TS 38.300: "NR and NG-RAN Overall Description".
5 3GPP TS 23.003: "Numbering, addressing and identification".
6 3GPP TS 33.501: "Security architecture and procedures for 5G System".
7 IETF RFC 4555: "IKEv2 Mobility and Multihoming Protocol (MOBIKE)".
8 ETSI GS MEC 003: "Mobile Edge Computing (MEC) Framework and reference architecture".
9 BBF SD-407: "Broadband Forum study on Fixed Mobile Convergence", https://www.broadband-forum.org/projects/5g/5g-wireless-wireline.
10 IEEE 802.11 Wireless LAN Medium Access Control (MAC) and Physical Layer (PHY) Specifications.
11 3GPP TS 29.281: "General Packet Radio System (GPRS) Tunnelling Protocol User Plane (GTPv1-U)".
12 3GPP TS 24.501: "Access-Stratum (NAS) protocol for 5G System (5GS); Stage 3".
13 3GPP TS 24.502: "Access to the 3GPP 5G Core (5GC) via non-3GPP access networks".
14 3GPP TS 29.500: "Technical Realization of Service Based Architecture".
15 3GPP TS 29.244: "Interface between the Control Plane and the User Plane Nodes; Stage 3".
16 3GPP TS 38.413: "NG-RAN; NG Application Protocol (NGAP)".
17 IETF RFC 3748: "Extensible Authentication Protocol (EAP)".

18 IETF RFC 4960: "Stream Control Transmission Protocol".

19 IETF RFC 7540: "Hypertext Transfer Protocol Version 2 (HTTP/2)".

20 IETF RFC 8259: "The Java Script Object Notation (JSON) Data Interchange Format".

21 IETF RFC 6824: "TCP Extensions for Multipath Operation with Multiple Addresses".

22 IETF RFC 7296: "Internet Key Exchange Protocol Version 2 (IKEv2)".

23 3GPP TS 23.204: "Support of Short Message Service (SMS) over generic 3GPP Internet Protocol (IP) access; Stage 2".

24 3GPP TS 23.041: "Technical realization of Cell Broadcast Service (CBS)".

25 3GPP TS 38.401: "NG-RAN; Architecture description".

26 3GPP TS 38.470: "F1 General Aspects and Principles".

27 3GPP TS 38.460: "E1 General Aspects and Principles"

28 3GPP TS 38.425: "NR User Plane Protocol".

29 3GPP TS 38.415: "PDU Session User Plane Protocol".

30 3GPP TS 38.410: "NG-RAN; NG general aspects and principles".

31 3GPP TS 38.420: "Xn General Aspects and Principles".

32 SP-170931: Presentation of TS 23.501 "System Architecture for the 5G System; Stage 2 (Release 15)" v2.0.1 for approval.

5

Access Control and Mobility Management

Devaki Chandramouli[1], *Subramanya Chandrashekar*[2], *Jarmo Makinen*[3], *Mikko Säily*[3] *and Sung Hwan Won*[4]

[1] *Nokia, Irving, TX, USA*
[2] *Nokia, Bangalore, India*
[3] *Nokia, Espoo, Finland*
[4] *Nokia, Seoul, South Korea*

5.1 General Principles

5.1.1 Mobility Management Objectives

The fifth generation (5G) access control and mobility management (MM) framework is intended to be adaptive and flexible, to cater for the disparate 5G mobility requirements coming from very diverse use cases.

Some main drivers for this MM framework are the following:

1) Support of common access control procedures for third generation partnership project (3GPP) and non-3GPP access. Enable development of a common protocol to perform access control for 3GPP access and non-3GPP access.
2) Frequent small data transmission. Smart phone applications have the need to send and receive small packets frequently to keep internet protocol (IP) connectivity open. This is referred to as "heart-beat" or "keep-alive." MM state machine should be conducive to support frequent small data transmission without additional signaling overhead.
3) Mobility on demand. Statistics from existing deployments showed that around 70% of the devices camping in cellular network are stationary thus, operators wanted to reduce the cost incurred due to support for MM framework by default resulting in additional cost overhead.
4) Ensuring that the MM state machine is split from session management (SM) state machine (Modular functional split). This is to minimize dependencies between logical function modules, remove overlaps in functional scope from network functions.

With the interest to increase convergence between 3GPP and non-3GPP access, features that are common to both access were considered "mandatory" and other access specific features were considered optional for the network to support. So, with the new system, registration procedure is part of the essential features for all access while features such as paging, reachability, etc. are 3GPP access specific thus considered optional rather

5G for the Connected World, First Edition.
Edited by Devaki Chandramouli, Rainer Liebhart and Juho Pirskanen.

it is necessary to support only for 3GPP access. Furthermore, need to support "mobility on demand" has also caused 3GPP to revisit the use of the term "mobility management" for MM framework in 5G system. The network function that supports registration, connection management (referred to as MM earlier) is called as AMF that stands for "access control and mobility management function."

In a broader sense "mobility management" includes all procedures used to support registration and mobility of a user equipment (UE), such as:

- Selecting a network;
- Registering and Deregistering from the network;
- Camping on a suitable cell and Cell Synchronization;
- Keeping the network informed about the present location of the UE to be reachable for paging even after the connection to the network has been released;
- Maintaining the connection to the network while the UE is moving (referred to as "handover");
- Re-establishing the connection between UE and network when the UE needs to send uplink (UL) signaling or user data or when the UE was paged by the network because the network wants to send downlink (DL) signaling or user data.

Furthermore, certain security related tasks are usually performed as part of the mobility management procedures: authentication, confidentiality protection of the subscriber's identity and confidentiality and/or integrity protection of signaling messages and user data. Detailed security aspects are specified in Chapter 7.

This chapter provides a detailed overview of access control and mobility management framework in 5G Systems. It also provides a detailed overview of radio resource control protocol (RRC) states, non-access stratum (NAS) states, handover and mobility procedure in various states, also handover procedures for Non-Standalone (NSA) mode and Standalone (SA) mode. In addition, it provides some insights on challenges due to high mobility and requirements to support ultra-reliable low latency with mobility. Furthermore, it outlines the procedures for inter-system mobility. Finally, it provides an outlook for supporting enhanced Mobility features in subsequent 3GPP releases.

5.1.2 Mobility Requirements for the 5G System

The mobility requirements for 5G services are significantly more stringent compared to 3G or 4G. The user plane and control plane latency, reliability and robustness to radio link failures, mobility and mobility interruption time requirements are defined to ensure the introduction of mobile systems which support the new capabilities of systems beyond IMT-2000 and IMT-Advanced. IMT-2020 [1] aims at more flexible, reliable and secure mobile system providing diverse services in the enhanced mobile broadband (eMBB), ultra-reliable and low-latency communications (URLLC), and massive machine type communications (mMTC). Further, 5G System is expected to operate in a wide range of frequencies (1–100 GHz) meaning that beamforming techniques are needed to compensate for the higher propagation loss at high frequencies. In this chapter, the 5G System mobility solutions are introduced which can address the 5G mobility requirements.

One of the key target agreements made so far in 3GPP is that handover interruption should be close to 0 ms with UE implemented with a single transceiver capability.

While this is challenging and depends on the exact definition of interruption-less mobility, this can be considered as a major improvement over 4G, which by default always involves a data interruption during handover due to legacy baseline handover procedure. Without going into more detailed solution, there are two basic principles on how the interruption-less handover can be achieved. With a single TX/RX solution the UE and network can deploy a normal handover with Make-Before-Break handover (connectivity to source cell is maintained until connectivity to target cell is established), which won't be interruption-less but can achieve close to 0 ms interruption with reasonable implementation cost. With more than one TX/RX solution at UE the connection to source cell can be maintained with uninterrupted user plane data while the connection to target cell is established, also known as multi-connectivity in 5G.

Performance-wise, the mobility solutions and supporting architectural decisions are having a paramount impact on the resulting capability of system performance. The user plane latency is targeted to be in the order of 4 ms for eMBB, while the URLLC services need far lower latency in the order of 1 ms. The 5G control plane needs to support both extreme power saving and low latency, e.g. transition time from battery efficient state to ready to data transfer. The minimum requirement for control plane latency is 20 ms and 3GPP has decided to target for 10 ms latency.

5.1.3 Mobility Support in the 5G System

Mobility supported in 5G for 3GPP access could be classified into several different categories. One classification is based on the RRC state of the UE:

a. Connected mode mobility.
b. Inactive mode mobility.
c. Idle mode mobility.

Further, in each of these states, depending on the type of mobility event, there is a further categorization:

a. Intra new radio (NR) mobility.
b. Inter radio access technology (RAT) mobility.

Then there is mobility in SA connectivity and NSA connectivity modes. This is applicable to both intra NR and inter-RAT use cases.

There are more flavors available based on type of deployments, namely classical gNB deployment and cloud gNB deployment. Cloud gNB is a gNB where the constituents of a gNB (CU-CP(Central Unit Control Plane), CU-UP(Central Unit User Plane), DU) are separated over F1 (see TS 38.473 [2]) and E1 interfaces (see TS 38.463 [3]). A classical gNB is a gNB where the F1 and E1 interfaces are internal. A cloud gNB may have more than one gNB-DU under its control, while a classical gNB has only one gNB-DU under its control.

For more details on RAN architecture, please refer to Chapter 4.

While the architecture of a classical gNB is realized by juxtaposing the logical entities of a cloud gNB, the 3GPP defined interfaces between those logical entities (F1 and E1) will simply become internal. Therefore, we describe the mobility events for a cloud gNB and explain how they are realized for a classical gNB.

We describe in this chapter the various kinds of mobility for different RRC states (e.g. Intra gNB, Inter gNB. Intra DU, Inter DU, and Inter RAT). RRC Connected, RRC Inactive and RRC_IDLE have their own mobility procedures.

5.2 Mobility States and Functionalities

5.2.1 NAS State Machine and State Transitions

The NAS layer can be categorized into two main sublayers (See TS 24.501 [4]): MM sublayer and SM sublayer. The main function of the MM sublayer includes registration management (RM), and the main function of the SM sublayer is to manage the protocol data unit (PDU) session context.

The NAS state machine is applicable for both 3GPP and non-3GPP access. In evolved packet system (EPS), 3GPP NAS SM protocol for non-3GPP access was emulated to support wireless local area network control plane protocol (WLCP) between UE and enhanced packet data gateway (ePDG) (see [5]). However, the NAS specifications and the actual protocol were still different and evolved separately for 3GPP and non-3GPP access. With 5G architecture, target was to specify a single NAS protocol that evolves together for both 3GPP and non-3GPP access. Although not all information elements within NAS are expected to be common and applicable for both 3GPP and non-3GPP access, it is expected that the overall mobility, authentication, and session management framework are applicable for both thus enabling support for converged core network. This also enables seamless mobility between 3GPP and non-3GPP access without the need to re-authenticate in all cases.

The simplified form of the state machine managed in the MM sublayer for RM is depicted in Figure 5.1. The state machine is managed per access type (i.e. 3GPP access or non-3GPP access).

In the RM-NULL state, 5GS services are disabled in the UE. No MM function is performed in this state. The UE in the RM-DEREGISTERED state can enter to this state by disabling the N1 mode, a mode of a UE allowing access to the 5GC via the 5G-AN. The UE in the RM-NULL state can transition to the RM-DEREGISTERED state if the N1 mode is enabled.

In the state RM-DEREGISTERED, no MM context has been established and the location of the UE is unknown to the 5GC and hence it is unreachable by the 5GC. To establish a MM context, the UE shall start the initial registration procedure. The UE in the RM-REGISTERED state can enter to this state by (locally) deregistering from the 5GS.

In the state RM-REGISTERED, a MM context has been established. Additionally, one or more PDU session context(s) may be activated at the UE. The UE may initiate

Figure 5.1 States in the MM sublayer.

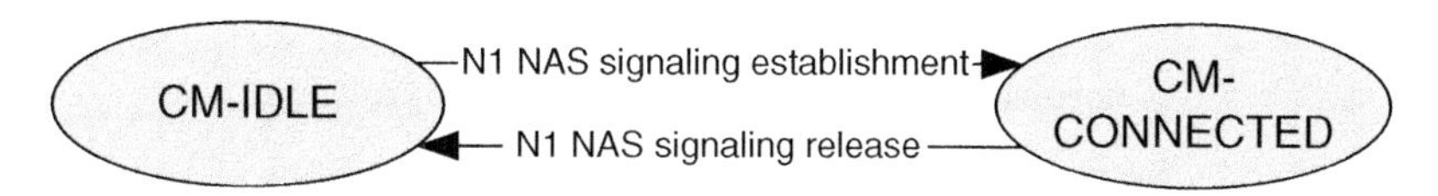

Figure 5.2 States for the CM.

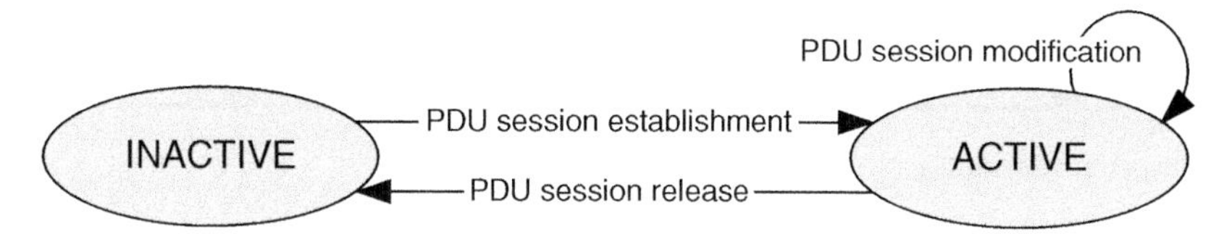

Figure 5.3 States in the SM sublayer.

the non-initial registration procedure (including the mobility registration update and periodic registration update).

Furthermore, in the state RM-REGISTERED, UE can either be in CM-CONNECTED state or CM-IDLE state (Figure 5.2). When the UE is in CM-IDLE state, the network must page the UE to be able to reach the UE in the case of the 3GPP access. Similarly, the UE needs to send an initial NAS message to transition to CM-CONNECTED state to transmit either signaling or user plane packets. When the UE is in CM-CONNECTED state, UE and network can communicate at the NAS layer level.

The simplified form of the state machine managed in the SM sublayer is depicted in Figure 5.3. The state machine is managed per PDU session.

In the INACTIVE state, no PDU session context exists. The successful PDU session establishment allows transition to the ACTIVE state.

In the ACTIVE state, PDU session context is active in the UE. The state of a PDU session context with the ACTIVE state changes to the INACTIVE state if the PDU session context is released by the PDU session release procedure. The PDU session context in this state can be modified by the PDU session modification procedure.

5.2.2 RRC State Machine and State Transitions

With 5G, there is a need to address the "smart" problem with "always on" applications:

- Multiple applications run on top of each other with independent heartbeats.
- "Always on" applications need to send and receive small packets frequently to keep IP connectivity open.
- The typical frequency of "keep alive" is once per minute, or once every few minutes, and the amount of data is very small (≪1 KB).
- Keep alive messages are typically generated by the application without any option for the network to control them (Figure 5.4).

Hence, to tackle this challenge, the 5G systems was expected to adopt connectivity with flexibility in its configuration to system access. In 5G, the system is expected to offer a solution to support "always connected" applications taking "always on" to the next level. Thus, the access to the network and state transition to connected state should be instantaneous from the application perspective, in the order of 10 ms [6].

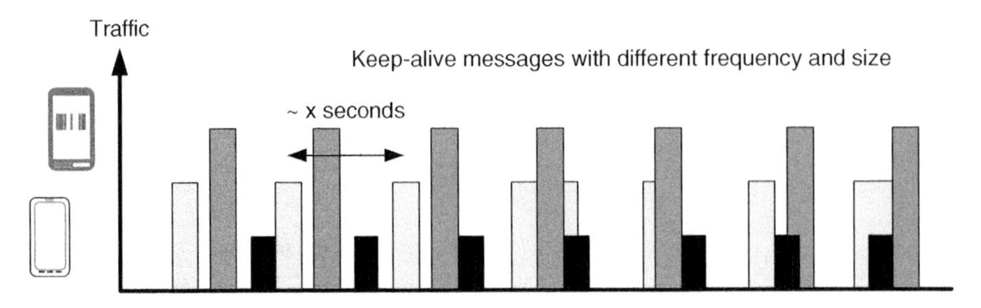

Figure 5.4 Traffic pattern illustration.

An overview of existing RRC states and state transitions specified by 3GPP for LTE is shown in Ref. [7], which also illustrates the mobility support between evolved universal mobile telecommunications system terrestrial RAN (E-UTRAN), universal terrestrial radio access network (3G RAN) (UTRAN) and GSM/EDGE RAN (GERAN). Details of mobility management for Evolved Packet System is described in TS 23.401[8] and TS 24.301[9]. The RRC states in LTE, namely RRC_IDLE and RRC_CONNECTED, try to minimize the UE power consumption, network resource usage and memory consumption while the RRC_CONNECTED was introduced for high UE activity and network controlled mobility respectively. The state transition from CONNECTED to IDLE and vice-versa requires a considerable amount of signaling to setup the UE's AS context in RAN and introduces delay and extra e2e system signaling. Smartphone applications and machine type communications (MTC) devices have resulted in the reconsideration of state handling efficiency in LTE systems. It is costly to keep all the UEs in RRC_CONNECTED state, as high user activity increases power consumption and mobility related signaling. On the other hand, the state transition from RRC_IDLE to RRC_CONNECTED, in the order of 50 ms, cannot meet the 10 ms latency requirements of 5G. This resulted in the need to reconsider RRC state machine for 5G System.

Although the UE is in CM CONNECTED state from the core network (CN) perspective, the RAN can suspend the RRC state of the UE during inactivity periods. From the RAN perspective, the UE can be in RRC connected, RRC Inactive or RRC Idle state. When inactivity is detected, the UE may request based on a configured timer that the network can suspend the RRC connection. Alternatively, the RAN may suspend the RRC connection from RRC_CONNECTED to RRC_INACTIVE after the data buffers are empty or if there is a temporary inactivity detected. The RRC_IDLE state may be rarely used, for example, as a recovery state when RRC resume fails. This resulted in the introduction of RRC_INACTIVE state for 5G and the RRC state machine comprises of RRC_IDLE, RRC_CONNECTED and RRC_INACTIVE [10]. Figure 5.5 illustrates the RRC State machine and expected transitions (for details see Ref. [11]):

Figure 5.6 shows how RRC state machine fits within the NAS states:

The UE mobility is controlled by the network during RRC connected state. However, during RRC Inactive state, the mobility of the UE is controlled by the UE with the assistance of the network. In RRC_IDLE state, i.e. in a kind of recovery state, the UE mobility using cell reselections is autonomously controlled by UE. Lightweight procedures called RRC suspend and RRC resume are respectively used to resume and suspend RRC connection. The RRC suspend message may contain service tailored configuration to address UEs with diverse service requirements.

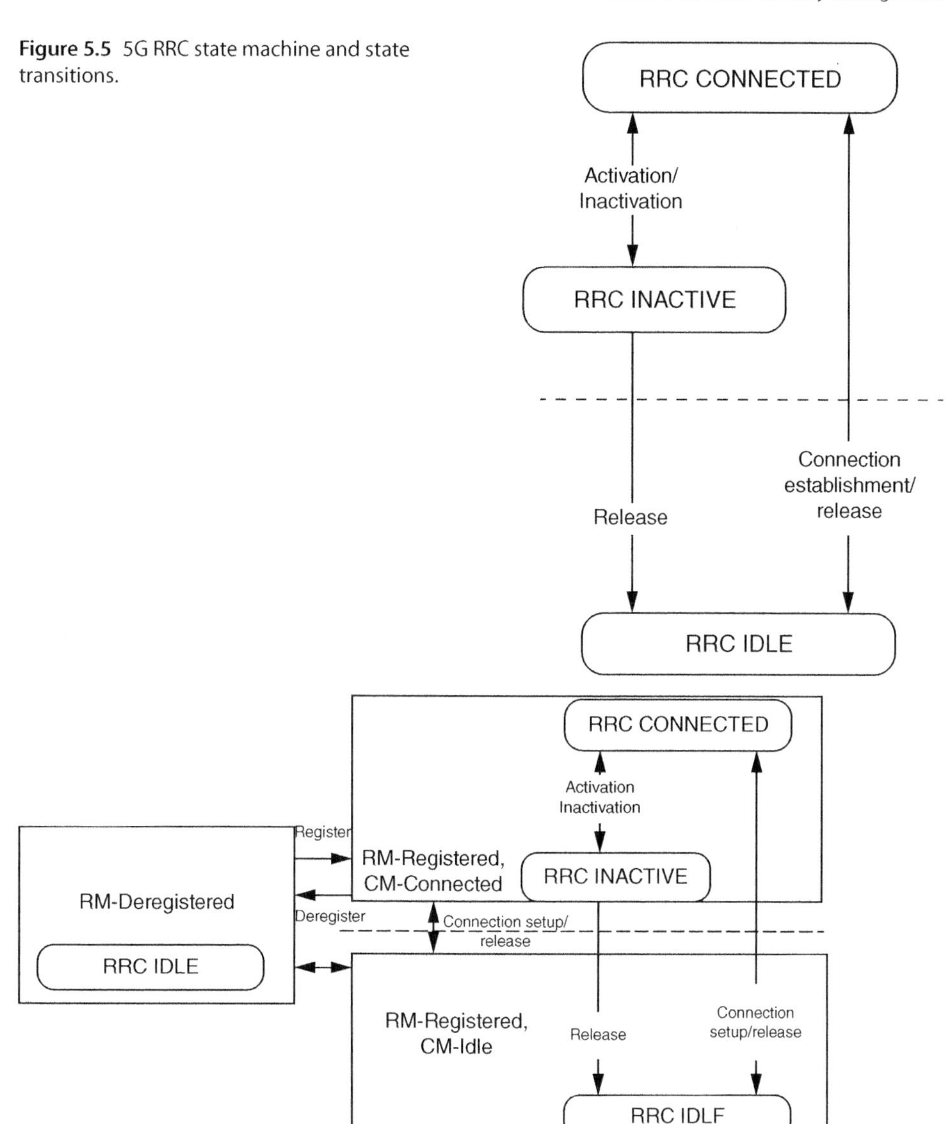

Figure 5.5 5G RRC state machine and state transitions.

Figure 5.6 5G RRC state machine embedded with NAS State machine.

The UE state transition happens when either the UE or the next generation RAN (NG-RAN) node detects a low activity condition and UE has no ongoing data traffic in the user-plane.

When the network commands the UE to Inactive state, the last serving NG-RAN node (gNB or ng-eNB) sends an RRC Connection Suspend message to the UE. The message that contains (at least) Resume Identification (ID, in this case the Last NG-RAN node (gNB or ng-eNB) ID), Inactive state related timing Information (e.g. Registration period), up-to-date list of TAs (Tracking Areas) in which the UE can move without TA update, and Security Information for UE identification while re-connecting to the network.

Active connection in RRC Connected state is needed again when an application needs to send data. The UE is already connected so it will reconnect to the network via its current NG-RAN cell and sends the RRC Connection Resume Request message to the NG-RAN node including (at least) UE ID, Resume ID, Inactive state related timing Information (e.g. time spent in inactive state), and Security Information to verify the UE context. The NG-RAN node responds to the UE with the RRC Connection Resume Complete message and UE is back to CONNECTED state.

A connection failure during the RRC Inactive can happen for example due to failed cell reselection or if the cell update to RAN was not acknowledged back to UE. Also RAN can detect and assume a connection failure or UE may have been powered off if RAN has not received any location update information from UE within the maximum reporting time period.

5.2.3 Inter-RAT Operation of RRC States

In legacy networks, when a UE in RRC_IDLE state measures and selects to camp in a better cell in a new RAT, it enters RRC_IDLE in the new RAT. Similarly, it is agreed in Release 15 that UE in the RRC_INACTIVE state can perform selection to another RAT and the UE enters the RRC_IDLE state in that RAT. The tight interworking between NR and LTE evolution during inactive state could be exploited more effectively for NG-RAN mobility. An NG-RAN node can either provide NR or E-UTRA control plane with RRC protocol terminations toward the UE. Therefore, it could be assumed that eLTE will support also RRC_INACTIVE and allow cell re-selection between the systems within the NG-RAN context. The RRC_INACTIVE state could have a common configuration with RAT specific details and a common RAN Notification area configuration shared within NG-RAN. The RRC States in the UE and the anchor eNB remain the same, even if the UE selects a cell that belong to another RAT and its RAN area. The procedure can be referred as NG-RAN RRC reactivation procedure to activate the RRC connection between the UE and the current anchor eNB, possibly located in another RAT. In addition to activating the RRC connection between the UE and the anchor NB, the anchor eNB may send the necessary UE context information to the new eNB.

As an example, let's consider a scenario where eLTE (E-UTRA) and NR are connected to 5GC (Figure 5.7). The UE is under the NR coverage and network suspends the RRC connection to RRC_INACTIVE. The UE AS context is stored in the anchor gNB and the UE. The stored UE context may include, e.g. data radio bearer (DRB) configuration, AS security context, etc. The UE is provided with a NG-RAN notification area that consists of cells that belong to eLTE and NR. If the UE reselects to eLTE, it does not need to update its location nor change RRC state to RRC_IDLE. When the UE has the data available for transmission, it initiates an NG-RAN RRC activation procedure to the current cell, which fetches the UE Context from the anchor gNB and resumes the RRC connection with the state transition to RRC_CONNECTED.

5.2.4 Benefits of the New RRC State Model

The new state model and proposed new RRC state together optimize the power consumption of mobile devices during the low activity periods while minimizing the latency for the first packet transmission from the UEs to the network. The identified

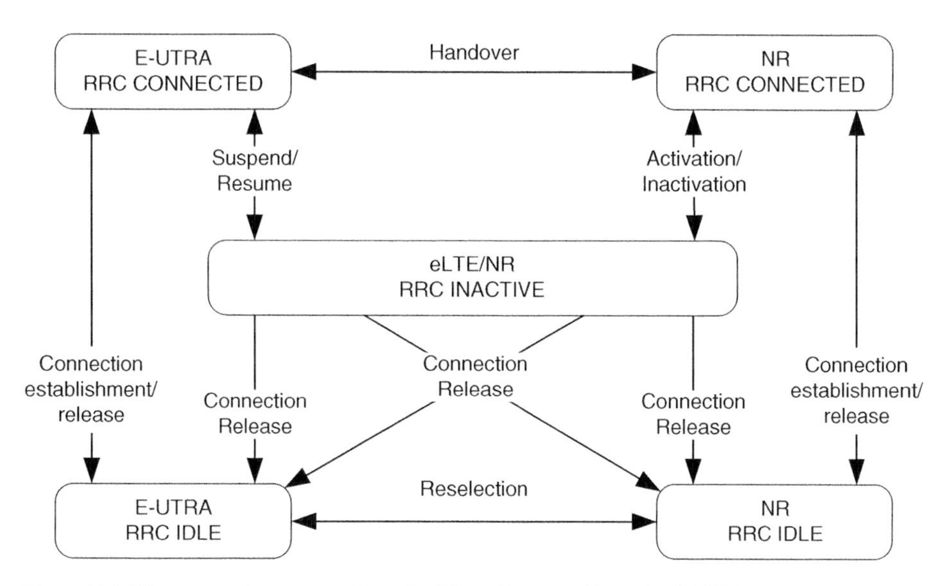

Figure 5.7 UE states and state transitions for NR and interworking with E-UTRA.

Table 5.1 Characteristics of RRC states in 5G.

5G state	Mobility	Paging	UE location information	Storage of UE context
RRC_CONNECTED	NW based cell selection & reselection	NR-RAN and 5GC	Cell level	UE & NG-RAN
RRC_INACTIVE	UE based cell selection & reselection	NR-RAN and 5GC	RAN Tracking Area	UE & NG-RAN
RRC_IDLE	UE based cell selection & reselection	5GC	CN Tracking Area	5GC

characteristics of the RRC state model are shown in the Table 5.1. The mobility and system access procedures of the new state model are configurable based on different aspects of use cases, device capability, access latency, power saving, security requirements and privacy.

The mobility state design for 5G has taken the learnings from the pros and cons of state handling design in the existing technologies and considered the 5G use cases and their requirements. Some benefits of the new 5G RRC state model using RRC_INACTIVE are that it:

- can keep the UEs always connected from 5GC perspective;
- enables minimum number of RRC states to avoid added complexity in the state model;
- provides fast and lightweight state transition between active data transmission and power saving, which can reduce CP signaling overhead for frequent state transitions;
- can fulfill the latency requirement of state transition for control plane;
- can support highly configurable procedures for contradictory requirements of various 5G use cases; and
- allows network slice specific state configuration.

5.3 Initial Access and Registration

UE(s) need to register to the network to receive services, e.g. enabling mobility and reachability of mobile devices. The registration procedure is used when the UE performs initial registration to the 5G system, a mobility registration update upon changing to a new Tracking area (TA) outside the UE's registration area in both CM_CONNECTED and CM_IDLE mode or when a UE performs a periodic registration update (due to a pre-defined time of inactivity), and additionally when the UE needs to update its capabilities or protocol parameters that are negotiated in the Registration procedure.

Initial registration is performed after the UE power-on, when the UE is not yet registered nor connected to the network. UE starts connection to the network using access control procedures which start from network selection. UE needs to determine to which public land mobile network (PLMN) to attempt registration and which access network to select. Initial access consists of functions where the cell search and acquisition are done at first allowing a UE to synchronize to the target cell. The 5G enables the synchronization channels to have different periodicity (5, 10, 20, 40, 80 or 160 ms) depending on, e.g. traffic load and power saving requirements, which is quite different from LTE which uses periodicity of 5 ms for synchronization signals. The reference signals (RS) are necessary for signal reception and are transmitted in 5G in the same sub-frame using the same bandwidth and beam as the corresponding user data.

The system information (SI) is distributed over each cell and this enables the UE to acquire the information needed to access the network. The SI includes the cell bandwidths and various configurations for downlink and uplink and is broadcasted using the Master Information Block (MIB) and the Secondary Information Blocks (SIBs). The MIB contains the information on how to read the SIB information. The SIB includes configurations for the UE, such as cell selection and reselection, random access (RA) parameterization and permissions such as access barring and configuration of broadcast services such as warning messages. In 5G the SI is divided into minimum SI and "other SI." Minimum SI is periodically broadcasted and comprises of basic information required for initial access, while the other SI is provisioned on-demand basis.

During the initial access over the radio interface, the UE attempts the connection to RAN. Initial access to RAN starts from Random Access procedure. There is a time-frequency location and identification, when the UE can access random access channel (RACH) but in principle the procedure happens in a random fashion. When the RAN has responded to RA, the connection setup over the radio interface triggers UE's sending an initial NAS message (the REGISTRATION REQUEST message in case of the initial registration), which is used to establish an NAS signaling connection between UE and AMF. After the security setup, the RRC connection can be enabled (i.e. the UE transitions to RRC_CONNECTED state) and a NAS signaling connection can be established.

The following figures depict interactions among UE and network functions in the 5GC upon initiation of the REGISTRATION REQUEST message (i.e. the initial registration procedure). Figure 5.8 shows the initial registration procedure where the AMF does not fetch UE context from any other AMF, and Figure 5.8 exhibits the initial registration procedure in which the AMF (new AMF) receives UE context from a different AMF (old AMF) that had served the UE previously: the new AMF attempts to receive UE context from the old AMF if the REGISTRATION REQUEST message contains the 5G globally

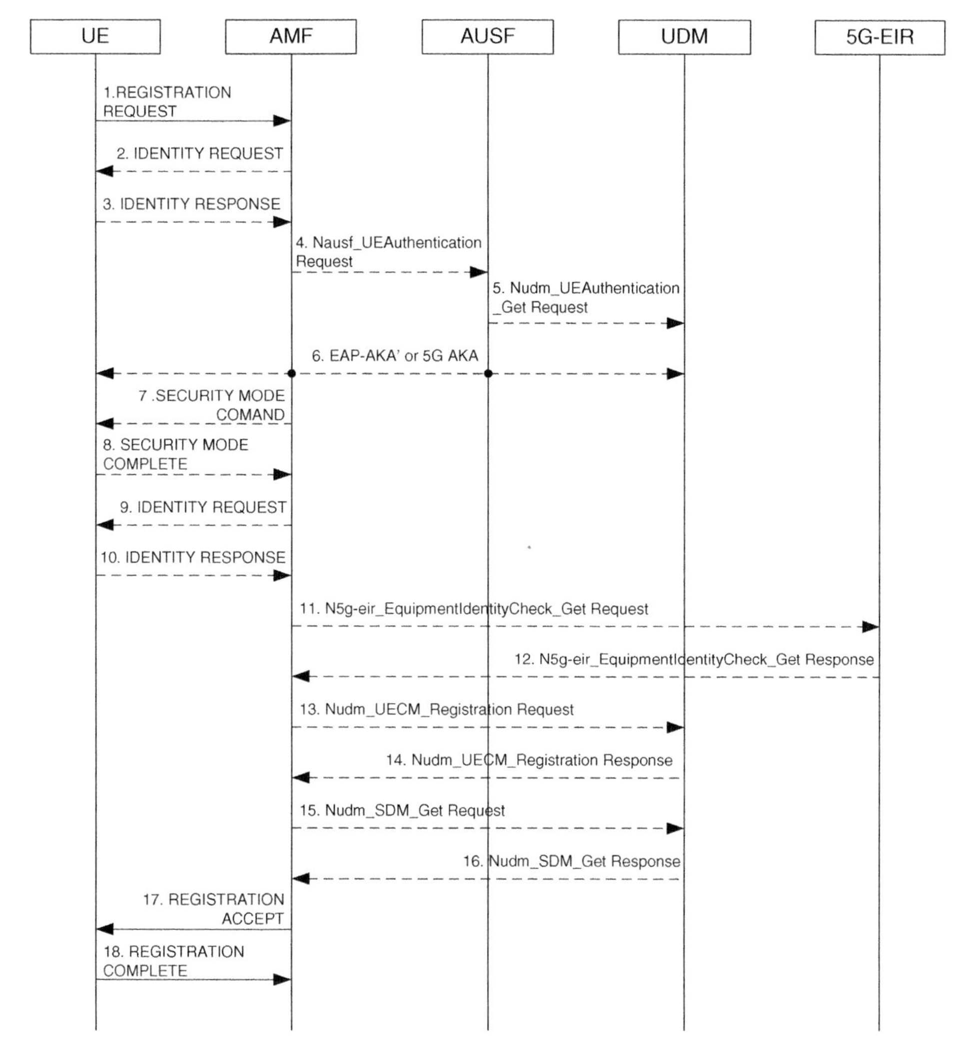

Figure 5.8 Initial registration procedure (UE context not fetched from another AMF).

unique temporary identifier (5G-GUTI) and the new AMF can contact the old AMF identified by the globally unique AMF identifier (GUAMI) derived from the 5G-GUTI. In both figures and corresponding step-wise descriptions, the procedure is simplified compared to the actual initial registration procedure. See 3GPP TS 23.502 [12] for further details.

1. The UE sends the REGISTRATION REQUEST message to register to the 5GS. The REGISTRATION REQUEST message is delivered via the RRC message (from the UE to the NG-RAN) and the next generation application protocol (NGAP) message (from NG-RAN to the AMF, see TS 38.413 [13]). To help the NG-RAN to make a proper AMF selection, the UE includes in the RRC message the GUAMI and/or requested network slice selection assistance information (NSSAI), if available. The

GUAMI is derived from a valid 5G-GUTI in the UE, if any. The REGISTRATION REQUEST message includes an identity of the UE. The identity is either the 5G-GUTI or the subscription concealed identifier (SUCI). The use of the 5G-GUTI is prioritized, i.e. if the UE has a valid 5G-GUTI, the UE should use the 5G-GUTI. In addition, the REGISTRATION REQUEST message may include, among many other parameters, the requested NSSAI. The requested NSSAI includes a list of S-NSSAIs that the UE requests to use in the PLMN (see Section 4.7). In case the UE has valid security parameters, the REGISTRATION REQUEST message is integrity protected.

2. If the AMF cannot retrieve SUCI using the 5G-GUTI provided by the UE and/or SUCI is not provided by the UE, the AMF sends the IDENTITY REQUEST message to the UE requesting the SUCI.
3. The UE responds with the IDENTITY RESPONSE message including the SUCI.
4. If the REGISTRATION REQUEST message (sent in step 1) does not successfully pass the integrity check, the AMF initiates UE authentication by invoking the Nausf_UEAuthentication service: the Nausf_UEAuthentication_Authenticate Request message is sent to the authentication server function (AUSF) selected by the AMF based on the SUCI. The message includes the SUCI. Otherwise, steps 4–16 can be skipped.
5. Upon receiving the Nausf_UEAuthentication_Authenticate Request message, the AUSF shall check that the requesting AMF is entitled to request the Nausf_UEAuthentication service. If the check is successful, the AUSF sends the Nudm_UEAuthentication_Get Request message to the unified data management (UDM). Upon reception of the Nudm_UEAuthentication_Get Request message, the UDM de-conceals the SUCI included in the message to gain the subscription permanent identifier (SUPI). Based on SUPI, the UDM chooses the authentication method, either EAP-AKA' [14] or 5G AKA [15], based on the subscription data available in the UDM.
6. Either EAP-AKA' or 5G AKA is performed involving the UE, the AMF, the AUSF, and the UDM.
7. The AMF sends the SECURITY MODE COMMAND message to initialize NAS signaling security.
8. The UE responds with the SECURITY MODE COMPLETE message. From this step, any NAS message is ciphered and integrity protected, if not done earlier.
9. If the permanent equipment identifier (PEI) cannot be retrieved, the AMF can request the UE to provide the PEI by sending the IDENTITY REQUEST message.
10. The UE sends the IDENTITY RESPONSE message including the PEI.
11. Optionally, the AMF initiates the PEI check by invoking the N5g-eir_Equipment IdentityCheck_Get service with the N5g-eir_EquipmentIdentityCheck_Get Request message which is sent to the 5G-equipment identity register (5G-EIR).
12. The 5G-EIR indicates to the AMF the PEI check result via the N5g-eir_Equipment IdentityCheck_Get Response message.
13. The AMF registers with the UDM by invoking the Nudm_UECM_Registration service, i.e. sending the Nudm_UECM_Registration Request message to the

UDM. The AMF is implicitly subscribed to be notified when it is deregistered in UDM.
14. The UDM responds with the Nudm_UECM_Registration Response message.
15. The AMF attempts to retrieve the subscription data using Nudm_SDM_Get service; the AMF sends the Nudm_SDM_Get Request message to the UDM. Upon receipt of the message, the UDM may retrieve the UE context from unified data repository (UDR).
16. The UDM responds with the Nudm_SDM_Get Response message. The message includes the subscription data.
17. The AMF sends the REGISTRAION ACCEPT message to the UE indicating that the registration is accepted. The message includes the 5G-GUTI if it is newly assigned to the UE by the AMF. The message also includes a registration area allocated for the UE. The allowed NSSAI can be included in the message as well; it indicates the S-NSSAIs that the UE can make use of while being registered to the PLMN. The S-NSSAIs included in the allowed NSSAI can be the subset of the S-NSSAIs contained in the requested NSSAI of the REGISTRAION REQUEST message (sent in step 1).
18. The UE sends the REGISTRATION COMPLETE message to the AMF to acknowledge if a new 5G-GUTI was assigned via the REGISTRAION ACCEPT message.

Figure 5.9 shows registration procedure with the AMF being able to retrieve UE context from old AMF:

1. First step is like step 1 of Figure 5.8. The only difference is that the UE includes the 5G-GUTI, not the SUCI in this case. Inclusion of the 5G-GUTI is a prerequisite for UE context fetch from the old AMF.
2. The new AMF invokes the Namf_Communication_UEContextTransfer service towards the old AMF by sending the Namf_Communication_UEContextTransfer Request message to the old AMF. The old AMF is the AMF identified by the GUAMI included in the 5G-GUTI. The message includes the complete REGISTRATION REQUEST message sent to the new AMF in step 1.
3. The old AMF verifies the REGISTRATION REQUEST message and responds with the Namf_Communication_UEContextTransfer Response message to the old AMF. If the verification is a success, the message includes the UE context. Otherwise, the new AMF cannot retrieve the UE context from the old AMF and should perform step 2 and onwards of Figure 5.8.
4. If the UE context from the old AMF does not include the PEI, the AMF can obtain the PEI from the UE. See steps 9 and 10 of Figure 5.8.
5. See steps 11 and 12 of Figure 5.8.
6. See steps 13 and 14 of Figure 5.8.
7. The Nudm_UECM_Registration service invoked by the new AMF triggers the UDM to initiate the Nudm_UECM_DeregistrationNotification service to the old AMF. The UDM sends the Nudm_UECM_DeregistrationNotification Notify message to the old AMF and the old AMF removes the UE context.
8. See steps 17 and 18 of Figure 5.8.

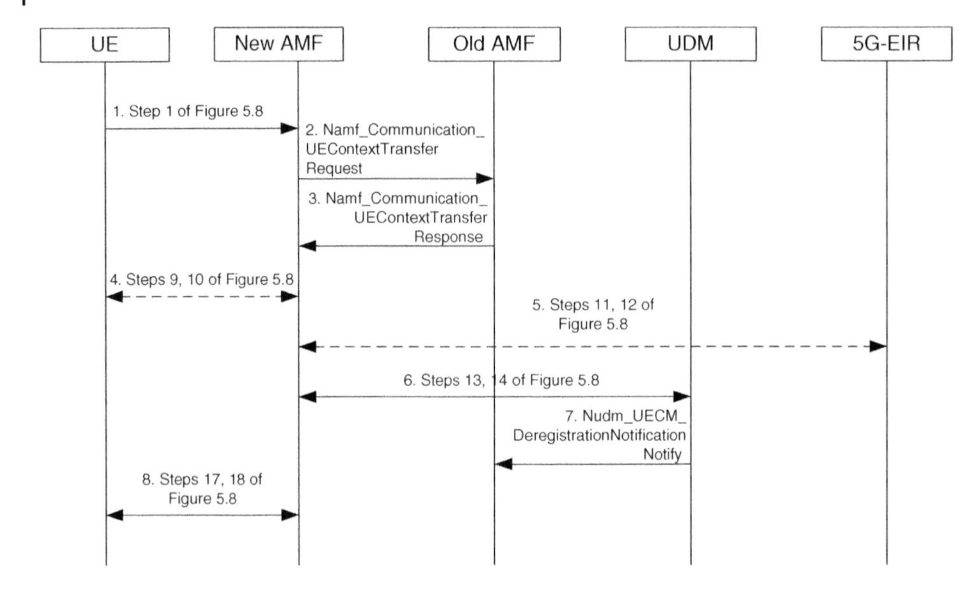

Figure 5.9 Initial registration procedure (UE context fetched from the old AMF).

5.4 Connected Mode Mobility

Connected mode mobility procedure differs depending on the type of deployment: NSA or SA.

5.4.1 NSA Mobility Scenarios

Here we consider mobility events related to MR-DC (LTE-NR Dual Connectivity). MR-DC (MultiRAT Dual Connectivity) is a generic term used to represent all the three variants of LTE-NR Dual connectivity, namely,

a. EN-DC (option 3 family): LTE eNB master and NR gNB secondary.
b. NE-DC (option 4): NR gNB master and eLTE ng-eNB secondary.
c. NGEN-DC (option 7): eLTE ng-eNB master and NR gNB secondary.

The Master Node (MN) and Secondary Node (SN) roles could be taken by any node according to the configurations listed above. The following are the mobility events that could occur while dealing with a UE in any of the MR-DC configurations. Here we describe the mobility events with a common MN and SN terminology:

a. Intra SN mobility;
b. Change of SN: Inter SN Handover (MN or SN initiated);
c. Inter MN Handover without SN change;
d. Master Node to eNB/gNB Change: SN Release.

5.4.1.1 Intra SN Mobility in a Cloud RAN Deployment

We describe mobility for the CU-DU split option, where the CU can be implemented, e.g. in a cloud, since it has use cases like inter-DU mobility to support that are not applicable for a gNB where CU and DU functions are collocated.

The intra SgNB mobility can be carried out over the SRB1 and SRB2 (eNB assisted) or over SRB3 (autonomous for gNB). Some topics that were considered:

a. The UE support for SRB3 is a major factor which influences if a UE will be configured with SRB3.
b. The support of SRB3 by a UE and network configuration for the same does not guarantee all the intra gNB mobility events to be executed over SRB3.
c. Whenever there is an inter MN cell change, the KeNB changes. This results in a S-KeNB change as well. Therefore, in all cases involving a security key change, a combined Pcell and PScell RRC re-configuration is performed. This is executed over SRB1/2 of MeNB.

However, it should also be noted that whenever SRB3 is used for RRC re-configurations or mobility events, there is a certain optimization in terms of latency of the executed procedure.

Intra CU, Inter DU Mobility using SCG SRB (SRB3) This scenario is applicable only for a cloud RAN gNB, since multiple DUs are not possible in a classical gNB.

This procedure is used for the case the UE moves from one gNB-DU to another gNB-DU within the same gNB-CU during NR operation. Figure 5.10 shows the inter-gNB-DU mobility procedure for intra-NR.

1. The UE sends a *Measurement Report* message to the source gNB-DU.
2. The source gNB-DU sends an Uplink RRC Transfer message to the gNB-CU to convey the received *Measurement Report*.
3. The gNB-CU sends an UE Context Setup Request message to the target gNB-DU to create an UE context and setup one or more bearers, which contains *Target Cell ID*.
4. The target gNB-DU responds to the gNB-CU with an UE Context Setup Response message, which contains *MobilityControlInfo*. In a successful case, the admission control check has passed and resources are allocated at the target.
5. The gNB-CU sends a UE Context Modification Request to the source DU, which indicates to suspend the data transmission for the UE, at the source gNB-DU. The source gNB-DU also sends a Downlink Data Delivery Status frame to inform the gNB-CU about the unsuccessfully transmitted downlink data to the UE. Downlink packets, which may include packet data convergence protocol (PDCP) PDUs not successfully transmitted in the source gNB-DU, are sent from the gNB-CU to the target gNB-DU.
6. S-gNB-DU responds with successful suspension of data transmission and UE scheduling.
7. PSCell change is triggered by gNB-CU over SRB3. Message contains RRC: RRCReconfiguration message, which contains the *CellGroupConfig* IE, and internally the SpCellConfig IE and the SCellConfig IE. SpCellConfig IE contains the reconfigurationWithSync IE, which indicates to UE that PSCell change request is in question. SpCellConfig IE defines also the target PhysCellId, new cell radio network temporary identifier (C-RNTI) and t304 for UE. TDCoverall timer is started after sending the RRC message.
8. UE sends an NR RRC: RRC Reconfiguration Complete message which confirms that UE has processed the message.

9. Random Access procedure is performed by the UE at the target gNB-DU.
10. Downlink packets are sent to the UE. Also, uplink packets are sent from the UE, which are forwarded to the gNB-CU through the target gNB-DU.
11. The gNB-CU sends an UE Context Release Command message to the source gNB-DU.
12. The source gNB-DU releases the UE context and responds the gNB-CU with an UE Context Release Complete message.

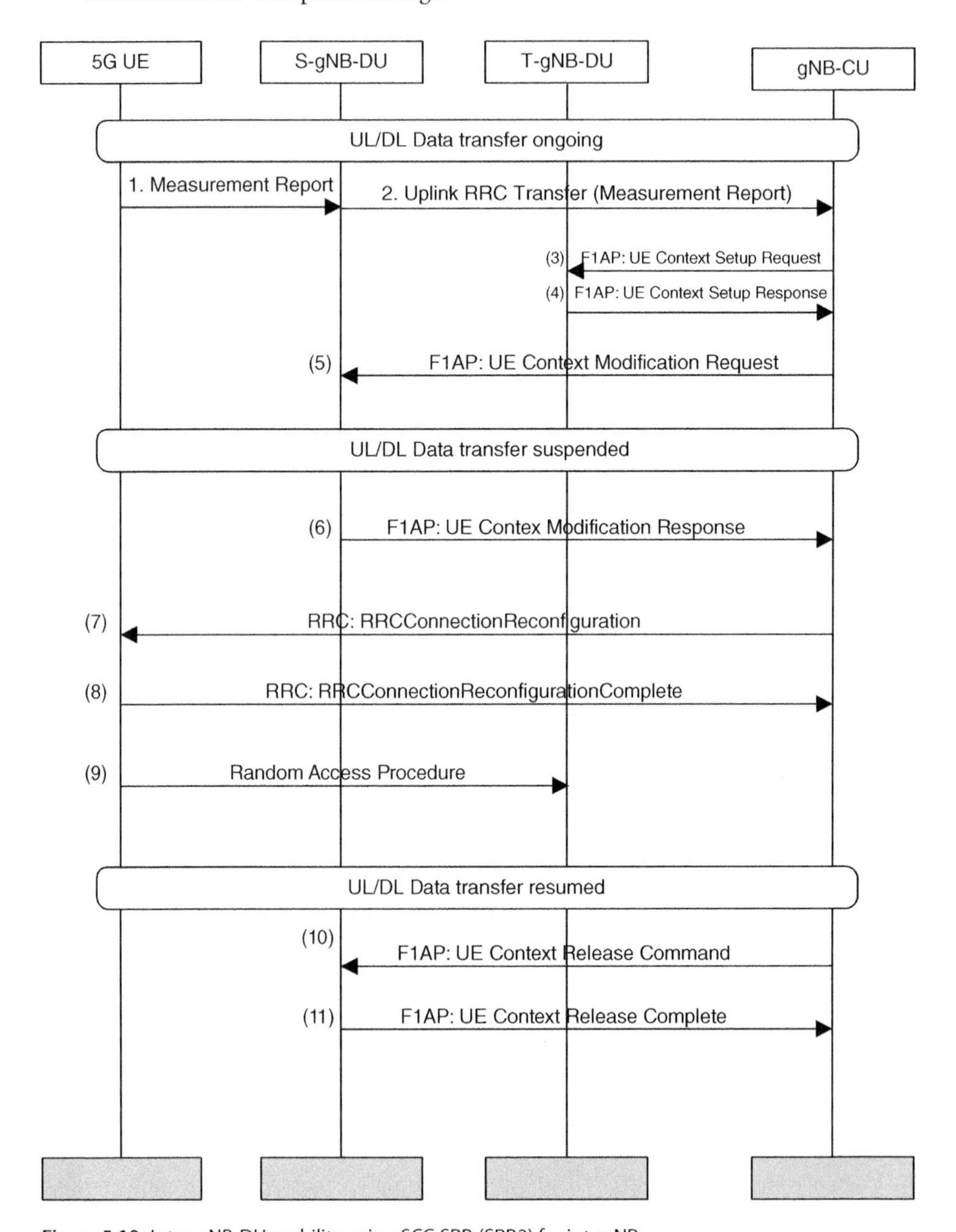

Figure 5.10 Inter-gNB-DU mobility using SCG SRB (SRB3) for intra-NR.

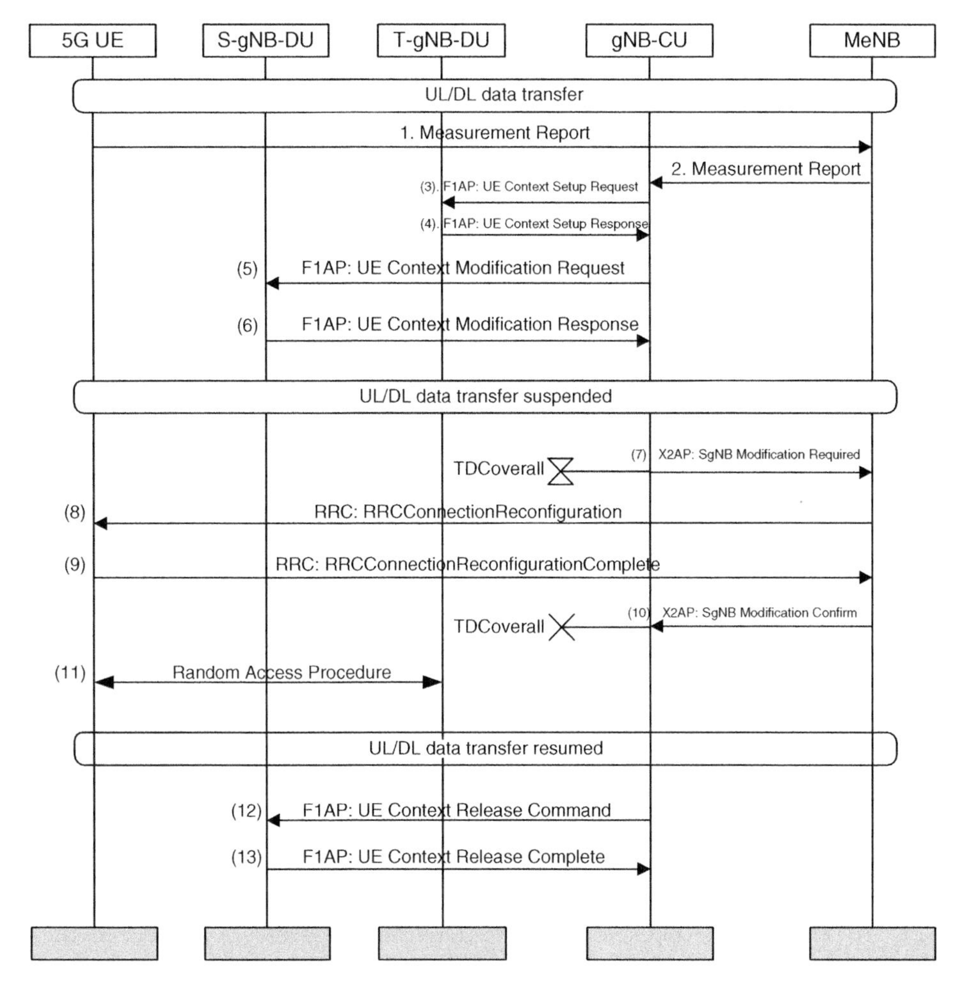

Figure 5.11 Inter-gNB-DU mobility using MCG SRB in EN-DC.

Intra CU, Inter-gNB-DU Mobility using MCG SRB This procedure is used for the case the UE moves from one gNB-DU to another gNB-DU within the same gNB-CU when only master cell group (MCG) SRB is available during EN-DC operation (see [16]). Figure 5.11 shows the inter-gNB-DU mobility procedure using MCG SRB in EN-DC.

As mentioned earlier, using MCG SRBs for mobility causes additional latency for the handover event. This is because of the additional X2 messages in DL and UL. Hence it is recommended to utilize SRB3 whenever supported by UE and available at gNB.

However, there could be several reasons why a gNB must turn to MCG SRBs despite having configured SRB3. Combined PCell and PSCell change in MN and SN, reliability of a handover event are some example use-cases.

It should also to be noted here that the UE will respond to the network in the same SRB route that it received the DL message.

1. The UE sends a *Measurement Report* message to the MeNB.
2. The MeNB transparently forwards this to the SgNB over X2 (see TS 36.423 [17]).

3. The gNB-CU sends an UE Context Setup Request message to the target gNB-DU to create an UE context and setup one or more bearers.
4. The target gNB-DU responds the gNB-CU with an UE Context Setup Response message after performing admission control and allocating required resources.
5. The gNB-CU sends a UE context modification request to suspend data transmission at DU for this UE.
6. gNB-DU responds with a successful suspension of data transmission and UE scheduling. At this stage, data transmission is suspended in both CU and DU.
7. The SgNB sends an X2: SgNB Modification Required to initiate a PSCell change. The RRC Re-configuration container inside this message contains the necessary configurations and information about the target cell.

8, 9. The MeNB and the UE execute the Handover event by performing the RRC Reconfiguration procedure.

10. The MeNB sends an SgNB modification confirm message to the gNB-CU. Downlink packets, which may include PDCP PDUs not successfully transmitted in the source gNB-DU, are sent from the gNB-CU to the target gNB-DU.
11. Random Access procedure is performed at the target gNB-DU. Downlink packets are sent to the UE. Also, uplink packets are sent from the UE, which are forwarded to the gNB-CU through the target gNB-DU.
12. The gNB-CU sends an UE Context Release Command message to the source gNB-DU.
13. The source gNB-DU releases the UE context and responds to the gNB-CU with an UE Context Release Complete message.

The procedure remains the same in NGEN-DC case, except for the fact that the interface between MN and MN is Xn.

5.4.1.2 Change of Secondary Node (MN and SN Initiated)

EN-DC The change of Secondary Node procedure is initiated either by MN or SN and used to transfer a UE context from a source SN to a target SN and to change the SCG configuration in UE from one SN to another.

The Change of Secondary Node procedure always involves signaling over MCG SRB towards the UE and it cannot be autonomously performed by SN over SRB3.

EN-DC: MN Initiated SN Change Figure 5.12 shows the signaling flow for the MN initiated Secondary Node Change:

1.-2. The MN initiates the SN change by requesting the target SN to allocate resources for the UE by means of the SgNB Addition procedure. The MN may include measurement results related to the target SN. If forwarding is needed, the target SN provides forwarding addresses to the MN. The MN may send the SgNB Modification Request message (to the source SN) to request the current SCG configuration before step 1.

3. If the allocation of target SN resources was successful (indicated in the SgNB Addition Request Acknowledge in step 2), the MN initiates the release of the source SN resources including a Cause indicating SCG mobility. The Source SN may reject the release. If data forwarding is needed the MN provides data forwarding addresses to the source SN. Reception of the SgNB Release Request

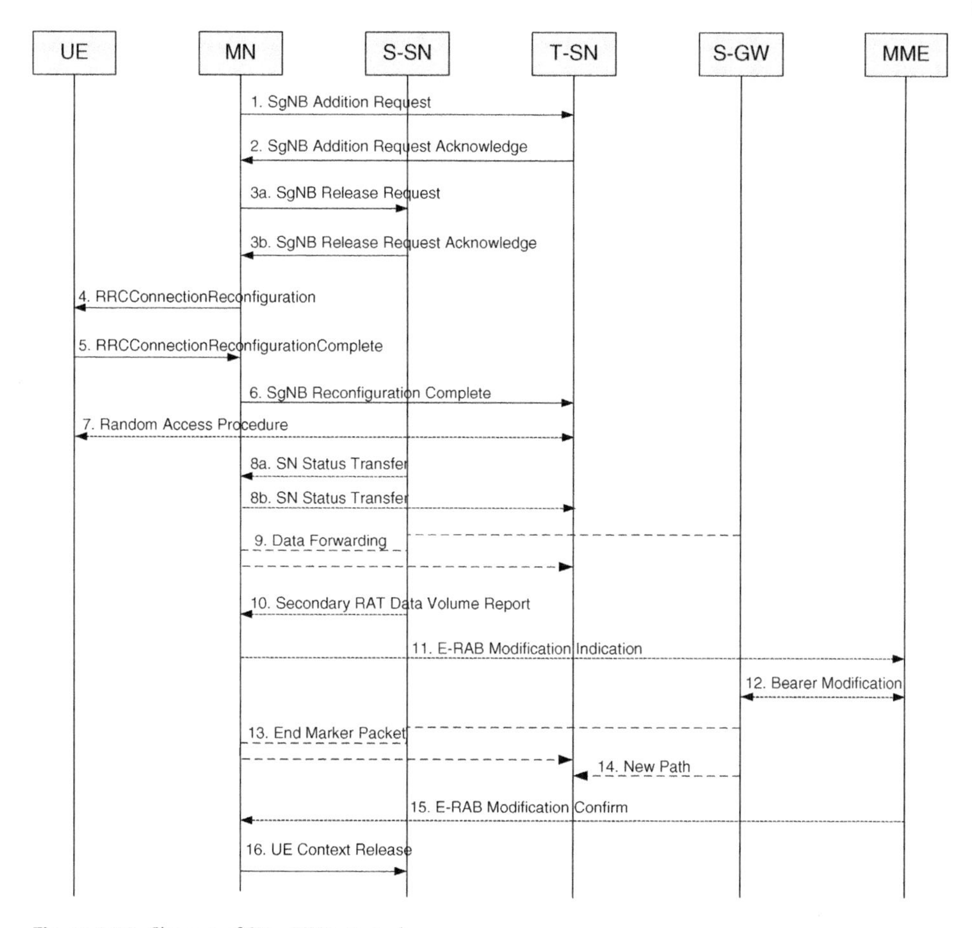

Figure 5.12 Change of SN – MN initiated.

message triggers the source SN to stop providing user data to the UE and, if applicable, to start data forwarding.

4-5. The MN triggers the UE to apply the new configuration by sending an RRC Re-configuration message. The MN indicates to the UE the new configuration in the *RRCConnectionReconfiguration* message including the NR RRC configuration message generated by the target SN. The UE applies the new configuration and sends the *RRCConnectionReconfigurationComplete* message, including the encoded NR RRC response message for the target SN. In case the UE is unable to comply with (part of) the configuration included in the *RRCConnectionReconfiguration* message, it performs the reconfiguration failure procedure.

6. If the RRC connection reconfiguration procedure was successful, the MN informs the target SN via *SgNBReconfigurationComplete* message with the encoded NR RRC response message for the target SN.

7. If configured with bearers requiring SCG radio resources, the UE synchronizes to the target SN.

8-9. UE performs Random Access procedure. If applicable, data forwarding from the source SN takes place. It may be initiated as early as the source SN receives the SgNB Release Request message from the MN.

10-11. The source SN sends the *Secondary RAT Data Volume Report* message to the MN and includes the data volumes delivered to the UE over the NR radio for the related E-RABs.

12-15. If one of the bearers was terminated at the source SN, path update is triggered by the MN.

16. Upon reception of the UE Context Release message, the source SN can release radio and C-plane related resource associated to the UE context. Any ongoing data forwarding may continue.

EN-DC: SN Initiated SN Change Figure 5.13 shows the signaling flow for the Secondary Node Change initiated by the SN:

1. The source SN initiates the SN change procedure by sending SgNB Change Required message which contains target SN ID information and may include the SCG configuration (to support delta configuration) and measurement results related to the target SN.

2.-3. The MN requests the target SN to allocate resources for the UE by means of the SgNB Addition procedure, including the measurement results related to the

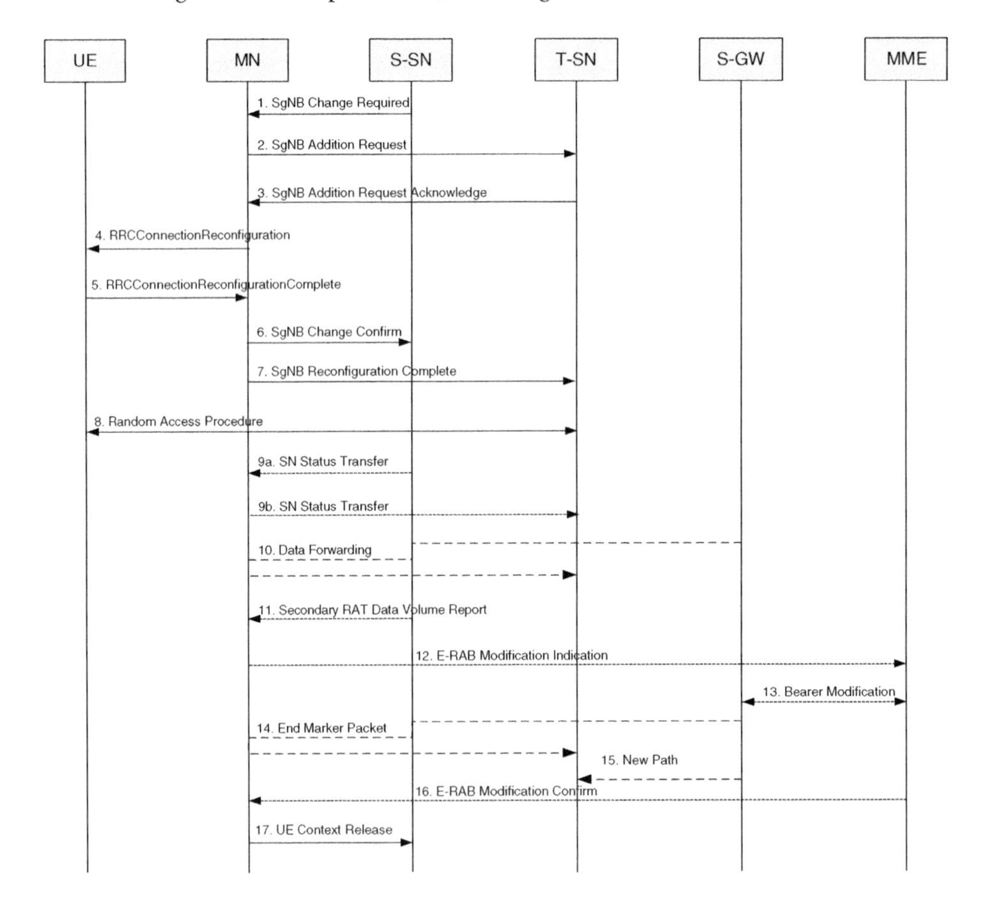

Figure 5.13 Change of SN – SN initiated.

target SN received from the source SN. If forwarding is needed, the target SN provides forwarding addresses to the MN.

4.-5. The MN triggers the UE to apply the new configuration. The MN indicates the new configuration to the UE in the *RRCConnectionReconfiguration* message including the NR RRC configuration message generated by the target SN. The UE applies the new configuration and sends the *RRCConnectionReconfigurationComplete* message, including the encoded NR RRC response message for the target SN. In case the UE is unable to comply with (part of) the configuration included in the *RRCConnectionReconfiguration* message, it performs the reconfiguration failure procedure.

6. If the allocation of target SN resources was successful, the MN confirms the release of the source SN resources. If data forwarding is needed the MN provides data forwarding addresses to the source SN. Reception of the SgNB Change Confirm message triggers the source SN to stop providing user data to the UE and, if applicable, to start data forwarding.

7. If the RRC connection reconfiguration procedure was successful, the MN informs the target SN via SgNB Reconfiguration Complete message with the encoded NR RRC response message for the target SN.

8. The UE synchronizes to the target SN by performing Random Access procedure.

9.-10. If applicable, data forwarding from the source SN takes place. It may be initiated as early as the source SN receives the SgNB Change Confirm message from the MN.

11. The source SN sends the *Secondary RAT Data Volume Report* message to the MN and includes the data volumes delivered to the UE over the NR radio for the related E-RABs.

12-16. If one of the bearers was terminated at the source SN, path update is triggered by the MN.

17. Upon reception of the UE Context Release message, the source SN can release radio and C-plane related resource associated to the UE context. Any ongoing data forwarding may continue

MR-DC This is described in a generic manner as it is applicable for both NEDC and NGEN-DC scenarios.

MR-DC: MN Initiated SN Change The MN initiated SN change procedure is used to transfer a UE context from the source SN to a target SN and to change the SCG configuration in UE from one SN to another.

The Change of Secondary Node procedure always involves signaling over MCG SRB towards the UE.

Figure 5.14 shows the signaling flow for the SN Change initiated by the MN:

1.-2. The MN initiates the SN change by requesting the target SN to allocate resources for the UE by means of the SN Addition procedure. The MN may include measurement results related to the target SN. If data forwarding is needed, the target SN provides data forwarding addresses to the MN. The MN may send the SN Modification Request message (to the source SN) to request the current SCG configuration before step 1.

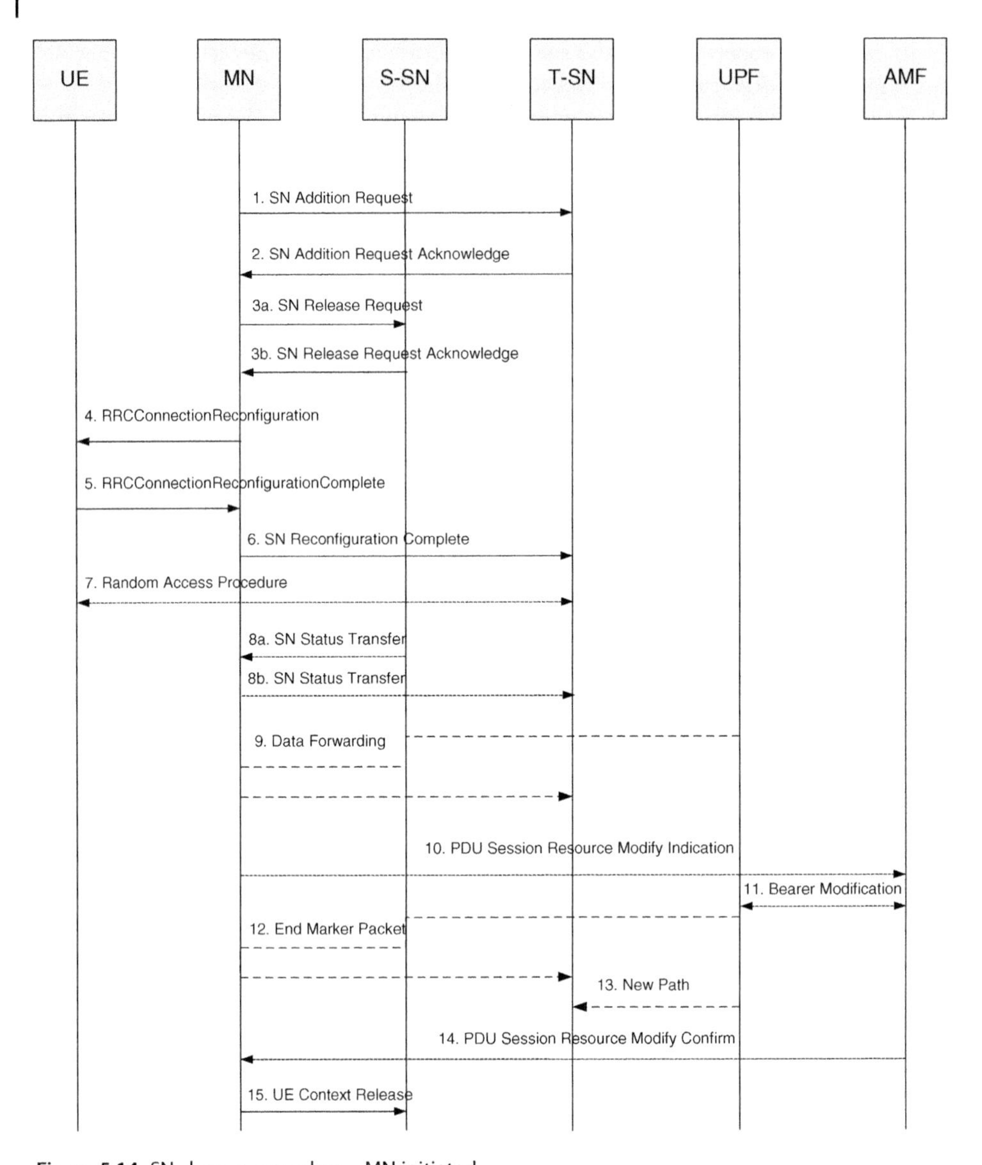

Figure 5.14 SN change procedure – MN initiated.

3. If the allocation of target SN resources was successful, the MN initiates the release of the source SN resources including a Cause indicating SCG mobility. The Source SN may reject the release. If data forwarding is needed the MN provides data forwarding addresses to the source SN. Reception of the SN Release Request message triggers the source SN to stop providing user data to the UE and, if applicable, to start data forwarding.

4.-5. The MN triggers the UE to apply the new configuration. The MN indicates the new configuration to the UE in the MN RRC reconfiguration message including the target SN RRC configuration message. The UE applies the new configuration and sends the MN RRC reconfiguration complete message, including the encoded SN RRC response message for the target SN. In case the UE is unable to comply with (part of) the configuration included in the MN RRC reconfiguration message, it performs the reconfiguration failure procedure.

6. If the RRC connection reconfiguration procedure was successful, the MN informs the target SN via SN Reconfiguration Complete message with the encoded SN RRC message for the target SN.
7. If configured with bearers requiring SCG radio resources the UE synchronizes to the target SN by performing Random access procedure.

8.-9. If applicable, data forwarding from the source SN takes place. It may be initiated as early as the source SN receives the SN Release Request message from the MN.

10-14. If one of the PDU session/QoS (quality of service) Flow was terminated at the source SN, path update procedure is triggered by the MN.

15. Upon reception of the UE Context Release message, the source SN can release radio and C-plane related resource associated to the UE context. Any ongoing data forwarding may continue.

MR-DC: SN Initiated SN Change The SN initiated SN change procedure is used to transfer a UE context from the source SN to a target SN and to change the SCG configuration in UE from one SN to another.

Figure 5.15 shows the signaling flow for the SN Change initiated by the SN:

1. The source SN initiates the SN change procedure by sending the SN Change Required message, which contains a candidate target node ID and may include the SCG configuration (to support delta configuration) and measurement results related to the target SN.

2.-3. The MN requests the target SN to allocate resources for the UE by means of the SN Addition procedure, including the measurement results related to the target SN received from the source SN. If data forwarding is needed, the target SN provides data forwarding addresses to the MN.

4.-5. The MN triggers the UE to apply the new configuration. The MN indicates the new configuration to the UE in the MN RRC reconfiguration message including the SN RRC configuration message generated by the target SN. The UE applies the new configuration and sends the MN RRC reconfiguration complete message, including the encoded SN RRC response message for the target SN. In case the UE is unable to comply with (part of) the configuration included in the MN RRC reconfiguration message, it performs the reconfiguration failure procedure.

6. If the allocation of target SN resources was successful, the MN confirms the change of the source SN. If data forwarding is needed the MN provides data forwarding addresses to the source SN. Reception of the SN Change Confirm message triggers the source SN to stop providing user data to the UE and, if applicable, to start data forwarding.
7. If the RRC connection reconfiguration procedure was successful, the MN informs the target SN via SN Reconfiguration Complete message with the encoded SN RRC response message for the target SN.
8. The UE synchronizes to the target SN by performing Random access procedure.

9.-10. If applicable, data forwarding from the source SN takes place. It may be initiated as early as the source SN receives the SN Change Confirm message from the MN.

11-15. If one of the PDU session/QoS Flow was terminated at the source SN, path update procedure is triggered by the MN.

16. Upon reception of the UE Context Release message, the source SN can release radio and C-plane related resource associated to the UE context. Any ongoing data forwarding may continue.

5.4.1.3 Inter-Master Node Handover without Secondary Node Change

EN-DC Inter-Master Node handover with or without MN initiated Secondary Node change is used to transfer context data from a source MN to a target MN while the context at the SN is kept or moved to another SN. During an Inter-Master Node handover, the target MN decides whether to keep or change the SN (or release the SN).

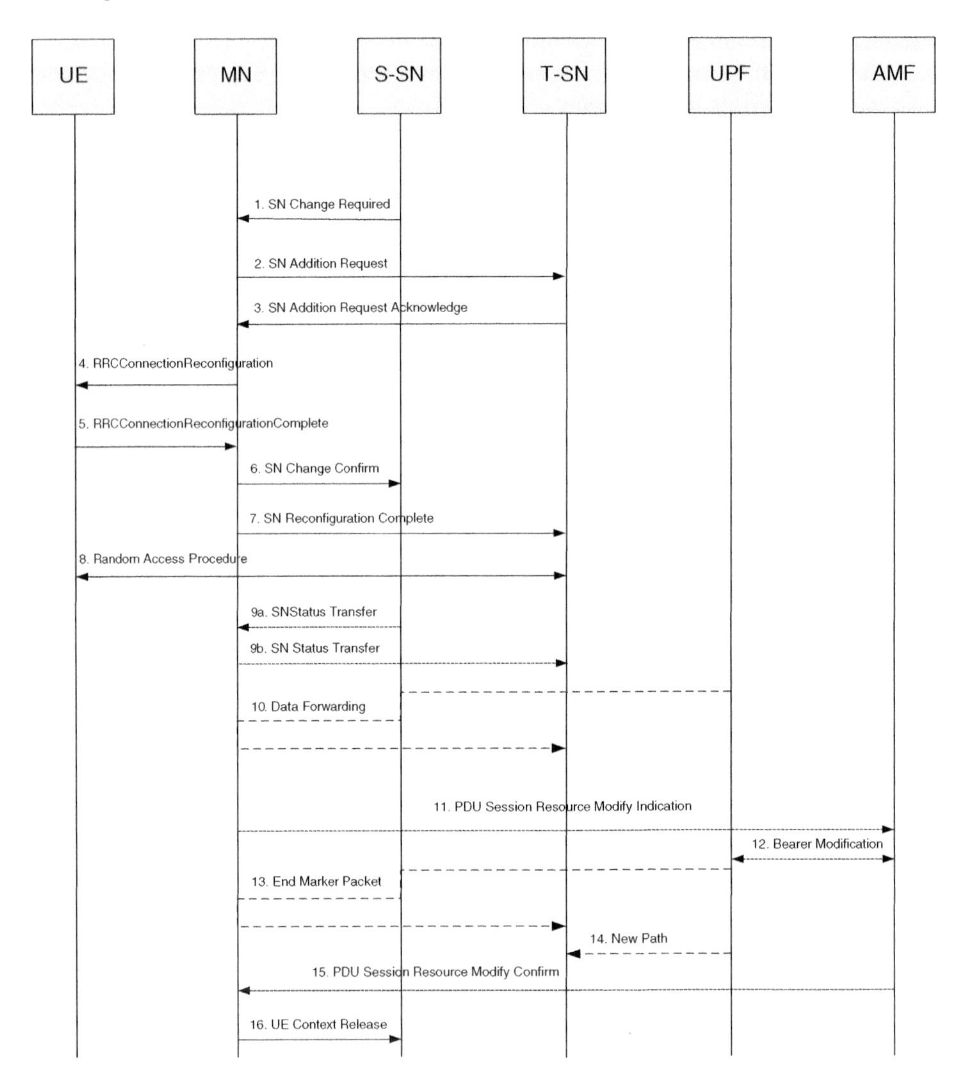

Figure 5.15 SN change procedure – SN initiated.

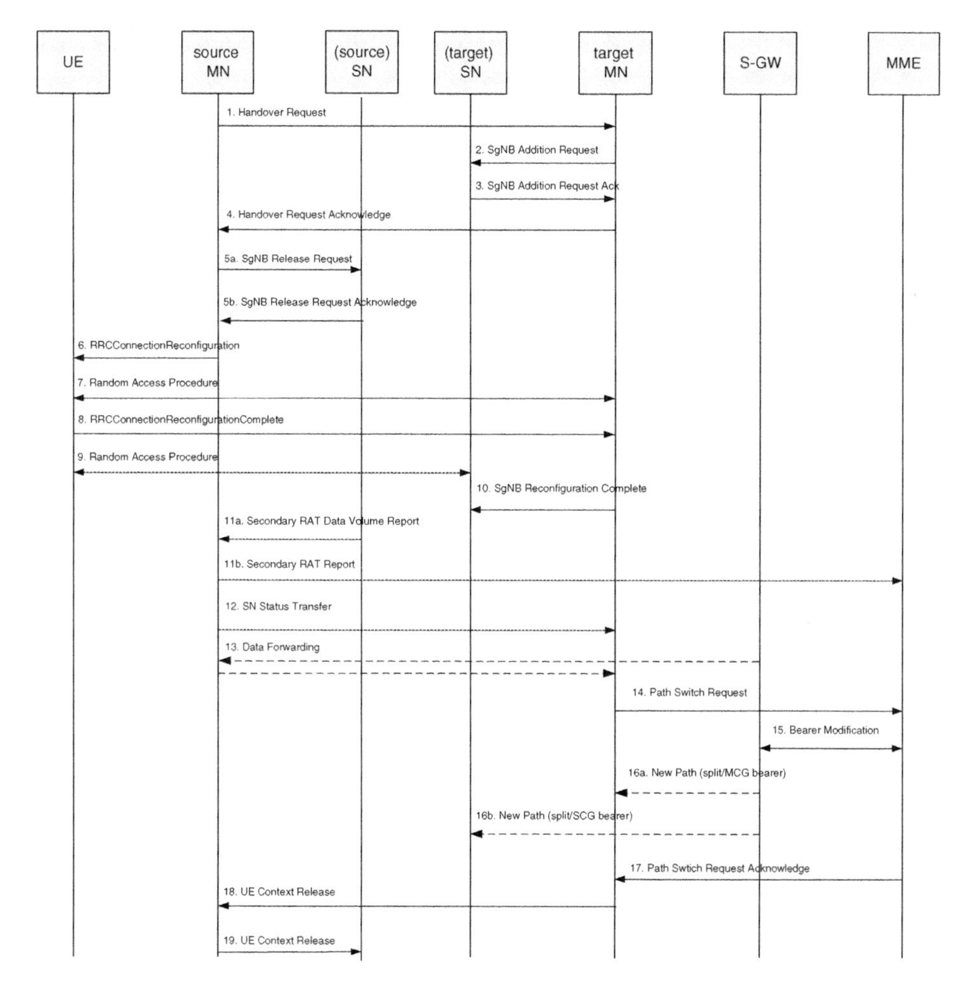

Figure 5.16 MN initiated inter-MN handover with or without SN change.

Figure 5.16 shows the signaling flow for inter-Master Node handover with or without MN initiated Secondary Node change. For an inter-Master Node handover without Secondary Node change, the source SN and the target SN shown in Figure 5.16 are the same node.

1. The source MN starts the handover procedure by initiating the X2 Handover Preparation procedure including both MCG and SCG configuration. The source MN includes the (source) SN UE X2AP ID, SN ID and the UE context in the (source) SN in the Handover Request message. The source MN may send the SgNB Modification Request message (to the source SN) to request the current SCG configuration before step 1.
2. If the target MN decides to keep the SN, the target MN sends SN Addition Request to the SN including the SN UE X2AP ID as a reference to the UE context in the SN that was established by the source MN. If the target MN decides to change the SN, the target MN sends the SgNB Addition Request to

the target SN including the UE context in the source SN that was established by the source MN.

3. The (target) SN replies with SN Addition Request Acknowledge.
4. The target MN includes within the Handover Request Acknowledge message a transparent container to be sent to the UE as an RRC message to perform the handover, and may also provide forwarding addresses to the source MN. The target MN indicates to the source MN that the UE context in the SN is kept if the target MN and the SN decided to keep the UE context in the SN in step 2 and step 3.
5. The source MN sends SN Release Request to the (source) SN including a Cause indicating MCG mobility. The (source) SN acknowledges the release request. The source MN indicates to the (source) SN that the UE context in SN is kept, if it receives the indication from the target MN. If the indication as the UE context kept in SN is included, the SN keeps the UE context.
6. The source MN triggers the UE to apply the new configuration.

7.-8. The UE synchronizes to the target MN and replies with RRCConnectionReconfigurationComplete message.

9. If configured with bearers requiring SCG radio resources, the UE synchronizes to the (target) SN.
10. If the RRC connection reconfiguration procedure was successful, the target MN informs the (target) SN via SgNB Reconfiguration Complete message.

11a. The SN sends the Secondary RAT Data Volume Report message to the source MN and includes the data volumes delivered to the UE over the NR radio for the related E-RABs.

11b. The source MN sends the Secondary RAT Report message to mobility management entity (MME) to provide information on the used NR resource.

12.-13. Data forwarding from the source MN takes place. If the SN is kept, data forwarding may be omitted for SCG bearers and SCG split bearers.

14–17. The target MN initiates the S1 Path Switch procedure. If new UL tunnel endpoint IDs (TEIDs) of the serving gateway (S-GW) are included, the target MN performs MN initiated SN Modification procedure to provide them to the SN.

18. The target MN initiates the UE Context Release procedure towards the source MN.
19. Upon reception of the UE Context Release message, the (source) SN can release C-plane related resource associated to the UE context towards the source MN. Any ongoing data forwarding may continue. The SN shall not release the UE context associated with the target MN if the indication was included in the SN Release Request in step 5.

MR-DC with 5GC Inter-MN handover with/without MN initiated SN change is used to transfer UE context data from a source MN to a target MN while the UE context at the SN is kept or moved to another SN. During an Inter-Master Node handover, the target MN decides whether to keep or change the SN (or release the SN).

Figure 5.17 shows the signaling flow for inter-MN handover with or without MN initiated SN change. Note that for an inter-Master Node handover without Secondary Node change, the source SN and the target SN shown in Figure 5.17 are identical.

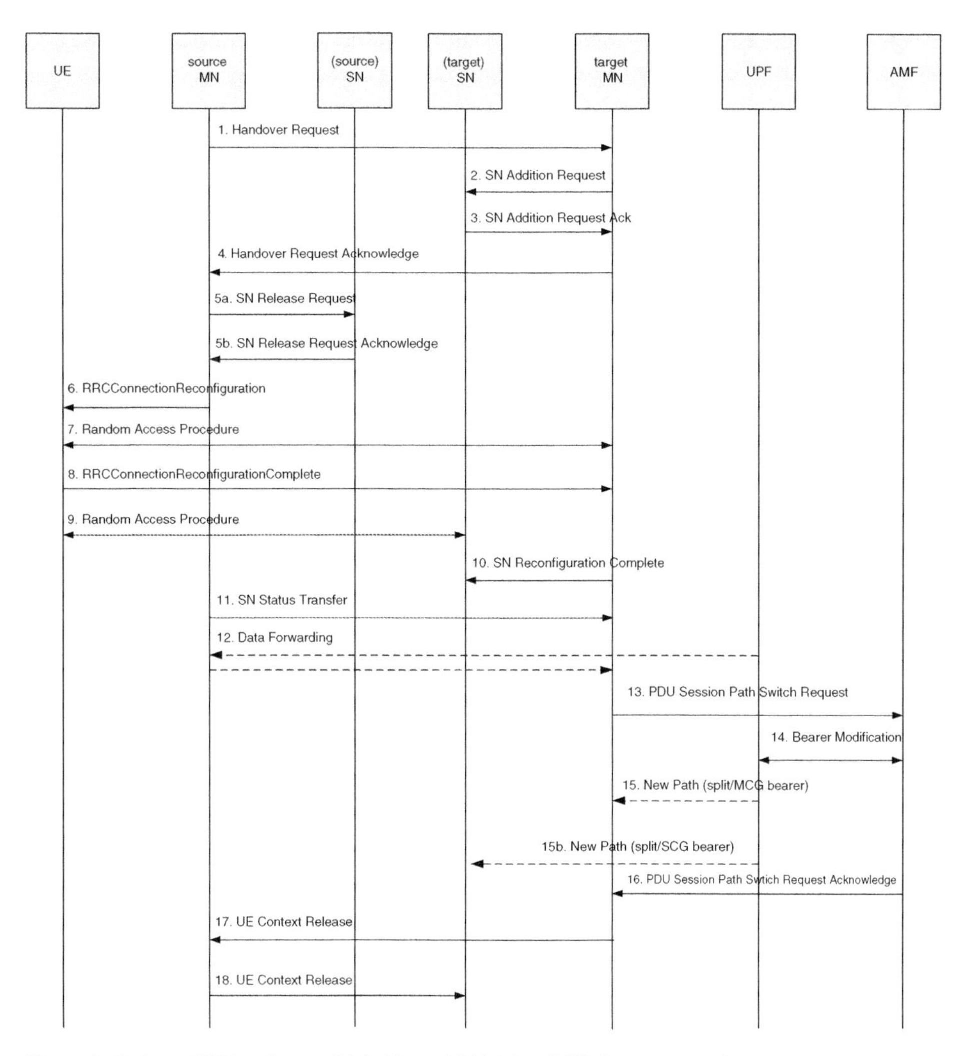

Figure 5.17 Inter-MN handover with/without MN initiated SN change procedure.

1. The source MN starts the handover procedure by initiating the Xn Handover Preparation procedure including both MCG and SCG configuration. The source MN includes the source SN UE XnAP ID, SN ID and the UE context in the source SN in the Handover Request message. The source MN may send the SN Modification Request message (to the source SN) to request the current SCG configuration before step 1.
2. If the target MN decides to keep the source SN, the target MN sends SN Addition Request to the SN including the SN UE XnAP ID as a reference to the UE context in the SN that was established by the source MN. If the target MN decides to change the SN, the target MN sends the SN Addition Request to the target SN including the UE context in the source SN that was established by the source MN.
3. The (target) SN replies with SN Addition Request Acknowledge.

4. The target MN includes within the Handover Request Acknowledge message a transparent container to be sent to the UE as an RRC message to perform the handover, and may also provide forwarding addresses to the source MN. The target MN indicates to the source MN that the UE context in the SN is kept if the target MN and the SN decided to keep the UE context in the SN in step 2 and step 3.
5. The source MN sends SN Release Request message to the (source) SN including a Cause indicating MCG mobility. The (source) SN acknowledges the release request. The source MN indicates to the (source) SN that the UE context in SN is kept, if it receives the indication from the target MN. If the indication as the UE context kept in SN is included, the SN keeps the UE context.
6. The source MN triggers the UE to perform handover and apply the new configuration.

7.-8. The UE synchronizes to the target MN and replies with MN RRC reconfiguration complete message.

9. If configured with bearers requiring SCG radio resources, the UE synchronizes to the (target) SN by performing Random access procedure.
10. If the RRC connection reconfiguration procedure was successful, the target MN informs the (target) SN via SN Reconfiguration Complete message.

11.-12. Data forwarding from the source MN takes place. If the SN is kept, data forwarding may be omitted for SCG bearers and SCG split bearers.

13–16. The target MN initiates the PDU Session Path Switch procedure. If new UL TEIDs of the UPF for SN are included, the target MN performs MN initiated SN Modification procedure to provide them to the SN.

17. The target MN initiates the UE Context Release procedure toward the source MN.
18. Upon receipt of the UE Context Release message from source MN, the (source) SN can release C-plane related resource associated to the UE context towards the source MN. Any ongoing data forwarding may continue. The SN shall not release the UE context associated with the target MN if the indication was included in the SN Release Request message in step 5.

5.4.1.4 Master Node to eNB/gNB Change

The Master Node to eNB/gNB Change procedure is used to transfer context data from a source MN/SN to a target eNB/gNB. Basically, this is about releasing the dual connectivity and getting back to single connectivity. This could happen either due to UE not requiring DC anymore or due to deteriorating NR coverage and UE cannot see any other gNB cells.

EN-DC Here the Master Node to eNB change procedure is explained in the context of EN-DC. This is the case when eNB is acting as the Master Node and gNB is the Secondary Node.

Figure 5.18 shows the signaling flow for the Master Node to eNB Change procedure:

1. The source MN starts the MN to eNB Change procedure by initiating the X2 Handover Preparation procedure, including both MCG and SCG configuration. The source MN may send the SgNB Modification Request message (to the source SN) to request the current SCG configuration before step 1.

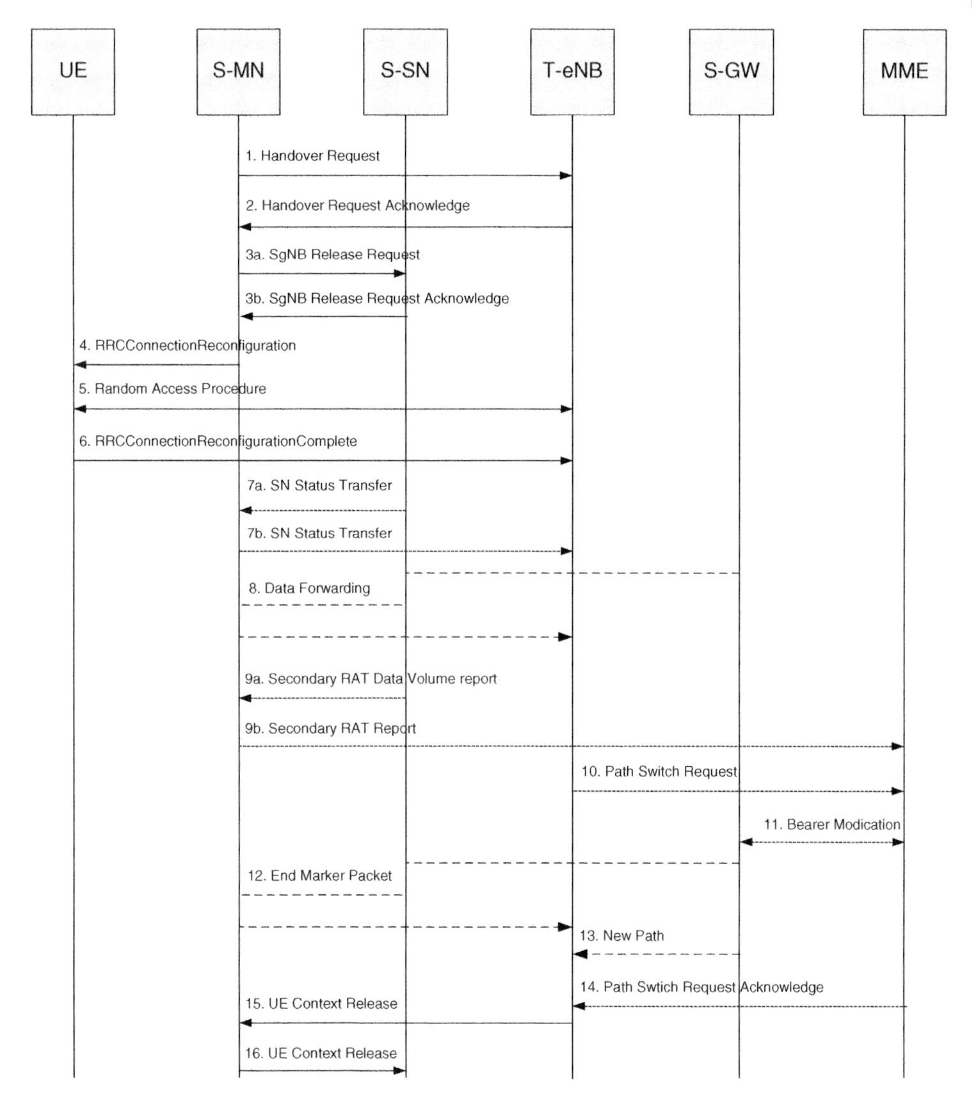

Figure 5.18 Master node to eNB change procedure.

2. The target eNB includes the field in HO command which releases SCG configuration, and may also provide forwarding addresses to the source MN.
3. If the allocation of target eNB resources was successful, the MN initiates the release of the source SN resources toward the source SN including a Cause indicating MCG mobility. The SN acknowledges the release request. If data forwarding is needed, the MN provides data forwarding addresses to the source SN. Reception of the SgNB Release Request message triggers the source SN to stop providing user data to the UE and, if applicable, to start data forwarding.
4. The MN triggers the UE to apply the new configuration. Upon receiving the new configuration, the UE releases the entire SCG configuration.

5.-6. The UE synchronizes to the target eNB by performing Random access procedure.
7.-8. If applicable, data forwarding from the source SN takes place. It may start as early as the source SN receives the SgNB Release Request message from the MN.
9a. The source SN sends the Secondary RAT Data Volume Report message to the source MN and includes the data volumes delivered to the UE over the NR radio for the related E-RABs.
9b. The source MN sends the Secondary RAT Report message to MME to provide information on the used NR resource.
10–14. The target eNB initiates the S1 Path Switch procedure.
15. The target eNB initiates the UE Context Release procedure towards the source MN.
16. Upon reception of the UE CONTEXT RELEASE message, the S-SN can release radio and C-plane related resource associated to the UE context. Any ongoing data forwarding may continue.

MR-DC with 5GC Here the Master Node to target eNB/gNB is explained as applicable to MR-DC in general. This is applicable for both cases:

1) Master Node to target ng-eNB
2) Master Node to target gNB

The change procedure is used to transfer UE context data from a source MN/SN to a target ng-eNB/gNB. Essentially, this is also about releasing the dual connectivity and getting back to single connectivity.

Figure 5.19 shows the signaling flow for the MN to ng-eNB/gNB Change procedure:

1. The source MN starts the MN to ng-eNB/gNB Change procedure by initiating the Xn Handover Preparation procedure, including both MCG and SCG configuration. The source MN may send the SN Modification Request message (to the source SN) to request the current SCG configuration before step 1.
2. The target ng-eNB/gNB includes the field in HO command which releases the SCG configuration, and may also provide forwarding addresses to the source MN.
3. If the resource allocation within target ng-eNB/gNB was successful, the MN initiates the release of the source SN resources toward the source SN including a Cause indicating MCG mobility. The SN acknowledges the release request. If data forwarding is needed, the MN provides data forwarding addresses to the source SN. Reception of the SN Release Request message triggers the source SN to stop providing user data to the UE and, if applicable, to start data forwarding.
4. The MN triggers the UE to perform HO and apply the new configuration. Upon receiving the new configuration, the UE releases the entire SCG configuration.

5.-6. The UE synchronizes to the target ng-eNB/gNB.
7.-8. If applicable, data forwarding from the source SN takes place. It may start as early as the source SN receives the SN Release Request message from the MN.
9–13. The target ng-eNB/gNB initiates the PDU Session Path Switch procedure.

14. The target ng-eNB/gNB initiates the UE Context Release procedure towards the source MN.
15. Upon reception of the UE Context Release message from MN, the source SN can release radio and C-plane related resource associated to the UE context. Any ongoing data forwarding may continue.

5.4.2 Standalone (SA) Mobility Scenarios

To support session continuity for UEs making inter-NG-RAN node mobility within the 5G system, the 5G system provides support for seamless handover procedures. If there exists a direct interface between NG-RAN nodes, i.e. the Xn interface, and the use

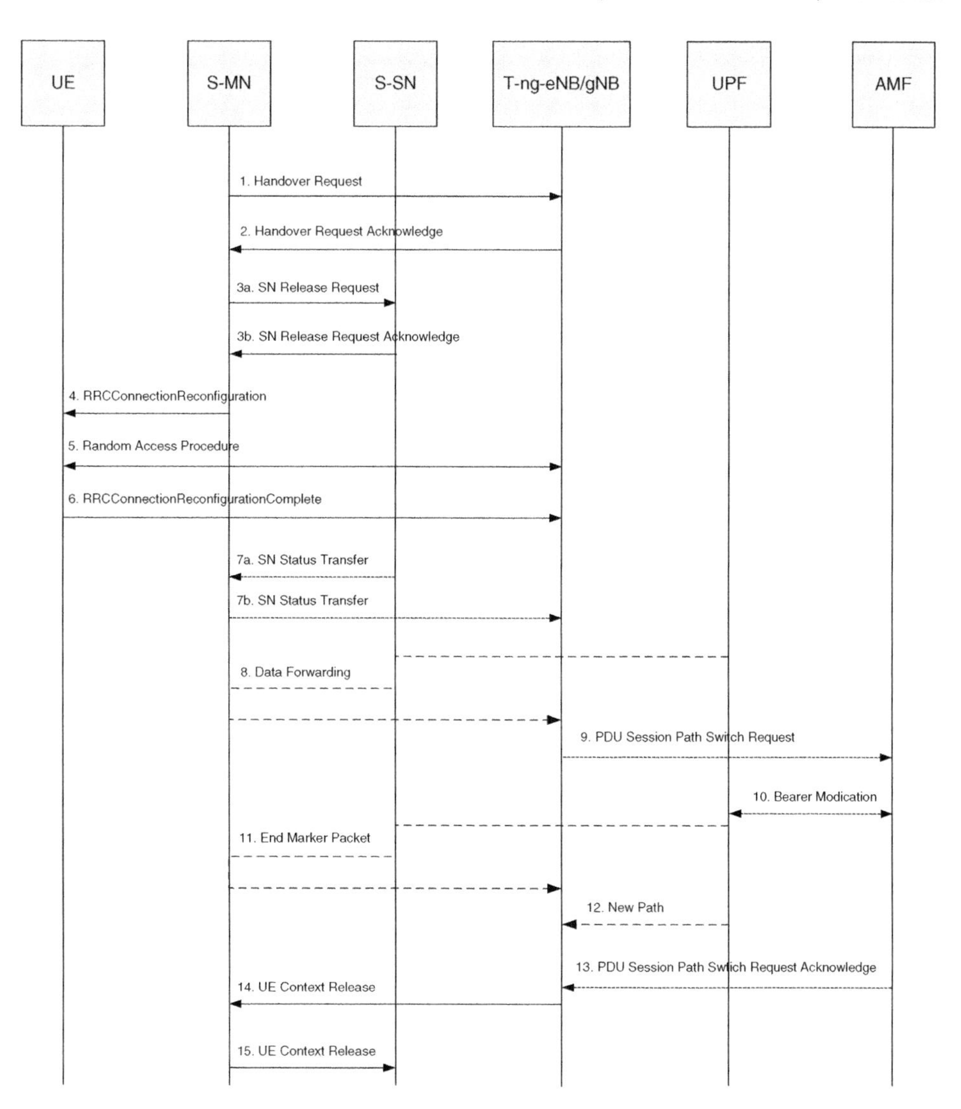

Figure 5.19 MN to ng-eNB/gNB change procedure.

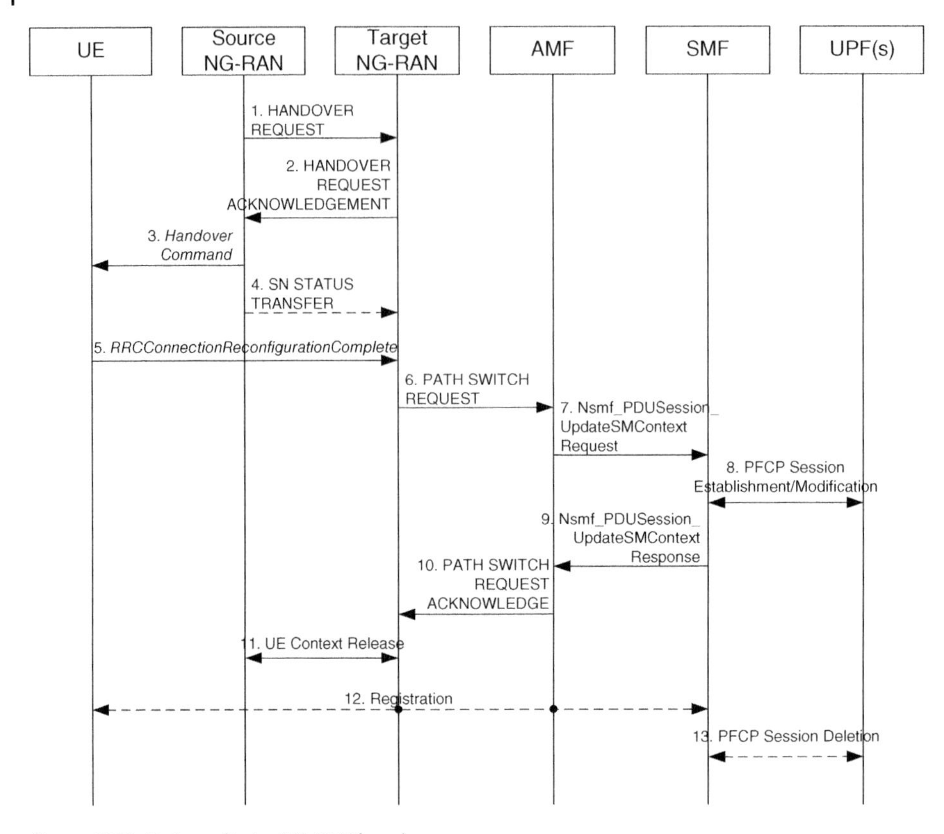

Figure 5.20 Xn based inter NG-RAN handover.

of it is not prohibited, the Xn handover procedure is performed. In the Xn handover procedure, the signaling for handover preparation and execution phases is exchanged between NG-RAN nodes and then the core network is involved. Otherwise, the N2 handover is performed. In the N2 handover procedure, the core network takes part from the handover preparation phase.

Relocation of the AMF is not supported by the Xn handover. If the AMF must be relocated, the N2 handover needs to be performed. On the other hand, both handover procedures support the intermediate UPF relocation.

5.4.2.1 Xn Based Handover

The following call flow (Figure 5.20) shows the Xn based inter NG-RAN handover.

1. The source NG-RAN node decides to hand the UE over to the target NG-RAN node, based on, e.g. measurement report from the UE. Then, the source NG-RAN node sends the HANDOVER REQUEST message to the target NG-RAN node. The message contains a transparent RRC container with necessary information to prepare the handover at the target side.
2. The target NG-RAN node performs the admission control and, if admitted, responds with the HANDOVER REQUEST ACKNOWLEDGE message to the source NG-RAN node. The message includes a transparent container to be sent

to the UE as an RRC message to perform the handover, i.e. *HandoverCommand* message.
3. The source NG-RAN node forwards the *HandoverCommand* message to the UE.
4. The source NG-RAN node also sends the SN STATUS TRANSFER message to the target NG-RAN node to transfer the uplink user data receiver status and the downlink user data transmitter status for each of the DRBs which require user data transfer and are accepted by the target NG-RAN node. The source NG-RAN node starts forwarding downlink user data from the core network to the target NG-RAN.
5. The UE completes the RRC connection reconfiguration with the target NG-RAN node. Now the target NG-RAN node can forward the downlink user data to the UE and the UE can send uplink user data to the target NG-RAN node.
6. The target NG-RAN node sends the PATH SWITCH REQUEST message to the AMF to inform of the handover and PDU session-related information including at least the downlink tunnel information to the target NG-RAN node.
7. The AMF sends the Nsmf_PDUSession_UpdateSMContext Request message to the SMF. The message includes the PDU session-related information.
8. If no UPF relocation is required, the SMF updates the UPF with the downlink tunnel information by exchanging the packet forwarding control protocol (PFCP) Session Modification Request and Response messages. One or more end markers are sent to the source NG-RAN node. The UPF starts sending downlink user data toward the target NG-RAN node. Otherwise, the SMF sends the PFCP Session Establishment Request message to a new intermediate UPF and the PFCP Session Modification Request message to the UPF working as a PDU session anchor, and receives response messages from them. Tunnel information is exchanged. One or more end markers are sent by the PDU session anchor UPF and are delivered to the source NG-RAN node via the old intermediate UPF. The PDU session anchor UPF starts sending downlink user data toward the new intermediate UPF. The source NG-RAN node forwards the end markers to the target NG-RAN node. The target NG-RAN node buffers the downlink user data coming directly from the core network until the target NG-RAN node receives the end markers from the source NG-RAN node. The path switch for downlink user data is completed.
9. The SMF responds with the Nsmf_PDUSession_UpdateSMContext Response message, which may include uplink tunnel information to the new intermediate UPF. Now the path witch for uplink user data is also completed.
10. The AMF acknowledges the path switch request. The resources for the inter-NG-RAN node data transfer are released.
11. The UE context in the source NG-RAN node is released.
12. The UE triggers mobility registration procedure if any of the trigger conditions are met (see Section 5.5.2).
13. In case the new intermediate UPF has been allocated, the SMF deletes the session in the old intermediate UPF.

5.4.2.2 N2 Based Handover

The following call flow (Figure 5.21) shows the N2 based inter NG-RAN handover.

1. The source NG-RAN node decides to hand the UE over to the target NG-RAN node, based on, e.g. measurement report from the UE. Then, the source

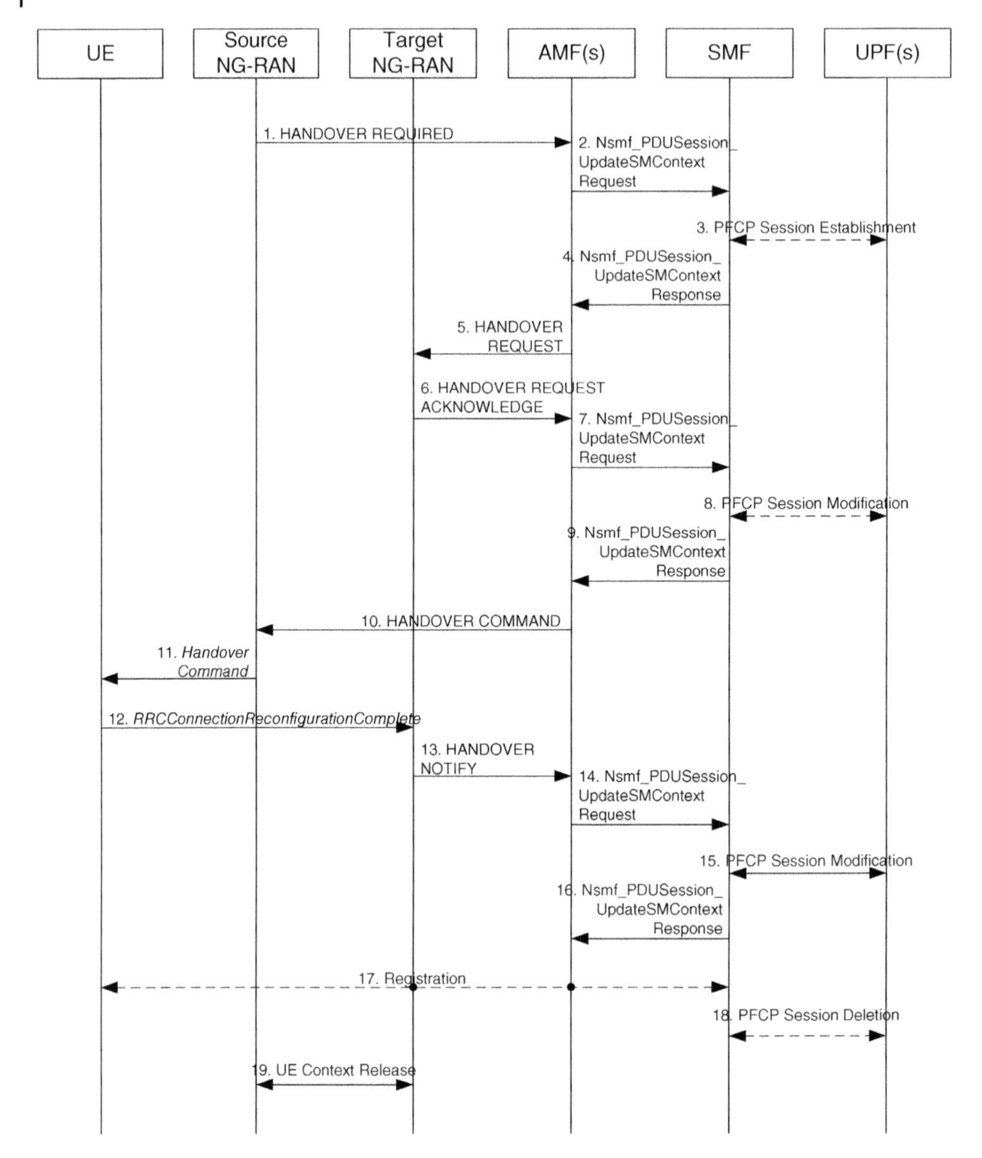

Figure 5.21 Inter NG-RAN node N2 based handover.

NG-RAN node sends the HANDOVER REQUIRED message to the AMF. The message includes the target NG-RAN node identity and a transparent container. If the AMF decides to relocate the UE's serving AMF, the AMF invokes the Namf_Communication_CreateUEContext service toward the new AMF passing the target NG-RAN node identity and the transparent container.

2. The AMF sends the Nsmf_PDUSession_UpdateSMContext Request message to the SMF including the target NG-RAN node identity.
3. Based on the target NG-RAN node identity, the SMF may decide to relocate the intermediate UPF, which leads to the exchange of the PFCP Session Establishment Request and Response message with a new intermediate UPF.

4. The SMF acknowledges the request from the AMF. In the acknowledgement message, the uplink tunnel information (toward the new intermediate UPF, if the SMF decided relocation in step 3) is included.
5. The HANDOVER REQUEST message is sent to the target NG-RAN identified by the target NG-RAN identity. The message includes the transparent container and the uplink tunnel information.
6. The target NG-RAN node sends the acknowledgement message back to the AMF. The message contains the *HandoverCommand* message and downlink tunnel information toward the target NG-RAN node. The tunnel information is associated with the PDU sessions that are accepted by the target NG-RAN node.
7. The AMF sends the Nsmf_PDUSession_UpdateSMContext Request message to the SMF. The message includes the downlink tunnel information.
8. Tunnel information for indirect data forwarding is exchanged. Refer to Ref. [12] for the details about tunneling for user data transfer.
9. The SMF responds with the Nsmf_PDUSession_UpdateSMContext Response message. In the case of the AMF relocation, the target AMF acknowledges to the source AMF with the Namf_Communication_CreateUEContext Response message including the *HandoverCommand* message.

10-11. The *HandoverCommand* message is forwarded to the UE. The source NG-RAN node starts forwarding downlink user date toward the target NG-RAN either directly or indirectly.

12. The UE completes the RRC connection reconfiguration with the target NG-RAN node. Now the target NG-RAN node can forward the downlink user data to the UE and the UE can send uplink user data to the target NG-RAN node.
13. The target NG-RAN node notifies that the handover on the radio side is completed toward the AMF by sending the HANDOVER NOTIFY message. In the case of the AMF relocation, the Namf_Communication_N2InfoNotify service is invoked.
14. The AMF sends the Nsmf_PDUSession_UpdateSMContext Request message to the SMF to inform of the handover completion.
15. The downlink tunnel information toward the target NG-RAN node is updated to the existing UPF or the new intermediate UPF. In the case of the new intermediate UPF allocation, the PDU session anchor UPF is also updated with the new tunnel information toward the new intermediate UPF. One or more end markers are delivered toward the old downlink user data path involving the source NG-RAN node, and now downlink PDUs are delivered via the new downlink user data path involving the target NG-RAN node and the new intermediate UPF (if any). The source NG-RAN node forwards the end markers to the target NG-RAN node. The target NG-RAN node buffers the downlink user data coming directly from the core network until the target NG-RAN node receives the end markers from the source NG-RAN node. The path switch for downlink user data is completed.
16. The SMF acknowledges the request.
17. The UE triggers the registration procedure if any of the conditions is met (see Section 5.5.2).

18. In case the new intermediate UPF has been allocated, the SMF deletes the session in the old intermediate UPF.
19. The UE context in the source NG-RAN node is released.

5.4.3 Conditional Handover

During connected mode mobility without dual/multi-connectivity, the UE is connected to one cell or more cells at the same time. The UE measures the neighboring cells to guarantee that the given factors, e.g. link power or quality, cell load, UE capability, access class, are above the configured thresholds to meet the QoS requirements. The UE is configured to report the measurement results back to network. The measurement reports can be triggered based on a specific mobility event configured by network, such as the neighbor cell becomes better than the serving cell by a configured quantity and this condition must be valid for a given period before the measurement report is sent to the network. Typically, the network sends the handover command to UE soon after receiving the measurement report and UE executes the handover procedure immediately to access the specified target cell. However, radio link failures may happen during the handover procedure due to UE missing the handover command or network not receiving the measurement report. The UE may also fail to access the target cell. The Conditional Handover (Figure 5.22) described in [18] is a mobility procedure where the UE receives the prepared handover command a bit earlier and is configured to evaluate and access the target cell a bit later to avoid the risk of the above-mentioned failures, for

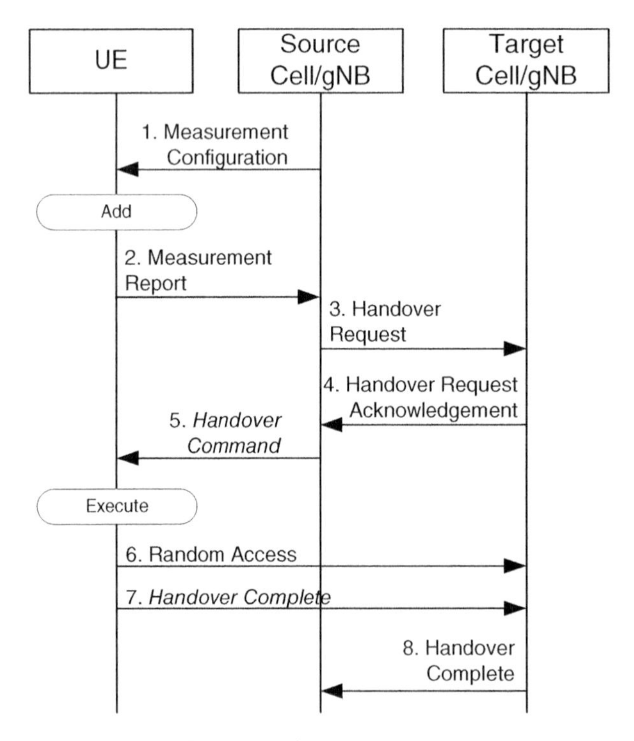

Figure 5.22 Conditional handover.

example. Thereafter, the Conditional Handover is a promising technique for improving the handover robustness. The following signaling flow indicates the difference to the baseline Xn handover:

1. The source NG-RAN node configures the UE for Conditional Handover measurements to identify at least one potential NG-RAN target node.
2. The configured condition is fulfilled and preparation of Conditional Handover is started with at least one cell being reported to network. The reported configuration is triggering the report earlier compared to baseline measurement report.
3. The source NG-RAN node decides to prepare the UE for conditional handover to target NG-RAN node candidate, based on, e.g. measurement report from the UE. The source NG-RAN node sends the HANDVOER REQUEST message to the target NG-RAN node. The message contains necessary information to prepare the handover at the target during the Conditional Handover execution.
4. The target NG-RAN node performs the admission control and, if admitted, responds with the HANDOVER REQUEST ACKNOWLEDGE message to the source NG-RAN node.
5. The source NG-RAN node prepares the *HandoverCommand* message to the UE with all the prepared candidate NG-RAN target nodes and the Conditional Handover Execution criteria.
6. The UE is measuring and evaluating the prepared target NG-RAN nodes against the HO execution condition. When the NG-RAN node is found which meets the target requirements, the Execute condition becomes valid and UE starts the synchronization and Random Access procedures to access the target NG-RAN node. The target NG-RAN node sends the PATH SWITCH REQUEST message to the AMF to inform of the handover and PDU session-related information including at least the downlink tunnel information to the target NG-RAN node.
7. The UE completes the Conditional Handover Execution with *HandoverComplete* message to the target NG-RAN node.
8. The target NG-RAN node informs the source node about successfully executed handover command.

5.5 Idle Mode mobility and UE Reachability

5.5.1 Overview

The need for an IDLE mode/state has been questioned in 5G as "Always-on Applications" used in the smart phones need to send and receive small packets frequently to keep IP connectivity open. Essentially the frequent connection requests are related to "push mail/content" services where the UE is frequently checking, if there are any new data available in the application server. Typically, this is a UE initiated event and happens in uncontrolled way from the radio network perspective. Another problem from network perspective are the "heart beat" or "keep alive" messaging events that may occur once per minute, or once every few minutes, and the amount of data is very small ($\ll$1 KB). These messages are used to keep the device towards the network in RRC Connected state.

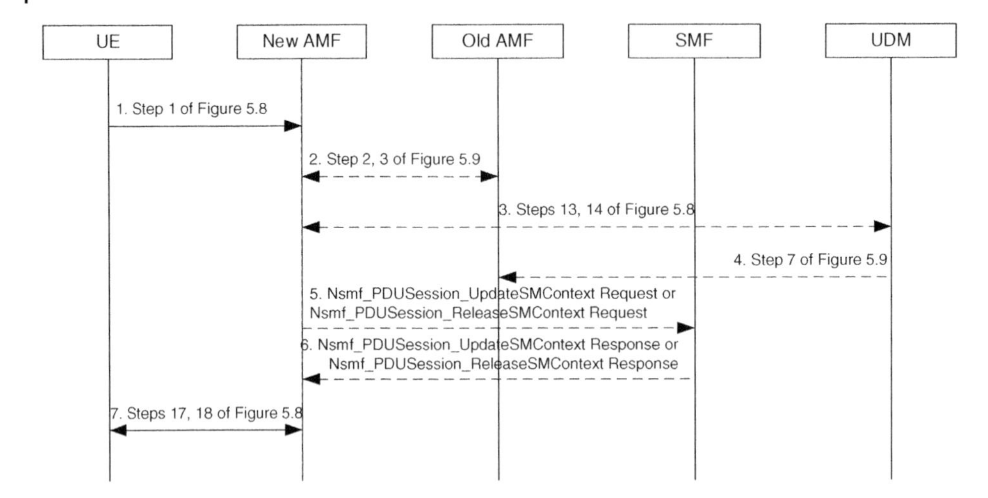

Figure 5.23 Registration procedure due to mobility or periodic update.

Despite the enhanced features and functionality, the IDLE state in 5G mobile systems is needed. One major reason is that the needed fault recovery mechanism will add complexity to the new RRC Inactive state. It is an essential requirement for the UE to be able to revert or fallback to a recovery state in case of sudden connectivity fault or network failure. The IDLE state in 5G can support the bootstrap procedures, initial PLMN selection, UE controlled mobility, contention based uplink transmission and core network based location tracking. The IDLE state allows also to reduce energy consumption in the UE, thus allows for longer re-charge cycles.

5.5.2 Mobility Registration and Periodic Registration

Mobility of a UE in IDLE mode is managed by the AMF with a granularity of a registration area, which is a list of tracking area identities. Therefore, if a UE moves out of the registration area assigned during the most recent registration, it performs a new registration, which is called the mobility registration update procedure. On the other hand, for the network to make sure that the UE located in the allocated registration area has not deregistered locally, i.e. deregistered without informing the network, the network controls the UE so that it performs a new registration periodically, which is called the periodic registration update procedure. If the UE does not perform a registration procedure more than a certain period (by default, four minutes longer than the periodicity of the periodic registration update procedure), the network starts a process to deregister the UE implicitly. Figure 5.23 shows the registration procedure due to mobility or periodic update. See Ref. [12] for further details.

1. In addition to step 1 of Figure 5.8, the REGISTRATION REQUEST message can include PDU session status or uplink data status. The uplink data status indicates the previously established PDU sessions(s) having uplink data pending. The PDU session status indicates the previously established PDU session(s) in the UE.

2–4. If the NG-RAN selected a new AMF, steps 2–4 are performed. See steps 13 and 14 of Figure 5.8 and steps 2, 3, and 7 of Figure 5.9.

5. For a PDU session included in the uplink data status, the AMF invokes the Nsmf_PDUSession_UpdateSMContext service towards the SMF with which the PDU session is associated by sending the Nsmf_PDUSession_UpdateSMContext Request message. This will trigger user plane resource reactivation in the NG-RAN. For a PDU session included in the PDU session status and not active in the AMF, the AMF releases the PDU session locally. The AMF invokes the Nsmf_PDUSession_ReleaseSMContext service towards the SMF with which the PDU session is associated by sending the Nsmf_PDUSession_ReleaseSMContext Request message.
6. The SMF responds to the request message.
7. See steps 17 and 18 of Figure 5.8.

5.5.3 Network Initiated Paging

In case user plane traffic (one or more downlink PDUs) or control plane signaling message(s) need to be delivered to the UE in IDLE mode, the AMF pages the UE. Since the AMF is not aware of the exact cell on which the UE is camping, the AMF may need to page the UE for the whole registration area in the worst case.

Figure 5.24 shows the paging procedure. See Ref. [12] for the further details.

1. When a UPF receives a downlink PDU for a PDU session for which there is no downlink tunnel information towards the NG-RAN stored in the UPF, the UPF sends the PFCP Session Report Request message to the SMF.
2. The SMF acknowledges the request. The UPF may forward the downlink PDU towards the SMF if the SMF is supposed to buffer downlink PDUs.
3. The SMF invokes the Namf_Communication_N1N2MessageTransfer service towards the AMF by sending the Namf_Communication_N1N2MessageTransfer Request message to the AMF. The message includes the *PDU Session Resource Setup List* to be delivered to the NG-RAN transparently in step 8.
4. The AMF acknowledges the request by sending the Namf_Communication_N1N2 MessageTransfer Response message with a cause "attempting to reach UE." The cause value implies that the AMF has started paging the UE.
5. The AMF sends the PAGING message to the NG-RAN node(s) belonging to the registration area in which the UE is registered.
6. Then the NG-RAN node(s) will page the UE.
7. The UE performs the RRC connection establishment procedure and transitions to the connected mode. In response to the *Paging* message, the UE sends the SERVICE REQUEST message to the AMF. Since there is no UE-specific association between the AMF and the NG-RAN, the UE provides the NG-RAN with the 5G-S-TMSI to ensure that the SERIVCE REQUEST message is routed to the correct AMF.
8. The AMF sends the INITIAL CONTEXT SETUP REQUEST message to the NG-RAN. The message includes the SERVICE ACCEPT message to be delivered

to the UE transparently, *PDU Session Resource Setup List* that the AMF received in step 3, and some UE context, e.g. security context. The *PDU Session Resource Setup List* indicates QoS profiles of QoS flows in the PDU session and the uplink tunnel information towards the UPF.

9-10. The RRC connection is reconfigured to reactivate the user plane radio resources for the PDU session. The SERVICE ACCEPT message is delivered to the UE.

11. The NG-RAN acknowledges the INITIAL CONTEXT SETUP REQUEST message. The message includes the *PDU Session Resource Setup Response List*. The *PDU Session Resource Setup Response List* contains the downlink tunnel information towards the NG-RAN and list of accepted QoS flows.
12. The AMF sends the Nsmf_PDUSession_UpdateSMContext Request message to the SMF. The message includes the *PDU Session Resource Setup Response List*.
13. The SMF updates the UPF with the downlink tunnel information towards the NG-RAN. Now the downlink PDU can be delivered to the UE via the NG-RAN.

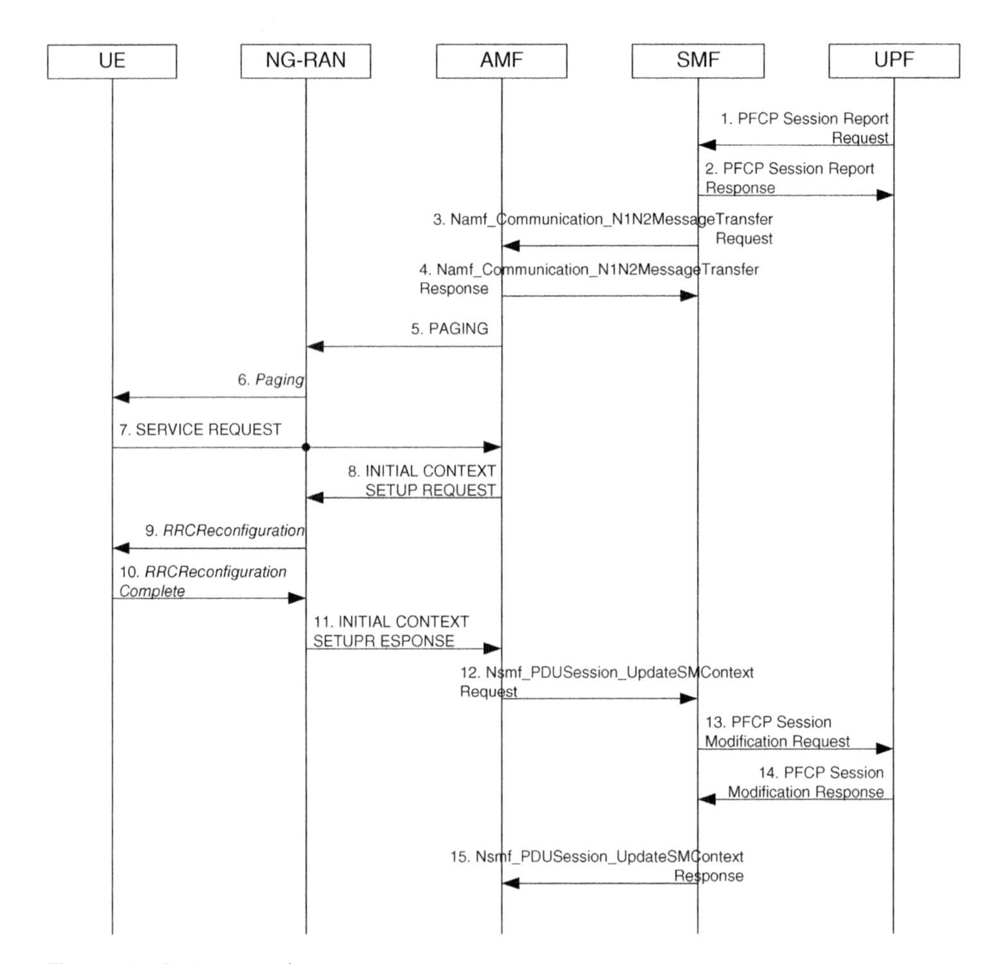

Figure 5.24 Paging procedure.

5.6 RRC Inactive State mobility and UE Reachability

5.6.1 Overview

This section on RRC Inactive state mobility covers the following topics:

- Cell selection and reselection
- Paging/Notification from RAN
- RAN Notification Area
- RRC Inactivation
- RRC Activation
- Need for configurability of RRC Inactive state
- Benefits of RRC Inactive state

To achieve a seamless UE state transition in the 5G system, a connectivity solution where the UE is kept "always-on" from Core Network (CN) perspective (control and user plane) was defined. Once the UE is registered to the network, the connection to the CN is kept alive. However, the RAN can suspend the RRC connection when there is inactivity. The RAN also has the opportunity to configure differently the behavior of UEs with different service requirements during times of inactivity.

The new RRC Inactive state has many features of the existing LTE_IDLE state, such as low activity towards network and UE based mobility using the cell reselection procedure. The need for configurability of the RRC Inactive state is motivated by 5G use cases which have highly diverse, and sometimes contradictory requirements in terms of reliability, mobility, latency, bandwidth, security and privacy, battery life etc. For example, the E2E latency requirement varies from <1 ms for use cases with ultra-low latency requirement such as industrial automation, to latencies from seconds to hours for use cases like massive (low-cost, long-range and low-power) MTC. The battery life requirement is irrelevant in some use cases like autonomous driving, where the device can get unlimited energy from the car, whereas the battery life requirement for battery operated devices ranges from three days for smartphones up to 15 years for a low-cost MTC device.

Allowing a device to use a specific RRC Inactive state configuration enables flexibility to the state handling mechanism. The requirements in Ref. [19] for state transition optimization for ultra-low complexity, power constrained, and low data-rate Internet of Things devices can be taken as an example of network slice specific state handling configuration. However, the relevant solutions may not necessarily be applicable for all use cases, e.g. autonomous driving of vehicle.

If parts of the RAN context are available in the network and UE, some of the potential RRC Inactive state configuration options include:

- Mobility/location tracking management configuration. RAN based mobility and location tracking, single/multiple cell-level tracking.
- Measurement configuration. Measurement configuration for cell reselection, camping, etc. considering the existence of multiple air interface variants.
- Camping configuration. Single/multiple-RAT camping, capacity based camping, etc.
- State transition/system access configuration. State transition and RACH access optimizations.
- Synchronization configuration. DL and/or UL synchronization.

5.6.2 Cell Selection and Reselection

During low activity periods the UE will evaluate the network conditions and select the suitable cell for camping based on UE measurements and according to rules given by the network. With the NR RRC state machine, the cell reselection is applicable in both RRC_IDLE state and RRC_INACTIVE state, while the cell selection is applicable only in RRC_IDLE state. The NAS controls the RAT selection and is looking for the selected PLMN, which is associated with the current RAT. The UE shall select a suitable cell based on RRC_IDLE state measurements and cell selection criteria. UE regularly evaluates the measurements against cell reselection criteria and searches for a better cell for camping. If a better cell is found, that cell is selected. During the normal service provisioning, the UE will decode the control channels to receive the system information of PLMN and receive the registration area information, as well as other AS and NAS Information. Identification of the cell where the UE should be camping is based on the reselection criteria, where the intra-frequency reselection is based on ranking of cells and inter-frequency reselection is based on absolute priorities. The cell reselection can be also UE speed dependent where the fast moving UEs are either preferred to stay at macro cell layer and/or perform the cell reselection faster.

Beam-mobility and multi-beam operation will also need support during low activity periods. For cell reselection in multi-beam deployment, the number of beams and the measurement quantity for cell reselection is computed within the cell beams of the same synchronization signal (SS)/physical broadcast channel (PBCH) block per cell. The linear average of the power values from one beam up to the maximum number of highest beam measurement quantity is derived with values above the threshold, thus indicating the cell reselection criteria.

5.6.3 Paging and Notification from RAN

In 5G, the UE can be paged from 5GC or NG-RAN. For paging occasions, the UE monitors the Paging Channel (PCH) only during the periods which are not configured with use of discontinuous reception (DRX) to reduce power consumption. When the UE is in RRC_INACTIVE state, the paging can be initiated by both 5GC and NG-RAN and for this purpose the UE needs to monitor both RAN-initiated and CN-initiated paging information from paging control channel (PCCH). To minimize complexity of paging from two different sources, the paging occasions of NG-RAN and 5GC should overlap and therefore the same paging mechanism can be used. As described in Figure 5.23, the RRC_INACTIVE state is configurable with service specific requirements thus allowing paging DRX cycle to be configurable either for power saving (long cycle) or low latency (short cycle). As the UE monitors both 5GC and NG-RAN initiated paging, the DRX configuration will be provisioned to UE with the paging initiator specific method. The 5GC-initiated paging and related DRX cycle is configurable via system information, while the NG-RAN can configure a UE specific DRX cycle to be used for RAN-initiated paging. During the paging occasions the UE periodically wakes up and monitors the physical downlink control channel (PDCCH) channel to check for the presence of a paging message. When the PDCCH indicates paging RNTI (P-RNTI) in the paging message, then the UE demodulates the PCH to see if the paging message was directed to it or someone else.

5.6.4 RAN Notification Area

The mobility procedure during the RRC_INACTIVE is optimized for the RAN based location tracking, which means that the UE in the RRC_INACTIVE state can be configured with a RAN Notification Area (RNA). The RNA is the area where the UE can move in RRC_INACTIVE state without notifying the network about the cell reselections. RNA is equivalent to RAN based paging, e.g. if the paging initiator is NG-RAN, then the UE is paged only in the configured RNA. This approach provides significant savings in terms of control plane signaling compared to paging always from core network over the whole tracking area, perhaps consisting of hundreds of cells. The RNA is flexible supporting stationary UEs and moderately fast moving UEs in low activity periods. Therefore, RNA can cover a single cell or multiple cells and the area falls within the registration area of 5GC. There are several ways to provision the UE with RNA. Typically, a stationary or slowly moving UE will get a configuration via list of cells, e.g. at least one cell. There is a trade-off between small and larger RNAs and for above pedestrian speeds the UE could be configured with a list of RNAs or a RAN area ID, where a RAN area ID is a subset of a CN Tracking Area or equal to a CN Tracking Area. If the UE is not configured with a list of cells, then it needs to read the system information and use the broadcasted a RAN area ID for RNA updates.

5.6.5 RRC Inactivation

RRC Inactivation will follow the expiration of inactivity timer, which indicates to network that the UE will benefit from configuration to RRC_INACTIVE low activity period and power saving. After the RRC inactivation, UE remains in CM-CONNECTED state and gets a configuration for RAN Notification Area where the UE can move without notifying NG-RAN network. The UE Context is stored in the last serving gNB with the serving AMF and UPF keeping their respective UE-associated N2 and N3 tunnels available. The NG-RAN node may configure the UE with a periodic RNA Update timer value at transition to RRC_INACTIVE.

The RRC inactive state of a UE can be configured based on the characteristics of the service(s) provided to the UE if such information is available at the network. According to Figure 5.25, RRC Inactive state is included in the Suspend message (see Ref. [20]). If the UE has multiple services or purposes, e.g. a device with multiple concurrent services, then the configuration might be done based on the service with the most stringent requirement state is included in the Suspend message. If the UE has multiple services, e.g. a device with multiple concurrent services, then the configuration might be done based on the service with the most stringent requirements.

5.6.6 RRC Activation

The RRC State transition from RRC_INACTIVE state to RRC_CONNECTED state can be triggered by UE or the network, e.g. when the UE low activity period is over and there is data available for uplink transmission or if the network must page the UE for incoming downlink data. The RRC activation described in Ref. [21] specifies the UE and network triggered transition from RRC_INACTIVE to RRC_CONNECTED.

In general, when the UE is Inactivated, the UE is provided by the last serving gNB with the I-RNTI identifier which is used to identify during the time of activation. This enables

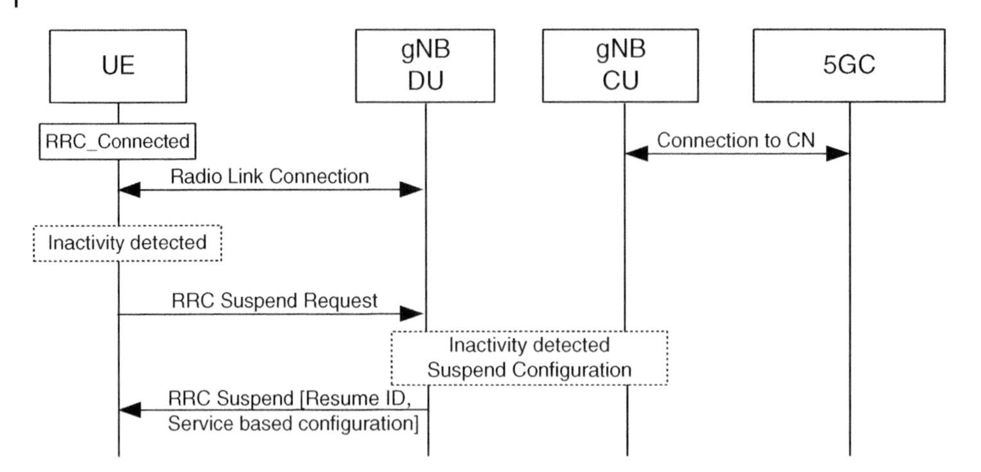

Figure 5.25 Configuration of a RRC inactive state.

the current serving gNB to resolve the anchor gNB and request for the UE Context data. After the RRC activation the path switch is requested from the AMF and the resources are released from the last serving gNB.

Network initiated state transition from RRC_INACTIVE to RRC_CONNECTED involves the RAN paging procedure with the relevant trigger, e.g. there is incoming DL user plane data, DL specific signaling from 5GC, update of critical system information, etc. The RAN based paging is distributed on the need basis in RAN so that the UE can be paged only from the cells under the last serving gNB footprint, or by distributing the paging over Xn interface to other gNBs, which belong to the same configured RNA.

When the UE has been successfully paged from RAN, the U attempts to activate the RRC connection by sending the RRC Connection Resume Request to the network.

5.7 Beam Level Mobility

5.7.1 Overview

The millimeter wave (mmWave) frequencies identified for 5G enable significant bandwidths for next-generation cellular mobile terminals, various new services in vertical applications, and industry deployments. However, radio links using mmWave frequencies are prone to quick channel variations and radio link failures due to deep free-space pathloss and absorption. These challenges can be addressed using beam forming directional antennas at the UE and gNB to improve the link budget for better cell coverage. The application of beam forming requires beam management framework to align the transmitter and the receiver beams, which calls for beam formed initial access, beam selection, beam tracking and fast beam failure recovery to identify and maintain the optimal selection of beams to connect the UE and gNB during active data transmission. The beam forming approach chosen for NR allows beam level mobility with low-cost analog antenna array beam-sweeping.

5.7.2 Beam Management

Beam level mobility, e.g. intra-cell mobility in NR, refers to mobility procedures within the same NR cell and can be done without higher level signaling procedures. That is, Beam level mobility does not require explicit L3 RRC signaling to be triggered and RRC is not required to know which beam is being used at a given point in time.

Multi-beam operation in NR uses synchronization signal block, SS block (SSB), which comprises a primary synchronization signal (PSS), a secondary synchronization signal (SSS), and a PBCH. The UEs are receiving these signals to operate the beam management functions and a given SSB is repeated within an SS burst set to define the gNB beam-sweeping transmission. Beam sweeping provisions the cell area with a set of beams in spatial domain. The transmitted beam interval and directions are specified and parameters are provided to UEs using system information and dedicated signaling. For example, during the initial cell selection, the UE assumes a default SS burst set periodicity of 20 ms which is repeated for gNB beam-sweeping transmissions. In general, an SSB is a group of four orthogonal frequency-division multiplexing (OFDM) symbols in time and 240 subcarriers in frequency where the Demodulation Reference Signal (DMRS) are associated with the PBCH to measure the SSB. The SSBs are grouped into the first 5 ms of an SS burst. Beam measurement will express the reception of SSB in beams in terms of Reference Signal Received Power (RSRP) of the received power with synchronization signals, Reference Signal Received Quality (RSRQ). Beam refinement is the procedure when the optimal set of beams is selected and configured for beamformed communication. Beam reporting sends the measurement information to network with the power and quality of the beamformed signals and the selection of beams is done during the beam refinement phase.

Narrow beam Channel State Information Reference Signal (CSI-RS) can be used for mobility management purposes in RRC_CONNECTED state. CSI-RS are configured to UE to achieve better measurement accuracy and better cell range when the same transmitted energy is applied to a narrower CSI-RS beam compared to wide beam SSB beam. The CSI-RS beam properties of serving and neighboring cells should include at least NR cell ID, slot configuration for CSI-RS and the periodicity e.g. 5, 10, 20, 40 ms, as well as configurable numerologies and association between CSI-RS and SSB for radio resource management (RRM) measurement. For beam management, it is possible to configure multiple CSI-RS to the same SS burst so the UE can first obtain cell synchronization using the SS bursts, and then use that as a reference to search and refine for CSI-RS resources.

Considering directional communications with beams, the best beams need to be periodically identified with radio link monitoring and beam management is used to maintain the alignment between the communicating nodes. For this purpose, SS- and CSI-based measurement results can be jointly used to measure if the best beam is selected for communication. The UE measurement configuration can include both SSB(s) and CSI-RS(s) for the reported cell(s) if both measurement types are available. Beam management is controlled by the Medium Access Control (MAC) protocol (configured by RRC) where the beam failure recovery procedure indicates to the serving gNB of a new SSB or CSI-RS when beam failure is detected. Beam failure is detected by counting beam failure instance indication from the lower layers to the MAC entity.

In RRC_CONNECTED, when the UE measures the beams of a cell, in practice, it measures a subset of the available beams because the beams not pointing to the UE are typically not measurable or not countable to the measurement result. Measurement results are filtered at first on physical layer for beam level quality and then at L3 for cell quality, potentially from multiple beams. Cell quality from beam measurements is derived in the same way for the serving cell and for the non-serving cells. Measurement reports may contain the measurement result of more than of the beams with the best value for measurement quantity, although using just the best beam typically represents the cell quality well when beam management is already done below RRC level. Network configures the beam measurements to be included in measurement reports. The possible reported values may contain only the beam identifier, the beam identifier with the measurement result, or no beam reporting.

For RRC_INACTIVE and RRC_IDLE users the beam management means selection of beamformed initial access allowing the UE to establish a physical link to access the network.

5.7.3 Beam Level and Cell Level Mobility

Just like in LTE, NR Release 15 specifies network controlled mobility to UEs in RRC_CONENCTED state, with the difference that the mobility is classified into cell level mobility and beam management, e.g. beam level mobility. Cell level mobility with beams refers to inter-cell mobility with handovers requiring L3 RRC signaling for inter-gNB handovers. To perform a cell level handover between two gNBs, the four-step procedure [21] is needed where the source gNB initiates handover and issues a Handover Request over the Xn interface. The target gNB performs admission control and provides the target RRC configuration as part of the Handover Acknowledgement. The source gNB provides the RRC configuration to the UE in the Handover Command as part of the connection reconfiguration. The prepared RRC Reconfiguration message contains the information required to access the target cell, which includes the target cell ID, the new C-RNTI, the target gNB security algorithm identifier. For beam mobility the information can include a set of dedicated RACH resources, the association between RACH resources and SSBs and the association between RACH resources and UE-specific CSI-RS configuration. Intention is to access the target without reading system information, so that the common RACH resources, target cell SIBs, etc. can also be included. When the Handover Command is executed, the RRC connection moves to the target gNB and the target gNB gets Handover Complete message from UE.

5.8 Support for High Speed Mobility

5.8.1 Overview

5G targets to address many different use cases. One of the more challenging ones is the case, where the user is moving at high velocity. Examples of this kind of situation are high speed trains and cars, where the passengers wish to keep seamless contact with the network and the critical vehicle communications should be maintained as well.

The 3GPP document TR 38.913 [6] specifies a scenario, where a train is moving at 500 km h^{-1} with up to 300 passengers connected to each macro cell. This leads to

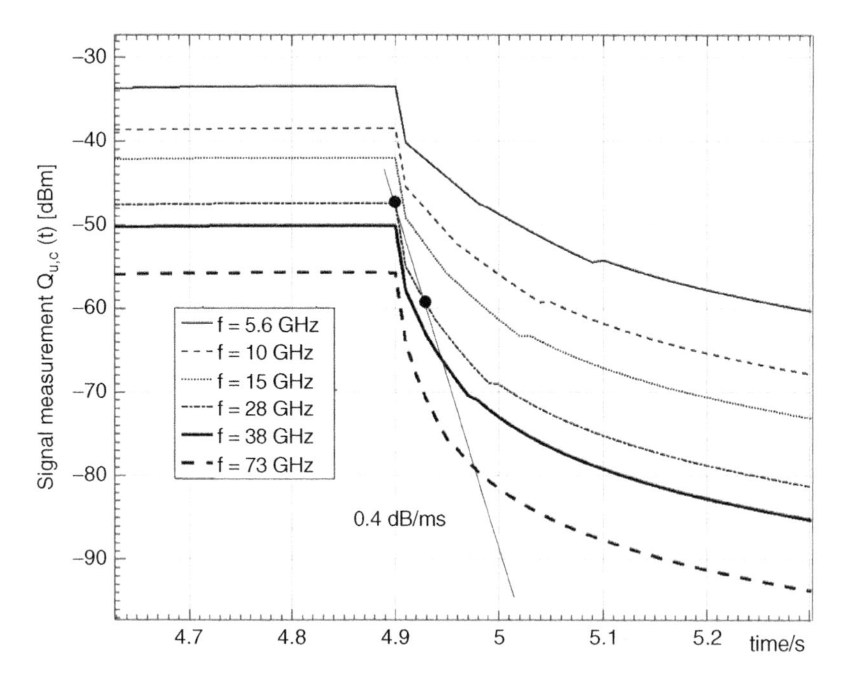

Figure 5.26 Received signal shadowing when the UE passes a street corner at 30 km h^{-1}.

frequent handovers for each user and a very high total number of handovers, even if we assume that only some of the users would be active. Furthermore, each handover should be executed within the short time the UE is within the coverage of both the source cell and the target cell. In this kind of high-speed scenario, the physical layer of 5G needs to be able to tolerate a very high Doppler shift as well.

Another viable scenario would be a car running along a street, which is covered by both <6 GHz macro cells and > 6 GHz small cells. Since the small cells can use frequencies up to several tens of GHz, their coverage areas will be small and their edges sharp. High frequency radio signals don't experience much diffraction and thus an obstacle between the transmitter and the receiver can cut the signal very abruptly. Thus, for these frequencies, a user velocity of 50 km h^{-1} would already be high speed.

Figure 5.26 illustrates how the received radio signal can fade away, when a UE in a slowly (30 km h^{-1}) driving car passes a street corner. The figure is based on simulation results presented in Ref. [22]. It shows for example that at 28 GHz the signal level can drop 12 dB in 30 ms, which can be sufficient to cut the connection in some situations.

5.8.2 Enablers for High Speed Mobility

In order, to facilitate a fast handover, where the UE can attach to the target cell before the connection to the source cell is lost, 5G System supports the following features:

1. The detection of the handover need is made as fast as possible. However, the decision to initiate a handover is a critical one. Handover is a heavy process and a decision to hand over to the wrong cell can lead to a long service interruption. Thus, careful

filtering of the cell measurement results is still needed, which limits the speed of detecting a handover need.

2. Handover process itself should be made as fast as possible, because a high velocity UE will experience very frequent handovers, when traveling across many small cells. One way to achieve this is a handover without random access procedure. Upon receiving a handover request from the source cell, the target cell defines the scheduling for the UE and sends it via the source cell to the UE. Thus, the UE can access the target cell directly, without random access procedure.
3. The multi-connectivity based intra-frequency make-before-break handover, specified for 5G, is a more robust solution than the normal single connected handover. This is because the UE can attach to one or more candidate target cells, while the traffic is still served by the source cell. Thus, new cells can be prepared and added to the multi-connectivity set of links as soon as they are available. After that, the switching of the traffic to one of the target cells can happen using a very fast and lightweight process. And if the selected target cell proves to be a bad one, the decisions can be reverted quickly. The traffic stuck in the wrong target cell will be retransmitted by another cell (or all cells) in the multi-connectivity set of links. The cell measurements used for steering the switching of traffic can be lighter filtered and have faster response times than those used for triggering handovers. Alternatively, the most critical low volume traffic (URLLC) can be continuously replicated to all available links.
4. Inter-frequency multi-connectivity enables a scenario, where the UE is continuously connected both to high frequency small cell layer and to macro cell layer. Figure 5.27 illustrates this scenario. Traffic is normally served by the small cells, which typically have higher capacity and this capacity is shared by less users that that of the large macro cells. In case new small cells can't be added to the set of links early enough for a "fast moving" UE, its traffic is seamlessly switched to the macro cell layer.

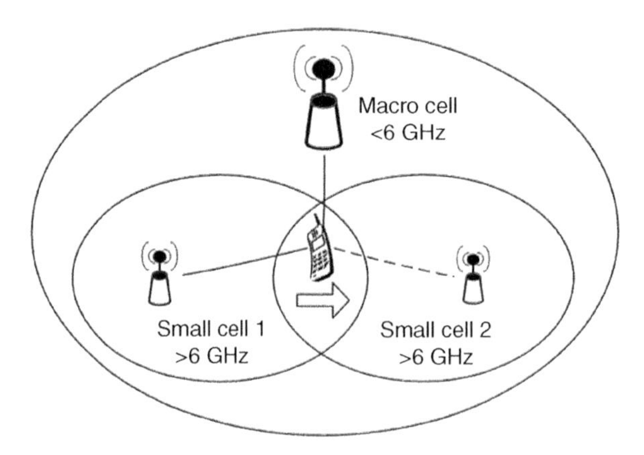

Figure 5.27 A multi-connected UE executing make-before-break handover while having a macro cell connection as back-up.

5.9 Support for Ultralow Latency and Reliable Mobility

5.9.1 URLLC Requirements

3GPP TS 22.261 [23] defines Ultra Reliable Low Latency Communications (URLLC) as a class of latency critical reliable communication service for e.g. the following use cases, which have 10 ms or less latency requirement:

- *Motion control.* Conventional motion control is characterized by high requirements on the communications system regarding latency, reliability, and availability. Systems supporting motion control are usually deployed in geographically limited areas but may also be deployed in wider areas (e.g. city- or country-wide networks), access to them may be limited to authorized users, and they may be isolated from networks or network resources used by other cellular customers.
- *Discrete automation.* Discrete automation is characterized by high requirements on the communications system regarding reliability and availability. Systems supporting discrete automation are usually deployed in geographically limited areas, access to them may be limited to authorized users, and they may be isolated from networks or network resources used by other cellular customers.
- *Automation for electricity distribution (mainly medium and high voltage).* Electricity is also characterized by high requirements on the communications service availability. In contrast to the above use cases, electricity distribution is deeply immersed into the public space. Since electricity distribution is an essential infrastructure, it will, as a rule, be served by private networks.
- *Intelligent transport systems.* Automation solutions for the infrastructure supporting street-based traffic. This use case addresses the connection of the road-side infrastructure, e.g. road side units, with other infrastructure, e.g. a traffic guidance system. As is the case for automation electricity, the nodes are deeply immersed into the public space.
- *Tactile interaction.* Tactile interaction is characterized by a human being interacting with the environment or people, or controlling a UE, and relying on tactile feedback.
- *Remote control.* Remote control is characterized by a device being operated remotely, either by a human or a computer.

Low latency and high reliability are basically conflicting requirements. High reliability of data delivery is usually achieved by retransmitting the lost data. However, re-transmissions take time and increase latency. This effect also impacts the data packets following the lost ones, since RAN is expected to rearrange the received packets and send them to the higher layers in the right order. Thus, the latency requirement should always be accompanied by a reliability requirement, which states the probability of a data packet being delivered within the stated latency.

3GPP TR 38.913 [6] specifies RAN latency as the time it takes to successfully deliver an application layer packet/message from the radio protocol layer 2/3 service data unit (SDU) ingress point to the radio protocol layer 2/3 SDU egress point via the radio interface in both uplink and downlink directions, where neither device nor Base Station reception is restricted by DRX. For URLLC, the target for average user plane latency should be 0.5 ms for UL, and 0.5 ms for DL. The target for maximum latency of a 32-byte packet is 1 ms with 99.999% reliability.

3GPP TR 38.804 [24] specifies that the latency target of 1 ms with 99 999% reliability specified in TR 38.913 [6] should be met in case no cell change (e.g. handover) is needed. If a cell change is required, e.g. because of mobility, some packets may be delayed more. There is no maximum latency specified for this case, but a general statement that the abovementioned latency should be the target.

5.9.2 The Challenges of URLLC Mobility

In the steady state (no cell change) the main sources of user data latency are:

1. Air interface latency, mainly driven by the radio frame structure.
2. Buffering latency (e.g. at radio link control protocol, RLC).
3. Processing latency of each node.
4. Retransmission latency. In case the delivery over air interface of a data packet or a part of it fails, a retransmission is needed. The retransmitted packet will experience an extra latency, since the receiving entity must first detect a faulty or missing packet, send a retransmission request to the transmitting entity and then the packet (or a part of it) must be transmitted again. Since RAN has the requirement of in order delivery, all the subsequent packets need to wait, until the retransmission is ready (or retransmission timer expires), before they can be processed in the receiver re-ordering function.
5. Transport network latency. Transport network latency especially impacts retransmissions, if the retransmission request and the actual retransmission travel long distances between RAN nodes.

The additional sources adding latency due to mobility:

1. Single connectivity handover – detach/attach. During a single connectivity handover, there will inevitably be a short service interruption, since the UE must first let go from the source cell, before it can access the target cell. This interruption will delay the delivery of the subsequent data packets.
2. Single connectivity handover – data forwarding. When the source cell initiates the handover, it may still have data in its buffers and new data may still be arriving from the core network UPF. Since the UE is no longer connected to the source cell, the source cell PDCP layer must forward the data to the target cell PDCP over the Xn (corresponds to X2 in LTE) until it stops receiving new data from the UPF and its buffers are empty. If the source cell PDCP and the target cell PDCP are not co-sited, the forwarded data will travel through the transport network to the source cell. During this time, it is not available for transmission to the UE. Any new data, that the UPF has sent to the target cell directly, must wait for the forwarding to complete first.
3. Multi-connectivity and make-before-break handover. A 5G multi-connectivity capable UE can communicate with both the source cell and the target cell simultaneously. Thus, there is no service break in the air interface. Similarly, if the UE using multi-connectivity links under the same PDCP instance, as presented in Figure 5.28a,

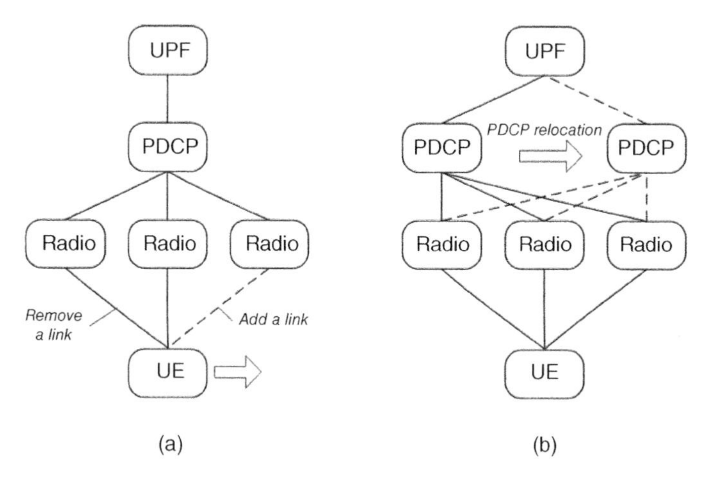

Figure 5.28 Mobility of a multi-connected UE consisting of two independent layers: radio mobility (a) and PDCP relocation (b).

there is no data forwarding either. But when the UE has traveled far enough, there will be a need to relocate the PDCP to a location providing better latency (and/or load conditions), as presented in Figure 5.28b. This relocation may introduce a service interruption depending on the selected relocation process.

5.9.3 Multi-connectivity as a Solution for URLLC Mobility

As described in the previous paragraph, the only way to avoid additional user plane latency at the air interface during mobility events (or any event, which includes a change in a serving cell) is to deploy make-before-break handover (e.g. soft handover). In a make-before-break handover, the UE is simultaneously connected to both the source cell and the target cell. Consequently, it is basically a form of temporary multi-connectivity. A UE carrying URLLC traffic is expected to deploy multi-connectivity also, when stationary, and thus it is natural that make-before-break handover should be based on multi-connectivity.

As described above, multi-connectivity can avoid interruptions in the air interface, but it may have a challenge, when relocating PDCP using legacy single connectivity principles. The two basic options for PDCP relocation process are (not yet discussed in 3GPP specifications):

a. The UE hosts only one PDCP counterpart per bearer as in a legacy single connected handover. A new (target) PDCP instance is established in the network and the connections from it to the same RLC instances, which the source PDCP is using, are built. When relocation starts, the gateway stops sending data to the source PDCP and sends an end-of-data mark (EM) to it instead. Any new data the UPF receives, it will send to the target PDCP. When the source PDCP receives the EM, it will send its status (e.g. last used PDCP PDU serial number) to the target PDCP using the Xn interface. After receiving this information, the target PDCP starts to process the data it has received from the gateway. During the PDCP status forwarding phase neither of the PDCP instances is processing data and thus there is a service interruption.

b. The UE hosts separate PDCP counterparts for each network PDCP instance (source and target) per bearer. Thus, when the target PDCP receives data from the gateway, it can process it immediately and there is no service interruption. Since the UE is now (temporarily) receiving data through two PDCP counterparts, the EM sent by the UPF shall now be forwarded to the UE for keeping the data packets in the correct order. However, the delivery of the EM may sometimes fail. In that case the UE needs to wait until reordering timer expires, before it can start processing the data coming from the target PDCP counterpart. Thus, a failure in EM delivery will introduce a short interruption. However, it should be noted, that if the source PDCP used multi-connectivity, when delivering the EM to the UE, its failure would be very rare.

Most often mobility events take place between cells on the same frequency. Consequently, a capability to communicate with two or more cells on the same frequency virtually at the same time is essential, to avoid transmission collisions and latency due to consequent retransmissions. This would benefit from MAC layer coordination to avoid interference problems between the links. However, this coordination doesn't need to be as strict as e.g. in CoMP [25]. It is sufficient that the MAC schedulers in each cell are given guidelines that prevent simultaneous UL and DL allocations on the same frequency band. In FDD and static TDD systems this is automatically the case. However, in a TDD system with dynamic DL/UL split, coordination would be beneficial. Secondly, simultaneous UL transmissions on the same radio frequency (RF) hardware would split the available transmit power and reduce maximum range. The MAC coordination function should avoid this as well.

5.10 UE Mobility Restrictions and Special Modes

5.10.1 Mobility Restrictions

Mobility restrictions are a set of functionalities exploited to restrict mobility handling or service access of a UE in the 5G system. They are supported by the UE, the NG-RAN and the 5G core network and consist of service area restrictions, forbidden area management, RAT restriction, and core network restriction.

The service area restrictions consist of either an allowed area, or a non-allowed area. The allowed area can contain up to 16 tracking areas or include all tracking areas in a PLMN. The non-allowed area can contain up to 16 tracking areas. The UE can initiate any MM and SM procedures while camped on a cell within a tracking area belonging to the allowed area or not belonging to the non-allowed area. On the other hand, while camped on a cell within a tracking area belonging to the non-allowed area or not belonging to the allowed area, the UE is prohibited from performing the mobility and periodic registration update procedure or the service request procedure involving user plane resource reactivation except for emergency services or for responding to core network paging. In addition, the SM procedures are also prohibited.

For the forbidden area management, the UE manages a list of "5GS forbidden tracking areas for regional provision of service," as well as a list of "5GS forbidden tracking areas for roaming." The UE updates one of the lists whenever a REGISTRATION

REJECT, SERVICE REJECT or DEREGISTRATION REQUEST message is received with the cause value "tracking area not allowed" or "roaming not allowed in this tracking area." In a tracking area whose identity is included in one of the lists, the UE is not permitted to initiate any communication with the network.

The RAT restriction defines the RAT which a UE is not allowed to access in a PLMN. In a restricted RAT, a UE is not permitted to initiate any communication for the PLMN. During the handover, the NG-RAN should determine the target RAT and target PLMN considering the RAT restrictions for PLMNs, if any.

Core network type restriction defines whether a UE can connect to 5G core network for a PLMN. During the registration, deregistration, or service request procedure, the UE disables capability to connect to 5G core network if the message includes the cause value "N1 mode not allowed."

5.10.2 MICO Mode

A UE may indicate preference for mobile initiated communication only (MICO) mode during the registration procedure. Based on the UE preference, UE subscription information, and network policies, or any combination of them, network determines whether MICO mode is applied for the UE and indicates it to the UE during registration procedure. The UE and core network can re-initiate or exit the MICO mode at subsequent registration procedure. The MICO mode needs to be applied explicitly in every registration procedure.

When the AMF applies MICO mode for a UE, the AMF considers the UE always unreachable while in idle mode. The UE in MICO mode is reachable for mobile terminated data or signaling only when the UE is in connected mode and for the PDU sessions that has been reactivated. The network rejects any request for downlink data delivery for an idle UE in MICO mode. The network also does not allow downlink transport over NAS for short message service (SMS), location services, etc.

A UE in MICO mode does not listen to paging while in the idle mode. An idle UE in MICO mode may stop any access stratum procedures, until the UE transitions from idle to connected mode due to one of the following triggers:

- a change in the UE (e.g. change in configuration) requiring the registration procedure;
- expiry of the periodic registration timer; and
- pending mobile-originated data or signaling.

5.11 Inter-System (5GS-EPS) Mobility

5G system provides a means to interwork with EPS, i.e. to support session continuity during the inter-system change (see Chapter 4, Section 4.11 for an overview on EPS interworking, migration from EPS). This section describes the inter-system change procedure for each of the interworking modes specified in 3GPP technical specifications [3, 12]:

1) UE operating in Single Registration mode
2) UE operating in Dual Registration mode

5.11.1 Inter-System Change in Single Registration Mode

The pre-condition for this scenario is that the UE is single registered in EPS or 5GS, maintains only NAS state at any time (Figure 5.29).

1. The E-UTRAN or NG-RAN decides to hand the UE over to the other type of RAN, i.e. NG-RAN or E-UTRAN and sends the HANDOVER REQUEST message to the MME or AMF, respectively. The message includes the target RAN identity and a transparent container.
2. In the case of handover from 5GS to EPS, after receiving the HANDOVER REQUEST message from the NG-RAN, the AMF invokes the Nsmf_PDU Session_ContextRequest service towards the PGW-C + SMF to obtain the SM contexts including the mapped EPS bearer contexts.
3. The source CN sends the Forward Relocation Request message to the target CN, selected considering the target RAN identity. The Forward Relocation Request message includes the MM context, the target RAN identity, and the transparent container.
4. The AMF or MME obtains the uplink tunnel information required for the target RAN. In the case of handover from EPS to 5GS, the AMF invokes towards the PGW-C + SMF the Nsmf_PDUSession_UpdateSMContext service. In the case of handover from 5GS to EPS, the MME exchanges the Create Bearer Request and Response messages with the S-GW.
5. The HANDOVER REQUEST message is sent to the target RAN identified by the target RAN identity. The message includes the transparent container and uplink tunnel information.
6. The target RAN sends the acknowledgement message back to the target CN. The message contains the *HandoverCommand* message and downlink tunnel information for the target CN. The tunnel information is associated with the PDU sessions or bearers that are accepted by the target RAN.
7. In the case of handover from EPS to 5GS, the AMF invokes towards the PGW-C + SMF the Nsmf_PDUSession_UpdateSMContext service to update the PGW-C + SMF with the downlink tunnel information. The PGW-C + SMF provides the downlink tunnel information to the PGW-U + UPF.
8. The Forwards Relocation Response message is sent to the source CN. The message includes the *HandoverCommand* message.

9, 10. The *HandoverCommand* message is forwarded to the UE.

11. The UE completes the RRC connection reconfiguration with the target RAN.
12. The target RAN notifies that the handover on the radio side is completed towards the target CN by sending the HANDOVER NOTIFY message.
13. The target CN confirms the completion of the relocation.
14. Then the source CN acknowledges the completion of the relocation.
15. The target CN informs that the handover is completed towards the PGW-C + SMF. In the case of handover from EPS to 5GS, the AMF invokes the Nsmf_PDUSession_UpdateSMContext service and in the case of handover from 5GS to EPS, the MME and S-GW exchange the Modify Bearer Request and Response messages respectively with the S-GW and the PGW-C + SMF. During the exchange of the Modify Bearer Request and Response message,

tunnel information is traded in addition. The PGW-C + SMF interacts with the PGW-U + UPF accordingly.

16. The mobility registration update (in the case of handover from EPS to 5GS) or tracking area update (in the case of handover from 5GS to EPS) is performed and unnecessary resources are released.

5.11.2 Inter-System Change in Dual Registration Mode

The pre-condition for this scenario is that the UE is dual registered in EPS and 5GS (Figure 5.30). The UE can maintain NAS states independently in EPS and 5GS. The network may or may not support N26. It is up to UE to decide which system to use for which session but it is up to the network to determine which session is admitted in the system based on regular SM procedures in the respective systems. It is up to the network and operator policy to determine for which PDU Session and/or packet data network (PDN) connection, service continuity is offered for inter system mobility. This information is expected to be provided to the UE as part of registration procedure and corresponding

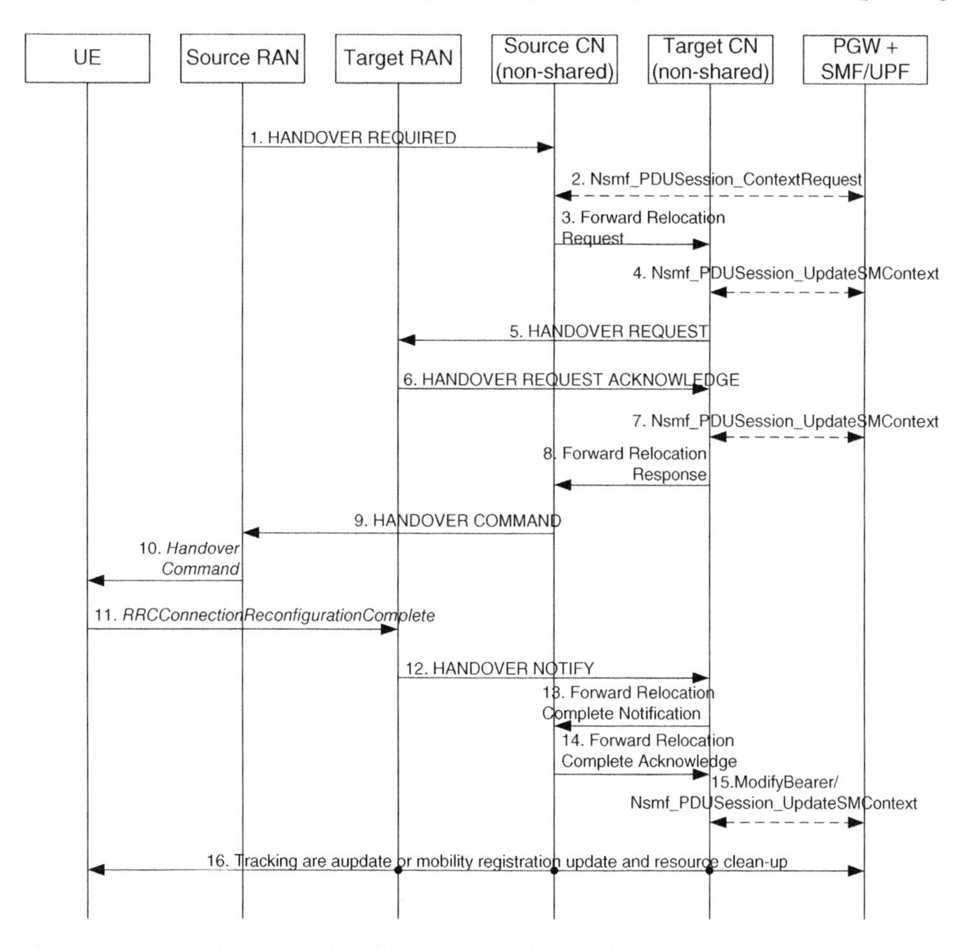

Figure 5.29 Inter-system change for a UE operating in single registration mode.

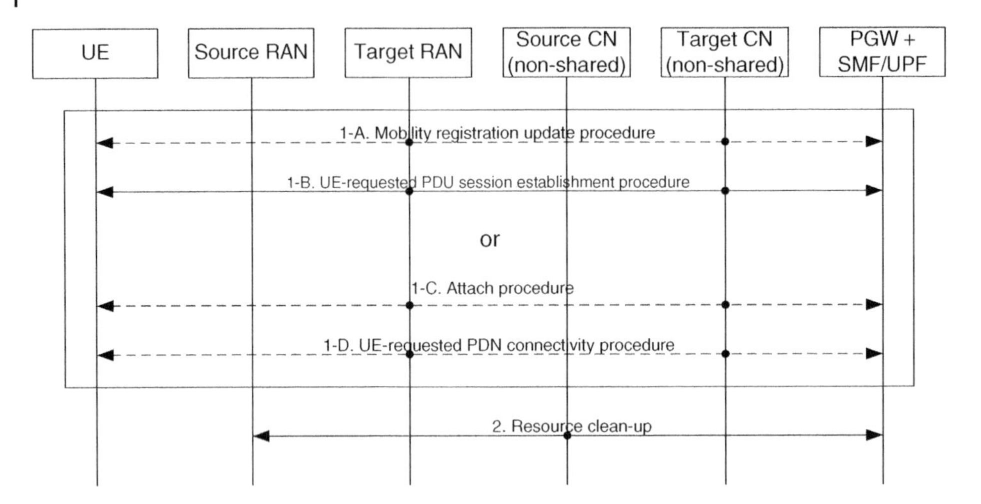

Figure 5.30 Inter-system change for a UE operating in dual registration mode.

session establishment procedures in the respective system. It is also up to the UE to decide which session to move from one system to another.

1. The UE decides to move a PDN connection or a PDU session from one system to the other. As shown in step 1-A, if the UE wants to move a PDN connection in the EPS to the 5GS, the UE performs the mobility registration update procedure (see Figure 5.23) in case the UE is not already registered to the 5GS. During the mobility registration update procedure, the home subscriber server, HSS + UDM does not send cancel location to the MME so that the UE can be kept registered to both systems. Then the UE initiates the UE-requested PDU session establishment procedure as shown in step 1-B. In the PDU SESSION ESTABLISHMENT REQUEST message, the UE sets the request type to "Existing PDU Session." Otherwise if the UE wants to move a PDU session in the 5GS to the EPS, the UE performs the attach procedure as shown in step 1-C in case the UE is not already registered to the EPS. During the attach procedure, the HSS + UDM does not send cancel location to the AMF so that the UE can be kept registered to both systems. The PDN CONNECTIVITY REQUEST message piggy-backed in the ATTACH REQUEST message includes the request type set to "Handover." If the UE is already registered to the EPS, the UE directly initiates the UE requested PDN connectivity procedure as depicted in step 1-D. In the PDN CONNECTIVITY REQUEST message, the UE sets the request type to "Handover."
2. The resources in the source system are cleaned up, i.e. a PDU session moved to the EPS or a PDN connection moved to the 5GS is released in the source system.

5.12 Outlook

It has been acknowledged that the evolution of LTE should be integrated to the 5G to possibly benefit from the widely deployed coverage of E-UTRA (LTE) in the 2020 timeframe. Tight integration of LTE (E-UTRA) and new 5G (New Radio/NR) air interfaces should not introduce additional core network signaling complexity, no extra RRC state transitions due to inter-RAT mobility and no extra load to RRC signaling. A moving UE RRC connection may be suspended or inactivated and UE is using the RRC Connected Inactive state during the low activity period. With tight integration, the connection resumption or activation back to RRC Connected could be done based on the system, which can provide better coverage or capacity. That is, a connection inactivation in 5G may be followed by resumption in LTE when 5G coverage is not available in the same geographical area.

In future releases, we expect support for mobility management in 5G System to be enhanced to support additional features for improved user experience and support new and emerging use cases in the market place. Therefore, 5G will raise new requirements to support new use cases and verticals, e.g.:

1. Handover with 0 ms interruption
2. Configurable reliability of handovers and link monitoring
3. Reduced jitter in the user throughput by optimized handovers based on service requirements
4. Configurable trade-off between power saving and latency
5. Service specific mobility configuration
6. Mobility support for multi-slicing networks

Following are some of the foreseen solutions that are under consideration:

1. Make before break handover
2. RACH-less handover
3. NR-NR Dual Connectivity
4. Seamless Inter-RAT mobility with Cloud RAN
5. Conditional handover

As and when new mobility management features are introduced, these new features may be applicable and used only in scenarios where necessary. This is to enable user centric flexible support of new features when and where essential, and at the same time not mandating support for such enhancements for use cases (e.g. stationary, low mobile, low cost device) where they are unnecessary. This approach goes well with the overall theme of the 5G System to be flexible, provide deployment tools for various use cases, and not mandate deployment and use of features but allow the actual deployment to decide when a certain feature is useful.

References

All 3GPP specifications can be found under http://www.3gpp.org/ftp/Specs/latest. The acronym "TS" stands for Technical Specification, "TR" for Technical Report.

1 Report ITU-R M.2410-0: "Minimum requirements related to technical performance for IMT-2020 radio interface(s) ", November 2017.
2 3GPP TS 38.473: "F1 Application Protocol (F1AP)".
3 3GPP TS 38.463: "E1 Application Protocol (E1AP)".
4 3GPP TS 24.501: "Non-Access-Stratum (NAS) protocol for 5G System (5GS)".
5 3GPP TS 23.402: "Architecture enhancements for non-3GPP accesses".

6 3GPP TR 38.913: "Study on scenarios and requirements for next generation access technologies".
7 3GPP TS 36.331: "Evolved Universal Terrestrial Radio Access (E-UTRA); Radio Resource Control (RRC)".
8 3GPP TS 23.401: "GPRS enhancements for E-UTRAN access".
9 3GPP TS 24.301: "Non-Access-Stratum (NAS) protocol for Evolved Packet System (EPS)".
10 3GPP TS 38.331: "NR; Radio Resource Control (RRC); Protocol specification".
11 Hailu, S., Säily, M., and Tirkkonen, O.: "Towards a configurable state model for 5G radio access networks", Global Wireless Summit 2016, Nov. 2016
12 3GPP TS 23.502: "Procedures for the 5G System".
13 3GPP TS 38.413: "NG-RAN; NG Application Protocol (NGAP)".
14 IETF RFC 5448: "Improved Extensible Authentication Protocol Method for 3rd Generation Authentication and Key Agreement (EAP-AKA')".
15 3GPP TS 33.501: "Security architecture and procedures for 5G system".
16 3GPP TS 36.300: "Evolved Universal Terrestrial Radio Access (E-UTRA) and Evolved Universal Terrestrial Radio Access (E-UTRAN); Overall description".
17 3GPP TS 36.423: "X2 application protocol (X2AP)".
18 R2-1706489, "Conditional handover basic aspects and feasibility in Rel-15", Nokia, June 2017.
19 3GPP TS 37.340: "NR; Multi-connectivity; Overall description".
20 5GPPP METIS-II project, Deliverable D6.1: "Draft Asynchronous Control Functions and Overall Control Plane Design", July 2016.
21 3GPP TS 38.300: "NR; Overall description; Stage-2".
22 Awada, A., Lobinger, A., Enqvist, A., et al. "A simplified deterministic channel model for user mobility investigations in 5G networks", IEEE International Conference on Communications (ICC), Paris, 2017, pp. 1–7.
23 3GPP TS 22.261: "Service requirements for next generation new services and market".
24 3GPP TR 38.804: "Study on new radio access technology Radio interface protocol aspects".
25 Won, S.H., Lee, H.-J., Oh, J., et al. "Coordination of multiple eNBs using short-term channel information", Proceedings of IEEE Globecom Workshop, pp. 765–770, 2014.

6

Sessions, User Plane, and QoS Management

Devaki Chandramouli[1], Thomas Theimer[2] and Laurent Thiebaut[3]

[1] *Nokia, Irving, TX, USA*
[2] *Nokia, Munich, Germany*
[3] *Nokia, Paris-Saclay,, France*

6.1 Introduction

UE(s) (User Equipment(s)) use the Fifth Generation System (5GS) to get data connectivity between applications on the UE and Data Networks (DNs) such as the 'Internet' or private corporate networks. Almost all applications running on the UE including voice require such data connectivity.

This chapter defines the data connectivity provided by the 5GS (Protocol Data Unit [PDU] Sessions, PDU Session continuity modes, traffic offloading, etc.). It then describes the 5GS Quality of Service (QoS) framework (QoS Flows, parameters of 5GS QoS, Reflective QoS, etc.). Finally, it gives an overview of session management including an overview on how applications can influence traffic routing and on policy control for PDU Sessions.

6.2 Basic Principles of PDU Sessions

6.2.1 What is a PDU Session?

The 'PDU Session' is the (new) abstraction for 5G user plane services like the concept of a Packet Data Network (PDN) connection in 4G Evolved Packet Core (EPC). A PDU Session provides connectivity between applications on a UE and a DN such as the 'Internet' or private corporate networks. A DN is identified by a Data Network Name (DNN) which is the equivalent of an Access Point Name (APN) in 4G EPC.

A PDU Session is associated with a single DNN and with a single slice (S-NSSAI).

PDU Sessions can provide different types of transport services corresponding to the nature of the PDU(s) carried over the PDU Session:

- Internet Protocol (IP) type PDU Sessions offer a Layer 3 service including IP address assignment and IP flow-based filtering and QoS. The 5GS provides the first hop router for the UE. An IP type PDU Session is expected to be the most commonly used PDU

5G for the Connected World, First Edition.
Edited by Devaki Chandramouli, Rainer Liebhart and Juho Pirskanen.

Session type for regular Internet connectivity and for operator offered services such as Internet Protocol Multimedia Subsystem (IMS) services (including voice).
  - An IP type PDU Session may offer either an IPv4 service or an IPv6 service or the combination of an IPv4 service and of an IPv6 service.
- Ethernet (ETH) type PDU Sessions offer a Layer 2 service including layer 2 (e.g. Media Access Control [MAC] address and/or Virtual Local Area Network [VLAN]) based QoS, forwarding and filtering of frames. ETH type PDU Session are expected to be the commonly used PDU Session for industrial applications, enterprise Local Area Network (LAN) services (e.g. when 5GS is used for connectivity within an enterprise).
- Unstructured type PDU Sessions provide a fully transparent data transport service that does not make any assumptions on the payload type and data format; traffic filtering and classification are not supported in this case. Typical use of unstructured type PDU Session would be for IoT (Internet of Things) traffic targeted towards Machine to Machine (M2M) applications.

One PDU Session carries PDU(s) of a single type only: either IP or ETH or Unstructured. 5G user plane delivery must support new use cases for a variety of vertical markets including for example industrial control applications relying on very low latency (Ultra Reliable Low Latency Communication [URLLC] services) and/or demanding non-IP connectivity services (e.g. ETH). URLLC services are defined in Section 6.3.

Other PDU Session types may be defined in future 3GPP releases.

The 5GS guarantees that applications on the DN and traffic forwarding on the DN are not bothered by UE mobility. For this purpose, the 5GC delivers and receives traffic of a PDU Session to and from the DNs via interfaces called N6 terminated on a User Plane Function (UPF) said to support a PDU Session Anchor (PSA) functionality (Figure 6.1).

The User Plane of PDU Sessions is supported by the Fifth Generation Access Network (5G-AN, e.g. the Fifth Generation Radio Access Network [5G-RAN] or the Non-3GPP Interworking Function (N3IWF) in case of Un-Trusted Non Third Generation Partnership Project [3GPP] access defined in Section 4.12) and by UPFs in the 5G Core. There can be multiple UPFs supported for a single PDU Session. An UPF that supports a PSA functionality interfaces with the DN via N6. At least one UPF is expected to act as a PSA for a given PDU Session.

3GPP Rel-15 specifications support service continuity for a PDU Session when the UE moves:

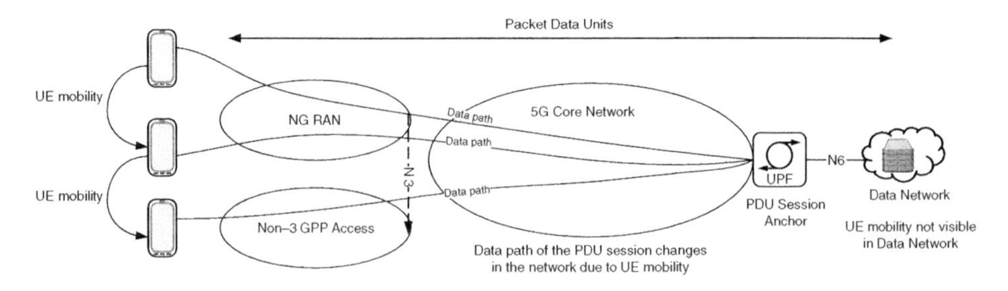

Figure 6.1 5GS user plane – a PDU Session topology example.

- within 3GPP access (within a Next Generation Radio Access Network [NG-RAN] and between NG-RAN(s)),
- between 3GPP and non-3GPP access to the 5G Core,
- between the 5GC and the 4G EPC.

This implies that seamless service continuity is offered for the PDU Sessions for a given UE (i.e. from an end user perspective, application continues to work without service disruption) when mobility occurs as mentioned above.

Whether a PDU Session may use 3GPP access or non-3GPP access or both is defined by policies configured by the operator in the UE.

In release 15 version of 3GPP specifications, a PDU Session cannot simultaneously use a 3GPP and a non-3GPP Access.

A UE may simultaneously establish multiple PDU Sessions (e.g. corresponding to different PDU types or to different devices within the UE). In that case the User Plane and the Control Plane of these PDU Sessions may be totally disjointed in the 5G Core (even though these PDU Sessions target the same DN).

SMF (Session Management Function) is the 5GC Control Plane Function responsible for establishing/maintaining the PDU Sessions (Figure 6.2). 5GC specifications define a generic UPF with capabilities that can be programmed by the SMF allowing flexibility for the deployment of 5G user plane features. UPF(s) may be geographically distributed or centralized.

The SMF (5GC CP) controls the UPF via the N4 interface.

6.2.2 Multiple Concurrent Access to the Same Data Network

End-to-End (E2E) network latency depends on multiple factors such as the air interface latency, packet switching and transport latency as well as processing of data flows in UPFs of the core. As air interface latencies are reduced to sub-millisecond level in 5G, propagation delay caused by the speed of light in the backhaul (i.e. optical fibre) is becoming the limiting factor. This leads to the need to distribute UPFs as well as application services to locations that are topologically and physically closer to the end user device.

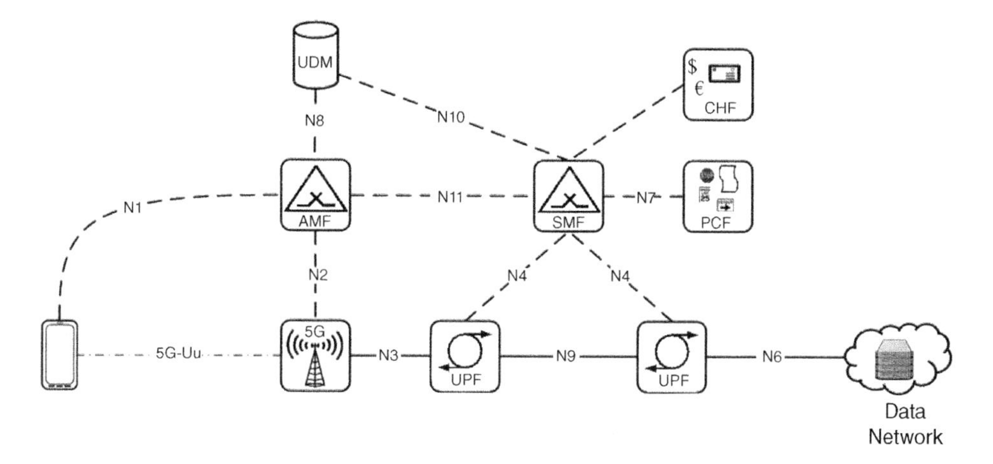

Figure 6.2 5G PDU Session management – impacted interfaces (as seen from SMF and UPF).

In 5GS, the UE can obtain services simultaneously from local/central access to the same DN. This allows the UE to obtain fast access to content/applications deployed locally but at the same time have access to any central server for all other content/applications: as opposed to 4G EPC where a PDN connection corresponds to a single interface to the DN, a 5GC PDU Session may support multiple PSAs and have multiple interfaces to the DN. The local access allows the end user enjoying application access with very low latency. The central access allows access to any application without service disruption due to mobility. This is to deal with the fact that due to specific application requirements (e.g. on end to end latency as for Augmented Reality) and due to network optimizations (e.g. for content caching in a content delivery network (CDN) for popular videos), different flows exchanged over a PDU Session may need to reach the DN at different locations (at different PSAs), some close to the UE and some more central.

Enabling seamless service continuity in a distributed network with full mobility required a major redesign of the 5G user plane.

Figure 6.3 shows a PDU Session example that illustrates key enhancements of 5G user plane capabilities. In this example, a 5G UE is concurrently served from different network locations ('Data Centers') hosting core network UPFs and cloud applications:

1. A *local serving location* selected to be near to the 5G UE, hosting low latency applications (like video caching and Augmented Reality) and a UPF acting as a temporary mobility anchor towards the access network. The local serving location may change

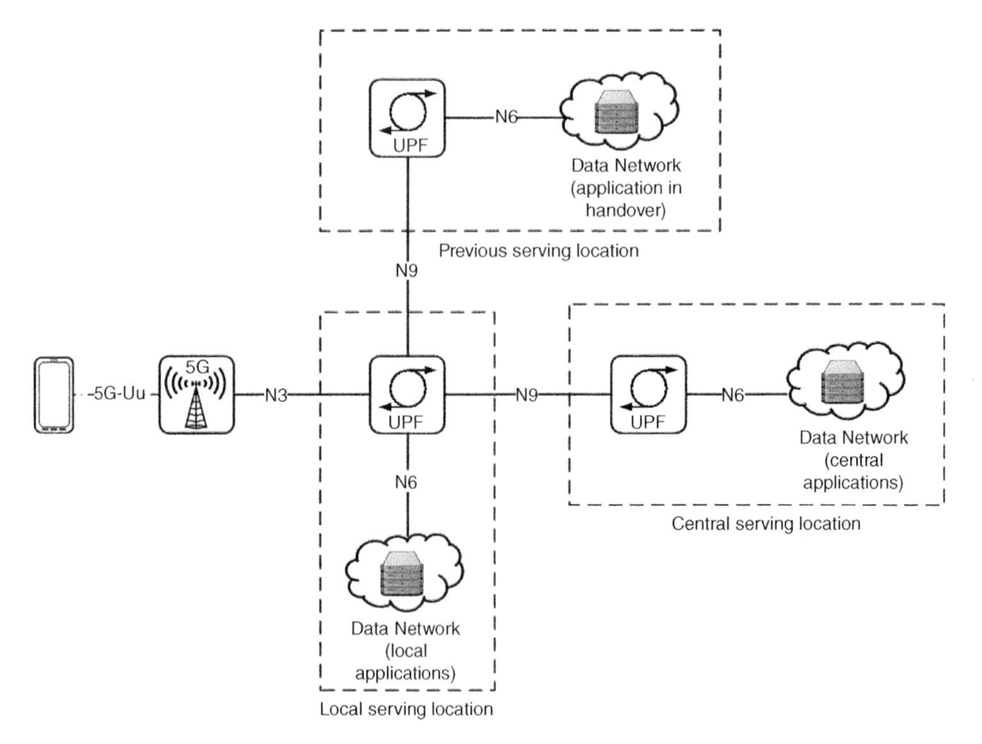

Figure 6.3 5G User plane – multiple concurrent (e.g. local/central) access to the same Data Network for a PDU Session.

when the UE is leaving the serving area of the 'local Data Center', and the challenge is to maintain service continuity during that change.

2. A *central serving location* providing a stable PDU Session termination for applications that do not tolerate changes of IP addresses, e.g. services like IMS voice. A central IP anchor is optional and would not be established in all cases.
3. A *previous local serving location*. Service continuity may require an application instance in the previous local serving location to be reachable on a temporary basis, until application-specific state and context information have been transferred to the new local serving location or until the application usage is over (in case the application does not support mobility). This allows the decoupling of application-level mobility from changes in primary serving locations (see Section 4.8 on Multi-Access Edge Computing [MEC]).

The example shows that a PDU Session may be configured to comprise multiple UPF typically arranged in a hub-and-spoke topology. The local serving location is hosting the hub UPF which acts as a branching point forwarding traffic flows to local applications or to other UPFs (acting as the spoke), e.g. connected to central applications. 3GPP has defined two User Plane solution architectures for controlling packet forwarding behaviour in the hub UPF:

(1) *Uplink Classifier* (UL CL) model (Figure 6.4) – With this model, an UPF supporting UL CL functionality can be inserted into the data stream to divert UL traffic matching destination-based filter rules. The UE is unaware of the existence of the UL CL and has a single network layer source address (e.g. IPv4 or IPv6 address). UL

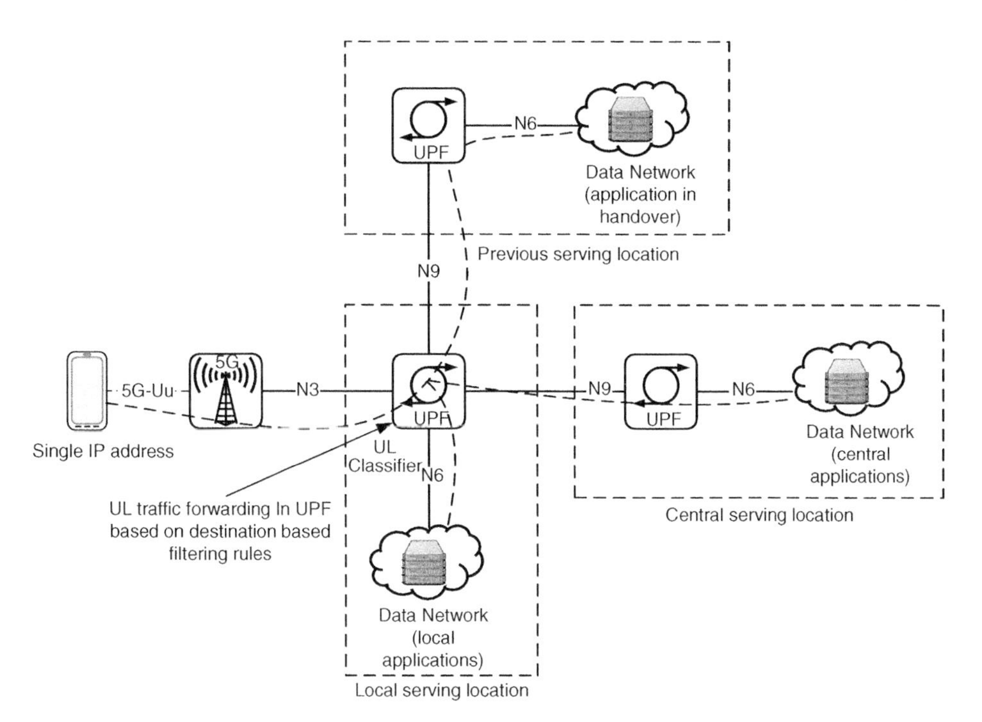

Figure 6.4 Uplink classifier solution for multiple concurrent access to the same Data Network.

Forwarding rules (in the UL CL) that are based on the destination-based filter rules ensure that UL traffic is sent via the correct PSA. When an UL CL is used, DL traffic from the DN (e.g. the Internet) is expected to be received via a PSA connected to the DN in the central serving location. Techniques like IP address translation (NAT) or local forced forwarding (tunnelling) may be used to ensure that DL traffic from local applications is sent via the PSA connected to the DN in the local serving location.

DL traffic received from all serving locations is sent by the UL CL towards the Access Network serving the PDU Session.

(2) Multi-homing – *IPv6 multi-homing* (Figure 6.5) is an alternative concept for IPv6 PDU Sessions with multiple IP anchors. It is based on multi-homing as defined in Request for Comments (RFC) 7157 [1]. The UE is assigned separate IPv6 prefixes for each PSA and needs to select one of those source prefixes based on address selection policies provided by the network as described in RFC 4191 [2]. A PDU Session with multi-homing capability requires an UPF that acts as a Branching Point UPF. The Branching Point UPF has UL forwarding rules associating each source IP prefix with a PSA. This solution depends on UE support for IPv6 multi-homing.

DL traffic received from all serving locations is sent by the Branching Point towards the Access Network serving the PDU Session.

IPv6 multi-homing can be employed in two scenarios:

1) Multi-homing as a ***permanent*** condition: a single PDU Session provides access to a local and to a centralized DN.
2) Multi-homing as a ***transient*** condition: this leverages IPv6 multi-homing for changing of PSA due to UE mobility while ensuring there is no service delivery gap:

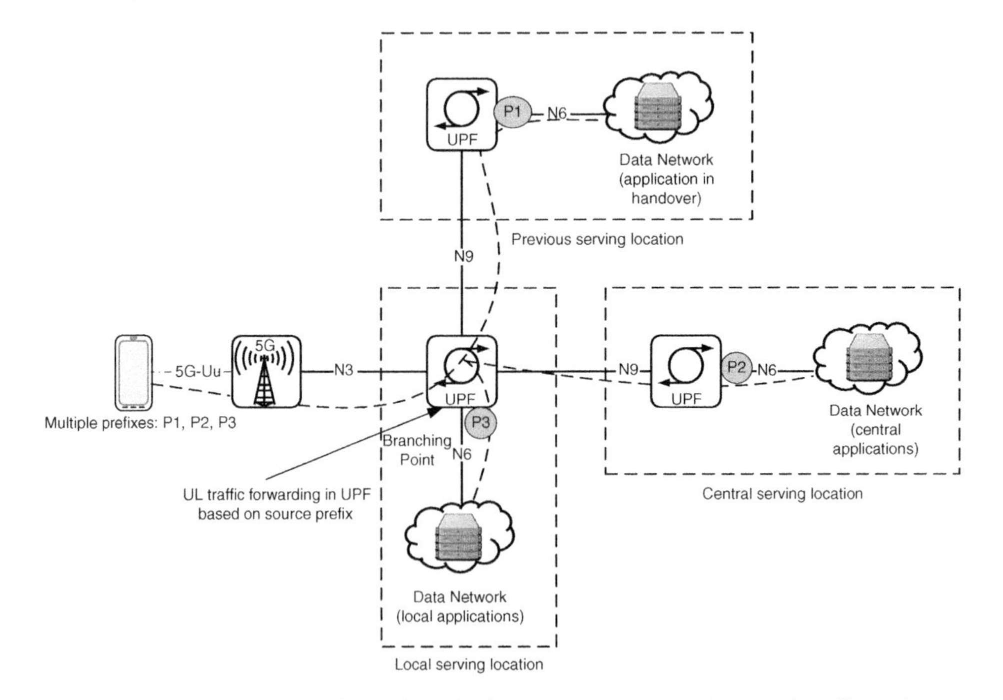

Figure 6.5 IPv6 multi-homing solution for multiple concurrent access to the same Data Network.

IPv6 Multi-homing allows a UE to send and receive data traffic via two PSAs simultaneously. This supports Session and Service Continuity (SSC) mode 3 defined in Section 6.2.3.

6.2.3 Support of SSC (Session and Service Continuity) Modes

When the UE moves it may be needed to change the PSA (i.e. interface to the DN) of a PDU Session. This change of PSA may be due to one of the following reasons:

- To ensure an end to end data transfer delay compatible with QoS requirements on the PDU Session.
- To optimize the backhaul resources of the network.

Changing the PSA may induce a change of IP address to be used by the UE on the PDU Session. Some applications like voice may not tolerate a change of IP address on the PDU Session while others may recover from a change of IP address or even recover from a small gap of IP connectivity.

Thus, 3GPP has defined different SSC modes to tackle this diversity of requirements about the loss of the UE IP address and the loss of IP connectivity.

5GS specifications support a wide range of SSC modes for PDU Sessions:

- For a PDU Session of SSC mode 1, the UPF acting as PSA at the establishment of the PDU Session is maintained regardless of the access technology (e.g. Access Type and cells) a UE is using to access the network. In case of a PDU Session of IP type, IP address continuity is supported regardless of UE mobility events. SSC mode 1 is well suited for applications like voice that are sensitive to the loss of IP address.
- For a PSA of SSC mode 2, the network may release the PSA when the UE mobility induces that a more optimum PSA should be allocated to the UE. In case of a PDU Session of IP type, the IP address/prefix corresponding to the released PSA is released. If the released PSA was the last remaining PSA for the given PDU Session, the PDU Session is released and the UE is 'invited' to request a new PDU Session ('break before make' mobility). SSC mode 2 is well suited for applications that can cope with the temporary loss of IP connectivity (e.g. mail programs or browsers).
- For PDU Session of SSC mode 3, the network allows the establishment of UE connectivity via a new PSA to the same DN before connectivity between the UE and the previous PSA is released ('make before break' mobility). This enables support for seamless service continuity when the UE mobility induces that a more optimum PSA should be allocated to the PDU Session. SSC mode 3 is well suited for applications that can cope with a change of IP address/Prefix while they cannot cope with a temporary loss of IP connectivity: for example, a video player may finish to get the current video chunk via the previous PSA while it will use the new PSA to get the next video chunks.

The support of SSC modes is documented in 3GPP TS 23.501 [3] section 5.6.9 and in 3GPP TS 23.502 [4] section 4.3.5.

6.2.4 Support of PDU Sessions Available/Authorized Only in Some Location (LADN)

5GS specifications support the concept of PDU Sessions available/authorized only in a set of locations called LADN (Local Area Data Network) service area. LADN may be used for applications that are only valid in some areas, for example a dedicated Augmented Reality application that is only available in a sport stadium.

LADN may be used e.g. to limit the access to some DNN(s) in the corresponding 'LADN service area':

- Tracking Areas serving the corporate premises for a corporate DNN; and
- Tracking Areas serving a 'stadium' for a DNN associated with applications dedicated to people participating to an event in the 'stadium'.

A LADN service area is defined as a set of Tracking Areas. In 3GPP Rel-15, LADNs apply only to 3GPP accesses.

For DNN subject to LADN restrictions, the LADN service area is configured in the Access and Mobility Management Function (AMF) on a per DNN basis, i.e. for different UEs accessing the same LADN, the configured LADN service area is the same. The AMF provides the UE with information on the LADN service area.

When the UE is out of the LADN service area of a DNN subject to LADN restrictions, the UE:

- shall not establish/modify a PDU Session for this DNN; and
- shall not request activation of the User Plane connection of a PDU Session for this DNN.

The support of LADN is documented in 3GPP TS 23.501 [3] section 5.6.5.

6.2.5 Data Network Authentication and Authorization of PDU Sessions

At the establishment of a PDU Session to a DNN subject to DN authentication/authorization, the DN-specific identity of a UE may be authenticated/authorized by a Data Network Authentication, Authorization and Accounting DN-AAA server (which is generally not operated by the Public Land Mobile Network [PLMN]). The DN-AAA server may belong to the 5G operator or to the DN owner (e.g. to an external provider).

This allows the DN owner (especially when it is an external provider) to control who enters the DN as well as to provide policies (for example, on the allowed throughput or on the allowed MAC addresses) restricting the access to the DN. This feature may be used for a PDU Session allowing access to a corporate network enabling the manager of the corporate network to control access to this network. This may be a requirement when the corporate network supports control of industrial processes (automated factories).

DN authentication/authorization of PDU Sessions works as follows:

- At the establishment of a PDU Session, if the SMF determines that authentication/authorization of the PDU Session establishment is required, the SMF triggers the authentication/authorization of the UE from the configured DN-AAA server. This is controlled by SMF policies associated with the DNN.
- The DN-AAA server may authenticate the PDU Session establishment. The SMF supports this UE authentication by relaying Extensible Authentication Protocol (EAP) exchanges between UE (via NAS SM signalling exchanged with the UE) and DN-AAA Server (via AAA signalling exchanged with the DN-AAA server).

- When DN-AAA server authorizes the PDU Session establishment, it may send to the SMF DN authorization data for the established PDU Session. The DN authorization data for the established PDU Session may include one or more of the following:
  - a reference to locally configured authorization data in the SMF;
  - a DN profile index to retrieve the SM or QoS policy from the Policy Control Function (PCF);
  - a list of allowed MAC addresses for the PDU Session; this may apply only for PDU Sessions of ETH PDU Session type; and
  - the PDU Session Aggregate Maximum Bit Rate (Session AMBR) (as defined in Section 6.4.4).

  The DN-AAA server may request receipt of information associating the UE identifier (Generic Public Subscription Identifier – GPSI) with the IP address(es) of the UE and/or with actual N6 traffic routing information to be used to reach the UE on the PDU Session; this information may be used by applications in the DN to communicate with the UE.

DN authentication/authorization of PDU Sessions is documented in 3GPP TS 23.501 [3] section 5.6.6 and in 3GPP TS 23.502 [4] section 4.3.2.3.

6.2.6 Allocation of IPv4 Address and IPv6 Prefix to PDU Sessions

A UE may correspond to a smartphone where the 3GPP layer ('the modem'), the 'Transmission Control Protocol (TCP)/Internet Protocol (IP) stack' and the applications are in a single piece of equipment.

A UE may also be made up of different entities such as:

- A smartphone containing the 3GPP layer ('the modem'); and
- Other devices like PC(s) containing the 'TCP/IP stack' and the applications but not supporting 3GPP technology.

In this latter case the UE needs to receive its IP addressing information via non 3GPP methods like Dynamic Host Configuration Protocol (DHCP), see [5].

For PDU Sessions associated with IPv4 (IPv4 or IPv4v6 PDU Session type), the 5GS may deliver the (unique) IPv4 address associated with the PDU Session using:

- *NAS (N1) SM signalling*. The IPv4 address associated with the PDU Session is provided to the UE within the dedicated 3GPP signalling indicating that the network accepts the establishment of the PDU Session,
- *DHCPv4*. When the UE indicates its willingness to have 'deferred IPv4 address allocation', no IPv4 address is provided to the UE within the dedicated 3GPP signalling indicating that the network accepts the establishment of the PDU Session. Once the PDU Session establishment is accepted, the UE (for example a PC behind a smartphone) is sending DHCPv4 [5] signalling in the User Plane to get the IPv4 address associated with the PDU Session.

For PDU Sessions associated with IPv6 (IPv6 or IPv4v6 PDU Session type), the 5GS may deliver the (possibly not unique) IPv6 Prefix associated with the PDU Session using SLAAC (IPv6 Stateless Address Autoconfiguration as defined in [6]):

– No IPv6 Prefix is provided to the UE within the dedicated 3GPP signalling indicating that the network accepts the establishment of the PDU Session. Once the PDU Session establishment is accepted, the network sends a Router Advertisement in the User Plane providing an IPv6 Prefix associated with the PDU Session
– When a new Prefix is allocated to a PDU Session supporting IPv6 multi-homing (as defined in Section 6.2.2), the network sends a Router Advertisement in the User Plane providing the new IPv6 Prefix associated with the PDU Session.

Other IP addressing allocation mechanisms will be defined in 3GPP Release 16 to support the specific requirements of Wireline access.

6.2.7 Support of PDU Sessions When Roaming

When the UE is roaming, its access is served by a Visited Public Land Mobile Network (VPLMN) that is different from its Home Public Land Mobile Network (HPLMN) (or different from a PLMN equivalent to its HPLMN).

In the case of roaming, the HPLMN may control whether for a PDU Session the access to the DN (the PSA) is located:

– In the VPLMN: this is called a LBO (Local Breakout) roaming mode.
– In the HPLMN: this is called a HR (Home Routed) roaming mode.

In HR mode (Figure 6.6) the HPLMN keeps control on the PSA, i.e. on the (N6) connectivity service of the PDU Session and on the corresponding Policy control; the VPLMN does not need to be aware of the specific data services provided on the PDU Session and the VPLMN responsibility is restricted to ensuring proper QoS, supporting the handovers, and handling the transitions between UP active and UP non-active state for the PDU Session.

The HR mode is well suited to cases where the HPLMN is providing specific data services to the users of a DN (for example, corporate Virtual Private Network [VPN]).

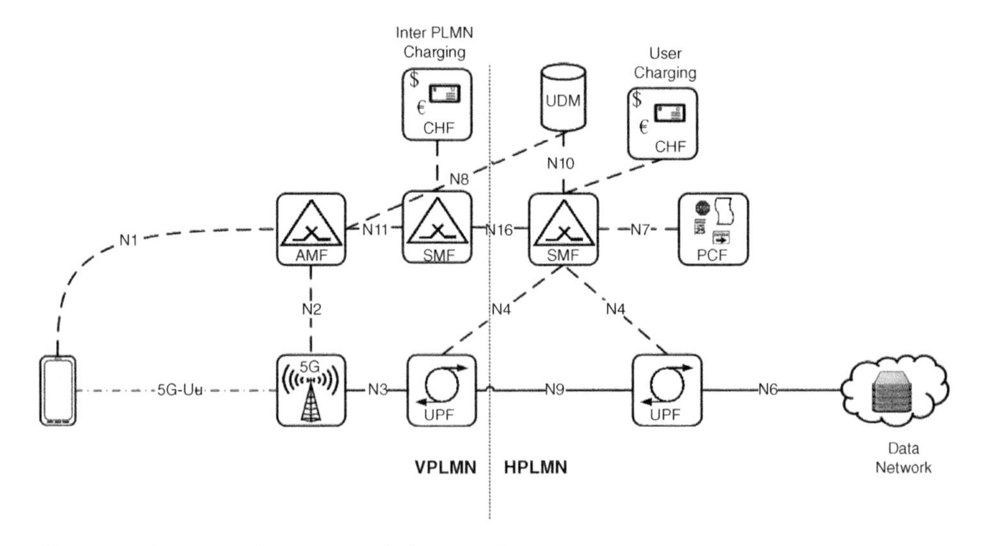

Figure 6.6 Home routed roaming mode for a PDU Session.

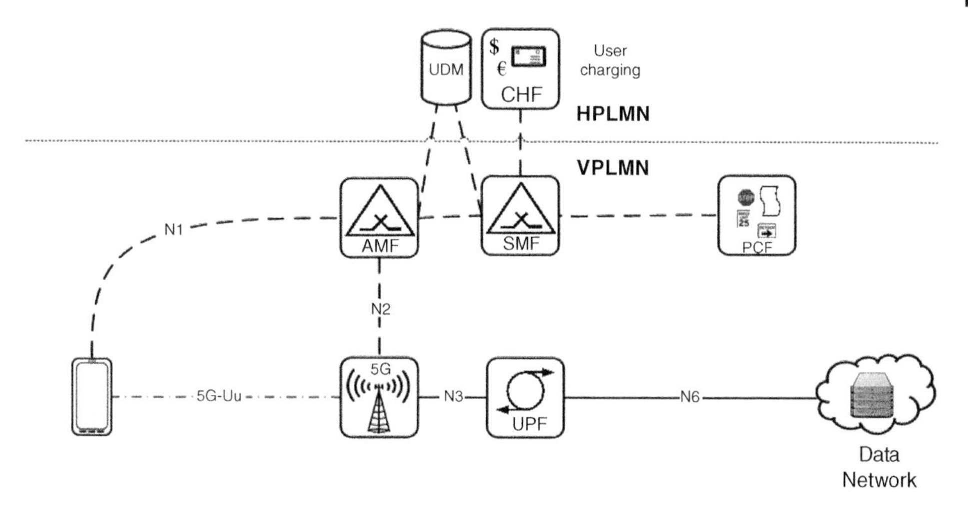

Figure 6.7 Local breakout roaming mode for a PDU Session.

In LBO mode (Figure 6.7) the VPLMN has full control over the PDU Session including the PSA (on the N6 connectivity service of the PDU Session) and on the corresponding Policy control.

Different PDU Sessions of the same UE may be established in different modes (LBO or HR). Subscription data tell the VPLMN on a per DNN and slice (S-NSSAI) basis whether the corresponding PDU Session may be established in LBO mode or in HR mode. The HPLMN may allow LBO usage for some ('friendly', e.g. belonging to the same operator group) VPLMN(s) while it only allows HR for other VPLMN(s).

6.2.8 Summary of the UPF Topology Options

As seen throughout this chapter, the user plane framework in 5GS offers a wide range of topology options. Figure 6.8 illustrates this fact:

Centralized UPF. For long term stable IP address assignment;

Cascaded UPF. For Multiple concurrent (e.g. local/central) access to the same DN (Section 6.2.2) or for HR roaming (Section 6.2.7); and

Parallel UPF. Transient usage of two PSA within a PDU Session at UE mobility (when the UE supports IPv6 multi-homing).

6.3 Ultra-reliable Low Latency Communication

One of the major new features brought by 5GS is the support of URLLC services. It corresponds to a combination of very low latency (1–10 ms) and high reliability (PDU Loss Rate $< 10^{-5}$) for PDU delivery (with a small expected throughput). Such a combination allows usage of 5GS for new applications like Intelligent Transport Systems (including Vehicle to Vehicle communication), Remote Control of industrial processes and Virtual/Augmented Reality.

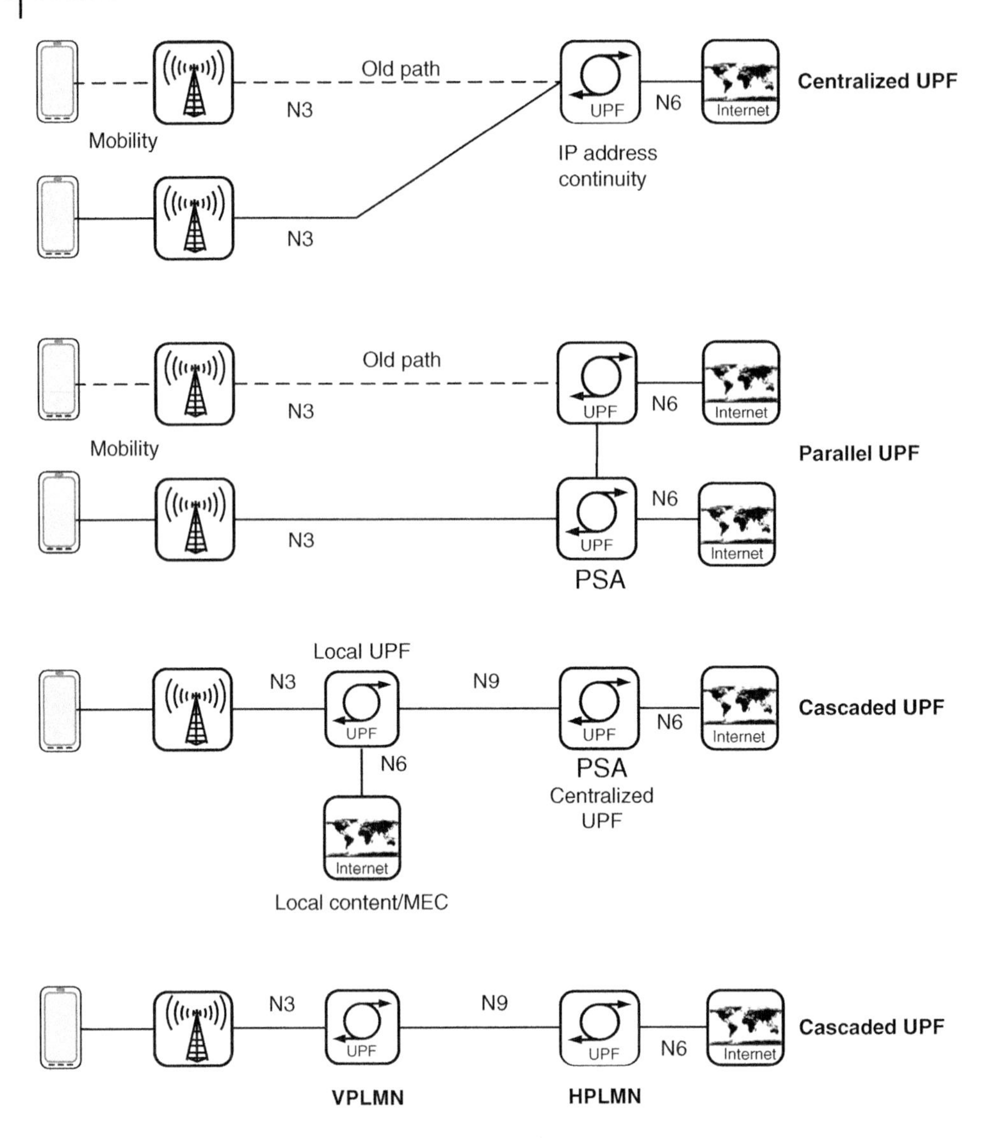

Figure 6.8 User plane architecture and user plane topology.

URLLC support requires following capabilities from 5GS:

- Specific QoS parameters, refer to Section 6.4.2.
- The location of the application 'near' to the UE: when the radio offers 1 ms or less TTI (Transmission Time Interval), it needs to be ensured that the delay due to the transport between NG-RAN and application is not becoming the main contributor to the end to end delay. Therefore, it is essential also for application to be located 'close' to the NG-RAN serving the UE.
- Thus, the SMF needs specific algorithms to select and allocate the PSA close to the NG-RAN:

- the usage of virtualization may allow deployments where the NG-RAN and the PSA are co-located
- The high reliability constraints require the NG-RAN to duplicate the DL traffic sent to the UE.
- In case of UE mobility, a change of PSA may be needed to meet the delay requirements of the flows within the PDU Session. In that case, SSC mode 3 service continuity (see Section 6.2.4) is required to allow a change of PSA without gaps in transmission/service delivery.

6.4 QoS Management in 5GS

6.4.1 What is a QoS Flow?

The 5G QoS model is based on QoS Flows. The QoS Flow is the finest granularity of QoS differentiation in a PDU Session. All PDU(s) within a QoS flow are given the same treatment as far as QoS is concerned (e.g. scheduling, admission control, PDU discard, etc.). Different PDU(s) within a QoS flow may nevertheless be charged differently due to different tariff plans or to an external entity subsidizing some flows, e.g. related with advertisements.

A QoS flow associates a QoS profile with a set of PDU(s) identified by traffic filters.

The 5G QoS model supports both QoS Flows which require Guaranteed Flow Bit Rate (GFBR) ('GFBR' QoS Flows) and also QoS Flows which do not require GFBR ('non-GBR' QoS Flows). A Guaranteed Bitrate (GBR) QoS Flow may be further qualified as delay critical GBR when end to end latency requirements are very tight (a PDU delayed more than the PDU Delay Budget is counted as lost if the transmitted data burst is less than the Maximum Data Burst Volume allowed within the period of PDU Delay Budget). GBR QoS Flows are typically used to support IMS voice service. Delay critical GBR may also be used to support industrial control applications that need support for low latency and ultra-high reliability.

The 5G QoS model also supports reflective QoS where the UE is required to reflect (i.e. mirror) the QoS for UL traffic based on the QoS applied for the DL traffic (see Section 6.4.3).

A QoS Flow Identifier (QFI) is used to identify a QoS Flow in the 5GS. The QFI is a scalar and is carried in an encapsulation header between the UPF and the 5G-AN (and between UPF(s)), i.e. without any changes to the e2e PDU header. QFI are used for all PDU Session Types. The QFI is unique within a PDU Session.

5GS QoS mechanism is documented in 3GPP TS 23.501 [3] section 5.7. Policy control related with QoS enforcement is documented in 3GPP TS 23.503 [7] section 4.3.

Figure 6.9 illustrates the 5G QoS framework:

6.4.2 QoS Parameters Handled in 5G System

A QoS Flow is characterized by:

- A QoS profile that defines the QoS parameters applied to the QoS flow;
- One or more QoS rule(s) which correspond to traffic filters associating PDU(s) to a QoS Flow.

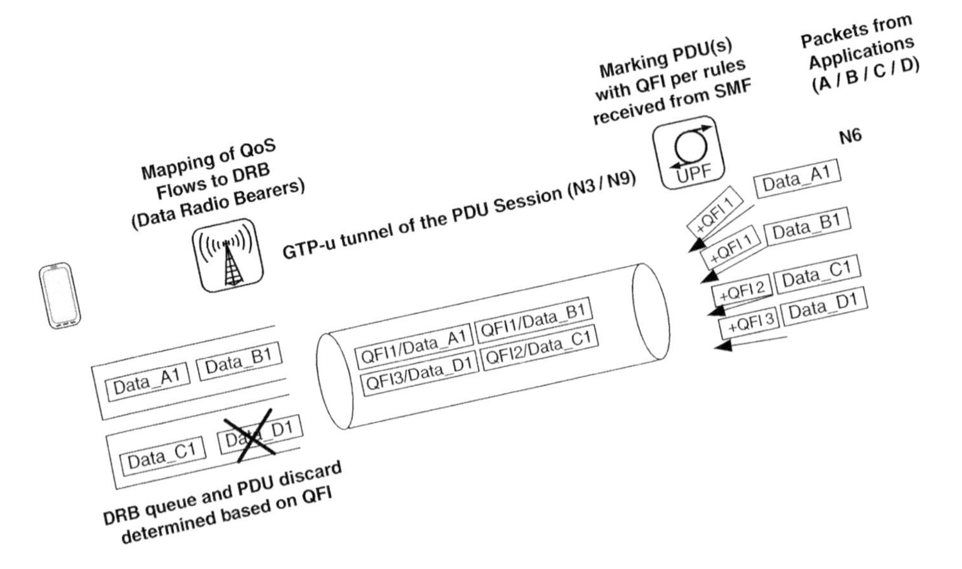

Figure 6.9 5G flow based QoS framework.

The QoS profile of a QoS Flow contains following QoS parameters:

- A 5G QoS Identifier (5QI):
  - A 5QI is a scalar that is used as a reference to 5G QoS characteristics, i.e. access node-specific parameters that control QoS forwarding treatment for the QoS Flow (e.g. scheduling weights, admission thresholds, queue management thresholds, link layer protocol configuration, etc.).
  - Standardized 5QI values have one-to-one mapping to a standardized combination of 5G QoS characteristics as specified in 23.501 [3] Table 5.7.4-1.
  - The 5G QoS characteristics for pre-configured 5QI values are pre-configured in the 5G-AN. The 5G QoS characteristics for dynamically assigned 5QI values are signalled as part of the QoS profile.
  - Standardized or pre-configured 5G QoS characteristics, are indicated through the 5QI value, and the default values are not signalled on any interface. 5GS provides means to override some QoS characteristics.
- An Allocation and Retention Priority (ARP):
  - The ARP contains information about the priority level, the pre-emption capability and the pre-emption vulnerability of the QoS Flow and is used to decide whether a new QoS Flow may be accepted or needs to be rejected in case of resource limitations (typically used for admission control of GBR traffic).
- For each Non-GBR QoS Flow, the QoS profile may also include a Reflective QoS Attribute (RQA) which indicates that certain traffic (not necessarily all) carried on this QoS Flow may be subject to Reflective QoS.
- For each GBR QoS Flow, the QoS profile also includes:
  - GFBR – UL and DL;
  - Maximum Flow Bit Rate (MFBR) – UL and DL;
- In case of a GBR QoS Flow only, the QoS parameters may also include:

- Notification control that indicates whether notifications are requested from the 5G-AN when the GFBR can no longer (or can again) be fulfilled for a QoS Flow.
- Maximum Packet Loss Rate – UL and DL.

Each QoS profile has one corresponding QoS Flow identifier (QFI), which is not included in the QoS profile itself.

Standardized 5QI values and their corresponding QoS characteristics as specified in 23.501 [3] Table 5.7.4-1 support a wide range of different traffic types/services:

- Very stringent combinations of delay (sub 10 ms end to end delay) and PDU loss rate (e.g. 10^{-5}) for services like Intelligent Transport Systems, Remote Control of industrial processes, etc.
- Low delay (50–150 ms) but higher acceptable PDU loss rate for conversational Voice/Video (Live Streaming); Different scheduling priorities and delays are defined for Mission Critical (firemen, etc.) and for regular voice services.
- Higher delay (300 ms) for web based, e.g. TCP-based applications like www, e-mail, chat, ftp, p2p file sharing, progressive video, etc. Different scheduling priorities are defined to support user/traffic differentiation.

6.4.3 Reflective QoS

As the name implies, Reflective QoS is to enable UEs to 'reflect' (i.e. mirror) the QoS for the UL traffic based on the QoS applied for the DL traffic. It allows support for symmetric QoS in the UL and DL.

Reflective QoS activated by the UPF enables the UE to map UL User Plane traffic, IP flows, to appropriate QoS Flows using implicitly derived (also referred to as UE derived) QoS rules (without using SMF signalled QoS rules via N1 signalling). Reflective QoS activated by the RAN enables the UE to map UL QoS Flows to appropriate DRBs.

This saves N1 signalling especially in case of very dynamic usage of IP flows by some applications requiring a specific QoS (e.g. in case of Internet browsing, TFTs (Traffic Flow Templates) may change so rapidly that it is impossible to perform signalling for every new TFT). This is achieved by the UE creating derived QoS rules (also referred to as implicitly derived QoS rules) based on the received DL traffic when reflective QoS is indicated by the 5GS for this DL traffic. When Reflective QoS is activated by the UPF, the UE shall use the derived QoS rules to determine mapping of UL traffic to QoS Flows. When the Reflective QoS is activated by the RAN, the UE shall use the derived QoS rules to determine mapping of UL QoS Flows to appropriate DRBs.

Reflective QoS is controlled by the 5GC (Figure 6.10) on per-packet basis by using the Reflective QoS Indication (RQI). The RQI is provided by the UPF in DL traffic together with the QFI in the (GTP-u) encapsulation header on N3 and then propagated by the 5G-AN on the interface with the UE.

Reflective QoS applies only for IP PDU Sessions and ETH PDU Sessions. It is possible to apply reflective QoS and non-reflective QoS concurrently within the same PDU Session.

The UE shall indicate whether it supports reflective QoS to the network (i.e. SMF) during the PDU Session establishment.

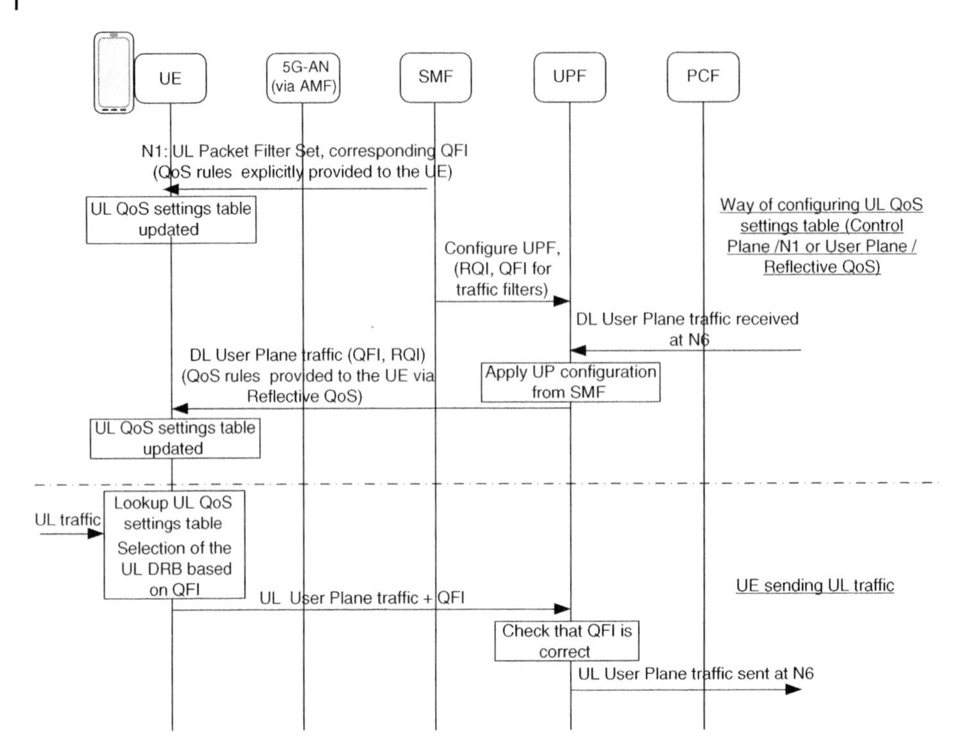

Figure 6.10 Controlling UL user plane QoS.

6.4.4 QoS Enforcement and Rate Limitation

The amount of traffic flowing through the 5GS for an UE is limited by

- Subscription for non-GBR traffic; this corresponds to AMBRs: Session-AMBR and User Equipment Aggregate Maximum Bit Rate (UE-AMBR);
- The GFBR – UL and DL – negotiated at the establishment of a GBR QoS flow;
- Service Level Agreements (SLA) between the HPLMN and the VPLMN in case of roaming.

Each PDU Session is associated with a per Session-AMBR. The Session-AMBR limits the aggregate bit rate that can be expected to be provided across all Non-GBR QoS Flows for a specific PDU Session.

The subscribed Session-AMBR is a subscription parameter which is retrieved by the SMF from the Unified Data Management (UDM).

The SMF may use the subscribed Session-AMBR or modify it based on local policies or use the authorized Session-AMBR received from PCF to determine the Session-AMBR to be applied to the PDU Session. The Session-AMBR to be applied to the PDU Session is signalled to the appropriate UPF(s), to the UE and to the 5G-AN (to enable the calculation of the UE-AMBR).

The Session-AMBR is measured over an AMBR averaging window which is a standardized value. In downlink direction, the Session-AMBR is enforced by the UPF. In uplink, the Session-AMBR is enforced by both the UE and the UPF.

Each UE is associated with a per UE-AMBR.

The UE-AMBR limits the aggregate bit rate that can be expected to be provided across all Non-GBR QoS Flows of all PDU Sessions of a UE. Each 5G-AN shall set the UE-AMBR policed value to the sum of the Session-AMBR values of all PDU Sessions with active user plane to this AN up to the value of the subscribed UE-AMBR. The subscribed UE-AMBR is a subscription parameter which is retrieved from UDM and provided to the 5G-AN by the AMF. The UE-AMBR is measured over an AMBR averaging window which is a standardized value. The UE-AMBR is not applicable to GBR QoS Flows. The 5G-AN enforces the UE-AMBR in both downlink and uplink.

6.4.5 QoS Control within 5GS

Within the 5GS, QoS Flows are controlled by the SMF, possibly enforcing policies received from the PCF and subscription information received from the UDM.

QoS profiles

- may either correspond to preconfigured QoS characteristics values in the 5G-AN; in this case, there is no need of any signalling with the 5G-AN to establish or release the corresponding QoS flows. This provides great flexibility to use many different QoS flows within a PDU Session. or
- may be dynamically established by the SMF in the 5G-AN (via the AMF) over the N2 reference point. The SMF may fine tune the 5G-AN delay requirements (referred to as 5G-AN part of the PDB – Packet Delay Budget) based on the transport delays between the DN and the 5G-AN; when transport delay requirements are low, bigger buffering time is available for the RAN thus enabling the RAN to optimize its resources.

QoS rule(s) may be provided by the SMF to the UE (via the AMF) using SM related signalling (PDU Session creation/modification) over the N1 reference point and/or may be derived by the UE by applying reflective QoS.

Rules provided by the SMF to the UPF over N4 may associate traffic filters with:

- A QFI setting, associating the PDU matching a traffic filter to a QoS Flow.
- A RQI when Reflective QoS is applied to this traffic.
- Transport (DiffServ Code Point [DSCP]) QoS setting determined based on the 5QI and ARP of the traffic.
- Usage monitoring (e.g. for charging) requests.
- Traffic gating.

This is further defined in Section 6.8.

6.4.6 Delay Critical GBR QoS Flows and Their Characteristics

To support applications with requirements for ultra-high reliable and low latency communications, GBR QoS flows of resource type ‘delay critical GBR’ were introduced. When there is a need to transmit packets with a latency in the order of 1–10 ms and to support reliability in the order of 10^{-5}, it was essential also to define the largest amount of data that the 5G-AN is required to transmit within a period of the 5G-AN PDB (i.e. the 5G-AN part of the Packet Delay Budget (PDB)). This is referred to as Maximum Data Burst Volume. Each Delay Critical GBR QoS flow is associated with a Maximum Data Burst Volume. The Maximum Data Burst Volume may be signalled with 5QIs to

Table 6.1 Delay critical 5QI QoS characteristics.

5QI value	Resource type	Default priority level	Packet delay budget (ms)	Packet error rate	Default maximum data burst volume (Bytes)	Example services
81	Delay critical GBR	11	5	10^{-5}	160	Remote control
82		12	10	10^{-5}	320	Intelligent transport systems
83		13	20	10^{-5}	640	
84		19	10	10^{-4}	255	Discrete automation
85		22	10	10^{-4}	1358	Discrete automation

the (R)AN, and if it is not signalled, default values apply for standardized 5QIs QoS characteristics.

For delay critical GBR QoS flows, a packet delayed more than the PDB is counted as lost if the transmitted data burst is less than Maximum Data Burst Volume within the period of PDB and the data rate for the QoS flow is not exceeding the GFBR. The PER (Packet Error Rate) also defines a target for reliability for the latency target set by the PDB.

Table 6.1 shows the delay critical 5QI values and the QoS characteristics:

6.4.7 Packet Filter Set

The Packet Filter Set is used in the QoS rules to identify one or more packet (IP or ETH) flow(s).

The Packet Filter Set may contain one or more Packet Filter(s). Every packet filter can be made applicable for the DL direction only, the UL direction only or both directions.

Packet Filter Sets can be of two types: IP Packet Filter Set and ETH Packet Filter Set, corresponding to those PDU Session Types.

For IP PDU Session type, the Packet Filter Set allows packet filtering based on a combination of the following aspects:

- Source/destination IP address or IPv6 prefix
- Source/destination port number
- Protocol ID of the protocol above IP/Next header type
- Type of Service (TOS) (IPv4)/Traffic class (IPv6) and Mask
- Flow Label (IPv6)
- Security Parameter Index (SPI)
- Packet Filter direction.

For ETH PDU Session type, the Packet Filter Set allows packet filtering based on a combination of the following aspects:

- Source/destination MAC address
- Ethertype as defined in IEEE 802.3

- Customer-VLAN tag (C-TAG) and/or Service-VLAN tag (S-TAG) VLAN Identifier (VID) fields as defined in IEEE 802.1Q
- C-TAG and/or S-TAG Priority Code Point (PCP)/Drop Eligible Indicator (DEI) fields as defined in IEEE 802.1Q
- IP Packet Filter Set, in the case that Ethertype indicates IPv4/IPv6 payload
- Packet Filter direction.

6.5 User Plane Transport

As different PDU Sessions may carry traffic of different types (IP, ETH, etc.), the traffic of PDU Sessions is transported in tunnels within the 5GS (Figure 6.11). This allows a common backbone to carry the traffic of different PDU Session types.

The same tunnelling mechanism applies:

- within the 5G Core (e.g. over N9 interface between a UPF in a local location and a UPF in a central location, or between a UPF in the VPLMN and a UPF in the HPLMN) and
- on the N3 UP interface between the 5GC and the access network regardless of the access type (3GPP, un-trusted Non 3GPP access).

This tunnelling corresponds to a GTP-u/User Datagram Protocol (UDP)/IP header (protocol stacks are defined in Chapter 4), reusing the same baseline tunnelling protocol as in EPC. As opposed to previous generations that defined one (GTP-u) tunnel per bearer (QoS level) within a PDN connection, 5GC defines one single (GTP-u) tunnel per PDU Session.

Over 5GS, the GTP-u header carries any per PDU information needed by 5GS entities to support proper 5GS QoS handling, e.g. QoS Flow ID (QFI) and the RQI defined in Sections 6.4.1 and 6.4.3.

Over N4 the SMF can associate traffic with a Network Instance possibly determined based on the target DNN and/or slice (S-NSSAI) of the corresponding PDU Session.

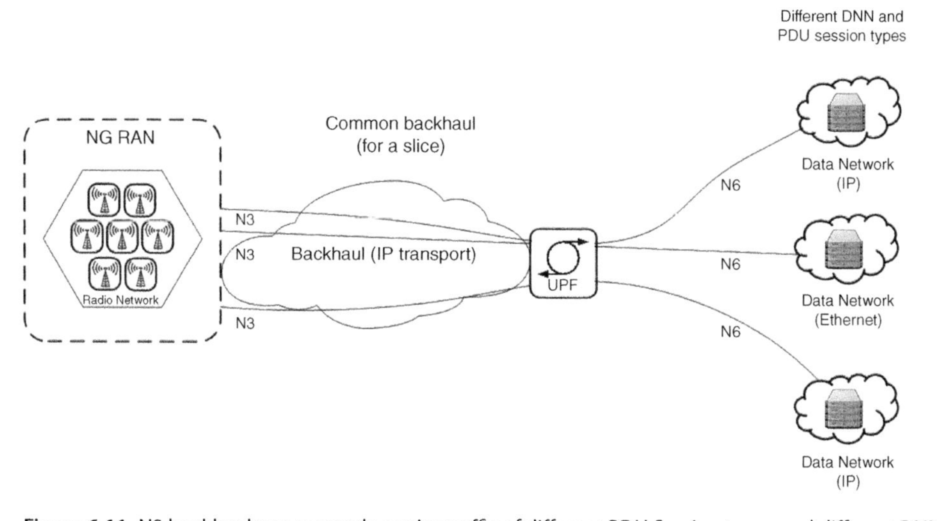

Figure 6.11 N3 backhaul transparently carries traffic of different PDU Session types and different DNN.

The UPF can use the Network Instance to determine the IP network (e.g. VPN) to transfer the corresponding traffic over N3/N9/N6.

Tunnel establishment requires exchange of (GTP-u) tunnel endpoints using control plane signalling. This enables the 5G-AN and UPF(s) to learn the (GTP-u) peer address for the tunnels to be able to send and receive traffic corresponding to a PDU Session. Handovers within 5GS and mobility between 3GPP and Non 3GPP access require updates to (GTP-u) tunnel endpoints using control plane signalling messages.

6.6 Policy Control and Application Impact on Traffic Routing

6.6.1 Overview

The PCF can send dynamically derived policies to the SMF; these policies may target:

- Either the PDU Session or
- a Service Data flow (SDF). An SDF refers to a set of PDU(s) (within a PDU Session) identified by traffic filters that are to be handled the same way in any dimension (QoS, charging, forwarding, gating, etc.).

Conversely the PCF may receive information from the SMF about the status of the PDU Session, e.g. information on the Access Type and the RAT Type, the roaming situation, the IP address/IPv6 prefix(es) allocated to the PDU Session, etc. This information can influence dynamic policies derived by the PCF.

The PCF is an optional function of the 5GC. Whether PCF control is applied to a PDU Session is defined by SMF policies based on the DNN and slice. When PCF control does not apply, locally configured SMF policies apply to the PDU Session (local policy rule-based control of the PDU Session).

PDU Sessions supporting IMS voice and emergency services are meant to support PCF control, allowing the IMS to request GBR QoS Flows via the PCF.

Policy control related with PDU Sessions and QoS enforcement is documented in 3GPP TS 23.503 [7] section 4.3.

6.6.2 Application Impact on UP Traffic Routing

Supporting necessary enablers for MEC is a key driver for 5GS (see Section 4.8 on MEC). How SMF and UPF enable support for MEC is defined in this section.

One of the key features of 5GS is to facilitate the deployment of UPF that provides the most optimum UP path between the UE (or the 5G-AN currently serving the UE) and 'critical' applications deployed at the edge of the network. There are different cases leading to an application being identified as 'critical', e.g.:

- The application has strict UP latency requirements (industrial control and Virtual Reality are typical examples).
- The application requires a big DL throughput while content caching at the edge may dramatically reduce the transmission cost for the operator and greatly improve end user experience (content caching for popular videos is an example).
- The application is used by a crowd of people located at the same place (stadium).

- Placing the application at the edge dramatically reduces the amount of data sent UL within the network (e.g. the application at the edge reports to a central server only when it has detected some configured pattern in a video: video surveillance).

Locating an UPF geographically 'close' to where the 5G-AN (radio site) of the UE is located and 'close' to where such applications have been deployed is thus a requirement in 5GS. UPF selection is further defined in Section 6.7.3.

But applications located at the edge will not handle all traffic as the biggest number of applications will remain reachable via central access to the DN (e.g. Internet). So, it should be possible to configure the local UPF so that it knows (i) which traffic is to be routed to local applications and (ii) which traffic is to be sent centrally (possibly to a central UPF as defined in Section 6.2.2). Furthermore, the UPF needs to be configured how (e.g. on which tunnel) traffic is to be routed to a given local application instance.

Some applications may support application mobility while other applications may not support application mobility. Support for application mobility implies that another instance of the application can be selected to serve the UE once the UE has moved beyond the serving area of the current application instance.

For applications that support mobility between application instances, the SMF needs to notify the application environment when a new local application instance is to be used. For applications known not supporting mobility, the SMF needs to ensure that despite of the mobility of a UE, the traffic can be exchanged with the same application instance if the application is running on the UE.

All points listed up to now show that the SMF needs some configuration to properly support local applications and MEC (see Section 4.6). Thus, the application controller (MEC system management), which is called an Application Function (AF) in 3GPP terminology, can issue requests to the 5GC that may contain:

(1) Information to identify traffic to be forwarded to a local application instance. The traffic can be identified in the AF request by:
    - Either a DNN and possibly slicing information (S-NSSAI) or an AF-Service-Identifier
        - When the AF provides an AF-Service-Identifier, i.e. an identifier of the service on behalf of which the AF is issuing the request, the 5G Core maps this identifier to a target DNN and slicing information (S-NSSAI).
    - An application identifier or traffic filtering information (e.g. 5-Tuple). The application identifier refers to an application handling UP traffic and is used by the UPF to detect the traffic of the application.
(2) Information on how to route locally traffic identified in bullet (1) above.
(3) Potential locations of applications towards which the traffic routing should apply. The potential location of applications is expressed as a list of Data Network Access Identifier(s) DNAI(s). This should be used as a guideline for SMF selection of a local UPF. This information is determined by the AF based on its knowledge of where (which Data Centers) application instances have been deployed (one of the roles of the MEC Controller is to manage instantiation of local instances of applications).

(4) Information on the target UE(s). This may correspond to:
   - Individual UEs identified using a GPSI which is a public identifier (like a Mobile Subscriber Integrated Services Digital Network [MSISDN]), or an IP address/Prefix;
   - Groups of UEs identified by a Group Identifier;
   - Any UE: the request applies to any UE accessing the combination of DNN, and slice (S-NSSAI).

(5) Indication that for the traffic related with an application, no DNAI change shall take place once selected for this application (the application does not support mobility);

(6) Temporal validity condition.

The AF sends its request to the PCF directly or via the NEF. When the AF request targets any UE or a group of UE(s), the AF request shall be sent via the NEF. In that case the AF request is likely to influence multiple PDU Sessions served by multiple SMFs and PCFs.

If the AF request is sent via NEF, the NEF may

- Validate the AF request
- Provides mapping between information from the AF and internal 5GS information (DNN, S-NSSAI, etc.)
- Determine the relevant PCF when the request targets an individual UE
- Store the requests in the UDR when the request targets multiple UEs and thus potentially multiple PCF(s):
  PCF(s) that need to receive AF requests targeting a DNN (and slice), and/or a group of UEs subscribe to receive UDR notifications about change on such AF request information. When PCF(s) receive a notification about such AF request, they take the received AF request information into account when making policy decisions for existing and future relevant PDU Sessions. In the case of existing PDU Sessions, the PCF's policy decision may trigger a PCC rule change from the PCF to the SMF.

When a PCC rule is received from the PCF, the SMF may take appropriate actions to reconfigure the User Plane of the PDU Session such as:

- Adding, relocating or removing a UPF in the data path to, e.g. insert an UL CL or a Branching Point UPF as described in Section 6.2.2.
- Allocate a new IPv6 Prefix to the UE (when IPv6 multi-homing applies).
- Update the UPF with new traffic steering rules (e.g. to forward relevant traffic to local application instances).
- Subscribe to notifications from the AMF to determine when it should insert a local UPF.

6.6.3 PCF Policies Targeting a PDU Session

The PCF can control following parameters of a PDU Session:

- The information to assist in determining the IP Address allocation method (e.g. from which IP pool to assign the address) when a PDU Session requires an IP address/Prefix – as defined in TS 23.501 [3] section 5.8.1.
- The default charging method for the PDU Session.

- The address of the corresponding charging entity.
- The conditions upon which the SMF shall fetch new policies for the PDU Session (called Policy Control Request Triggers).
- The Authorized Session-AMBR.
- The default 5QI and ARP of the QoS Flow associated with the default QoS rule.
- Applicability of Reflective QoS for QoS Flows within a PDU Session.
- Usage Monitoring Control related information.

Multiple Authorized Session-AMBR and default 5QI and ARP may be provided by the PCF with conditions where they are to apply (e.g. time condition).

6.6.4 PCF Policies Targeting a Service Data Flow

The PCF can instruct the SMF to ensure actions on PDU(s) identified by SDF filtering information (also called SDF templates). PCC rules associate SDF filtering information with actions to apply on the corresponding traffic.

Such policies don't apply for PDU Sessions with unstructured PDU Session Type.

A PCC rule may be predefined (locally configured on the SMF) or dynamically provisioned during the lifetime of a PDU Session. The latter is referred to as a dynamic PCC rule.

It is possible to take a PCC rule into service, and out of service, when a specific condition is met, e.g. at a specific time of day without any PCF-SMF interaction at that point in time.

Actions defined on traffic identified by an SDF template may provide instructions to the SMF in order to:

- Apply QoS as defined in Section 6.4
  - All traffic associated with the same QoS parameters build a QoS Flow. A QoS flow may thus correspond to multiple SDF template (when these SDF templates are associated with the same QoS requirements but could correspond to different requirements in terms of traffic steering/charging/gating, etc.)
- Perform Gating Control, i.e. discard selected traffic upon PCF control
  - This may be used to prevent Voice traffic before IMS Voice charging has been started
- Discard traffic that don't match the SDF template of any active PCC rule
  - wild-carded SDF filters allow exchange of traffic that does not match any SDF template of any other active PCC rules.
- Monitor the amount of traffic for charging or for other purpose such as service traffic limitation (per day, month, etc.)
- Steer traffic: this may be used to control forwarding of traffic towards different N6 interfaces of the same DN identified by a DNAI; DNAI(s) are global (valid for any DNN) and may be used to identify different local/central Data Centres. This may also be used to steer traffic towards Service Functions deployed at N6 on the DN.
- Apply relevant User Plane security (integrity protection) (this is enforced by the RAN).

These actions are executed locally by the SMF or result in N4 commands towards UPF, see Section 6.8.

An SDF template may identify:

- A PDU direction (UL, DL, both)
- A Precedence (to disambiguate different SDF templates that would identify the same traffic).

An SDF may be identified by the IP or ETH Packet filters (see Section 6.4.7) depending on the type of PDU Session.

The SMF is responsible of any network signalling towards UPF and 5G-AN that are required to adapt a PCC rule (see also Section 6.7).

6.7 Session Management

6.7.1 Overview

The SMF is responsible for any network (5GC) signalling required to control a PDU Session and to set the User Plane handling within this PDU Session. Following are some of the functionalities performed by the SMF:

- Selection of the UPF(s) supporting the PDU Session (see Section 6.7.3).
- N4 Signalling to control User Plane handling in the UPF (see Section 6.8).
- Exchange N2 session management (SM) Signalling via AMF to set QoS parameters in the 5G-AN, e.g. to reserve resources for GBR traffic. This may take place when receiving a new or a modified PCC rule from the PCF, when applying a local policy at PDU Session establishment but also when the User Plane of a PDU Session becomes active (e.g. when the CM state of the UE becomes CM-CONNECTED, see Section 5.2).
- Exchange N1 SM signalling (via AMF) with the UE, e.g. to:
  - establish, modify and release the PDU Session
  - provide the UE with QoS rule(s), i.e. a mapping between traffic filter(s) and the QoS that has to be apllied to the corresponding traffic.
- Interactions with the AMF to
  - exchange N1 SM signalling with the UE via the AMF,
  - exchange N2 SM signalling with the 5G-AN serving the UE via the AMF,
  - receive from the AMF requests to activate the User Plane of a PDU Session,
  - get event notifications from the AMF when the UE has moved out of an area identified in the event subscription from the SMF; this may be used to handle LADN (see Section 6.2.4).
- When PCF control is to apply, selection of the PCF and then support of any signalling exchange with the selected PCF (see Section 6.6).

6.7.2 PDU Session Establishment

Only UE(s) initiate the PDU Session Establishment procedure. This procedure may be started in following cases:

- to establish a new PDU Session:
  - the PDU Session may be established if the UE is registered with the 5GS (e.g. 'always-on' PDU Session for IMS Voice),

- the PDU Session Establishment may be, due to an internal event in the UE (e.g. the user opens a new application that triggers the request for a new PDU Session) or in response to a device trigger (Short Message Service [SMS]), received by the UE from a network application.

- to establish a new PDU Session for IMS Emergency services,
- to support a UE initiated PDU Session handover between 3GPP and non-3GPP access,
- to support a UE initiated PDU Session handover from EPS to 5GS.

Upon reception of a UE request for a new PDU Session (received via the AMF), the SMF may (see Figure 6.12):

- Contact the UDM to retrieve subscription data and register itself in the UDM for this PDU Session.
- Check whether the UE request is compliant with the user subscription.
- Check whether PCF control is to apply to this PDU Session. If it applies, the SMF determines whether to use the same PCF as the AMF or whether local policies (e.g. related with slicing) require a dedicated PCF.
- Contact the PCF to retrieve policies (for the PDU Session and for SDFs within this PDU Session as described in Section 6.6).
- Select UPF(s) to serve this PDU Session (see Section 6.7.3).
- In case of IP PDU Session Type allocate or request the allocation of an IP address/IPv6 prefix.
- Interact with the charging entity for this PDU Session.
- Configure the UPF(s) over N4 (see Section 6.8).

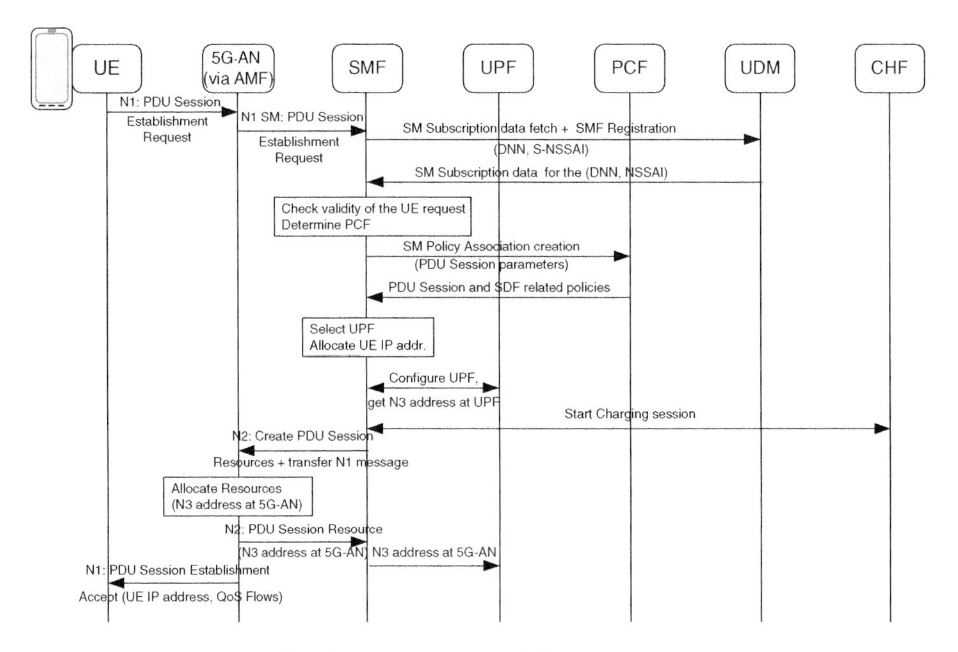

Figure 6.12 High-level call flow for PDU Session establishment.

– Configure (via the AMF) the 5G-AN with QoS flows for this PDU Session.
– Answer (via the AMF and the 5G-AN) to the UE providing QoS Rules (and an IP address/IPv6 prefix) for this PDU Session.

6.7.3 UPF Selection

The selection and reselection of the UPF are performed by the SMF considering, e.g.

- UPF deployment scenarios such as centrally located UPF and distributed UPF located close to or at the Access Network site (Section 6.2.4),
- the parameters of the PDU Session (DNN, slice).

For home routed roaming case, the UPF in HPLMN is selected by the SMF in H-PLMN, and the UPF(s) in VPLMN are selected by the SMF in V-PLMN.

The UPF selection involves two steps:

– *Step 1.* Discovery of available UPF(s). This step may take place while there is no PDU Session to establish and may be followed by N4 Node Level procedures where the UPF and the SMF may exchange information such as the support of optional functionalities and capabilities.
– *Step 2.* Selection of an UPF for a particular PDU Session; this is followed by N4 session management procedures.

6.7.3.1 Discovery of Available UPF(s)

The SMF may be locally configured with the information about the available UPFs, e.g. by the Operations, Administration and Maintenance (OA&M) system when an UPF is instantiated or removed.

The UPF selection functionality in the SMF may optionally utilize the NRF to discover UPF instance(s).

The NRF may be configured by OA&M with information on the available UPF(s) or the UPF may register itself in the NRF.

6.7.3.2 Selection of an UPF for a Particular PDU Session

Parameter(s) considered by the SMF for the UPF selection may include:

– parameters of the PDU Session such as the DNN, the S-NSSAI and PDU Session Type (IP, ETH, etc.);
– the service expected from the UPF (central PSA, Intermediate UPF connecting a 5G-AN, UL CL as defined in Section 6.2.2, etc.);
– for selecting a UPF that should serve as a PSA requirements on the IP address to allocate to a PDU Session (static address, requirement received from the PCF or related with a DNN policy, etc.);
– UPF location and UE location information;
  - this is especially important when there is a need to meet stringent delays associated with URLLC (Section 6.3) or to comply to Application requests (MEC) provided via a DNAI parameter included in a PCC rule (see Section 6.6.2);
– Local operator policies;
– UPF load/capacity.

6.8 SMF Programming UPF Capabilities

All UPFs involved in a PDU Session are controlled by the SMF via the N4 interface. In 3GPP Release 15, N4 is not defined as a Service Based Interface and the UPF is not expected to offer services via REST Application Programming Interfaces (APIs). However, UPF can consume services from NFs, e.g. NRF for UPF registration.

The functionalities and policies enabled in each UPF may be different and depend on the specific use case and PDU Session topology. UPF(s) are programmable allowing the SMF to select and tailor the UPF functionalities.

Predefined user plane entities such as 'Serving Gateway' or 'PDN Gateway' as defined for 4G EPC did not offer the flexibility required for handling the large variety of connectivity models and session topology options of 5GC.

The following functions are offered by a UPF:

- Internal 5GS (N3/N9) and external (N6) interface termination.
- Traffic detection based on Packet Detection Rules (PDRs) sent by the SMF; the SMF may associate a PDR with a FAR (Forwarding Action Rule), a QER (QoS Enforcement Rule), and a URR (Usage Reporting Rule) defined here after.
- Packet forwarding based on FARs sent by the SMF. This includes:
  - Over N4 the SMF can associate traffic with a Network Instance possibly determined by the SMF based on the target DNN and/or slice (S-NSSAI) of the corresponding PDU Session. The UPF can use the Network Instance to determine the IP network (e.g. VPN) to transfer the corresponding traffic over N3/N9/N6.
  - Usage of proper tunnelling mechanisms over N6. Such tunnelling is beyond the scope of 3GPP specifications and may depend on business agreements between the operator and the DN.
  - Support of traffic steering for service chaining on N6. Upon control of the SMF, the UPF may apply traffic marking and traffic forwarding rules to support selective introduction of Service Functions (e.g. NAT, TCP optimizers, etc.) at N6.
  - Replication of user plane traffic for lawful interception.
  - Mobility anchor and support for access network handover (change of address where to send DL traffic to the access network after a handover).
  - DL Data buffering (when the UE is not reachable as it is IDLE, i.e. no User Plane transport is yet set-up towards the UE for the PDU Session) and data forwarding to ensure data integrity during handover.
  - Sending one or more 'end marker' to the source access network node during a handover procedure: after data switching to the new data path an 'end marker' PDU is sent on the old data path to indicate that no further PDU is expected on the old data path.
- QoS enforcement based on QERs sent by the SMF. This includes:
  - Rate enforcement of session AMBR. Refer to Section 6.4.4 on QoS.
  - Tunnel-level packet marking for QoS enforcement: the UPF may add a QoS Flow ID (QFI), Section 6.4.2, or RQI, Section 6.4.3, to DL traffic based on instructions from the SMF.
  - Transport-level packet marking, e.g. setting the transport IP header DSCP.

– Usage Reporting based on URRs sent by the SMF. This includes:
  - Counting of resource usage (e.g. for Charging record generation or for credit control).
  - UPF reporting usage monitoring data towards the SMF via N4.
– Event reporting to SMF based on various trigger conditions.

Future releases may also include traffic steering capabilities when a UE is simultaneously connected via multiple access networks (e.g. 5G radio access and untrusted Wi-Fi access).

References

All 3GPP specifications can be found under http://www.3gpp.org/ftp/Specs/latest. The acronym 'TS' stands for Technical Specification, 'TR' for Technical Report.

1 IETF RFC 7157: "IPv6 Multihoming without Network Address Translation".
2 IETF RFC 4191: "Default Router Preferences and More-Specific Routes".
3 3GPP TS 23.501: "System Architecture for the 5G System; Stage 2".
4 3GPP TS 23.502: "Procedures for the 5G System; Stage 2".
5 IETF RFC 2131: "Dynamic Host Configuration Protocol".
6 IETF RFC 4862: "IPv6 Stateless Address Autoconfiguration".
7 3GPP TS 23.503: "Policy and Charging Control Framework for the 5G System; Stage 2".

7

Security

Peter Schneider

Nokia, Munich, Germany

7.1 Drivers, Requirements and High-Level Security Vision

7.1.1 5G Security Drivers

There is broad consensus that a Fifth Generation (5G) network must be secure, in the sense that it protects the privacy of its users and the confidentiality and integrity of the traffic it transports, but also in the sense that it is protected against any kind of cyberattack that may affect the availability and integrity of the network or the confidentiality of the data it stores. This chapter outlines the 5G security drivers and requirements, and presents a 5G security vision as well as an architecture, building on material from [1]. It gives a description of the 5G security features specified by Third Generation Partnership Project (3GPP) for the 5G System, with a focus on showing the differences to Evolved Packet System (EPS). Finally, it provides a closer look at security issues that are currently not very much in the focus of 3GPP, but highly relevant for 5G networks: Secure Software Defined Networking (SDN), building on material from [29], and secure Network Function Virtualization (NFV).

Inherently, wireless communication is vulnerable and needs specific protection against interception and tampering. Consequently, since the second generation of mobile networks, Global System for Mobile communication (GSM), encryption is used on the radio interface to secure the user communication. In the next two generations of mobile networks, Universal Mobile Telecommunications System (UMTS) and Long Term Evolution (LTE), the security architecture was significantly enhanced. Besides encryption of user traffic, UMTS and LTE also provide mutual authentication between mobile terminals and the network, as well as integrity protection and encryption for the control traffic. It is common practice in modern mobile networks to secure network management communication between network element and management system, using protocols such as Secure Shell (SSH) [27] or Transport Layer Security (TLS) [13]. The security features of UMTS and LTE not only ensure a high level of security and privacy for the subscribers, but, very importantly, also make such networks robust against various forms of attacks against the integrity and availability of the services they provide.

Specified nearly 10 years ago, the EPS security concepts have not shown any major flaws, leading to the question why new security concepts may be required at all for

5G for the Connected World, First Edition.
Edited by Devaki Chandramouli, Rainer Liebhart and Juho Pirskanen.

the next mobile network generation. As a short answer, it is the support of a variety of new use cases on the one hand, and the adoption of new networking paradigms on the other hand that make it necessary to reconsider some elements of the current security approach.

While EPS was designed primarily to support the mobile broadband use case (i.e. broadband access to the Internet), 5G targets a variety of additional use cases with a variety of specific requirements. Some use cases, such as vehicular traffic or factory control, place the highest demands on the dependability of the network. Human lives may depend on the availability and integrity of the network service used by such applications.

To support each use case in an optimal way, security concepts will need to be more flexible. For example, security mechanisms used for ultra-low latency, mission-critical applications may not be suitable in massive Internet of Things (IoTs) deployments where mobile devices are inexpensive sensors that have a very limited energy budget and transmit data only occasionally.

To efficiently support new levels of performance and flexibility required for 5G networks, it is understood that new networking paradigms must be adopted, such as SDN and NFV. At the same time, these new techniques also bring new threats. For example, when applying NFV, the integrity of virtual network functions (VNFs) and the confidentiality of their data may depend to a large degree on the isolation properties of a hypervisor, or more generally, on the integrity and correctness of the whole cloud software stack. Vulnerabilities in such software components have surfaced quite often in the past. In fact, it remains a major challenge to provide a fully dependable, secure NFV environment. SDN, for its part, bears the threat that control applications may wreak havoc on a large scale by erroneously or maliciously interacting with a central network controller.

Another driver for 5G security is the changing ecosystem. Current networks are dominated by large monolithic deployments, each controlled by a single network operator owning the network infrastructure while also providing all network services. By contrast, 5G networks may see many specialized stakeholders providing end-user network services, as illustrated in Figure 7.1.

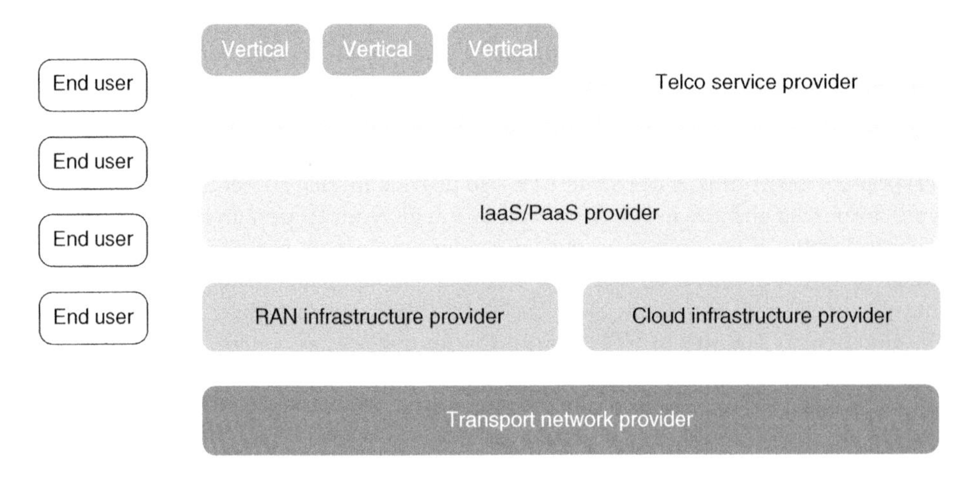

Figure 7.1 Stakeholders providing 5G services.

There may be dedicated infrastructure providers (InPs) decoupled from telco service providers that host several service providers as tenants on a shared infrastructure. Telco service providers in turn may not only offer end-user communication services, but may also provide complete virtual networks or network slices specialized for specific applications, such as industrial IoT applications. These network slices may be operated by verticals, i.e. non-telco enterprises like logistics enterprises or manufacturing companies. A manufacturing company could run a virtual mobile network specialized for industrial IoT applications on its own plants. The security aspect here is the building and maintenance of new trust relationships between the various stakeholders to ensure trusted and trouble-free interaction, resulting in secure end user services.

7.1.2 5G Security Requirements

When specifying a security concept, typically one of the first steps is to define the security requirements. This section does not aim at assembling a complete list of security requirements, but rather highlights some high-level requirements and their sources.

A major source of 5G requirements is the Next Generation Mobile Networks (NGMN) Alliance's 5G White Paper [2]. This paper emphasizes that in 5G networks, the "enhanced performance is expected to be provided along with the capability to control a highly heterogeneous environment, and capability to, among others, ensure security and trust, identity, and privacy." So, on the one hand, the security mechanisms need to comply with the overall 5G requirements, including extremely fast control plane procedures, extremely low user plane latency, and highest energy efficiency. On the other hand, [2] explicitly requires many security improvements compared to today's networks, for example:

- 'Improve resilience and availability of the network against signaling based threats';
- 'Improve system robustness against smart jamming attacks of the radio signals and channels'; and
- 'Improve security of 5G small cell nodes.'

Subsequently, the NGMN Alliance has driven a working group on 5G security which has delivered three dedicated documents (called "packages") [3] with 5G security recommendations. These relate to topics such as potential security improvements of the access network, Denial of Service (DoS) protection, network slicing, mobile edge computing, low latency and consistent user experience.

Clearly, the EPS security concepts are a starting point as well as a benchmark for 5G security, with respect to the mobile broadband use case which will remain of major importance. 5G must obviously be able to provide at least the same level of protection where needed as EPS, thus EPS security features are a natural security baseline for 5G networks. On top of this, it is rewarding to revisit security features that have been discussed but were not adopted for EPS. This discussion is captured in [5] and comprises features such as International Mobile Subscriber Identity (IMSI) catching protection, user plane integrity protection, or ensuring non-repudiation of service requests. IMSI catching means to trick mobiles into revealing the identity of the subscriber, the IMSI, by carrying out an active attack involving usage of a faked base station. Non-repudiation of service requests means that users cannot reasonably deny that they made a service request, as the request origin can be proven, typically by use of digital signatures.

Since Release 15, 3GPP SA1 specifies the service requirements for the 5G system in [6], with a dedicated section on security. There are various security requirements listed that extend what has been required for previous mobile network generations. For example, it is now required to be able protecting the subscriber identity against active attacks, i.e. prevent IMSI catching. Further, it is required that the 5G system supports an access independent security framework and includes a mechanism for the operator to authorize subscribers of other networks to receive temporary service (e.g. for mission critical services), even without access to their home network.

With respect to authentication for User Equipments (UEs) attaching via Evolved Universal Mobile Telecommunications System Terrestrial Radio Access Network (E-UTRAN), EPS only supports EPS AKA (EPS Authentication and Key Agreement). For 5G, it is a requirement that alternative authentication methods with different types of credentials are supported, aiming at IoT devices in isolated deployment scenarios, e.g. on factory plants. Moreover, the network is required to support a resource efficient mechanism for authenticating groups of IoT devices.

Document [6] also covers the area of network slicing, and requires slice isolation, in the sense that traffic as well as management operations within one slice have no significant impact on other slices. In this context, [6] also considers the involvement of third parties, and requires the network to support slice management (creation, modification, deletion) by such third parties. Moreover, in private 5G networks, the use of identities, credentials and authentication methods controlled by a third party must be supported.

As mentioned earlier, by the adoption of NFV, new security threats arise and need to be mitigated. Respective security requirements can be found in specifications provided by the European Telecommunications Standards Institute (ETSI) Industry Specification Group (ISG) NFV. As early as 2013, [7] included a list of security requirements, with the intention to address the new NFV threats. Among others, this list mentions protection measures against security vulnerabilities in the virtualization layer, protection for data stored on shared storage or transmitted via shared networking resources, protection of the new interfaces exposed by adopting the ETSI NFV architecture, including the management and orchestration components, and mechanisms to isolate different sets of VNFs running on the same NFV infrastructure.

Subsequently, specifications of the ETSI NFV's security group further elaborated on threats and respective security requirements. For example, [8] lists "potential areas of concern," which include validation and enforcement of the topology (of the graph consisting of the NFVs and their interconnections), a secure boot of the NFV platforms, AAA (Authentication, Authorization and Accounting) mechanisms for NFV users (tenants), private keys in cloned VNF images, or backdoors via virtualized test and monitoring functions. Another specification, [9], provides "security and trust guidance" and lists the high-level NFV security and trust goals.

A very comprehensive list of security requirements is given by [10], specifying what is required to allow the secure execution of sensitive VNF components in an NFV environment. A basic requirement is the implementation of a tamper-resistant, tamper-evident Hardware-Based Root of Trust (HBRT), like a Trusted Platform Module (TPM), see [11]. The HBRT is meant to support a secure boot as well as a highly secure management of cryptographic keys (key generation, key storage) and execution of crypto operations making use of the securely stored keys. Various other requirements are specified, including authentication of users of the NFV environment, attribute-based access control as specified in [12], communication security by means of TLS [13], verification of the provenance and integrity of all software components before installing and executing them, secure logging, and many more.

The brief overview on security requirements given in this section focuses on the NGMN Alliance, 3GPP SA1 and ETSI NFV as sources of requirements, and it clearly shows that there is already a significant body of diverse and somewhat challenging 5G security requirements. This body will be further increased by security requirements relating to the variety of additional mechanisms and protocols specified by other organizations, such as the Internet Engineering Task Force (IETF), that are expected to become relevant for future 5G networks.

7.1.3 5G Security Vision

Considering the 5G security drivers and requirements, three main characteristics of "what 5G security should be" emerge: Supreme built-in security, flexibility, and automation, as visualized in Figure 7.2.

7.1.3.1 Supreme Built-in Security

5G networks must support a very high level of security and privacy for their users (not restricted to humans) and their communication. At the same time, networks must be highly resistant to all kinds of cyber-attacks. To address this twofold challenge, security cannot be regarded as an add-on only; instead, security must be considered as part of the overall architecture and be built into the architecture right from the start. Based on the secure network architecture, also secure implementations of each of the individual network functions (NFs) are essential. Network operators need assurance methods, i.e. means to assess the security properties of the NFs they deploy, to ensure the required high degree of security of their networks.

7.1.3.2 Flexible Security Mechanisms

5G security must be flexible. Instead of a one-size-fits-all approach, the security setup must optimally support each of a variety of services. This entails the use of individual virtual networks – network slices – for individual applications, as well as the adjustment

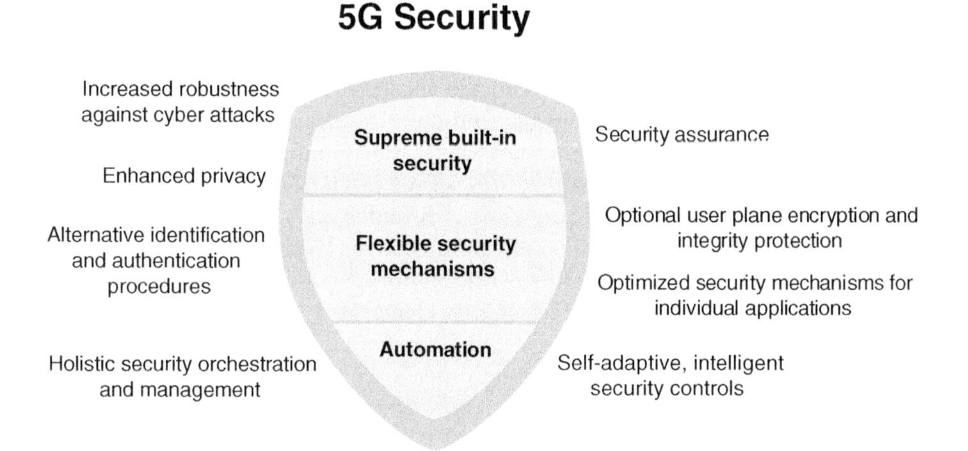

Figure 7.2 5G high-level security vision.

of the security configuration per network slice. Security features subject to this flexibility may comprise the mechanisms for identifying and authenticating mobile devices and/or their subscriptions, or for determining the way user traffic is protected. For example, some applications may rely on traffic protection offered by the network. These applications may require not only encryption, as in EPS case, but also user plane integrity protection. However, other applications may use end-to-end security on the application layer. They may want to opt out of network-terminated user plane security because it does not provide additional protection for the application layer payload but rather increases the processing time and energy consumption of mobile devices.

7.1.3.3 Security Automation

As one aspect of security automation, automated holistic security orchestration and management will be required to cope with the complexity of managing security efficiently and consistently throughout a network that spans multiple, possibly independent infrastructure domains, and that comprises a plethora of VNFs that may be allocated dynamically in these different domains. The holistic security management includes the task of specifying and distributing security policies to virtual and physical security functions and maintaining their consistency in a dynamic network setup.

The other important automation aspect is the automation of security controls. Intelligent, adaptive security controls will be needed to detect and mitigate the yet unknown threats – those that will try to exploit any weaknesses and flaws that may be found despite all efforts to rule them out. These kind of predictive security controls may leverage analytics and machine learning to be able to adapt themselves to cope with ever changing and evolving security threats.

7.2 Overall 5G Security Architecture

It may still be too early to describe the complete security architecture for future 5G networks. However, many elements of this architecture have already emerged, or can be anticipated – they are discussed in the present and the subsequent sections.

It can be fairly assumed that in 5G mobile network most of the NFs will be deployed in cloud environments. These cloud environments are not restricted to central clouds, but will comprise also highly distributed edge cloud deployments to facilitate edge computing close to the mobile devices. Outside the clouds, base station equipment will be deployed to provide radio coverage to mobile devices. Realtime layers of the radio stack can be implemented in dedicated base station equipment outside the edge clouds, close to the remote radio head. Non-Realtime layers such as the Packet Data Convergence Protocol (PDCP) layer providing radio interface security can be implemented as a VNF in the edge cloud. This split architecture is leveraged by the high-level split of the radio protocol as specified by 3GPP. Figure 7.3 provides a schematic view of this architecture and depicts possible security elements. It is beyond the scope of this book to investigate each of those in detail. Therefore, an overview description is given in this section, and the most important aspects are discussed in more detail in subsequent sections.

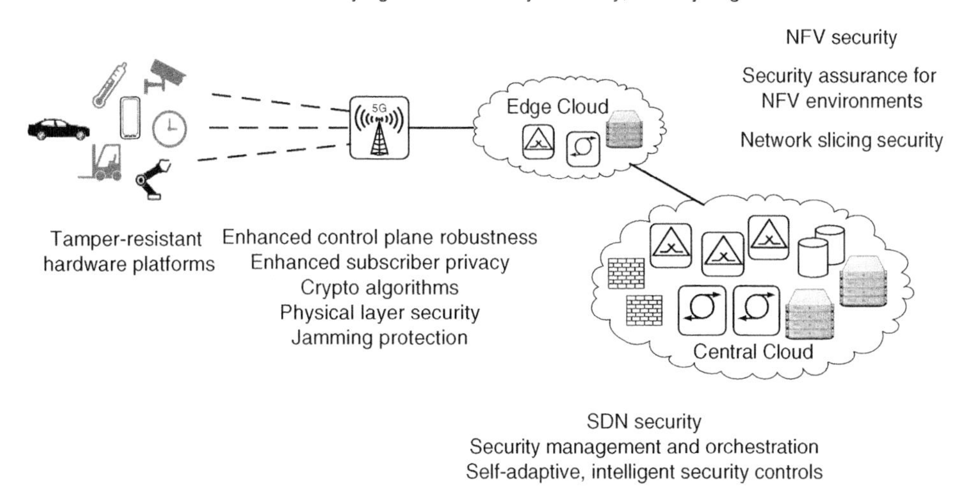

Figure 7.3 Elements of a 5G security architecture.

7.2.1 Security Between Mobile Devices and the Network

Security features relating to the interface between mobile devices and network are subject to standardization in 3GPP's Working Group SA3. Section 7.3 describes what is expected to be specified by 3GPP for 5G Phase 1 at the time of writing this book, while this section provides a more general overview, considering also mechanisms for future standardization.

Security between mobile devices and (radio and core) network includes authentication, and thus includes the types of identities (for subscriptions, devices or human users), the credentials that can be used, as well as the options available to store them on mobile devices. The current approach with the Universal Subscriber Identity Module (USIM) application on a Universal Integrated Circuit Card (UICC), commonly called a "Subscriber Identity Module (SIM) card," will continue to play a major role, particularly for the mobile broadband use case. In addition, other types of tamper-resistant hardware platforms for secure storage of identities and credentials and for secure execution of algorithms may be added to optimally support use cases such as massive IoT deployments.

To support more flexibility in the security setup, the security negotiation between the mobile device and network must be flexible enough to allow, for example, a mobile device to "opt out" of network terminated user plane traffic security. More flexibility may also be required with respect to crypto-algorithms. While the algorithms used in EPS are still very sound and not endangered by known attacks, new algorithms may need to be added, particularly so-called lightweight crypto algorithms. These refer to algorithms that minimize energy consumption for crypto operations without sacrificing the security. They are very important in the context of IoT devices. Other reasons for revisiting the selection of crypto algorithms may be the emerging of new computational techniques, such as quantum computing, or the detection of new attacks or more effective crypto analysis methods for the selected algorithms.

For control traffic, cryptographic integrity protection is vital to prevent spoofing, connection hijacking or impersonation attacks. Encryption is also important in this case to protect private user information transmitted in the control plane. The control plane protocols must be designed and implemented to anticipate all kinds of malicious behavior of mobiles – even of those that have been successfully authenticated as legal subscriber devices. It is no option to rely on standard-conformant behavior of mobile devices, as these, often without knowledge of their owners, can be corrupted by malware and may attack the network.

Like in EPS, security on the radio interface will rely on crypto protocols located rather at the top of the layer 2 protocol stack. It is an interesting field of research to investigate to what degree physical layer security mechanisms could enhance the security architecture, for example by providing additional security for communication on the lowest layers, where in current mobile networks no cryptographic protection is available. In EPS for example, this communication comprises control messages on the Media Access Control (MAC) layer, by which mobile devices inform the base station about the amount of data they want to transmit at a given time, and by which base stations inform mobile devices which radio resource blocks they must use to transmit the data. The individual physical properties of a transmission channel between a base station and an authenticated mobile device may be monitored to detect when an incoming unprotected lower layer message was sent by some attacking device instead of the legal, authenticated mobile device.

7.2.2 Security in the Telco Cloud

As most NFs are expected to run in NFV environments, NFV security mechanisms will play a key role in the 5G mobile network security architecture. They are discussed in Section 7.5. Connectivity within and between data centers is expected to be controlled via SDN, calling for a sound SDN security approach. SDN-specific security mechanisms are explained in Section 7.4. In the context of network slicing, security mechanisms must ensure a strict isolation between different network slices running on shared infrastructure. This is necessary to prevent virtual machines in one slice from impacting those in other slices. It is also required to prevent information from leaking between slices on side channels, e.g. via physical memory sequentially used by different slices, where information stored by one slice may not properly be wiped out when another slice gets access to the memory. More details on network slicing security can be found in Section 7.6.

When NFs are no longer bound to specific hardware but may be instantiated on different hardware platforms, it might be more difficult to ensure their proper, secure operation. In this sense, NFV may have a significant impact on security assurance methods that will need to take this dynamic allocation of software functions to different hardware infrastructures into account.

An automated holistic security orchestration and management will become crucial in large, cloud-based 5G mobile network deployments. There will be many different security functions, including virtualized or physical firewalls and intrusion detection systems, and security functions within hypervisors or other elements of the virtualization software. All these must be configured consistently, with all individual policies meaningfully contributing to the overall defense concept, and must be dynamically adapted to changes of the network topology, like migration of VNFs onto other physical resources.

Finally, there will be unpredictable threats that try to exploit weaknesses that may be present despite the care involved in designing and implementing the network. Moreover, DoS attacks may be carried out even in the absence of specific vulnerabilities, by simply flooding network elements or the overall network with traffic. Such attacks are not of a theoretical nature. Mobile botnets pose a very real threat to mobile networks. A botnet is a potentially large set of corrupted devices controlled by an attacker that can carry out large scale attacks against specific servers or networking services. To cope with such new threats, smart, automatic and self-adapting security controls are required. By closely monitoring what is happening within the network, i.e. analyzing the traffic flows and the behavior of mobile devices and network nodes, and, in a machine learning approach, considering which kind of anomalies preceded past security incidents, smart security controls will be able to detect malicious activities – even previously unknown ones – at an early stage and trigger mechanisms to mitigate such threats and maintain the integrity and availability of the network.

7.3 3GPP Specific Security Mechanisms

As standardization aims at ensuring interoperability, the 3GPP security mechanisms focus on securing the 3GPP-specified interfaces. Of those, the radio interface may be considered as the one that is most endangered, because it is inherently accessible by local attackers. The method to secure the interface between UE and network is therefore a major area of the 3GPP security architecture, and it comprises not only the cryptographic algorithms used to secure traffic on this interface, but also the mutual authentication between mobile device and network, including the means to conceal the subscription identity to prevent tracking.

3GPP specifies 5G security in [18]. 5G security re-uses many of the mechanisms specified for EPS in [15] and [16]. An excellent description of EPS security is given in [4].

7.3.1 Security Between UE and Network

7.3.1.1 EAP-Based Authentication

A major change compared to EPS is the introduction of the Extensible Authentication Protocol (EAP, see [20]) framework for mutual Authentication and Key Agreement (AKA) between UE and network. EAP is extensible in the sense that different authentication methods can be used within this framework. In EPS, EAP with the authentication method EAP-AKA′ [21] is specified for non-3GPP access only. Aiming at access independence, in 5G it can be used over any kind of access. EAP-AKA′ is the EAP authentication method that is mandatory to be supported in 5G, but other methods, such as EAP-TLS, may also be implemented for some scenarios, e.g. in private networks.

On the network side, EAP is processed by the Security Anchor Function (SEAF), acting as EAP authenticator in the serving network (which can be the visited or home network depending whether the terminal is roaming or not), and the Authentication Server Function (AUSF), acting as authentication server, in the home network (see Chapter 4 for a detailed discussion of the 5G network architecture). In 5G phase 1, the SEAF is assumed to be collocated with the Access and Mobility Management Function (AMF). The AUSF communicates with the Authentication Credential Repository and Processing Function (ARPF) inside the Unified Data Management (UDM) that holds the long-term authentication credentials of the UE and performs all processing that requires access to these credentials. It is the UDM/ARPF which decides on the authentication procedure (EAP-AKA′ versus 5G AKA), and depending on this decision, suitable authentication material is produced and sent to the AUSF. In case of EAP-AKA′, this is an Authentication Vector (AV) including keys called Cyphering Key (CK′) and Integrity Key (IK′) that are derived from the long-term credential.

The authentication exchange is depicted in Figure 7.4.

During the EAP processing, UE and home network, represented by the AUSF, authenticate each other and derive a common key, called EMSK (Extended Master Session Key), in a way that the key remains hidden to any intermediate nodes. From EMSK, the AUSF derives a key called K_{AUSF} and from this key another key called K_{SEAF} which is passed from the AUSF in the home network to the SEAF in the serving network. This key is bound to the serving network identity, by including the Serving Network Identifier (SN id) into the input of the key derivation algorithm. Moreover, a constant "5G:" is part of the input, binding the key to an EAP-AKA′ run in a 5G system (as opposed, e.g. to an EPS AKA run). In case of EAP-AKA′, the serving network identity is also considered when deriving the keys CK′ and IK′ in the ARPF. The AUSF receives these keys before starting the authentication process. The SEAF derives from K_{SEAF} a key K_{AMF} and passes it to the AMF. From K_{AMF}, keys K_{NASenc} and K_{NASint} for encryption and integrity protection of Non-Access Stratum (NAS) signaling traffic are subsequently derived.

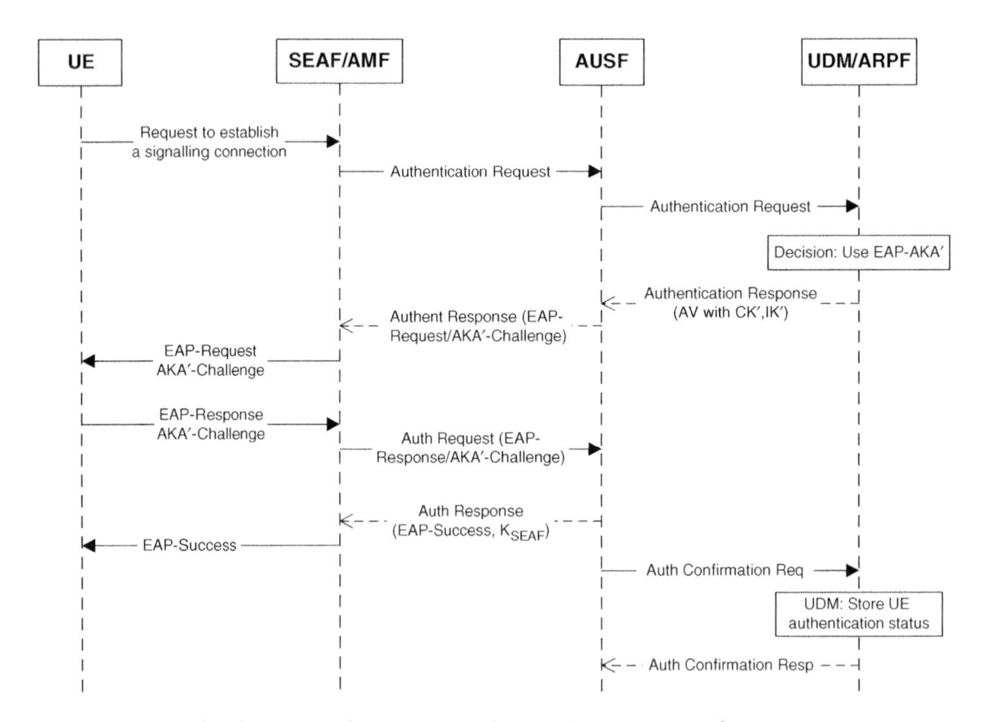

Figure 7.4 Mutual authentication between UE and network using EAP-AKA′.

The UE is also aware of the SN id and that this is a 5G authentication run, and it uses the same derivations as the network to generate K_{SEAF}, K_{AMF}, K_{NASenc} and K_{NASint}. When the UE can successfully check the integrity of a message from the network protected by K_{NASint}, the UE obtains assurance from the home network that the UE shares K_{SEAF}, K_{AMF}, K_{NASenc} and K_{NASint} with the specific serving network whose identifier was used as input to the derivation of K_{SEAF}.

It should be noted that the UE always authenticates with its home network in the EAP authentication run, as opposed to EPS AKA, where authentication is done between UE and serving network, based on authentication material the home network provides to the serving network.

Moreover, EAP between UE and AUSF in the home network allows UE and AUSF to retain a key shared only between UE and AUSF but not known to the visited network. Although currently not specified by 3GPP, such a key may be used to establish security associations directly between UE and home network.

EAP-AKA′ based authentication must be supported in the UE and in serving network entities (i.e. AMF/SEAF), but not necessarily in Home Public Land Mobile Network (HPLMN) entities such as AUSF and ARPF/UDM. So, an operator not using EAP-AKA′ for its own subscribers does not have to implement it in AUSF and ARPF/UDM. Still, foreign subscribers roaming in this operator's network will not be prevented from using EAP-AKA′, based on the mandatory support in the AMF/SEAF.

7.3.1.2 5G AKA

Besides EAP, also an enhanced version of EPS AKA, called 5G AKA, must be supported by the UE and serving network entities. Different to EPS AKA, 5G AKA can be used over 3GPP as well as non-3GPP access. The part of the Mobility Management Entity (MME) and the Home Subscriber Server (HSS) in EPS-AKA are taken by the SEAF/AMF and the AUSF/ARPF/UDM in 5G AKA. Like in EAP based 5G authentication, 5G AKA binds the key it generates to the serving network identity and to the constant "5G:." Different to EPS AKA, the expected result (called XRES*) on the AKA challenge is not part of the Authentication Vector passed from the home network to the serving network, but only a one-way hash of it, called HXRES*. This is sufficient for the visited network to verify whether a UE provided the correct answer (called RES*) on the AKA challenge, because the serving network can apply the same hash to RES* and compare the result to HXRES*. 5G AKA specifies an authentication confirmation message, in which the serving network passes RES* as received from the UE to the home network. This proves to the home network that the UE is indeed present in the serving network, as the serving network is not able to calculate RES* itself. Depending on its policies, a home network may not acknowledge that the UE is present in the serving network without successful authentication confirmation. For example, it may reject registration requests from the UE in a serving network, if no sufficiently recent confirmed authentication via the serving network has happened. This is an advantage compared to EPS AKA, where a fraudulent serving network can falsely claim a UE is connecting to it and possibly charge the home network, by first requesting an authentication vector and then sending an Update Location request.

The 5G AKA authentication exchange is visualized by Figure 7.5.

After passing the Authentication Vectors to the SEAF, 5G AKA requires an additional roundtrip between SEAF/AMF and AUSF. The reason for this is that in 5G, the serving

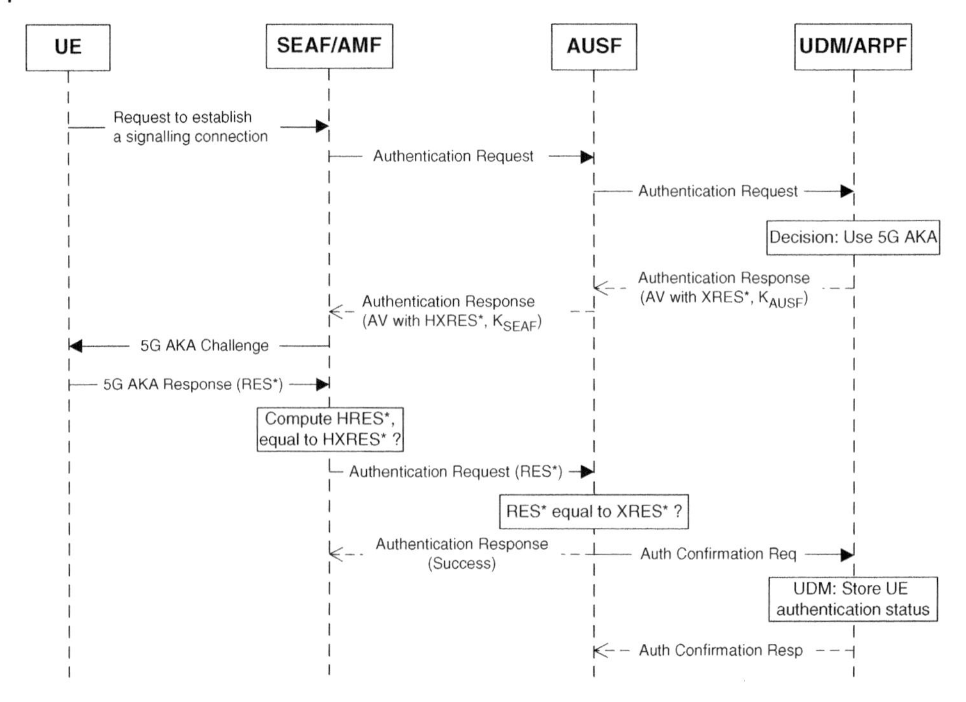

Figure 7.5 Mutual authentication between UE and network using 5G AKA.

network only receives the unconcealed subscription identifier (see section on Subscription Identity Privacy) after successful authentication confirmation by the home network, and it needs this identifier in deriving the key K_{AMF}.

7.3.1.3 Access Stratum (AS) Security

A key K_{gNB} is computed from K_{AMF} by the AMF and passed to the gNB (5G base station) as the root key for Access Stratum (AS) security, i.e. allowing protection of traffic between UE and gNB. In handovers between different cells of the same gNB, this key need not be changed. In handovers to other gNBs, a new K_{gNB} must be derived, preferably using fresh input not known to the old gNB. This is called a "vertical" key derivation and requires involvement of the AMF. Like in EPS, there is also a "horizontal" key derivation without involvement of the AMF. This variant is faster, but has the slight disadvantage that a compromised K_{gNB} at the old gNB would mean that the new K_{gNB} is also compromised.

As in EPS, integrity protection and encryption is specified for Radio Resource Control (RRC) signaling, using respective keys derived from K_{gNB}. In 5GS, it is possible to secure AS signaling even when there is no Data Radio Bearer (DRB) setup for the following reasons:

1. Support secure AS signaling connection for the network to perform redirection even when the UE is registered without DRB. For instance, not ciphering and integrity protecting the RRC release message with redirection may result in attackers sending the UE to a "wrong" cell. Redirection in LTE was not secured. Thus, attacks were possible.

2. AS security setup is essential in case of non-3GPP access to establish an Internet Key Exchange Version 2 (IKEv2) association even before Protocol Data Unit (PDU) Sessions/DRBs are setup. Thus, enabling support for AS security setup for signaling only connections is essential to enable a common registration procedure for 3GPP and non-3GPP access.

Different from EPS, for user plane traffic not only encryption, but also integrity protection must be supported – although both are "optional to use." It depends on the policies of the network what kind of protection is used. For example, for access to a specific data network, the policy may be not using encryption, because this data network establishes a protected tunnel toward the UE anyway, and the data network operator and the 5G network operator have agreed on not using AS user plane protection for this reason. In general, user plane protection is used, and keys for user plane encryption and integrity protection are derived from K_{gNB}.

A gNB may be split into a central and distributed unit (CU-DU). In this case, the termination for AS security will reside in the CU, which is more likely to have physical protection (like being locked within a solid building). Also, RRC and user plane traffic are thus protected by AS security on the CU-DU interface (called F1). However, as not all traffic at this interface will be protected this way, confidentiality, integrity protection and replay protection is mandated on this interface, as well as on the interface between the CU control plane and the CU user plane (called E1), if such interfaces are implemented. This can be achieved by the means described for non-service-based interfaces in Section 7.3.2.

7.3.1.4 Key Hierarchy and Crypto Algorithms

An overview of the 5G key hierarchy is given by Figure 7.6.

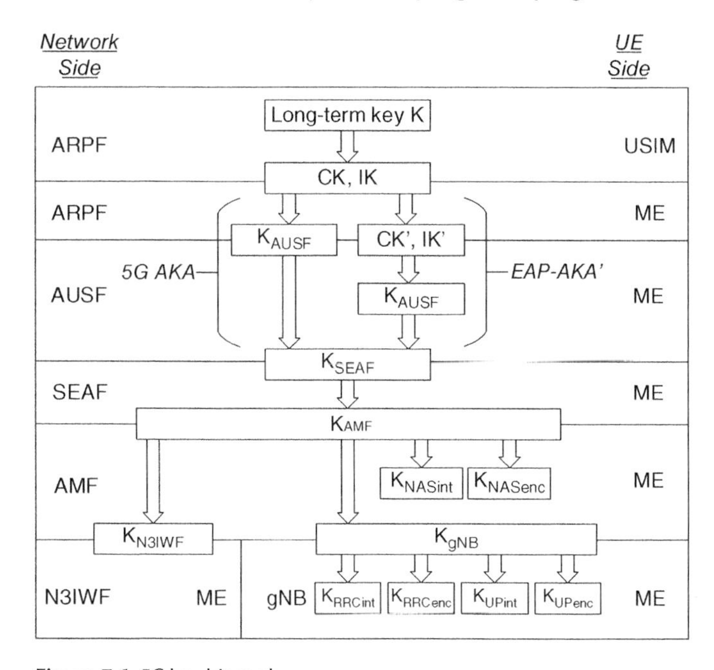

Figure 7.6 5G key hierarchy.

In the first step, keys called CK and IK are derived from the long-term key K, in the network and on the USIM. 5G AKA and EAP-AKA′ use different ways to derive K_{AUSF} from CK and IK. The subsequent derivations are common to both methods. On the UE side, CK and IK are passed from the USIM to the ME (Mobile Equipment) part of the UE, where the subsequent derivations are carried out.

For both NAS and AS security, 5G phase 1 specifies the same three pairs of crypto algorithms as in EPS (each pair consisting of an encryption and an integrity protection algorithm), i.e. the three pairs of algorithms based on the three crypto algorithms AES (Advanced Encryption Standard), SNOW 3G and ZUC, where AES and SNOW 3G are mandatory to support. In future, it may be reasonable to add further algorithms, for example lightweight crypto algorithms, as mentioned already in Section 7.2, or algorithms with keys longer than 128 bits. The specification allows for easy inclusion of new algorithms once this is required.

7.3.1.5 Security for Untrusted Non-3GPP Access

Untrusted non-3GPP access requires an IPsec [23] security association to be established between UE and an entity in the core network. In 5G, this entity is the Non-3GPP Interworking Function (N3IWF). A protocol stack has been specified that allows exchanging NAS signaling messages during the establishment of the IPsec security association by means of IKEv2 [22]. This way, both EAP-AKA′ and 5G AKA can be used for authentication, with the key derivations visualized in Figure 7.6. However, instead of the key K_{gNB} used in 3GPP access, a key K_{N3IWF} is derived and passed to the N3IWF. This key is subsequently used as the MSK (Master Session Key) in the IKEv2 handshake to authenticate the setup of the IPsec association.

7.3.1.6 Security for Non-Standalone Architecture with EPS (EN-DC)

The security concepts for EN-DC (Evolved Universal Mobile Telecommunications System Terrestrial Radio Access Network – New Radio – Dual-Connectivity) are very similar to those in the EPS DC. They are described in [15], starting with Release 15. For each UE session using EN-DC, the MeNB (Master eNB) generates a key for the SgNB (Secondary gNB) based on K_{eNB} and a counter. It passes the key to the SgNB, and the counter to the UE, together with an indication which algorithms are to be used between UE and SgNB. As EN-DC supports also a signaling radio bearer terminated at the SgNB, keys for RRC integrity protection and encryption are derived at the SgNB and the UE, as well as a key for encrypting the DRBs. Integrity protection for DRBs is not activated in EN-DC.

7.3.1.7 Subscription Identifier Privacy

Previous generations of mobile networks protect the subscription identifier, the IMSI, by usage of temporary identifiers. However, in certain situations, like a failure of an MME holding the current temporary identifier, it is required to use the IMSI and send it in clear form. This allows passive IMSI catching attacks, but more critically, an active attack using a purposefully designed fake base station ("IMSI catcher"), pretending to be the legal network, can trick the UE into revealing the IMSI. Proposals how to overcome this issue were made – but not agreed upon – already when UMTS security and EPS security were specified.

5G changes this situation by introducing the Subscription Permanent Identifier (SUPI) and the Subscription Concealed Identifier (SUCI). The UE computes the SUCI by encrypting the individual part of the SUPI, with a public key provisioned on the UE. The Mobile Country Code (MCC) plus Mobile Network Code (MNC) part of the SUPI is not encrypted, thus a serving network can identify the home network of the subscription. The home network is in possession of the respective private key and can determine the SUPI by decrypting the SUCI based on the private key. Subsequently, a 5G-GUTI (Fifth Generation Globally Unique Temporary Identifier) is allocated by the network and it is used to identify the UE context and protecting subscriber confidentiality. The AMF allocates a 5G-GUTI for the UE immediately after the UE is successfully registered. The UE is required to provide the 5G-GUTI and not the SUPI or SUCI, if it is trying to register in the same or an equivalent Public Land Mobile Network (PLMN). To avoid tracking based on the GUTI, it is mandatory to frequently change it in an unpredictable way. NAS signaling procedures support concealed transmission of a newly assigned GUTI to the UE after usage of the GUTI in clear form. There is no need to use the GUTI in clear form more than one time, i.e. tracking via the GUTI can be prevented. Paging messages use the Fifth Generation S-Temporary Mobile Subscriber Identity 5G-S-TMSI derived from the GUTI to uniquely identify the UE. The UE uses the 5G-S-TMSI in RRC messages to allow the Radio Access Network (RAN) selecting an appropriate AMF for delivering the NAS message that is encapsulated within a RRC message. Again, after such an exposure of the 5G-S-TMSI in clear format, it is required to immediately allocate a new, unpredictable 5G-GUTI.

7.3.1.8 Usage of the UICC

In EPS, the use of a UICC within the UE to securely store the long-term credentials and to perform certain security algorithms is mandatory. According the current state of the specifications, it is mandatory to use the UICC in 5G, but the standard mentions that the use of another hardware solution, the SSP (Smart Secure Platform) currently under standardization by ETSI, may also be specified, given that certain security criteria will be fulfilled.

7.3.1.9 Secondary Authentication

Previous mobile network generations specify a method to transport information for authentication between a UE and an external packet data network (connected via the Gi or SGi reference points) within the Protocol Configuration Options (PCOss), i.e. inside 3GPP specified signaling messages. This PCO-based authentication is rather limited. In 5G, it is replaced by what is called a "secondary authentication," based on EAP. EAP messages can be exchanged in session management messages between UE and SMF (Session Management Function), which acts as EAP authenticator. The SMF can relay the EAP messages to a AAA server acting as authentication server inside the external data network. Usage of EAP-based secondary authentication is a significant improvement over PCO-based authentication in terms of flexibility and security.

7.3.2 Security for Network Interfaces

7.3.2.1 Non-Service-Based Interfaces

In EPS, network interfaces are secured using IKE/IPsec ([22, 23]). This is mandatory, if the interfaces span different security domains, for example if an interface is between two different PLMNs. As a rule, signaling traffic must be integrity protected, and if sensitive information (such as keys or subscription identifiers) are transported, also encrypted.

Notably, this kind of protection has not been deployed for inter-PLMN traffic very much so far, leaving PLMNs vulnerable to various kinds of attacks launched via interconnection networks. IKE/IPsec-based network interface security has been adopted in 5G. On interfaces between gNBs or between RAN and Core Network, integrity protection and encryption is mandatory for all types of traffic, unless these interfaces are protected otherwise, e.g. "physically protected," a condition that can hardly be met for a classical RAN deployment. However, in edge cloud deployments, indeed other protection mechanisms may become effective, e.g. "wholesale" protection of traffic between edge clouds and central clouds.

7.3.2.2 Service-Based Interfaces

In the Service-Based Architecture (SBA), see Chapter 4 for a detailed description of SBA, Network Functions (NF) communicate using Hypertext Transfer Protocol (HTTP)/2 [24], and as HTTP usually relies on TLS [13] for security, support of TLS based on server- and client-side certificates has been mandated for all NFs. Like IKE/IPsec, also TLS maybe replaced by other forms of protection. Mutual authentication (either explicitly via TLS, or implicitly, e.g. based on the fact that the entities communicate within a single security domain), is mandatory for Service Registration and Service Discovery requests of NFs toward the NRF (Network Repository Function).

For authorization of service requests of a consumer NF toward a producer NF, an approach based on Open Authorization 2.0 (OAuth 2.0) [26] has been adopted, where the consumer NF (the OAuth Client) requests an authorization token from the NRF (the OAuth Authorization Server) and presents it to the producer NF (as OAuth Resource Server) when requesting a service. The consumer NF must authenticate the producer NF when requesting the service, while the producer NF may rely only on the token and the authentication of the consumer NF that must have happened before the token was passed to the consumer NF.

7.3.2.3 Interconnection Security

In roaming cases service-based interfaces are used between NFs in different PLMNs. To secure the PLMN interconnection, a new entity called Secure Edge Protection Proxy (SEPP), has been specified. All control plane traffic between two PLMNs must be exchanged via the N32 reference point between two SEPPs belonging to the two PLMNs. A SEPP serves as single entry point into a PLMN. It hides the details of the topology of the PLMN it belongs to and performs the required security mechanisms for the N32 interface. The latter include usage of TLS either directly to the other SEPP, or toward the next hop intermediary within the interconnection network. However, there is currently consensus in 3GPP SA3 to specify security mechanisms on the application layer, too. For service-based interfaces, information elements are carried in the HTTP message body. Some of these information elements may need to be read or even be modified by intermediate nodes within the interconnection network. Thus, selective protection of the different information elements can be useful, e.g. applying integrity protection but no encryption to information elements that need to be read but not modified by intermediaries, and providing encryption only to information elements that are not relevant for routing decisions by intermediaries. Similar principles may also apply to any Diameter-based interfaces (see [25] for more details on the Diameter protocol).

7.3.3 Summary and Outlook

The 3GPP 5G security architecture builds strongly on the well-proven EPS security architecture, but also introduces many new features such as:

- Introduction of EAP as an access independent authentication framework;
- Introduction of integrity protection for the user plane between UE and network (mandatory to support, optional to use);
- "IMSI-catching" prevention by use of an encrypted subscription identity;
- Security for the SBA;
- Improved support for inter-PLMN security.

Additional 5G security enhancements have already been discussed during the 3GPP 5G security study work [19]. Future enhancements may include the option to terminate user plane security between UE and network not in the RAN, but in a UPF within the core, aiming at scenarios where different core network slices may be operated by different parties sharing the RAN, in this case, user plane security terminated in an individual core slice would not be affected by attacks on the shared RAN. More security features related to network slicing scenarios involving multiple third parties may also need to be specified. Another area for enhancements is optimized security for new use cases, such as massive IoT and Ultra-Reliable Low Latency Communication (URLLC) applications. This may include new lightweight crypto algorithms, new authentication methods, and new ways to store subscription credentials on mobile devices.

7.4 SDN Security

With SDN network node control functions can be separated from forwarding functions. Control functions of network nodes like switches, routers or gateways can be implemented at a centralized location. Moreover, SDN comprises the concept of network programmability, i.e. the centralized control functions provide interfaces that can be used by other software applications to control the network. SDN is applicable to the cloud infrastructure, namely to the networks providing connectivity between racks and server blades inside data centers, as well as the networks interconnecting data centers. SDN is also applicable to virtual switches implemented in hypervisors to provide networking between virtual machines running on the same hypervisor instance. A specific flavor of SDN, called OpenFlow (OF), has been specified by the Open Networking Foundation (ONF). Their specifications are publicly available via the organization's web site. The work of the ONF also included a "Project Security," which produced some specifications, in particular [14], which provides a good overview of OpenFlow security issues. While [14] has a clear emphasis on OpenFlow protocols, in the following, a more general discussion of SDN security aspects is given.

7.4.1 Separation of Forwarding and Control Plane

Due to the separation of forwarding and control plane, SDN introduces an interface between SDN controller and SDN switch. This interface increases the attack surface of the overall system. It could allow attacks on the integrity and confidentiality of the controller to switch communication, DoS attacks, or attacks aiming to get control over switches and controllers by exploiting vulnerabilities in the protocol software or the interface configuration. However, securing such an interface is a well-known task and suitable means are readily available, such as usage of TLS [13] to perform mutual authentication and protect cryptographically the legal communication, thus protecting against any interception and excluding all communication faked by malicious third parties.

7.4.2 Centralized Control

The separation of forwarding and control plane allows to (logically) centralize the control of a network. While centralized control can contribute to unifying security policies and thus improve the overall network security, centralized control will also increase the impact of certain attacks, namely attacks that succeed in crashing or compromising such central controllers. Therefore, secure design and implementation minimizing the risk of vulnerabilities is essential for controllers.

7.4.3 Controllers Running in Cloud Environments

SDN allows to implement controllers on virtual machines in cloud environments. In this case, controllers will be exposed to threats present in such an environment and may be compromised via this environment. See Section 7.5 for a detailed discussion of NFV security. On the positive side, cloud environments can help to overcome DoS attacks by dynamically allocating additional resources to controllers.

7.4.4 Agile and Fine Granular Control

SDN allows an agile and fine granular control of the flows in a network. This facilitates the deployment of security solutions that may apply dynamic policies to flows, such as diversion, rate limiting or blocking. Care must be taken that fine granular policies do not introduce unnecessary complexity and the potential for errors into the network configuration.

7.4.5 Network Programmability via the Northbound Controller Interface

SDN introduces the so-called northbound interface, where applications have access to SDN controllers. This allows implementation of new security solutions that make use of the possibility to execute central, and at the same time fine granular and agile control over the network via an SDN controller.

On the other hand, the concept of possibly several different applications, including third party applications (that may not be under the full control of the network operator) executing control over a network raises security issues, including authentication of applications, authorization of requests and resolving conflicting requests. These issues are certainly resolvable, but solutions must be designed and implemented with great care to avoid erroneous behavior caused by multi-party network control.

As discussed in the context of centralized control, secure design and implementation is essential for SDN controllers. This applies to the northbound interface, to prevent

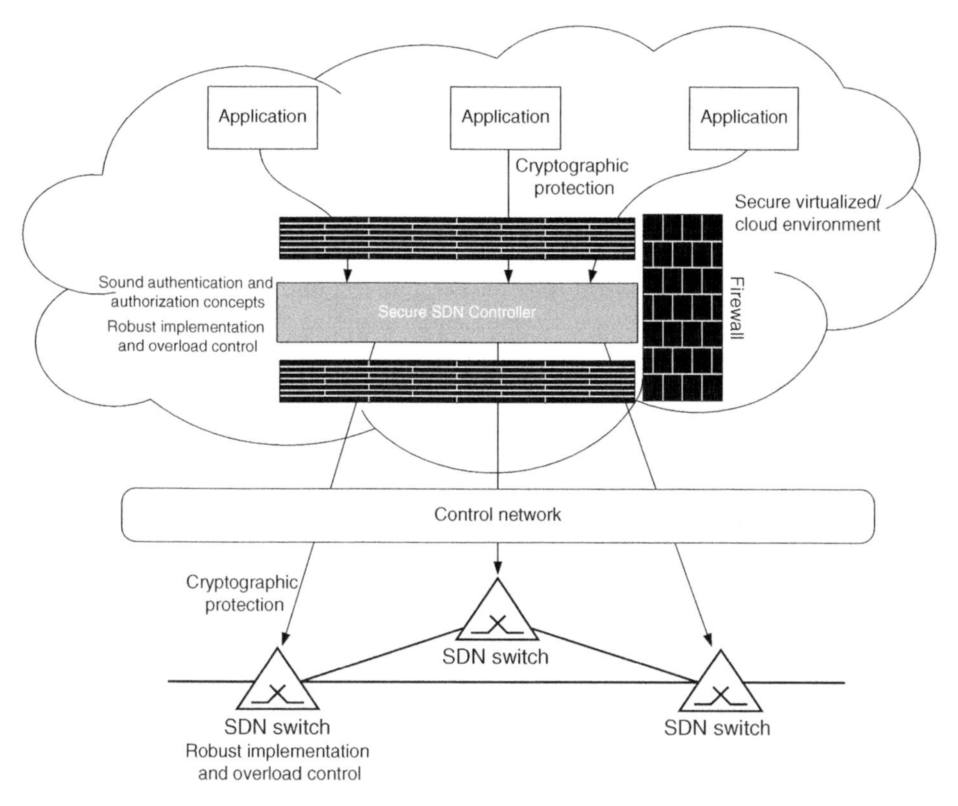

Figure 7.7 SDN security mechanisms.

malicious applications being able to compromise a controller via this interface and subsequently to exhibit unauthorized control over network resources.

Figure 7.7 visualizes the recommended SDN security measures. Of these, the most essential one is an SDN controller that is robust against overload and attacks, and implements sound authentication and authorization mechanisms at its northbound interface. An additional firewall around the controller may be used to fend off typical attacks that can happen in TCP/IP networks, for example so-called TCP SYN flood attacks, where an attacker in the network tries to exhaust certain networking resources of a victim, here the SDN controller. Robust implementation and overload control also applies to the nodes that implement the data layer, the SDN switches. Traffic between applications and the controller and between the controller and the switches can be cryptographically protected by state-of-the-art techniques such as TLS. While the actual effort for applying the cryptographic protocols may be negligible in this context, it must be noted that a key management solution is also required, which comprises the creation, distribution and maintenance of long-term keying material – private/public key pairs and certificates in case of TLS. However, such a key management solution may be required independently of the use of SDN, for securing the management of network elements.

To summarize, SDN brings new security challenges to networks, when multiple, diverse applications are admitted to “program the network,” i.e. control network resources via northbound interfaces. However, SDN may enable more flexible and efficient deployment of security solutions, if those solutions can be implemented as software applications making use of SDN controllers without relying on traditional security devices. Building on this, SDN has the potential to make networks more secure.

7.5 NFV Security

As pointed out already, adoption of NFV raises significant new threats. With the virtualization layer and the specific new software entities such as the virtual infrastructure manager, VNF managers and NFV orchestrators, the overall amount of software that may be vulnerable to attacks increases significantly. Even more critical, the concept of sharing physical infrastructure may bring together different applications in a rather unpredictable way. Thus, malicious applications may try to attack other applications using the same physical resources, for example physical memory. The goal and purpose of the virtualization software is to make the infrastructure sharing transparent to all applications. However, errors in this software may allow successful breaches of this type of isolation.

7.5.1 Providing Highly Secure Cloud Software

It is therefore imperative that the virtualization layer, e.g. the hypervisor, is implemented with the highest care to minimize the likelihood of vulnerabilities. Likewise, all the virtualization software (sometimes called the cloud stack) must be sound and robust, and anticipate erroneous or even malicious behavior of all application software that may access cloud software application programming interfaces. The result must be a highly secure NFV platform.

Like the software of non-virtualized network elements, VNFs must be designed and implemented with security in mind, following product creation processes that comprise steps such as a threat analysis for the respective function, the specification of a security concept, secure coding, hardening, security testing and auditing. Such steps will not guarantee error-freeness, but they are supposed to drastically reduce the number of potentially exploitable errors. They also can give network operators a degree of assurance that the system will be secure against compromises by attackers. This so-called security assurance clearly must consider the split between NFV platform and application software, and consider that VNFs may run on different hardware and/or software platforms.

7.5.2 Assuring NFV System Integrity

Relying on a sophisticated and well-implemented security-aware creation process may not be sufficient for NFV-based 5G deployments – stronger trust in the integrity of a concrete deployment may need to be established. Based on a TPM (see [11]) acting as HBRT, integrity of the NFV platform can be ensured at boot time. Later, e.g. before launching a certain VNF, platform integrity may be verified by remote attestation mechanisms. Clearly, VNF images also need to be verified – an attacker may have been able to tamper with an image before the VNF is launched. Further, it must be ensured that a VNF can be bound to the verified platform (or set of verified platforms) – otherwise it may be migrated to a compromised one that may successfully attack the VNF, e.g. read out the VNF's confidential data. Means are available to achieve this goal. The interested reader is referred to [17].

7.5.3 VNF and Traffic Separation

Natively, a secure cloud provides separation of VNFs, but in a shared infrastructure, it is a logical separation only. There may be cases where a specific VNF or set of VNFs is considered so sensitive, that any side-channel attacks, e.g. via a vulnerable hypervisor must be excluded. In this case, dedicated physical resources may be assigned to a VNF or a set of VNFs, not shared by other VNFs. This comes at the cost of reduced resource efficiency. It may be acceptable in large central data centers with abundant resources. When it comes however to small, distributed edge cloud deployments, resources may be so scarce that setting aside specific hardware resources for use by a specific application is not a valid option.

Complex networks normally use traffic separation as a security mechanism. For example, Operation and Maintenance (O&M) traffic may be fully separated from user plane traffic, using different virtual networks. Thus, malicious user plane traffic cannot disturb O&M of the network. This concept is fully applicable to NFV-based networks. Hypervisors can support different virtual switches within a single hypervisor instance, dedicated to different traffic types. Virtual and physical switches can support different Virtual Local Area Networks (VLANs), and Virtual Private Networks (VPNs) can be created in wide area transport networks, thus allowing the creation of fully separated end-to-end networks for different traffic types such as user plane traffic, control plane traffic and O&M traffic.

7.5.4 Security Zones

Another well-known network security measure is perimeter security and network-internal traffic filtering to create different security zones within a network. For example, one zone may contain all network elements that need to communicate with peers outside the network, and another zone may comprise nodes with only network-internal communication relationships. This concept can be transferred into NFV environments. Instead of physical firewall devices, virtualized firewalls will typically be applied, and virtualized intrusion detection systems may inspect the traffic for potential threats. In terms of functionality, such virtualized security devices are equal to physical ones. They have the advantage of flexible scalability, allowing them to cope better with floods of malicious traffic. On the other hand, one could argue that they are more endangered, as they may be attacked via the shared infrastructure they use. The separation between different security zones created via network internal firewalls is only a logical one. However, as pointed out above, on the cost of reduced resource efficiency, a physical separation may be implemented for a particularly sensitive security zone.

7.5.5 Cryptographic Protection

To ensure confidentiality and integrity of data in transit or on storage, cryptographic mechanisms may be used. For the traffic over 3GPP-specified reference points, between mobile device and network, this type of protection has already been discussed in Section 7.3. Many interfaces inside an NFV-based network will however not be standardized. Such interfaces do exist between VNFs, but also between different VNF-components that may be implemented as different virtual machines and could be running on different physical platforms, interconnected by a network. Applying cryptographic protection on all these interfaces seems hardly feasible – not only in terms of computing effort and delay, but also in terms of key management. Moreover, it must be noted that attackers inside the NFV environment need not necessarily attack the traffic in transit – it may be much more rewarding for them to try attacking the virtual machines themselves, for example by trying to access their memory while relevant data is present in cleartext. Consequently, virtualized networks may rather rely on networking security provided by the cloud infrastructure, which ensures that a VPN specified to connect a specific set of virtual machines cannot be accessed by any other virtual machine or outside attacker. The cloud infrastructure may in turn apply cryptographic protection where it is needed. For example, the fibers interconnecting distributed data centers are physically exposed, and could become subject to various forms of tapping. Therefore, encryption of all traffic via these fibers must be applied, most likely to the aggregated traffic, thus reducing significantly the complexity of key management compared to individual encryption of individual traffic flows.

Concerning long-term storage of data, e.g. on shared disk arrays, relying on general infrastructure security could be sufficient. For enhanced security, individual encryption with keys maintained within the VNFs seems a plausible option, as it would require an attacker, in addition to gaining illegal access to the stored data, to also successfully attack the VNF and retrieve the keys.

7.5.6 Secure Operation and Management

Secure operation and management is crucial for non-virtualized as well as for virtualized networks. In the latter case, the dynamic nature of the network and the complexity of having potentially many different network slices sharing the same infrastructure pose a specific challenge. As discussed in Section 7.2, a high degree of automation can help to cope with this challenge.

7.6 Network Slicing Security

7.6.1 Isolation

When slicing a network, i.e. creating multiple virtual networks on a common physical infrastructure, the typical intention is that each of these virtual networks is independent of the other virtual networks. If we assume network multi-tenancy, where different parties operate different slices, network slice isolation becomes the crucial security aspect. We can distinguish two isolation aspects – resource isolation and security isolation. The former refers to the fact that resources assigned to a network slice (such as computing, storage and networking resources) cannot be "hijacked" by any other slice. This does not mean that each slice always has the same, fixed amount of resources available. Rather, a Service Level Agreement (SLA) between slice operator and InP may specify for example a minimum amount of resources that is always guaranteed, and may further specify terms and conditions for the usage of additional resources. Resource isolation ensures that the SLA is fulfilled, even in situations where other slices may suffer from overload situations or DoS attacks. Security isolation refers to the property that information in one slice cannot be accessed or modified by other slices sharing the same infrastructure.

As pointed out in Section 7.5, sound NFV security can ensure both types of isolation within NFV environments. For the transport between distributed cloud deployments, VPN techniques can be applied to ensure isolation, where SDN allows for highly dynamic and flexible control of different VPNs sharing the same transport infrastructure. Slices may also share non-virtualized equipment, for example base station equipment handling the lower protocol layers. Here, isolation needs to be assured by respective equipment-specific mechanisms. For example, when several slices share a single cell, a common radio scheduler may be configurable to implement scheduling policies that ensure that each of the slices gets radio resources in this cell according to the SLA valid for the slice.

Obviously, isolation works better the less common resources are used. Not sharing physical infrastructure would therefore be the best isolation approach, but this is clearly out of question, considering the huge advantages expected from infrastructure sharing. Still, in some cases it may be reasonable to use even physical separation, like for example physically separating different subscription databases used by different slices.

If we assume a model where an InP rents out slices to third party organizations, such as industry verticals, then with sound isolation mechanisms in place, a slice used by a third party organization may be isolated from other slices. However, there is no native isolation against the InP itself, who is in control of the infrastructure and can access all the information that is processed on it. So third parties operating slices must trust the InP not only to take suitable measures to ensure isolation between slices, but also to ensure that InP access capabilities are not abused to access private third party data. Among others, this means that also the threat of malicious insiders within the InP organization must be considered and suitably mitigated.

7.6.2 Enhanced Slice Isolation Infrastructure

An industry vertical may need to provide 5G connectivity for, as an example, Industry 4.0 (I 4.0) applications on the vertical's factory plants, and for this purpose rent a suitable network slice from a network operator providing the infrastructure along with the NFs that constitute the network slice. The vertical becomes a tenant of the network operator. The slice provides connectivity between the I 4.0 devices and a Data Network (DN) operated by the vertical that hosts the I 4.0 application servers. If the applications are highly sensitive, the vertical may not simply trust the network operator to ensure isolation, but desire better, enforceable isolation. As described in [28] and outlined in the following, there are different ways how this can be achieved.

7.6.2.1 Over-the-Top Security

An obvious option for the vertical in this case is the use of Over-The-Top (OTT) security, where end-to-end security associations between mobile devices and application servers are established on top of the network protocol layers. A common example for OTT security applied by subscribers of a mobile data service is the usage of TLS [13] for secure web browsing. In the industry vertical scenario, OTT security requires that the vertical operates an own database holding identifiers and credentials for all the connected devices, and provisions these identifiers and credentials on the mobile devices. This database must be implemented on a platform controlled by the vertical itself – running it in some InP's cloud would defeat the desired isolation.

OTT security does not contribute to resource isolation, but it can provide confidentiality and integrity protection of the tenant traffic even against the network operator. However, the network operator will still be able to retrieve a lot of meta-data, e.g. which subscribers get access to the vertical's data network, location, communication times and volumes, communication relationships and so on.

7.6.3 Other Aspects

The slicing concept is meant to provide virtual networks optimized for specific applications. This optimization may also apply to security mechanisms. Moreover, slices may have individual security assurance levels, focusing security assurance efforts on those applications where it is most meaningful. The assurance level of a slice is limited by the assurance level of the underlying infrastructure.

While a single slice may be lean and easy to operate and control, a whole network hosting many slices is much more complex than a traditional, un-sliced network. The attack surface will therefore be larger in a network comprising multiple slices, in particular in network multi-tenancy scenarios, where third parties like industry verticals operate individual slices. Interfaces and procedures provided by the network operator to third parties for managing slices must be carefully secured and management operations must be authorized. Similarly, when slices controlled by verticals have interfaces to common parts of the network controlled by the operator, these interfaces must be secured and access must be authorized. The same holds for inter-slice communication in the sense of communication between (virtual) networks operated by different organizations (network operators or third parties). Other slicing specific procedures that may increase the attack surface, and must therefore be secured, diligently comprise slice selection procedures and slicing-specific authentication and authorization procedures.

With sound slice isolation, a DoS attack on a single slice will not affect other slices in the same network. However, a slice that is engineered for a "small" application, like for a small number of mobiles and/or a low amount of traffic may rather easily be overwhelmed by a flooding attack, compared to a large, un-sliced network. DoS attacks against specific slices can probably be carried out even if the slices are not explicitly made visible by the network, because the existence of specific slices may be hard to conceal in the longer term.

A threat that may arise in case of simultaneous connectivity of a UE to several slices is the forwarding of malicious traffic between different slices via the UE. However, this would mean the UE acts maliciously, probably because of some malware having been installed. As a network or slice needs anyway be protected against malicious or erroneous UE behavior, no specific security measures may be required against this threat.

7.7 Private Network Infrastructure

To achieve a higher degree of isolation as provided by OTT security, an industry vertical may deploy its own 5G network infrastructure. In principle, a complete private 5G network may be deployed in a factory, at a harbor or airport, if the issue of spectrum usage is solved, e.g. unlicensed spectrum is used or licensed spectrum is allocated to the vertical in a certain region or area. It allows such private networks to avoid deployment of USIMs, also allows them not to use 3GPP (AKA) security mechanisms but any suitable EAP method. While some of the I 4.0 devices may never leave a plant, others may also need connectivity outside the plant and may require service continuity for some services in use.

References

1 Nokia White Paper, "Security challenges and opportunities for 5G mobile networks", 2017, available at https://resources.nokia.com/asset/201049https://www.ngmn.org/5g-white-paper/5g-white-paper.html.
2 Next Generation Mobile Network Alliance, "5G White Paper", Version 1.0 Feb 17, 2015, available at https://www.ngmn.org/5g-white-paper/5g-white-paper.html.
3 Next Generation Mobile Network Alliance, "5G Security Recommendations Package #1", "5G Security Recommendations Package #2: Network Slicing", "5G Security – Mobile Edge Computing/Low Latency/Consistent User Experience", (available at https://www.ngmn.org/de/publications/technical.html).
4 Forsberg, D., Horn, G., Moeller, W.D., and Niemi, V. (2013). *LTE Security*, 2e. Wiley.
5 3GPP TR 33.821: "Rationale and track of security decisions in Long Term Evolved (LTE) RAN/3GPP System Architecture Evolution (SAE)".
6 3GPP TS 22.261: "Service requirements for the 5G system".
7 ETSI GS NFV 004: "Network Functions Virtualization; Virtualization Requirements".
8 GS NFV-SEC 001: "Network Functions Virtualization; NFV Security; Problem Statement".
9 GS NFV-SEC 003: "Network Functions Virtualization; NFV Security; Security and Trust Guidance".
10 GS NFV-SEC 012: "Network Functions Virtualization; Release 3; Security; System architecture specification for execution of sensitive NFV components".
11 ISO/IEC 11889-1:2015: "Information technology – Trusted platform module library – Part 1: Architecture".
12 NIST SP 800-162: "Guide to Attribute Based Access Control (ABAC) Definition and Considerations".
13 IETF RFC 5246: "The Transport Layer Security (TLS) Protocol".
14 Open Networking Foundation TR 511: "Principles and Practices for Securing Software-Defined Networks".
15 3GPP TS 33.401: "3GPP System Architecture Evolution (SAE); Security architecture".
16 3GPP TS 33.402: "3GPP System Architecture Evolution (SAE); Security aspects of non-3GPP accesses".

17 Nokia White Paper: "Trusted NFV systems", available at https://resources.ext.nokia.com/asset/201400.
18 3GPP TS 33.501: "Security architecture and Procedures for 5G System".
19 3GPP TR 33.899: "Study on the security aspects of the next generation system".
20 IETF RFC 3748: "Extensible Authentication Protocol (EAP)".
21 IETF RFC 5488: "Improved Extensible Authentication Protocol Method for Third Generation Authentication and Key Agreement (EAP-AKA')".
22 IETF RFC 7296: "Internet Key Exchange Protocol Version 2 (IKEv2)".
23 IETF RFC 4301: "Security Architecture for the Internet Protocol".
24 IETF RFC 7540: "Hypertext Transfer Protocol Version 2 (HTTP/2)".
25 IETF RFC 6733: "Diameter Base Protocol".
26 IETF RFC 6749: "OAuth2.0 Authorization Framework".
27 IETF RFC 4251: "The Secure Shell (SSH) Protocol Architecture".
28 Schneider, P., Mannweiler, C., and Kerboef, S. (2018). Providing strong 5G mobile network slice isolation for highly sensitive third-party services. In: *Proceedings of the IEEE WCNC*. ISBN: 978-1-5386-4068-5.
29 Nokia, "Building Secure Telco Clouds", 2014, available at https://resources.nokia.com/asset/200289.

8

Critical Machine Type Communication

Zexian Li[1] *and Rainer Liebhart*[2]

[1] *Nokia, Espoo, Finland*
[2] *Nokia, Munich, Germany*

8.1 Introduction

As outlined in Chapter 1, key differentiators between long-term evolution (LTE) and 5G are the capability to provide ultra-low latency and extreme high reliability connectivity service to applications. In this chapter we will describe how these features can be used for divergent use cases and most importantly what kind of solutions in the radio and core network will leverage low latency and high reliability.

With low latency we mean usually a few milliseconds (in the extreme case 1 ms) end-to-end (E2E) latency between client and server on the user plane. With ultra-high reliability of at least 99.999%, 99.999 9% or even higher we mean reliability of the connection interface between device and network. Thus, reliability is an issue for keeping a wireless connection active to reach the device (note that also fronthaul and backhaul connections are relevant for device reachability) allowing for a very high packet reception rate. On the other hand, network availability is also crucial for many applications. This refers to the downtime of the network and its components such as base stations (BSs), core network elements, registers, authentication and application servers (ASs).

8.1.1 Industrial Automation

The most important and well-known use cases requiring ultra-low latency and extreme high reliability are remote health-care and surgery, some industrial applications like collaborative, mobile robots, remote control and autonomous driving. Table 8.1 (see also Table 1.3 in Chapter 1) depicts the relevant key performance indicators (KPIs) of these use cases.

Industrial applications in a limited area or region, e.g. in a factory, harbor, airport, campus and the like seems most promising to achieve a positive business case with low latency and high reliable 5G connectivity. Deploying a nationwide Ultra-Reliable Low Latency Communication (URLLC) network is too costly as it requires a high density of radio sites and many edge clouds hosting URLLC sensitive applications.

5G for the Connected World, First Edition.
Edited by Devaki Chandramouli, Rainer Liebhart and Juho Pirskanen.

Table 8.1 Low latency and high reliability use cases.

Scenario	End-to-end latency	Jitter	Survival time	Communication service availability (%)	Reliability (%)
Tactile interaction	0.5 ms	—	—	99.999	99.999
Discrete automation: motion control	1 ms	1 μs	0 ms	99.9999	99.9999
Electricity distribution: high voltage	5 ms	1 ms	10 ms	99.9999	99.9999
Remote control	5 ms	—	—	99.999	99.999
Discrete automation	10 ms	100 μs	0 ms	99.99	99.99
Intelligent transport systems: infrastructure backhaul	10 ms	20 ms	100 ms	99.9999	99.9999
Process automation: remote control	50 ms	20 ms	100 ms	99.9999	99.9999
Process automation: monitoring	50 ms	20 ms	100 ms	99.9	99.9
Electricity distribution: medium voltage	25 ms	25 ms	25 ms	99.9	99.9

8.1.1.1 Discrete Factory Automation/Motion Control

These are closed-loop control applications requiring ultra-low latency and high reliability, e.g. use of collaborative robots in a factory:

- *Latency.* From <1 to 10 ms.
- *Data rate.* Low as messages typically small (<50 bytes).
- *Reliability.* Ultra high, ranging up to $<10^{-9}$.
- *Accuracy.* Can be ultra-high, down to cm range.
- *Jitter.* Variable though ultra-high, linked to isochronous operation and telegram delivery – from <1 μs to 10 ms.

8.1.1.2 Process Automation

These are supervisory and open-loop control applications, as well as process monitoring and tracking operations requiring URLLC for a potentially high number of connected devices.

- *Latency.* From a few ms to 1 s.
- *Data rate.* Low as each transaction is typically <100 bytes.
- *Reliability.* Packet loss rate $<10^{-5}$.
- Massive number of low power sensors.

8.1.1.3 Tele Operation

These applications require URLLC like remote control of machines and vehicles.

- *Latency.* ~50–150 ms, influenced by presence of haptic/tactile feedback.
- *Data rate.* Low in downlink (DL) for small control commands, potentially very high in uplink (UL) for video.
- *Reliability.* Variable from 10^{-3} to 10^{-5}.
- *Accuracy.* Can be ultra-high, down to cm range.

8.1.2 V2X

Another future technology where 5G (but also LTE) can play a key role is Vehicle-to-X communication (V2X), especially Vehicle-to-Vehicle (V2V) and Vehicle-to-Network/Vehicle-to-Infrastructure (V2N/V2I). While in case of V2V vehicles are broadcasting information directly to their neighbors, e.g. once the vehicle is breaking or detects an obstacle (pedestrian or another vehicle) in front, V2N/V2I allows exchanging information in a wider area, e.g. regarding an accident or obstacle around a corner or junction. For V2V following limitations apply:

- Sensors provide only a local environment support, and cannot "look" beyond other vehicles, obstacles and corners.
- Some sensors can only offer decreased performance or fail in adverse weather conditions.
- Distance meters are limited by a considerable inherent delay.

In contrary, network assisted information sharing

- can take a wide area into account;
- might provide information faster than via sensor measurements;
- can lead to a higher reliability when combining information received via sensors and wireless networks.

The expectation is that V2X communication in general leads to higher road efficiency as cars can maintain lower distances at higher speeds, increased road safety as cars get more, enriched and faster information about their environment and potential hazards, and improved passengers' comfort.

Use cases where application servers may be involved to receive and distribute information to vehicles are e.g.: emergency vehicle priority, traffic light optimal speed advisory, traffic sign in the car, traffic jam warning and Software/map updates. These kind of use cases benefit from Edge Computing capabilities where application servers are hosted in Edge Clouds close (typically 5–25 km) to the radio sites (see Figure 8.1).

When talking about full autonomous driving (automation level 5, see Chapter 1.5) wireless technologies like 5G can support and complement in-board systems like sensors and cameras in controlling the vehicle, but not replace them. This means that

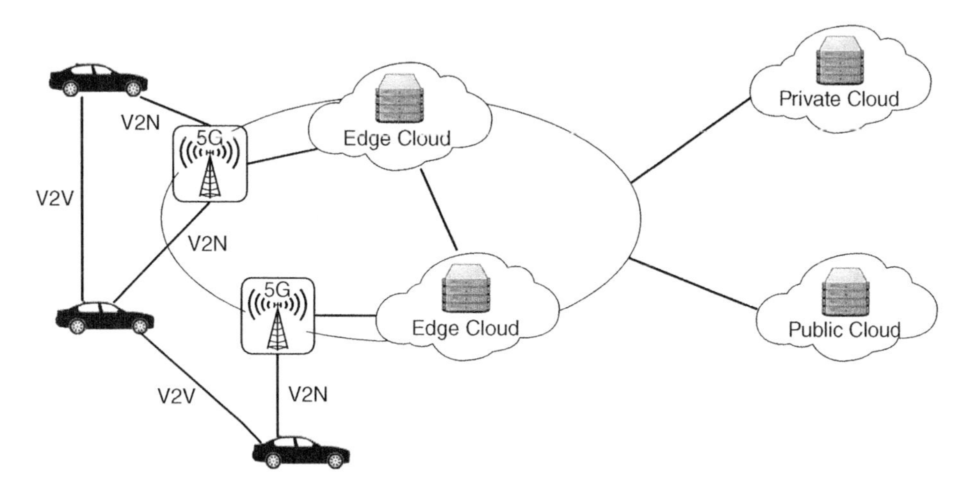

Figure 8.1 V2X communication and Edge Clouds.

controlling a vehicle must be possible within and outside network coverage, without declining the safety of passengers and vehicles. The network can allow usage of very dense platoons of vehicles of less than 1 m, e.g. several trucks following each other where on the first truck, the platoon leader, is equipped with a human driver, while outside radio coverage the platoon density is automatically increased to several meters. The smaller the distance in such a platoon, the more vehicles can be "packed" together, the less drivers are needed, the less fuel is needed. Very dense platoons require however extreme high reliable networks and ultra-low latency (V2V and V2N) communication between the platoon members, which comes only with 5G.

8.1.3 Public Safety

Mission critical communication in the context of Public Safety has several flavors: mission critical push-to-talk, mission critical data and mission critical video. Public Safety communication over LTE was specified by 3rd Generation Partnership Project (3GPP) from Rel-12 onwards. It contains enablers like Group and Device-to-Device communication over LTE, as well as Mission Critical Push to Talk (MCPTT) application based on the Internet Protocol Multimedia Subsystem (IMS).

Mission critical communication has more stringent requirements regarding packet delay than normal voice or data communication, e.g. mission critical voice user plane requires 75 ms, mission critical signaling 60 ms for packet delay. With 5G, in general, we can expect much lower latency on the user plane, faster connection setup times, which is important when granting the floor in case of push-to-talk, and higher reliability. 3GPP may specify a 5G broadcast solution only in Rel-16. Based on the experience with LTE based Multimedia Broadcast/Multicast Service (MBMS) the 5G broadcast solution can be designed from the very beginning allowing for more flexible and efficient deployments (e.g. deployments in smaller areas).

Last but not the least 5G Network Slicing allows operators to build dedicated slices for Public Safety networks, fulfilling the specific requirements Public Safety authorities might have. Such a slice is separated from other slices, thus provides not only higher reliability and lower latency for Public Safety services but protects the Public Safety slice against attacks from users of other slices and can provide a higher degree of isolation (e.g. guard the Public Safety slice from user plane traffic generated by other slices, also guard the slice from attacks to other slices etc.) and security by implementing special cryptographic algorithms.

8.2 Key Performance Indicators

Use cases and their requirements regarding throughput, latency, jitter, reliability, availability, etc. are extensively discussed in Chapter 1 of this book (see Sections 1.3–1.5). KPIs for the next generation radio access network were studied by 3GPP in TR 38.913 [1]. Regarding latency and reliability TR 38.913 provides the following target values:

8.2.1 Control Plane Latency

The target for control plane (CP) latency should be 10 ms. Control plane latency in this context means the needed time to move the device from idle to active state.

8.2.2 User Plane Latency

User plane latency refers to the time it takes to successfully deliver an application layer packet from the radio protocol layer 2/3 ingress point to the radio protocol layer 2/3 egress point via the radio interface in both UL and DL directions. For URLLC, the target for user plane latency should be 0.5 ms for both UL and DL transmission. For enhanced Mobile Broadband (eMBB), the target for user plane latency should be 4 ms for both UL and DL transmission.

8.2.3 Latency for Infrequent Small Packets

For infrequent application layer small packet transfer, latency refers to the time it takes to successfully deliver an application layer packet from the radio protocol layer 2/3 ingress point to the radio protocol layer 2/3 egress point, when the mobile device starts from its most "battery efficient" state. This kind of latency shall be no worse than 10 seconds on the UL for a 20-byte application packet measured at the maximum coupling loss of 164 dB.

8.2.4 Reliability

Reliability refers to the success probability of transmitting several bytes within a certain delay. A general URLLC reliability requirement for one transmission of a packet is $1-10^{-5}$ for 32 bytes with a user plane latency of 1 ms.

In case of eV2X (enhanced Vehicle-to-X communication), communication availability, resiliency and user plane latency for delivery of a 300 bytes packet, are as follows: Reliability is $1-10^{-5}$, user plane latency is 3–10 ms, for direct communication via side link and communication range of a few meters, and the same when the packet is relayed via the base station.

8.3 Solutions

8.3.1 System Design Challenges

Up to now, spectral efficiency has been the main design target for all previous wireless cellular communication systems, including LTE and earlier generations like 2G and 3G. URLLC, which has been widely envisioned as the enabler for supporting various newly emerging vertical services and applications, brings new challenges for the overall 5G system design. As discussed in Chapter 8.2, in 3GPP Rel-15 New Radio (NR) [1], the design target is to support communication reliability corresponding to Block Error Rate (BLER) of 10^{-5} and up to 1 ms radio protocol layer latency for delivering short packets with a packet length of 32 bytes. The requirements will become more stringent when considering the coming 3GPP Rel-16 as discussed for example in TR 22.804 [2]. From design point of view, the difficulty does not come with one stringent requirement, either high reliability or low latency. However, with the simultaneous requirements on ultra-high reliability and low latency, we are facing much more challenges. As can be easily understood, the stringent requirements will impact the design of data channels, e.g. retransmission schemes, building necessary redundancy and so on. Furthermore,

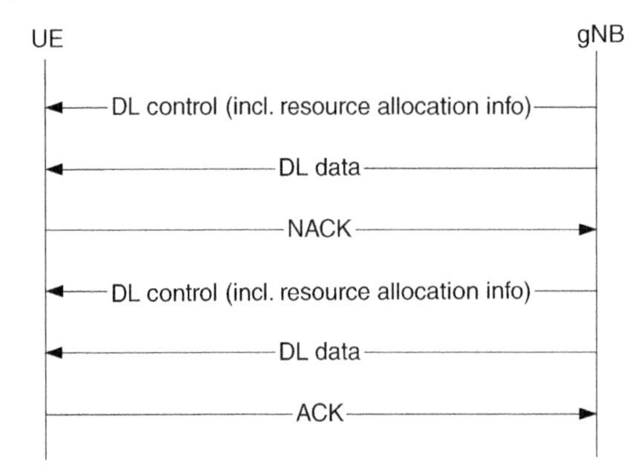

Figure 8.2 Example procedure for DL data transmission.

URLLC is putting more strict requirements on the control channel design as well. In this section, we will focus on the requirements of control channels.

Figure 8.2 illustrates a simplified system model for DL data transmission after an URLLC UE (User Equipment) has established the link with the gNB (5G NodeB). In Figure 8.2, it is assumed that with two transmissions the target BLER of 10^{-5} can be achieved.

Potentially all steps in the figure can impact the finally achievable reliability rate. Before data transmission, gNB sends DL control information including resource grant to the UE. Only after correctly decoded control information, the UE can try to decode the data packet sent within the allocated resource and provide feedback to the gNB. The feedback can be either an ACK (Acknowledgment) or a NACK (negative Acknowledgment) to indicate the success or failure in the data packet reception. In the figure it is assumed that NACK is transmitted after the first transmission due to failure decoding at UE side. After receiving NACK, gNB starts to retransmit the same data packet with the same or different transmission format. The gNB again instructs the UE to monitor the allocated resources for the data retransmission by sending a new resource allocation command. In case UE can get the data packet correctly, an ACK is sent.

As discussed in [3], errors in decoding the control channel can significantly reduce the overall E2E communication performance, especially the reliability. To be more specific, in case the UE is not able to decode the control channel information, e.g. resource allocation information, correctly, the UE is not aware of the present transmitted data packet and therefore it will not even try to decode the data packet. Consequently, there is no feedback from the UE, i.e. neither ACK nor NACK will be transmitted and at gNB side, such state could be recognized as a Discontinuous Transmission (DTX) and results in retransmission of the same data packet to the UE. In addition, there is the possibility that the gNB falsely recognizes DTX or NACK feedback (in case the first data packet transmission is not successfully decoded at the UE) as ACK. In this case, gNB will not take any further action including retransmission as it assumes the data packet was correctly received. However, in both cases, UE has not managed to decode the received packet correctly. Another way to improve reliability in such cases is to implement retransmission of the same physical data unit (PDU) at higher layers, e.g. using Radio Link Control (RLC) layer Automatic Repeat Request (ARQ) retransmission. However, this leads to increased latency.

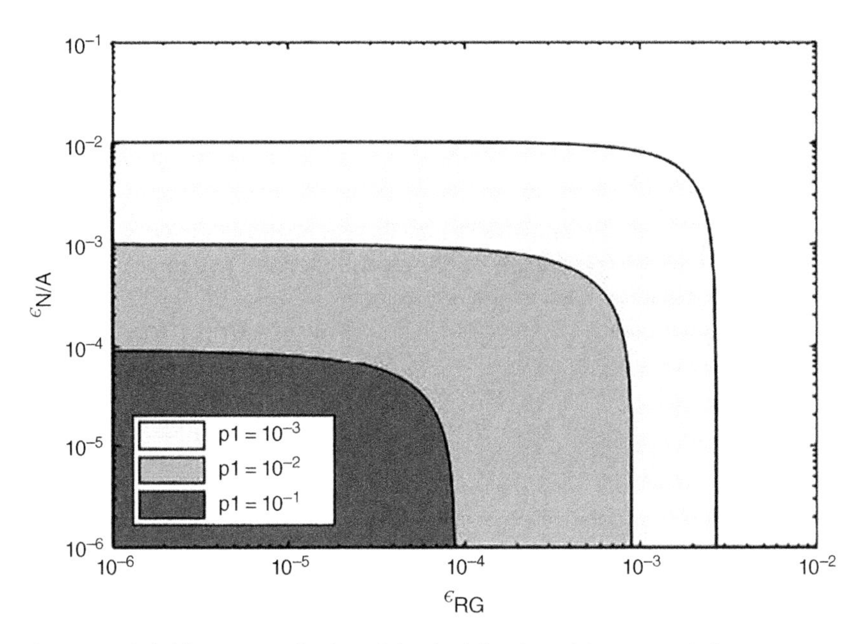

Figure 8.3 Reliability regions for downlink scheduling-based data transmissions.

Figure 8.3 illustrates the reliability regions for supporting reliability of $1 - 10^{-5}$ in case of cellular communications. In Figure 8.3 we denote ε_{RG} as the error rate of decoding the resource allocation information, p_1 as the BLER of the initial data transmission, and $\varepsilon_{N/A}$ as the error rate of decoding NACK as an ACK and decoding DTX as an ACK. It is assumed that there is sufficient time for two transmissions, i.e. in addition to the first transmission, the second transmission can be utilized to further increase reliability within the latency constraints. A residual BLER of 10^{-5} upon the data retransmission is assumed. In case NACK is detected, retransmission is done in a way that the residual BLER target is achieved upon soft combining the received information from both the initial transmission and the retransmission. In case of detecting a DTX, the retransmission is performed with more robust transmission to compensate the lack of soft information from the initial transmission at the receiver. Of course, in case NACK is received, it is also possible that a more robust Modulation and Coding Scheme (MCS) is selected to guarantee that the data packet can be decoded correctly after combining the first and second transmission. It is shown that a target with BLER $= 10^{-1}$ for performing the initial transmission entails at most 10^{-4} BLER for ε_{RG} and $\varepsilon_{N/A}$. Such control channel reliabilities are not supported by LTE and might cause a large signaling overhead in the system. The control channel error rate constraints can be relaxed by performing the initial data transmission using a more conservative MCS. For instance, the BLER constraint for the control signals, i.e. ε_{RG} and $\varepsilon_{N/A}$, can be reduced to 10^{-3} in case of adopting a 10^{-2} BLER target for the initial data transmission. In other words, there is a certain tradeoff between the reliability of control and data channels, suggesting that the link adaptation for data channels needs to consider the errors of control channels.

In addition to error cases regarding the DL control channel, another type of error which can reduce the overall reliability performance is related to UL ACK/NACK signal detection at the gNB. The erroneous decoding of an ACK as a NACK triggers unnecessary data retransmissions, resulting in waste of resources without any damage from reliability aspect. On the contrary, the erroneous decoding of a NACK as an ACK signal leads to absence of a necessary retransmission, which results in reduced reliability performance. It is easy to understand that the errors of ACK/NACK affect only the retransmission round.

One aspect related to DL data transmission, which has not been included in Figure 8.3, is how the gNB can determine the suitable MCS selection (corresponding to different resource sizes) for the DL data transmission. In the current LTE design, the eNB scheduler can use a certain method to select the MCS index based on the Channel Quality Indicator (CQI) report. For URLLC CQI reporting, the error of decoding a given CQI value wrongly as a lower value results in a lower level of MCS selection, which is not optimal from spectral efficiency point of view, however, there is no penalty on latency and reliability as the success probability rate is increased. On the other hand, the gNB might wrongly decode the received CQI as a higher value. Such an error leads to the selection of a higher MCS level for data transmission and hence a reduced communication reliability. This error type brings much more damage toward reliability.

Similarly, for UL data transmission, there are requirements on various control channels as well, for example scheduling request (SR) in case with scheduling based UL data transmission, resource grant information transmission and feedback. For instance, one source of error is missing the resource grant resulting in not sending the data packet in UL as the UE has no idea about which resource blocks (RBs) can be used for UL data transmission. This error might happen during the initial transmission round and/or retransmission rounds. In UL, the gNB potentially can notify this event when it does not receive any data signal from the UE. In case the gNB identifies the missing of resource grant for the initial transmission round, it can allocate additional radio resources for the retransmission round to compensate the loss of initial transmission.

In the following sections, we will discuss various technology components to improve system performance especially in terms of latency and reliability, which is kind of a tool box for URLLC support. Although, we are discussing latency and reliability aspects separately for an easier understanding, these two aspects are not independent, but rather tightly correlated. For example, the technology components mainly targeted for low latency support could bring more retransmission opportunity as well, which brings gains in terms of reliability.

8.3.2 Low Latency

8.3.2.1 Architecture

Edge Clouds and Edge Computing Ultra-low latency on the user plane requires the two endpoints of the communication path, e.g. application client on the device and Application Server (AS) in the network, being close together. Speed of light is limited to 300 000 km s^{-1} and in fiber it is only 200 000 km s^{-1}. Consequently, (ultra) low latency applications cannot be hosted in the Internet like most of today's applications, but in so-called Edge Clouds (sometimes referred to as Edge Computing). Assuming a maximum of 1 ms latency on the air interface between device and 5G base station for UL and DL, mainly the distance between Edge Cloud and base station determines the overall latency (an additional factor is the processing time in the evolved nodes). An

Edge Cloud in this context consists of one or several small data centers with potentially few racks, which can host local User Plane Functions (UPFs), Multi-Access Edge Computing platforms (MEC), local Application Servers (e.g. content cashing or video streaming servers) and 5G Radio Central Units (CUs). From 3GPP point of view the MEC platform is considered as an Application Function (AF).

A CU implements non-realtime higher layers of the radio protocol stack (Packet Data Convergence Protocol [PDCP] and Radio Resource Control [RRC]), while the Distributed Units (DUs) are implementing the realtime lower layer parts of the radio stack (up to the Medium Access Control (MAC) layer). This DU/CU split was specified by 3GPP in Rel-15 with F1 as interface between DU and CU (see Chapter 4). Core network elements that are part of the control plane are usually hosted in central Telco Clouds. Also, UPFs can be deployed in these central clouds, if they are used to access centralized services in the Internet or IMS where ultra-low latency plays no role. The next Figure 8.4 provides a logical network view including Edge Clouds and central Telco Clouds and their inter-connection.

These Telco Clouds can consist of hundreds or thousands of racks. Depending on the country size only very few or up to tens of them exist. Figure 8.5 shows the typical distances between radio site, DU, CU and central Telco Cloud. Distances may vary depending on transmission capabilities, latency requirements and number of sites.

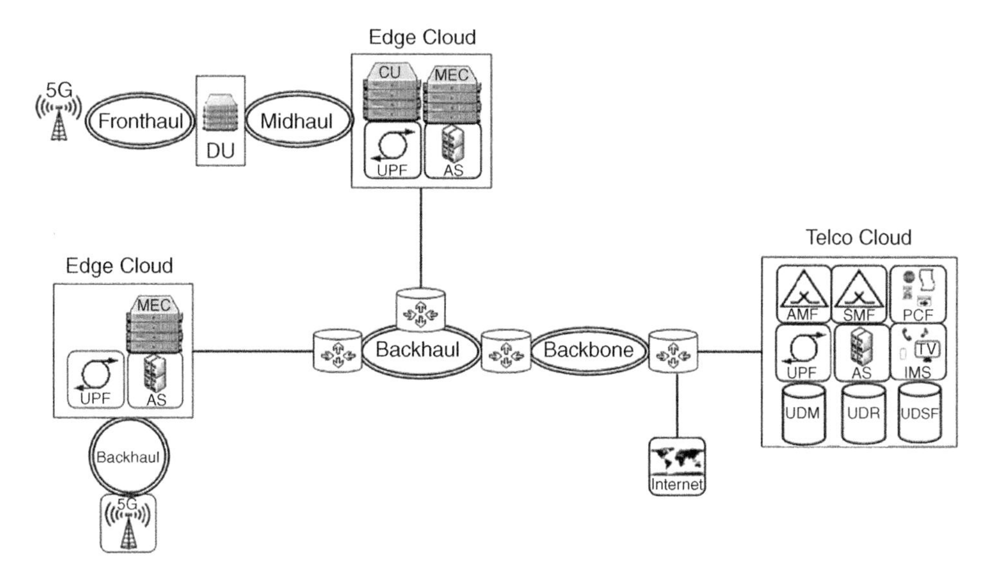

Figure 8.4 Logical network view with Edge and Telco Clouds.

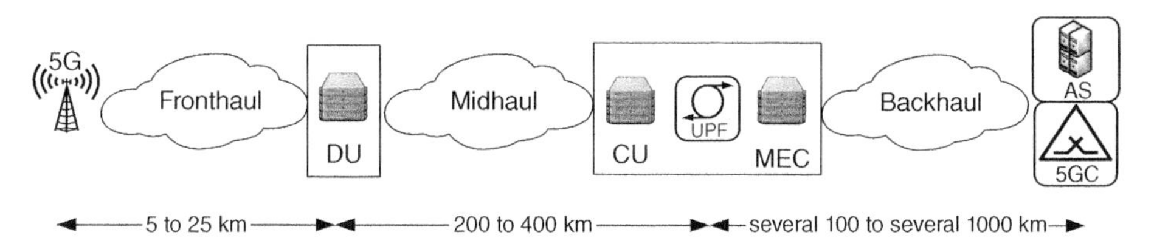

Figure 8.5 Network distances.

To allow for 1 ms latency between device and application server a different configuration is necessary where DU, CU are collapsed and hosted with a local UPF, potentially MEC and a local application server in the same Edge Cloud. This deployment is shown in Figure 8.6.

Micro-Operator Networks Deployment of Edge Clouds is one way to reduce latency. Another way is deployment of local or private networks (in a factory, at a harbor or airport) by micro-operators providing URLLC capabilities as e.g. required by safety critical applications. One example illustration of a micro-operator for vertical usage is illustrated in Figure 8.7.

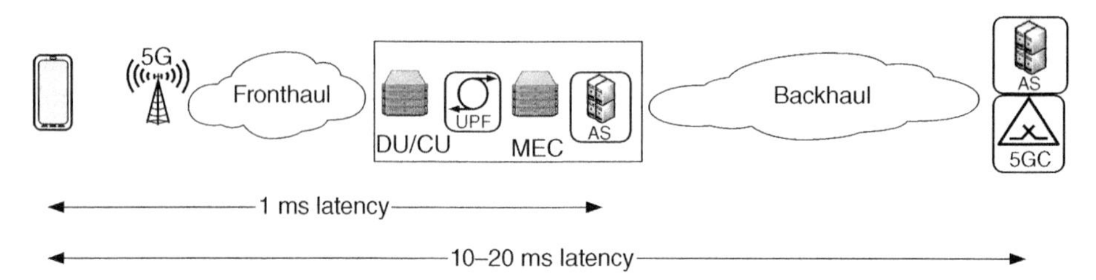

Figure 8.6 Low latency configuration.

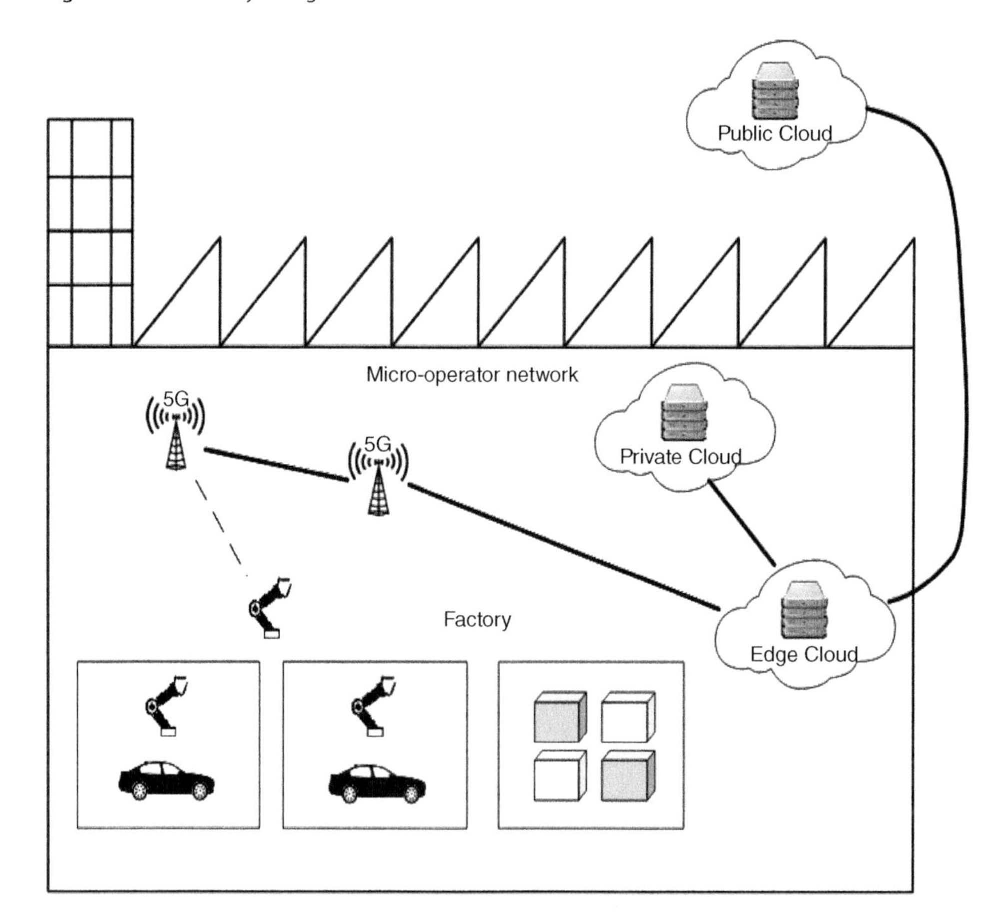

Figure 8.7 Micro-operator network for verticals.

As shown, the micro-operator can deploy a dedicated network to support specific services that might have very strict requirements in terms of latency and reliability. At the same time, there could be data traffic, for example maintenance traffic (SW updates), that is destined to a wide area or public network. Using the concept of micro-operators, the overall E2E latency among devices within the restricted area can be significantly reduced by moving the application layer processing close to end users. Micro-operator networks can be deployed and maintained by different parties: the public mobile network operator, the factory owner, specialized networking companies, 3rd parties running the factory infrastructure, etc.

Multi-Access Edge Computing MEC in general is meant to bring computing and IT capabilities to the edge of the (mobile) network, close to the consumers, thus becoming the "cloud" for real-time and personalized mobile services and improving user experience. However, from deployment perspective MEC is not limited to the network edge, it can also be deployed at Enterprise premises or at central locations (central Telco Cloud), depending on the application needs. MEC deployments allow for reduced latency and support local context delivery, which is key to provide critical communication capabilities, robust and highly responsive connectivity to Internet of Thing (IoT) applications. The MEC platform consists of Application Enablement, enabling authorized applications to access the network, Application Programming Interfaces (APIs) and management capabilities to run applications as software-only entities within a network or part of it. APIs can provide access to radio network information (allowing relocation of applications in case of user mobility), location information, bandwidth management data and UE identities. Figure 8.8 shows the principle MEC platform components.

8.3.2.2 Radio Network

In this section our objective is to discuss the latency reduction technology components, which have been accepted or are discussed within 3GPP with regards to the radio network. Table 8.2 summarizes the key technology components, which have been specified in 3GPP Rel-15 or are currently under discussion.

New Numerology and Short TTI Duration 3GPP has introduced a very flexible frame structure for 5G NR which offers different possibilities to shorten the Transmission

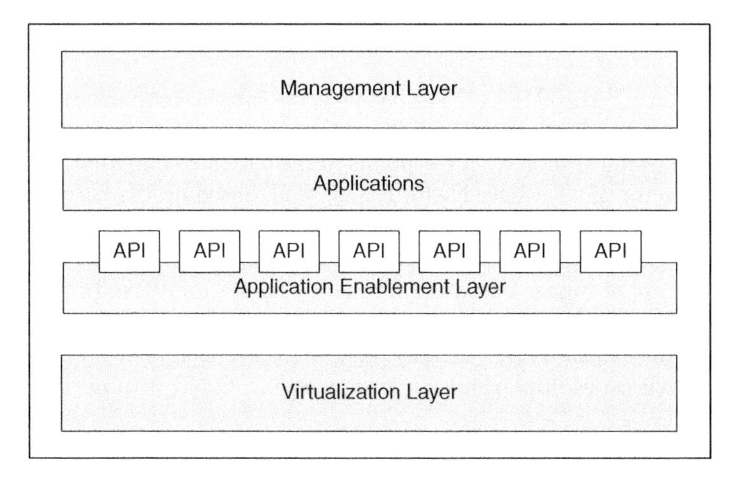

Figure 8.8 MEC platform architecture.

Table 8.2 List of technology components for latency reduction.

Technology components	Short description
New numerology, short Transmission Time Interval (TTI)	Short TTI size to ensure fast transmission of data packets. 3GPP NR supports reduced slot length with flexible subcarrier spacing (SCS) and further non-slot transmission of 1–13 symbols.
Maximum data burst size	Largest amount of data the 5G-AN is required to transmit within a period of 5G-AN PDB, the 5G-AN part of the packet delay budget.
Frame structure with bi-directional slots	By introducing one or multiple switching points within one slot, the time division multiplexing of DL/UL control symbols and data symbols in one slot is possible, allowing fast and energy efficient pipeline processing of control and data in the receiver.
Scheduling policy	Non-slot based scheduling: minimum scheduling unit is not on slot level, but on symbol(s) level (mini-slot level).
	Preemption scheduling: Prioritized scheduling of data packets with low latency constraints to minimize gNB queuing delays. Pre-emption of eMBB traffic for URLLC data packet transmissions.
Grant-free UL transmission	Grant-free UL transmission scheme avoids regular handshake delays between UE and gNB including at least scheduling request and UL resource grant allocation, and moreover relaxes the stringent requirements on the reliable resource grant.
UE and gNB processing time	Reduced UE and gNB processing time to ensure fast creation of transport blocks for transmission as well as fast processing at the receiver end for feedback transmission.

Table 8.3 OFDM numerologies for 5G NR (normal CP length).

Subcarrier spacing (kHz)	15	30	60	120	240
Symbol duration (μs)	66.7	33.3	16.7	8.33	4.17
Nominal normal CP (μs)	4.7	2.3	1.2	0.59	0.29
Min scheduling interval (symbols)	14	14	14	14	28
Min scheduling interval (slots)	1	1	1	1	2
Min scheduling interval (ms)	1	0.5	0.25	0.125	0.125

Time Interval (TTI) duration compared to LTE. For instance, the subcarrier spacing can be flexibly configured to support 5G operation in different frequency bands. The 15 kHz Sub-Carrier Spacing (SCS) (same as in LTE) corresponds to the baseline configuration, and can be scaled with a factor of 2^N, where $N \in [0, 1, 2, 3, 4, 5]$. Based on the current agreement sub-carrier spacings of between 15 and 120 kHz are defined in Release 15. In later releases, sub-carrier spacing greater than 480 kHz could be accommodated as well, especially in the context of higher operating frequency like in mmWave bands. With higher sub-carrier spacing, more symbols can be accommodated in a sub-frame, resulting in lower acquisition time. Table 8.3 illustrates the current agreed OFDM (Orthogonal Frequency-Division Multiplexing) numerologies for 5G NR with normal CP length.

On top of this, the number of OFDM symbols per TTI can also vary. Within NR, both slot based and non-slot based scheduling mechanisms are supported. To be more specific, the users can be scheduled with resource on slot level of 14 OFDM symbols, and/or on non-slot level (mini-slot based scheduling) where the length of a mini-slot can be configured between 1 and 13 symbols. Short TTIs can therefore be built by reducing the symbol duration (increasing the SCS) and/or reducing the number of symbols per TTI. For example, a TTI duration of 0.125 ms can be obtained by scheduling users on a slot resolution in case of 120 kHz subcarrier spacing (N = 3). As another alternative, it is also possible to use 15 kHz SCS (N = 0) and schedule users with a non-slot based resource of one to three symbols (~71–222 μs), leaving room for processing time and HARQ (Hybrid Automatic Repeat Request) feedback transmissions within 1 ms (see [4]). The non-slot based scheduling is therefore useful for time-critical services, while other less time-critical services can still be served with longer TTI durations.

Maximum Data Burst Volume When there is a need to transmit packets with latency in the order of 1 ms to 10 ms and to support reliability in the order of 10^{-5}, it is essential to define the largest amount of data the 5G-AN is required to transmit within a period of the 5G-AN Packet Delay Budget (PDB), the 5G-AN part of the PDB. This is referred to as Maximum Data Burst Volume. Each Delay Critical Guaranteed Bitrate (GBR) Quality of Service (QoS) flow is associated with a Maximum Data Burst Volume. The Maximum Data Burst Volume may be signaled together with 5G Quality of Service Identifiers (5QIs) to the (R)AN, and if it is not signaled, default values apply for standardized 5QIs QoS characteristics.

For delay critical GBR QoS flows, a packet delayed greater than the PDB is counted as lost packet, if the transmitted data burst is less than the Maximum Data Burst Volume within the period of PDB and the data rate for the QoS flow is not exceeding the Guaranteed Flow Bitrate (GFBR). The PER (Packet Error Rate) also defines a value for reliability of the latency target set by the PDB.

Frame Structure with Bi-directional Slots As shown in Table 8.3, 5G NR frames can be built on the configurable slot types on a scalable numerology designed for a wide range of scenarios and use cases fulfilling a variety of requirements. So far three types of slots were specified: DL only slot, UL only slot and bi-directional slot (see Figure 8.9) providing low latency support for both Frequency Division Duplex (FDD) and Time Division Duplex (TDD). Bi-directional slots are a viable option to achieve low latency in TDD where multiple switching points can be configured within one slot. The time division multiplexing of DL/UL control and data symbols in one slot allows fast and energy efficient pipeline processing of control and user data in the receiver. Decoupling the control and user data provides flexible and rapid UL/DL scheduling and enables very low latency of HARQ acknowledgements without dependency on the control channels from predetermined TDD UL/DL configurations. Depending on UE capabilities and the support of aggressive processing time requirements, a very short time delay in symbols for the gap between DL assignment and DL data transmission, the gap between DL data and acknowledgement for DL data and the gap between UL assignment and UL data transmission could be supported.

In addition to the low latency slot structure, dynamic TDD would allow for flexible assignment of UL and DL resources, further reducing the end user experienced latency.

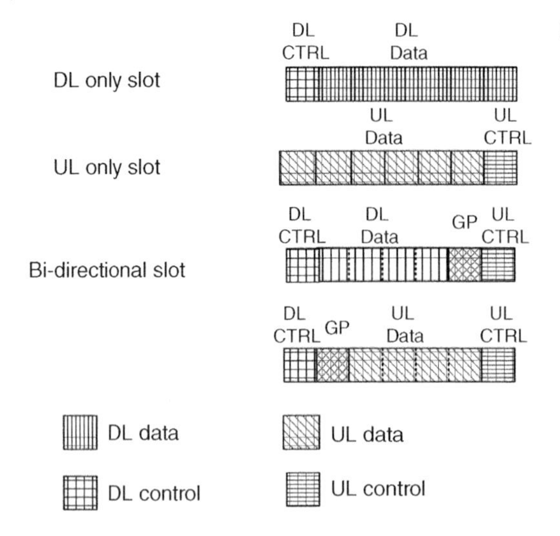

Figure 8.9 Flexible slot structure in 5G NR.

Scheduling Policy During the design of 5G NR, the stringent URLLC requirements were considered at the very beginning of the design phase. Scheduling is another key element for latency reduction and below we will discuss a group of technology components especially for latency reduction based on scheduling algorithms.

Non-slot Based Scheduling As mentioned before, non-slot based scheduling (mini-slot) is a key enabler within 5G NR which is useful in various scenarios including low latency transmission, multi-beam operation in the millimeter wave spectrum and transmission in unlicensed spectrum. It is widely envisioned that non-slot based scheduling is an essential enabler to fulfill the challenging latency targets in low frequency bands of TDD and FDD NR. Providing better delay spread performance with a low subcarrier spacing in wide area use cases, non-slot based transmission is preferred over slot based transmission.

As the smallest scheduling unit, mini-slot supports the transmissions with a duration in symbols, providing in-built low latency as processing time at both UE and gNB side are reduced significantly. The mini-slot length of two, four or seven symbols is part of the 3GPP recommendation in Rel-15. Supporting of front-load Demodulation Reference Signal (DMRS) in mini-slot scenarios allows the DMRS based channel estimation being performed earlier, upon which the UE and gNB data reception processing is dependent on, while the support of code blocks interleaving over frequency in mini-slots allows processing of the symbol as soon as it is received over-the-air.

As shown in Figure 8.10, there are at least two different approaches for the scheduling of a mini-slot. The self-scheduling of a mini-slot (as illustrated in case a of the figure) is made from the Physical Downlink Control Channel (PDCCH) location at the beginning of each DL mini-slot, which enables immediate transmission without waiting for the start of a slot boundary and has no constraints from DL/UL switching point alignment in a slot of static TDD system on additional PDCCHs. The cross-scheduling of a mini-slot (as illustrated in case b of the figure) is made from the PDCCH location at the beginning of the slot, which is attractive for UE power consumption savings.

Figure 8.10 Mini-slot PDSCH scheduling.

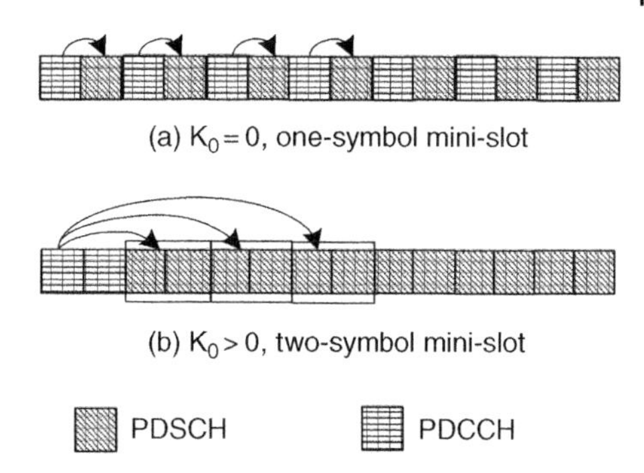

Efficient URLLC and eMBB Multiplexing Dynamic multiplexing between URLLC and non-URLLC traffic including traditional eMBB traffic is preferred due to the flexibility of resource usage and better spectral efficiency. However, this is a challenging task given the very different requirements of the services. Taking URLLC traffic and eMBB traffic as an example, URLLC traffic is typically bursty and sporadic and must be immediately scheduled preferably with short TTI to fulfill the strict latency requirements. On the contrary, eMBB traffic transferring large data packets would benefit from the usage of 1 ms or even longer TTIs leading to smaller control signaling overhead and avoiding fragmentation of packets into multiple smaller ones. One solution for multiplexing is to statically reserve certain radio resources for URLLC data transmissions exclusively. However, this results in waste of radio resources when there is no URLLC traffic to transfer. A more efficient solution is to use dynamic multiplexing, for example preemptive scheduling following [5, 6]: eMBB traffic is scheduled on all the available radio resources with a long TTI, e.g. 1 ms. When URLLC data arrives at the gNB, it is immediately transmitted to the corresponding UE by overwriting part of an ongoing eMBB transmission, using mini-slot transmission, as illustrated in Figures 8.11 and 8.12. This has the advantage that the URLLC data packet is transmitted right away without waiting for ongoing scheduled transmissions to be completed and without the need for pre-reserving radio resources for URLLC traffic. The drawback of the preemptive scheduling scheme is the performance of the impacted eMBB service

Figure 8.11 Resource allocation framework for URLLC and eMBB multiplexing: downlink multi-service multiplexing example.

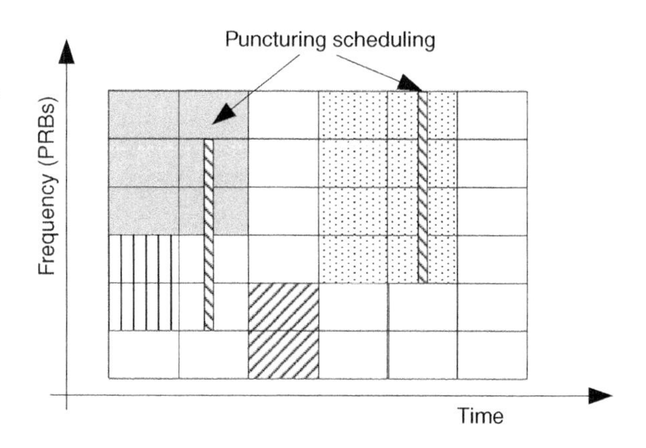

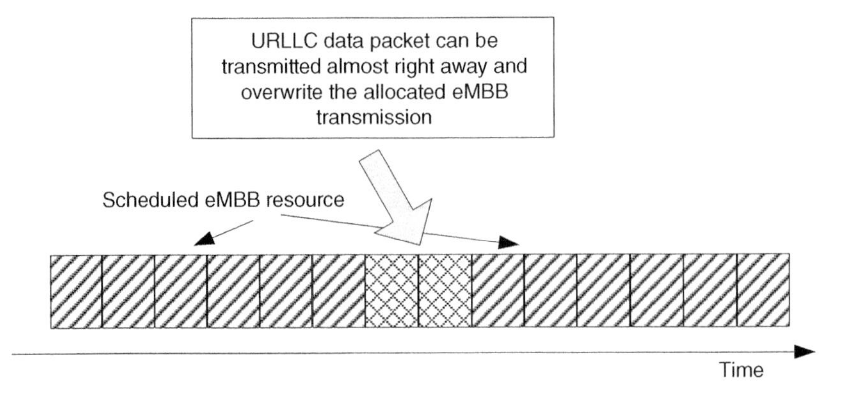

Figure 8.12 Example of preemptive scheduling.

whose transmission is partly removed, hence the decoding performance will suffer. To minimize this negative impact, the study in [5] proposes including an indication to the "victim" UE to give the information that part of its transmission has been overwritten by high priority URLLC traffic. This enables the eMBB UE to take this effect into account when decoding the transmission, i.e. it knows that part of the transmission is corrupted. To further improve the performance, various enhancements have been proposed, e.g. applying smart HARQ retransmission mechanisms, where only the overwritten part of the initial transmission is retransmitted.

Similarly, the concept can be extended to cover the UL direction as well. The UL scheduling is conducted by the gNB sending scheduling grants in the DL to the UEs. Among others, the scheduling grant includes pointers to time-frequency UL resources that the UEs shall use for their UL transmission. The gNB can choose to schedule UEs with different TTI sizes; e.g. on slot or mini-slot resolution level, given the options and constraints offered by the 5G NR flexible frame structure. For UL multiplexing between eMBB and URLLC services, one scheme which has been proposed and discussed in 3GPP is the pause-resume scheme (see [7]). To be more specific:

A gNB can selectively configure certain eMBB users that are scheduled in the UL over one or multiple slots to still monitor DL physical control channels carrying the scheduling grants for URLLC UEs at the beginning of every mini-slot (or a sub-set of those) during the ongoing UL eMBB transmission.

In case the capacity becomes an issue and there is need for urgent scheduling of URLLC traffic for UL data transmission, the gNB will send a pause-resume signaling message to one or multiple eMBB users that have an ongoing UL transmission and where already allocated resources are overlapping with the resources the gNB intends to allocate to URLLC UEs for UL transmission.

The pause-resume signaling message informs the eMBB UEs to put ongoing UL transmission on pause for a short duration, where-after it shall continue (resume) its UL transmission.

This approach requires that these eMBB UEs have the capability to monitor the DL control channel more frequently on mini-slot level, and the processing time for the DL control is comparable to that of URLLC UEs. In this case, the pause-resume signaling message can be sent in the same symbol(s) as the UL grant for URLLC. The pause-resume signaling is sent in parallel with the scheduling grant to the URLLC UE.

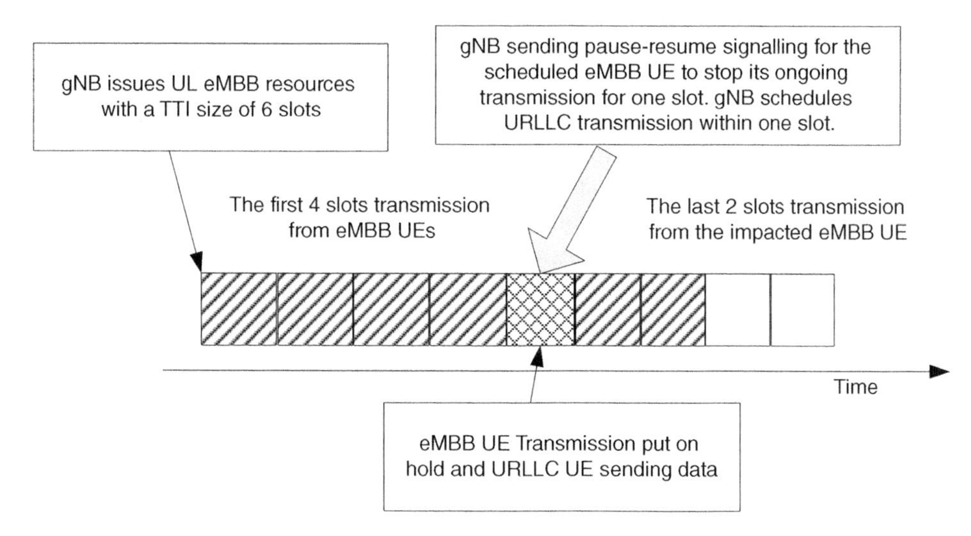

Figure 8.13 Pause-resume scheduling mechanism in uplink.

The pause-resume signaling message is assumed to be sent on PDCCH, being either (group-)common or UE-specific.

In the example shown in Figure 8.13, the gNB first schedules an eMBB UE to transmit with a TTI size corresponding to six subframes (or slots) in the UL. The eMBB UE starts UL transmission with the allocated resource based on the corresponding scheduled transmission. During UL transmission, the eMBB UE continuously monitors the DL control channel on mini-slot level. In case the UE receives a pause-resume message, it will stop the ongoing eMBB transmission for a period specified in the DL control channel, e.g. one or multiple mini-slots, while afterwards resuming the eMBB transmission to transmit the last two subframes of the TTI. During the mini-slot (or slot) where the ongoing eMBB transmission is put on pause, the gNB schedules the latency critical URLLC transmission. Alternatively, the URLLC transmission can puncture the eMBB transmission. Another possibility is that a eMBB UE does not need to be configured to monitor the URLLC resource allocation, and it transmits as in normal operation. The URLLC UE will transmit its UL data packet using the same resource as the eMBB UE, but possibly with a higher Tx power. This scheme is referred to as superposition transmission. This may lead to interference between URLLC and eMBB UEs and more advanced receivers such as interference cancelation receivers are necessary to achieve an acceptable reliability performance.

In addition to inter-UE multiplexing, intra-UE multiplexing between URLLC and eMBB services could be a potential problem as well. It is possible that the UE could prioritize different logical channels (LCHs) between URLLC and eMBB in case URLLC data arrives before eMBB data transmission. Below we are mainly discussing the case that one URLLC UE has ongoing UL transmission of eMBB data when URLLC data arrives or in another case the UE has an upcoming scheduled UL transmission but does not have sufficient time to prepare URLLC data for this transmission. There are different situations which should be studied further in the coming 3GPP Rel-16 or even future releases because UL URLLC data transmission can use either granted resources or configured UL grant-free resources or both.

Logical Channel (LCH) Based Scheduling Request (SR) For grant-based transmission, when scheduling URLLC UL data, the gNB needs to provide proper amount of spectrum resources and MCS. In Rel-15 it is already supported that the time interval between SR resources configured for a UE can be smaller than a slot, which means, depending on the latency requirement, it is possible that there is SR resource available for the UE per mini-slot. Furthermore, multiple SR configurations can be configured for one UE. Which SR configuration is used depends on the LCH that triggers the SR. This enables the gNB to be aware of the critical transmission and to make e.g. proper scheduling decisions. Different ways including multi-bit SR or multiple SR configurations to indicate information for critical latency transmissions were discussed in 3GPP. Following 3GPP agreements, the SR format can be kept similar as in LTE. But the gNB can configure multiple resources with different periodicity and resources for different services, for example one set of dedicated resources for URLLC services and another set for eMBB services which have relaxed requirement on latency. In this way, the gNB can get information of service type based on the resources where SR is received. The additional information is carried implicitly.

Consider the scenario that different SR resources are configured to request resources for UL transmission with different QoS requirements. One simple example is to configure two set of resources, one for latency critical services and another one for non-latency critical service as illustrated in Figure 8.14 (see [8]).

This figure assumes that two sets of resource blocks are configured, one for low latency services (per-slot) and another one for non-URLLC services which takes place at every 10 slots. In this way, the gNB can learn the required service type easily.

URLLC Transmission with UL Grant Free Resource Assuming frequent UL grant free resources are available (e.g. in every mini-slot) and if allowed, the UE could transmit both URLLC and eMBB data, as shown in Figure 8.15a, if there is no issue with Tx power since the grant free resource and the granted resource are overlapping in time, but not in frequency. In this case, UE transmission power might become a bottleneck. When the UE Tx power becomes as an issue or if simultaneous transmission is not supported, one straightforward solution can be that UL grant free transmission has the highest priority and eMBB data transmission is stopped or punctured as shown in Figure 8.15b. This is like the DL puncturing case where puncturing indication or some similar signaling to the receiver (i.e. gNB) would be necessary to reduce negative impacts on decoding and possible retransmission of packets.

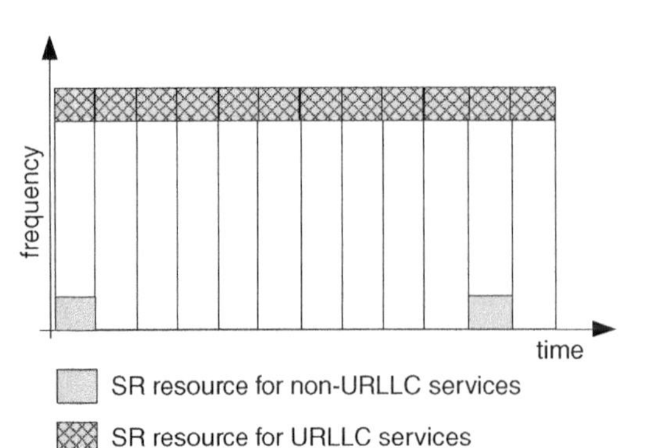

Figure 8.14 Example of multiple SR resource configuration.

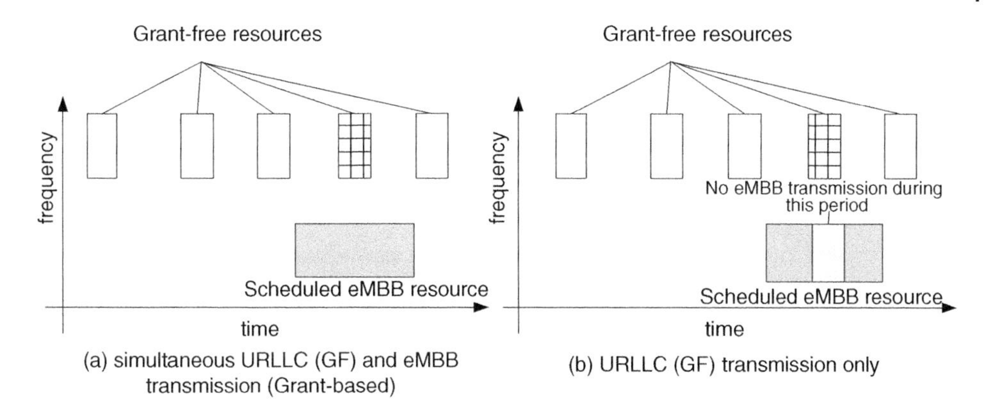

Figure 8.15 URLLC transmission with UL grant free resource.

Figure 8.16 Intra-UE puncturing.

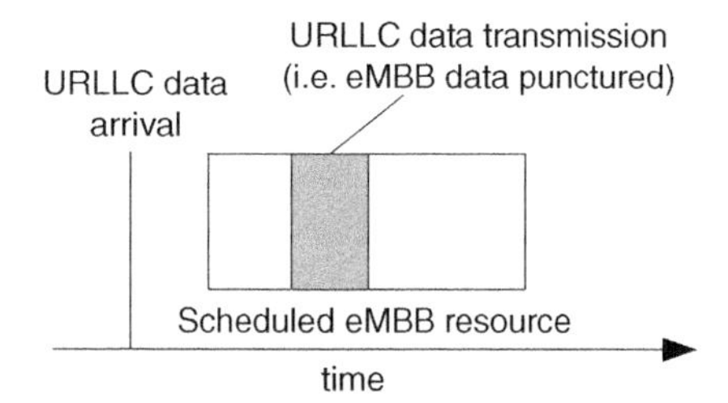

URLLC Transmission with UL Granted Resource In this case, a straightforward solution is the UE is requesting dedicated resources for URLLC UL data transmission which certainly will bring more latency due to the involved scheduling procedure: SR transmission, DL control channel reception and all necessary processing time.

Another option is sending URLLC data packet without waiting while using the already allocated eMBB resource as shown in Figure 8.16. This operation is like inter-UE punctured scheduling in DL. The benefit of this operation is reduced latency since URLLC data packets can be transmitted right away with the already allocated resources for eMBB transmission without any collision possibility. The potential problem related to this mechanism is the corrupted eMBB data. Similar as for DL preemptive operation, sending puncturing information to the gNB might be necessary to minimize the impact on eMBB performance. The problem to be solved in this case is how to send the puncturing information, e.g. together with URLLC data as in-band control signaling or as dedicated control signaling.

UL Grant-Free Transmission Reserving radio resources for delivering the SR is not efficient for applications that generate sporadic data traffic. For the extreme low latency and reliability requirements, it is desirable to support a grant-free UL transmission scheme that avoids the regular handshake delay between SR and UL grant allocation and relaxes the stringent requirements on the reliable grant.

There are two types of configuration schemes supported in 3GPP Rel-15 ([9]). For the UL grant-free type 1, UL data transmission without grant is based on RRC (re-)configuration without any L1 signaling which is suitable for deterministic URLLC traffic patterns whose properties can be well matched by appropriate resource

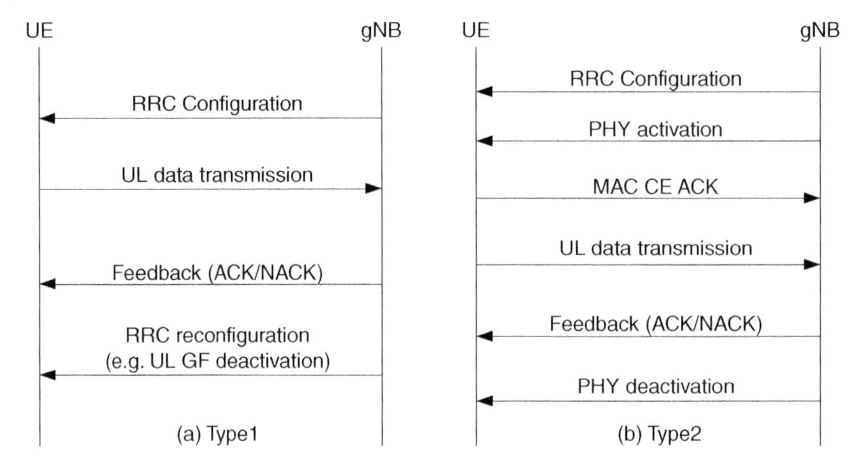

Figure 8.17 UL grant free transmission.

configuration. The UL grant-free type 2 allows additional L1 signaling for a fast modification of semi-persistently allocated resources, which enables flexibility of grant-free UL operation in terms of URLLC traffic properties including but not limited to the variable packet arrival time and/or packet size.

The resources configured for grant-free UL transmission can be shared or not among UEs of both type 1 or type 2 depending on network decision. Supporting a non-contention based scheme or a contention based scheme of resource allocation depends whether to match the requirements of spectrum efficiency and performance requirements for the given use cases. For predictive traffic, non-contention based allocation provides efficient resources dedicated to UEs with short access delay requirements, while keeping the high reliability gain from orthogonality. For sporadic small packet transfer and aperiodic transmissions in light load conditions, configuring grant-free UL transmission on shared resources in a contention-based manner is more efficient, but advanced Multiuser Detection (MUD) receivers would be required to ensure high reliability and to cope with resource collision of multiple users. The concept of contention-based access is shown in Figure 8.17.

UL grant-free transmission enables low latency access to data channels. To improve the reliability in case of collisions, diversity transmission, e.g. repetitions over time, can be utilized where packet replicas are sent multiple times to reduce the collision probability. It offers prominent gains and lower implementation complexity in comparison to multi-user detection. However, how to set the optimal size of the grant-free resource is a key problem to be solved. Depending on the network scenario, e.g. UE population and traffic pattern, the network may adopt various strategies to minimize or resolve collisions in the contention resource as indicated in [10].

Grant-free based transmission is in general more efficient for small packet UL transmission in terms of lower latency and lower overhead. While for medium to large packets, due to limited flexibility of link adaptation and power control for grant-free, it might result in higher latency to fully use grant-free for transmitting such kind of packets. One promising way to solve this problem is supporting efficient mode switching from grant-free to grant based, adaptively meeting requirements of incoming packets with different size.

Table 8.4 Reduced processing time for slot-based scheduling.

	HARQ timing parameter	Units	15 kHz SCS	30 kHz SCS	60 kHz SCS	120 kHz SCS
Front-loaded DMRS only	N1	Symbols	8	10	17	20
Front-loaded and additional DMRS	N1	Symbols	13	13	20	24
Frequency-first re-mapping	N2	Symbols	10	12	23	36

Reduced UE Processing Time The baseline UE processing time capability in NR Release 15 for slot-based scheduling, including Carrier Aggregation (CA) case with no cross-carrier scheduling and with single numerology for PDCCH, Physical Downlink Shared Channel (PDSCH), and Physical Uplink Shared Channel (PUSCH) and no Uplink Control Information (UCI) multiplexing, is given by Table 8.4.

Within 3GPP, there are ongoing discussions about the non-slot based processing. Compared to slot-based processing, the processing time can be further reduced and there is a strong trend that the value of N1 would be three symbols only.

8.3.3 High Reliability and Availability

8.3.3.1 Core Network

Many services like Mission Critical Public Safety services require (extreme) high-availability and reliability. Also, other services, e.g. traditional voice, chat, web browsing, video streaming, benefit from features that guarantee a high degree of network resilience. While network resilience can be increased by proper SW design or following certain Hardware (HW) deployment options (e.g. hot and cold stand-by of nodes), 5G comes also with some built-in network resilience and load balancing features.

One such feature is the concept of Access and Mobility Management Function (AMF) Sets and the enablers introduced for AMF resiliency, such as enablers for state-efficient AMF deployments, planned AMF maintenance (e.g. frequent SW upgrades in case of micro-services support) and AMF auto-recovery. The AMF Region concept is like the concept of Mobility Management Entity (MME) pools in the Evolved Packet Core (EPC). Pooling of elements (hot and cold standby) and respective load balancing between core network elements improves system reliability and guarantees network service availability, even if a single Network Function (NF) instance fails. An AMF Set (see [11]) consists of AMFs that serve a given area and network slice. An AMF Region consists of one or more AMF Sets. AMF Region and AMF Set identifiers are encoded in the Globally Unique Access and Mobility Management Function Identity (GUAMI), which identifies the AMF assigned to a UE.

The AMF Set concept is used to allow an AMF storing its UE context and session data in the Unstructured Data Storage Function (UDSF) or in other backup AMFs belonging to the same AMF Set (called "backup AMF"). The backup AMF is determined based on the GUAMI of the failed AMF. The GUAMI assigned to the UE indicates the corresponding AMF Region and Set. Once an AMF fails and it peer nodes (e.g. 5G base stations, other AMF) detect the AMF failure, the connected 5G base stations are informed, release the UE specific signaling connection toward this AMF and mark the AMF as failed. However, the UE's RRC connection and possible user plane connection are maintained. In a failure case the base stations direct subsequent messages to a backup AMF, if available. If no backup AMF is available, the base stations select an AMF from the same AMF Set. The selected AMF must inform its peer nodes that it is now serving the UEs of the failed AMF. The selected AMF can process UE's transaction. Thus, the failure of the AMF instance does not impact services offered to the UE.

While the Region concept is not new (it is like the MME pool concept in EPC), 3GPP has newly introduced many features in Rel-15 enabling auto-recovery of an AMF, allowing easier software upgrades, load balancing of AMFs without any impact to the UEs. In Evolved Packet System (EPS), MME restoration requires release of RRC and S1 connections for all connected UEs and it requires UEs in idle mode to be paged almost simultaneously to perform load balancing tracking are update (TAU). This is both service impacting and may cause excessive signaling load in the network. However, with the new enablers specified for AMF resiliency, AMF planned maintenance and AMF failures are handled in a completely seamless manner to the UEs (see Chapter 4 for more details). One of the important enablers includes the ability for an UE to provide the AMF Set Identity (ID) as part of the 5G S-Temporary Mobile Subscription Identifier (5G-S-TMSI) in the RRC signaling to the 5G base station. This allows the base station to dynamically select an AMF from the given AMF Set. Furthermore, an AMF pointer can point to multiple AMF instances allowing the base station to select any AMF instance and forward Non-Access Stratum (NAS) messages to this AMF. In EPC, the UE provides only the S-Temporary Mobile Subscriber Identity (S-TMSI) including the MMEC (pointing to a single MME) in RRC signaling sent to the eNB. Thus, it is not possible for an eNB to know which other MME is able to process the UE transaction without requiring a re-registration. Manual configuration of MMEGI in the eNB to map to an MMEC could be done in a proprietary manner however this may result in constraints on the number of unique MMECs available for a deployment (especially for big networks) as the MMECs must be unique across all MMEGIs in the network. Additional enablers allowing support for AMF scalability, load balancing and resiliency are explained in Chapter 4 of this book.

Instead of using another AMF from the same AMF Set as backup, the AMF can also store the context data in the UDSF (see Chapter 4) allowing any AMF in the AMF Set to retrieve the UE's context data and process UE's transactions. The UDSF is part of the data layer (like the Unified Data Repository UDR) and allows storing and retrieval of context and session data, or more generally of any unstructured data, by any NF from the control plane of the 5G Core. UPFs and application servers can optionally also store their session data in the UDSF. Figure 8.18 shows how the UDSF fits into the 5G Core architecture.

The UDSF can be a standalone function or e.g. collocated with the UDR. The interface between UDSF and NFs N18/Nudsf is not specified by 3GPP in Rel-15. Introduction of the UDSF for storing/fetching context and session data follows the data driven model known from cloud architectures with a clear separation between compute logic and data layer. Context storing and retrieval between similar NFs (e.g. AMFs of one AMF Region and Set) is no longer needed. The concept of storing session data in a data base external to the NFs can impose high demands regarding performance, e.g. very low delays when writing/reading in the data base, and availability on the data base. Thus, deployments and actual NF implementation should ensure that the frequency of storage and retrieval

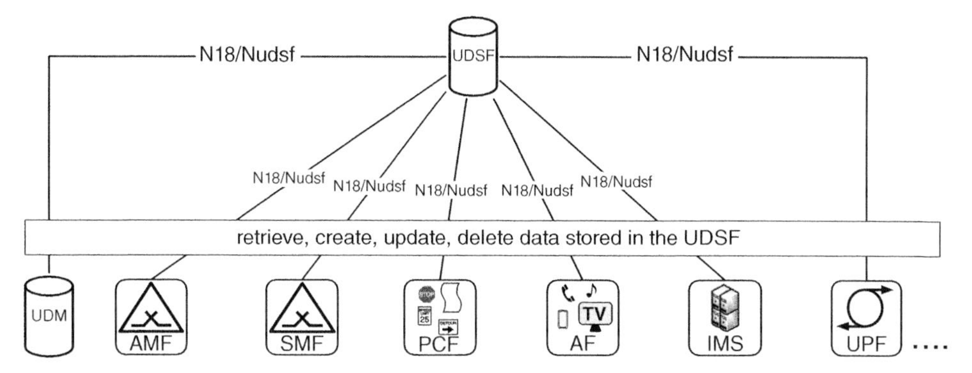

Figure 8.18 UDSF as part of the 5G core architecture.

of data is balanced to enable high availability and resiliency of the NFs, at the same time ensure that there is no performance impact due to storage of UE context in the (external) UDSF. If the context is stored at the end of a procedure, this will not impact the performance of any real-time and dynamic procedure and transaction ongoing in the network for a given UE, but allows for improved resiliency, scalability and availability (allowing any compute instance to process a UE transaction). For an AMF, it can be expected that a state-efficient implementation concept is followed, i.e. storing a stable version of the UE context at the end of a certain procedure, and not storing UE context at the end of every single transaction.

Load balancing between AMFs of a certain Set is achieved by assigning a weight factor to each AMF according its capacity. The AMF provides this weight factor to connected 5G base stations so that the base station can select an AMF based on its weight factor. Without going into details, it should be noted that also overload control mechanisms were specified by 3GPP for AMF (e.g. N2 interface overload control and NAS level mobility management congestion control) and Session Management Function (SMF) (NAS level session management congestion control). These control mechanisms can lead to rejection of signaling messages and push back of UEs with a back-off timer, i.e. a UE is only allowed connecting to the network after a certain time. Overload control mechanisms help avoiding failure and restart of a network element, thus leading to higher service availability.

In addition, from user plane perspective, 3GPP has standardized several new 5G QoS values and characteristics allowing delay and reliability critical applications to be delivered over 5G Core and Radio network. A detailed description of these 5G QoS values can be found in Chapter 6. Table 8.5 provides an overview of the standardized 5G QoS values for delay critical services.

8.3.3.2 Radio Network

It is obvious that the quality of the radio link between transmitter and receiver directly affects the overall system reliability. In most of the research related work, if not all, Signal to Interference plus Noise Ratio (SINR) is used to measure the quality of the radio link. The higher the SINR, the lower is the packet error probability, and hence the higher is reliability and the lower is latency. It is therefore important that URLLC UEs experience a SINR above a certain threshold with very high probability. On the high-level, there are two ways to increase SINR:

Table 8.5 Standardized delay critical 5G QoS values.

5G QoS identifier	Default priority level	Packet delay budget (ms)	Packet error rate	Default maximum data burst volume (Bytes)
81	11	5	10^{-5}	160
82	12	10	10^{-5}	320
83	13	20	10^{-5}	640
84	19	10	10^{-4}	255
85	22	10	10^{-4}	1358

Table 8.6 List of technology components for enhancing reliability.

Micro-diversity	Higher order SU-MIMO diversity to reduce the probability of experiencing deep fades.
Macro-diversity: multi-connectivity	Data duplication macroscopic transmission from multiple cells to the same UE to improve reliability. Robustness against shadow fading and cell failures.
Hybrid ARQ	Enhanced time-diversity and frequency-diversity (in case of frequency hopping) by using HARQ with soft combining. Adding robustness toward potential link adaption imperfections. Blind repetition.
Enhanced control channel reliability	High reliability of downlink control carrying resource grant, as well as uplink control carrying feedback information such as ACK/NACK and CSI. Achieved by using stronger coding, power boosting and asymmetric signal detection techniques.
Interference mitigation	Advanced UE interference mitigation receivers (e.g. MMSE-IRC or NAICS). Network-based inter-cell interference coordination.

(1) enhancing the signal power, for example with increased redundancy, diversity, and
(2) reducing the interference power via interference management, e.g. interference aware scheduling, interference cancelation receivers and so on.

Table 8.6 lists example technology components for enhancing reliability in the radio system.

Micro-diversity Having multiple antennas at either the transmitter or the receiver side or both is a resource efficient way to improve the SINR via spatial diversity. Particularly, single-user (SU) single-stream (i.e. Rank 1) transmission modes are the most relevant for URLLC use cases, as the goal is to improve performance on the lower tails of the SINR distribution. As reported in [12], URLLC devices should operate with at least 2×2 (preferably 4×4) SU single-stream transmission schemes, i.e. maximizing the diversity order of the wireless link. After discussion in 3GPP, 4×4 scheme was selected to get sufficient diversity orders. To be more specific, with micro-diversity:

- Multiple antennas can be deployed at both transmitter and receiver side with single-stream single-user transmission modes to achieve high diversity order and related power gain, and decrease the SINR outage probability.

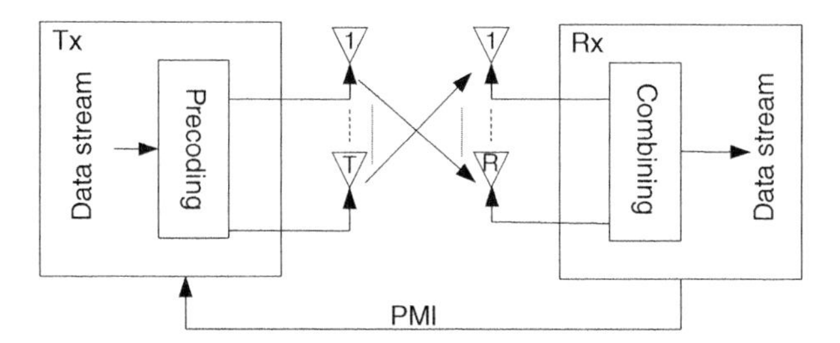

Figure 8.19 Example of micro-diversity operation.

- Support for both closed-loop and open-loop schemes. Closed-loop typically offers the best performance but it is sensitive to feedback errors and out-of-date Channel State Information (CSI) reports. Therefore, the most appropriate transmission mode should be selected on a per-user basis, depending on its speed, reliability requirements, among other relevant parameters.
- Microscopic diversity is applicable to both data and control channels.

From concept level, the following high-level Figure 8.19 illustrates how it works in general. With the assumption of independent fading channels among different antenna pairs, high diversity gain can be achieved.

Macro-diversity There is a variant of multi-connectivity schemes providing a macroscopic diversity for increasing reliability. Macroscopic diversity, i.e. data duplication and redundant transmission/reception from multiple cells/TRPs (Total Radiated Power), is also required to combat the slow fading effects (or shadowing and/or blocking) and to provide mobility robustness during handovers as discussed later. In this regard, data duplication at the PDCP layer has been agreed for NR in 3GPP Rel-15 [13]. At the lower layers, inter-cell non-coherent joint transmission is among the candidate transmission schemes. Macroscopic diversity also provides benefits in terms of resilience against failures of the cellular infrastructure.

Multiple-cells

Different cells can allocate different resources to the same UE for the same packet, i.e. independent schedulers are used among cells. At the receiver side, soft combining of received data packets from multiple cells is performed. This applies to both UL and DL data.

Multiple-TRPs

Use of multiple TRPs has been agreed with a focus on providing high data rate and diversity. For URLLC, multi-TRP communication can be one of the potential enablers of high reliability of URLLC services in low frequency and high frequency bands. In multiple TRPs, data or control information can be shared among multiple TRPs to URLLC UEs by applying joint transmission schemes, where different versions of the same data packet or control information can be received jointly. The UE can combine them in the physical layer. Therefore, the diversity gain is achieved.

As an example, inter-cell non-coherent joint transmission can be implemented in diverse ways as discussed in [14]. Three well-known techniques to increase robustness

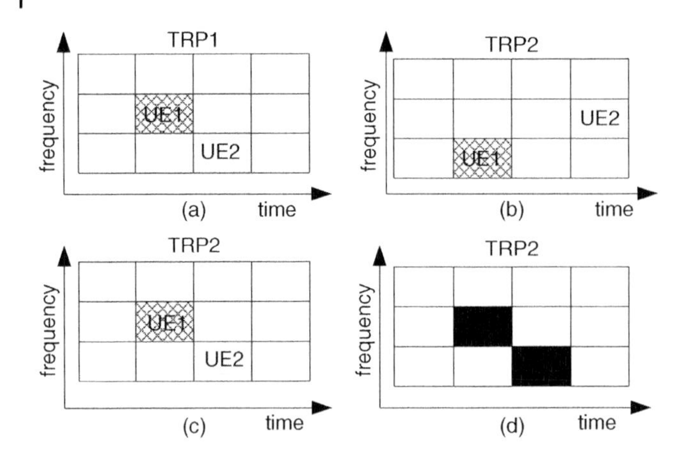

Figure 8.20 Example of baseline transmission and joint-transmission: (a) baseline; (b) macro-diversity/soft combining; (c) SFN; and (d) Narrow-band muting.

of a communication link are revisited for applications in the addressed scenario, namely Single-Frequency Network (SFN), narrowband muting and macro-diversity with soft combining as shown in Figure 8.20. Based on [14], it can be observed from simulation results that in a dense indoor deployments, inter-cell coordination is a powerful approach to increase the reliability of the transmissions without incurring longer delays as it is the case with retransmissions.

We will now look at the performance for outdoor scenarios. In case of noise-limited scenarios, soft combining is an appropriate solution candidate. Therefore, we will focus on the scheme with soft combing. In this scheme, non-coherent transmission of the desired packet is done by cooperating TRPs, as depicted in Figure 8.20 b. The same data packet is sent from multiple TRPs independently. The UE applies soft combining on the received data packets. Such non-coherent transmission can be done independently, such that each TRP performs independent scheduling and with multiple PDCCHs, one per cooperating TRPs.

One benefit of this technique is that it gives redundancy against some effects of the channel which are correlated over time and frequency, like blocking or radio link failure. Because the same packets are transmitted from multiple sources independently, it is less likely that all of them will be lost. Also, multiple independent simultaneous transmissions reduce the need for retransmission and thus the probability of long delays.

The following Figure 8.21 shows the performance difference between regular single TRP transmission and the studied multiple TRP transmission with soft combing at UE side. The 3GPP outdoor simulation environment is adopted, and it is assumed that 20% of the UEs are indoor. From the simulation results, we can see that with soft-combing, the offered URLLC load is about 2.6 Mbps and in case with regular transmission, the value is about 1.8 Mbps which means more than 40% capacity gain due to the joint transmission from multiple TRPs.

High Reliable Control Channels URLLC service requires tight latency and high reliability simultaneously. The data can be correctly received only, if the control channel is decoded correctly. Therefore, the control channel should have even higher reliability for URLLC

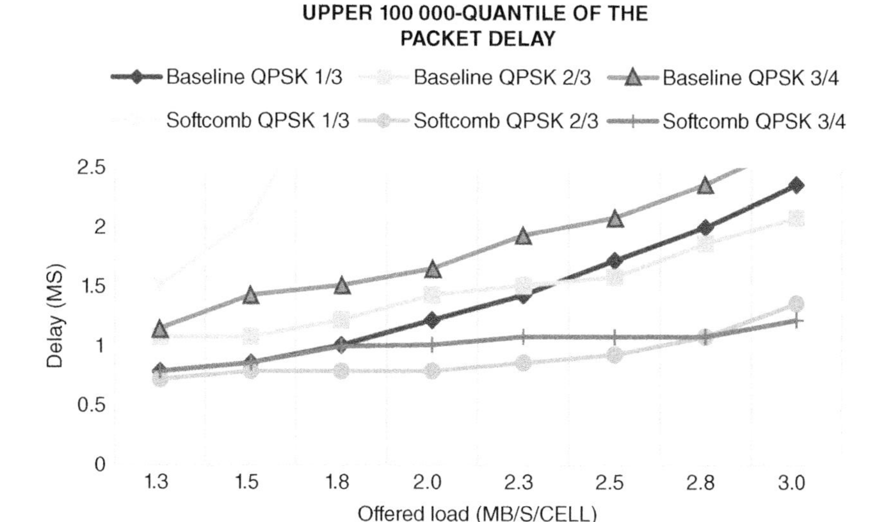

Figure 8.21 Performance comparison between baseline (regular unicast based transmission) and soft combining scheme.

Table 8.7 SINR gain due to higher AL with 40-bit DCI.

SINR gain (dB)	4GHz, 4Rx		700 MHz, 2Rx	
	TDL-A 30 ns	TDL-C 300 ns	TDL-A 30 ns	TDL-C 300 ns
AL 8 vs. AL 16	~2.4	~2.8	~3	~3.5

services or at least on the same level as the data channel. There are many methods for improving the reliability of control channels. Due its importance, we will discuss various ways to increase the reliability of different control channels or control messages.

High Aggregation Level With higher aggregation level (AL), the control information can be transmitted with more physical control channel resource elements, resulting in lower code rate coding schemes and lower order of modulation schemes which can be adopted for reducing the bit/symbol error rate. High aggregation levels up to 16 have been already supported in 3GPP Rel-15 for URLLC control channels. Performance gain due to higher AL is obvious as indicated in Table 8.7.

Compact Downlink Control Information (DCI) Reducing unnecessary control bits in control information for URLLC services allows to further improve the reliability of control channels on top of high aggregation level or relax the bandwidth requirements for transmitting the control information.

Link level simulations are used to evaluate the potential performance gains coming from compact Downlink Control Information (DCI) design [15] with the following simulation parameters in Table 8.8 as agreed in 3GPP:

Table 8.8 Simulation parameters for compact DCI.

Parameter	Value
DCI payload (excluding 24 bits Cyclic Redundancy Check [CRC])	40 bits, 30 bits
System bandwidth	20 MHz
Carrier Frequency	4 GHz, 700 MHz
Number of symbols for CORESET (Control Resource Set)	2
CORESET BW (contiguous physical resource block [PRB] allocation)	20 MHz
Subcarrier spacing	30 kHz
Aggregation level	Compact DCI study: 8, 16
Transmission type	Interleaved
REG bundling size	6
Modulation	QPSK (Quaternary Phase-Shift Keying)
Channel coding	Polar code (DCI)
Transmission scheme	1-port precoder cycling
Channel estimation	Realistic
Channel model	TDL-A (delay spread: 30 ns) TDL-C (delay spread: 300 ns)
UE speed	$3\,\mathrm{km\,h^{-1}}$
Number of BS antennas	2Tx
Number of UE antennas	4Rx for 4G, 2Rx for 700 MHz
Residual target BLER	10^{-5}
Deployment	Urban macro as listed in 3GPP 38.802

Main objective of the system level simulation is comparing the performance of normal DCI (assumed to be 40 bits) and compact DCI (assumed to be 30 bits) with two-symbol length CORESET (Control Resource Set).

The performance gain due to the 30-bit compact DCI is summarized in Table 8.9. The comparison is based on the SINR at BLER = 10^{-4} as the curves have better statistical convergence at this level. We do not expect significant difference at BLER = 10^{-5} as the curves have a similar slope. Table 8.9 shows that less performance gain is observed for higher AL.

Based on simulation results it can be observed that 0.4~0.8 dB gain can be achieved with compact DCI (30 bits versus 40 bits) with AL 8 at BLER = 10^{-4}. And for AL 16, even less gain around 0.3~0.4 dB can be achieved with compact DCI with AL 16 at BLER = 10^{-4}.

The potential impact on the overall system design, especially regarding flexibility, should be considered before making decisions of whether specifying compact DCI in 3GPP or not. In other words, both the performance gain and restrictions due to compact DCI need to be considered. To be more specific, depending on which information field(s) will be modified in compact DCI format comparing to regular DCI format,

Table 8.9 SINR gain with 30-bit DCI versus 40-bit DCI.

SINR gain (dB)	4GHz, 4Rx		700 MHz, 2Rx	
	TDL-A 30 ns	TDL-C 300 ns	TDL-A 30 ns	TDL-C 300 ns
AL 8	~0.4	~0.8	~0.4	~0.8
AL 16	~0.3	~0.3	~0.3	~0.4

various adverse impacts can be expected. For example, reducing the size of the resource allocation fields in either time domain, frequency domain or both would limit the scheduling flexibility, which may result in reduced spectral efficiency. Similar impacts can be observed from other fields, e.g. MCS, HARQ ID, Redundancy Version (RV), etc.

One more aspect that need to be considered is the impact on the number of DCI format sizes and the number of blind decoding attempts when a UE is expected to monitor other DCI formats, e.g. when a UE supports both eMBB and URLLC services. A new compact DCI format will increase the UE complexity, introduces additional constraints on the monitored DCI formats and affects the blind decoding budget for the different formats.

Considering only up to ~0.4 dB gain for AL16 (most likely to be used for URLLC services) with compact DCI and the possibly severe performance degradation due to various restrictions, it may not worth the effort to specify compact DCI unless more clear use cases and benefits can be identified which is an open topic for the coming Rel-16 work in 3GPP.

Asymmetric Detection of ACK/NACK As discussed earlier, protecting the NACK signal is more important than protecting the ACK signal, this is because erroneous NACK detection degrades the communication reliability (see [16, 17]). On the other hand, wrongly decoding an ACK as a NACK message will not result in performance degradation in terms of reliability, only reducing the spectral efficiency due to unnecessary retransmission. This leads to the idea of using enhanced NACK protection by applying e.g. asymmetric signal detection. For this purpose, the threshold ν for the binary hypothesis testing can be set in a way that the correct detection of NACK is favored as shown in Figure 8.22. Of course, the drawback of this approach is the higher probability of wrong detection of an ACK as a NACK compared to the case of symmetric signal detection, in which the same error probability is achieved for the miss detection of ACK and NACK. This results in unnecessary retransmissions. The performance gains due to asymmetric ACK/NACK detection can be found in [18].

Adaptive Configuring CQI Report The high reliability of existing transmission schemes of CQI report is not satisfied at low SINR. When a reported CQI value is decoded wrongly as higher value, it will result in selecting a higher level of MCS by the base station leading to reduced communication reliability. Different ways to increase the reliability of CQI report itself have been proposed, e.g. changing the allocated radio resource for sending CQI report or reducing the content of CQI report.

To improve the overall reliability, CQI report for URLLC services should be transmitted in a very reliable manner. In the operating principle specified for LTE (NR related

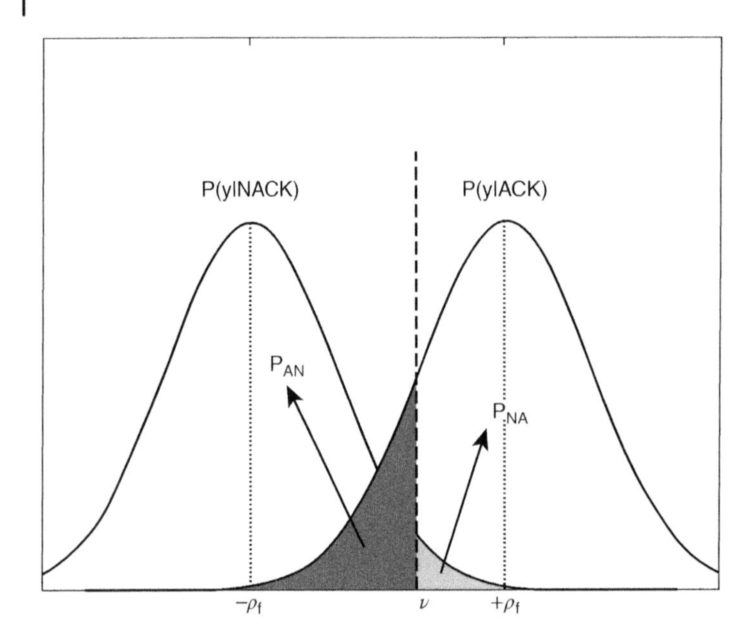

Figure 8.22 Illustration of decoding ACK/NACK signals.

discussion in 3GPP is ongoing), a fixed amount of radio resources is used for delivering CQI report over PUCCH. However, the achievable reliability with such a method is not sufficient, especially considering that there is a high chance that a reported wideband CQI value is decoded wrongly as higher value, which results in higher MCS selection and the before mentioned reduced communication reliability. Obviously, one way to increase the reliability is to repeat the same CQI information over time, however, the increased latency might be a problem. Other potential enhancements are the following:

- Increased physical resource for URLLC UE CQI reporting while keeping the same CQI payload size as for eMBB UEs, e.g. 4 bits: with the increased resource, the effective coding rate can be reduced which leads to higher reliable CQI decoding at the gNB.
- Define a smaller CQI payload while keeping the same resources between URLLC UEs and eMBB UEs: in this case the number of CQI values for URLLC UEs is smaller compared to eMBB UEs which leads to a reduced granularity of reported channel quality. With the same resource for CQI reporting, the reliability performance for CQI decoding can be improved.

In Figure 8.23 we compare the error probability of decoding CQI values as higher CQI values with a smaller CQI payload (Figure 8.23a: 3-bit versus 4-bit CQI, normal resources) or with utilizing double resources for CQI transmission (Figure 8.23b: 4-bit CQI). The simulation assumptions from TS 36.104 [19] are used and the SINR is set at −3.9 dB. It can be seen from Figure 8.23 that double the resource will give more benefit as expected, while the gain from reduced CQI payload size (3-bit versus 4-bit) is not neglectable either considering the URLLC use case. In addition, it should be noted that the current LTE CQI performance requirement in 3GPP (1%) is not sufficient for

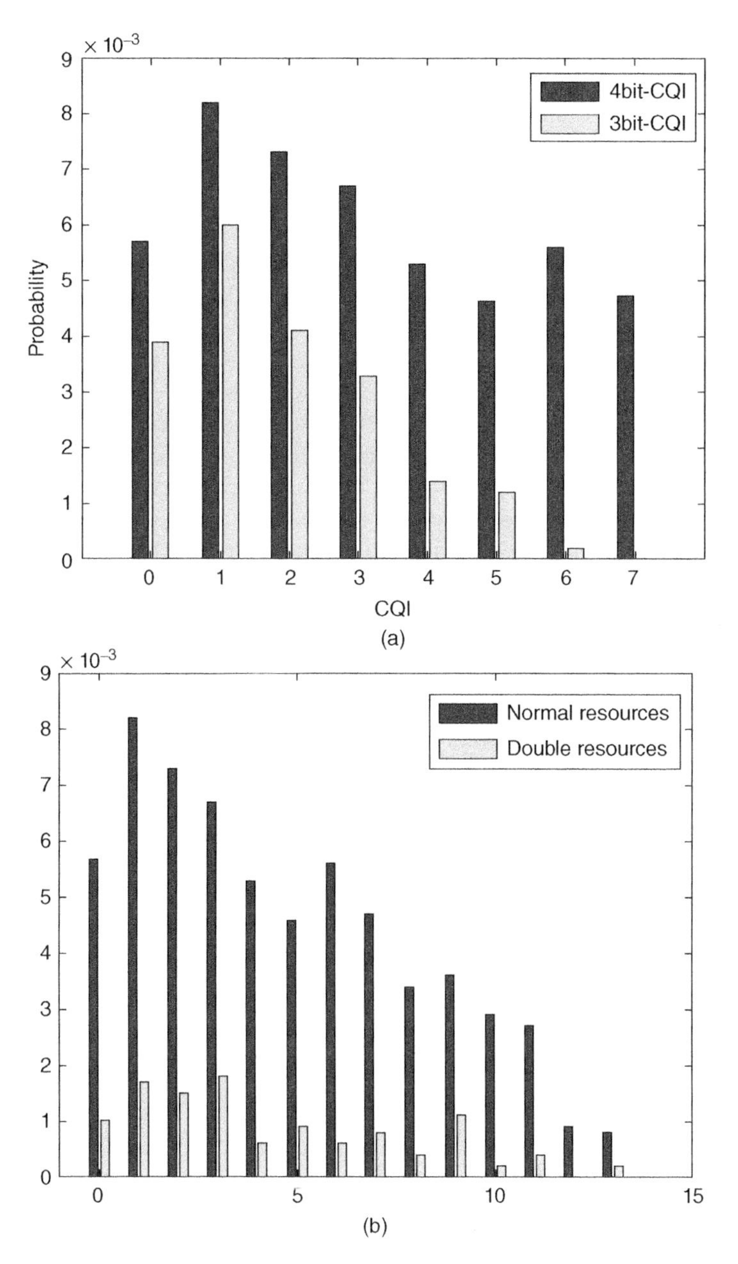

Figure 8.23 Performance of CQI report.

NR URLLC. As a summary, it would be beneficial to consider ways for improving the reliability of CQI report itself from both standard and implementation point of view.

Repetition of Scheduling Information There are several ways to repeat scheduling information, e.g., repeating within the same CORESET, across different CORESETs, across different monitoring occasions and so on. We are focusing on the scenario where

PDSCH is repeated. For this scenario, reliability of resource assignment messages can be increased by including the resource allocation information of the current transmission and sub-sequent retransmissions, in case data repetition is supported as currently agreed in 3GPP Rel-15, into the messages. In this case, even if a UE is missing one assignment message, the allocated resource could be identified with the subsequent assignment message or the previous assignment message.

For example, in case of blind repetition (i.e. continuous transmission without waiting for feedback) without time domain PDCCH repetition, if the UE missed a single multi-TTI DCI (DCI scheduling K-repetitions), it will miss the PDSCH transmission in all the scheduled TTIs for the data packet since the UE has no idea that the gNB has already started to send a data packet. In Figure 8.24 DL data transmission is taken as one example for illustration, with the assumption of $K = 4$. This scheme is called Option 1. With Option 1, a single DCI is used to indicate the PDSCH resource information over multiple concessive TTIs. Option 1 has smaller signaling overhead compared to other options discussed below where several types of repetitions are considered. However, from reliability point of view, it suffers from high error probability as it depends on a single shot transmission only. To be more specific, in case the UE misses the PDSCH scheduling information at the very beginning, it will miss all the following data packet repetitions as well.

A solution to avoid this situation is using the independent DL assignment message to schedule the resource for the current TTI only as shown below in Figure 8.25 (Option 2). Option 2 has at least two advantages: the higher PDCCH reliability as having several, separate DCIs and the ability to enable some type of frequency hopping or dynamic resource allocation. This comes with the drawback of increased signaling overhead due to several DL DCIs for the same data transmission.

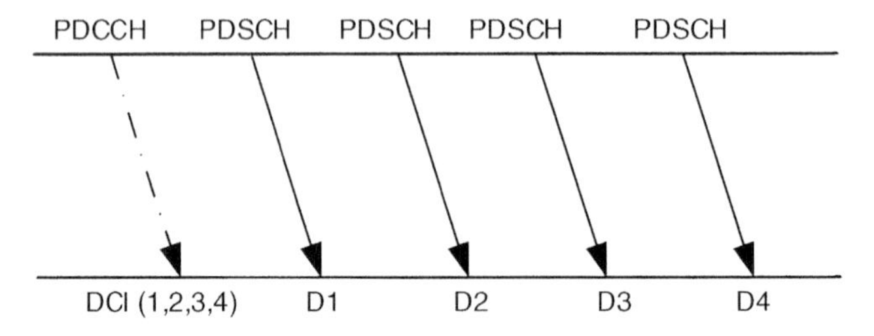

Figure 8.24 Example of DL HARQ processing with multi-slot scheduling (Option 1).

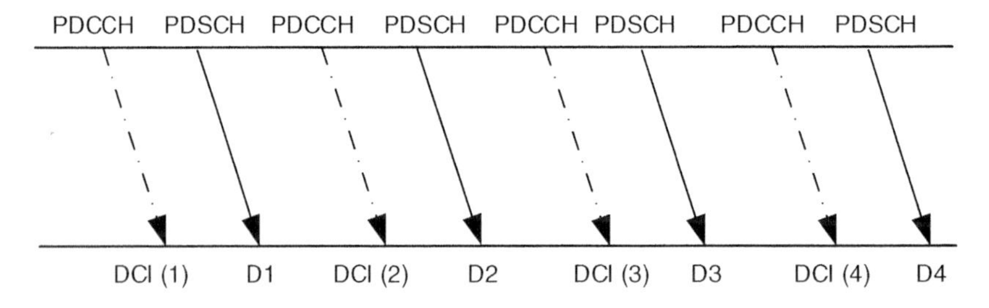

Figure 8.25 Example of individual scheduling for each blind repetition independently (Option 2).

One way to further improve reliability is to enable a gNB combining Option 1 and Option 2 in a way that DCI indicates the number of repetitions but also sends a DL assignment together with the repetition as illustrated in Figure 8.26 (Option 3).

Taking the example shown in the Figure 8.26 above where it is assumed to have maximal four transmissions and no ACK received during this period, the first assignment message includes the resource allocation information for all four transmissions. In the second slot, the resource assignment information is updated with three transmissions only, i.e. from the 2nd to the 4th transmission. Following the same principle, the last scheduling message includes the resource information for the last transmission only, i.e. the 4th transmission. The benefit of this method is the increased reliability of assignment messages. In case a UE misses one assignment message, the allocated resource could still be identified with the subsequent or the previous assignment message. The same principle can be applied for scheduling K transmissions for PUSCH. For this option, the gNB does not necessarily have to transmit PDCCH with every repetition. The gNB can choose the number of repetitions for PDCCH depending on the reliability target. Furthermore, the number of repetitions can be flexibly configured for the different UEs.

Another alternative of PDCCH repetition is shown in Figure 8.27 (Option 4) where the PDSCH resource assignment information is repeated before the first data transmission occurs. Under the assumption that the content of the repeated DCI is the same, it is sufficient if the UE can successfully decode only one PDCCH. Basically, the gNB can transmit the same DCI message multiple times using multiple PDCCH candidates in the search space.

Regarding Option 4, there are a few factors that may limit the number of PDCCH repetitions in a TTI. Firstly, there could be PDCCH capacity and/or increased blocking probability, if PDCCH with large AL must be repeated a few times within the limited control resource. Secondly, there is a limit on the number of Control Channel Elements

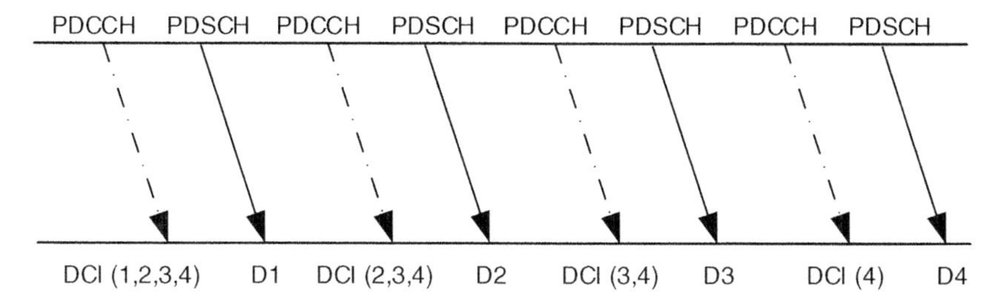

Figure 8.26 Reliable transmission of DL assignment information (Option 3).

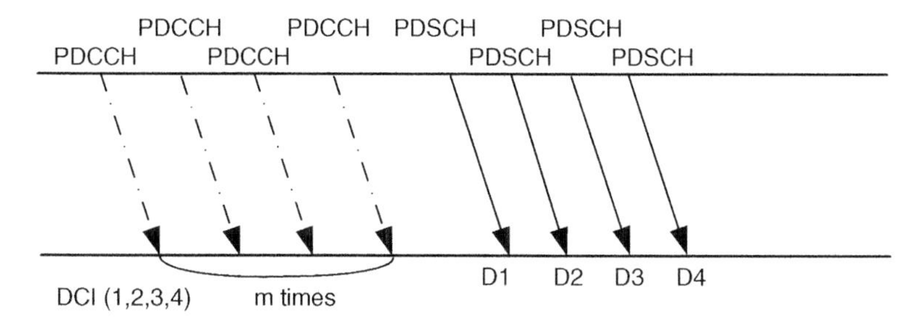

Figure 8.27 PDCCH repetition before data transmission (Option 4).

(CCEs) per slot where a UE can do channel estimation for PDCCH. According the current agreements in 3GPP Rel-15, the maximum number of CCEs one UE can monitor within one slot is 56 in case of 15 and 30 kHz SCS, 48 in case of 60 kHz SCS and 32 in case of 120 kHz SCS. This would practically allow at most two repetitions for AL = 16 (or no repetition for 120 kHz SCS) considering that the UE also needs to monitor the CSS (Common Search Space) and possibly other candidates.

The successful PDSCH decoding probability of all four options can be derived as shown in Table 8.10.

From the analysis results shown in the above Figure 8.28 where K = 4 ($\varepsilon_{D,1} = 10^{-2}$, $\varepsilon_{D,2} = 10^{-3}, \varepsilon_{D,3} = 10^{-4}, \varepsilon_{D,4} = 10^{-5}$) is assumed, Option 1 (i.e. the option without any repetition) results in the worst performance putting strict requirements on control channel reliability. For the other options, in case the error probability of DCI decoding is low, there is no clear performance difference. When the error probability of DCI increases, for example to 10^{-2}, Option 3 and Option 4 (with m = 3 and m = 4) still

Table 8.10 Overall error probability for different options of PDCCH repetition.

Option 1: $P = (1 - \varepsilon_{DCI})(1 - \varepsilon_{D,K})$

Option 2: $P = \sum_{n=1}^{K} \binom{K}{n} (1 - \varepsilon_{DCI})^n \varepsilon_{DCI}^{K-n} (1 - \varepsilon_{D,n})$

Option 3: $P = \sum_{n=1}^{K} \varepsilon_{DCI}^{K-n} (1 - \varepsilon_{DCI})(1 - \varepsilon_{D,n})$

Option 4: $P = (1 - (\varepsilon_{DCI})^m)(1 - \varepsilon_{D,n})$

ε_{DCI}: Probability of missing a DCI; $\varepsilon_{D,n}$: BLER of decoding data with n receptions; K: Maximum number of data transmissions; P: Overall PDSCH reliability; and m: number of PDCCH repetitions in Option 4.

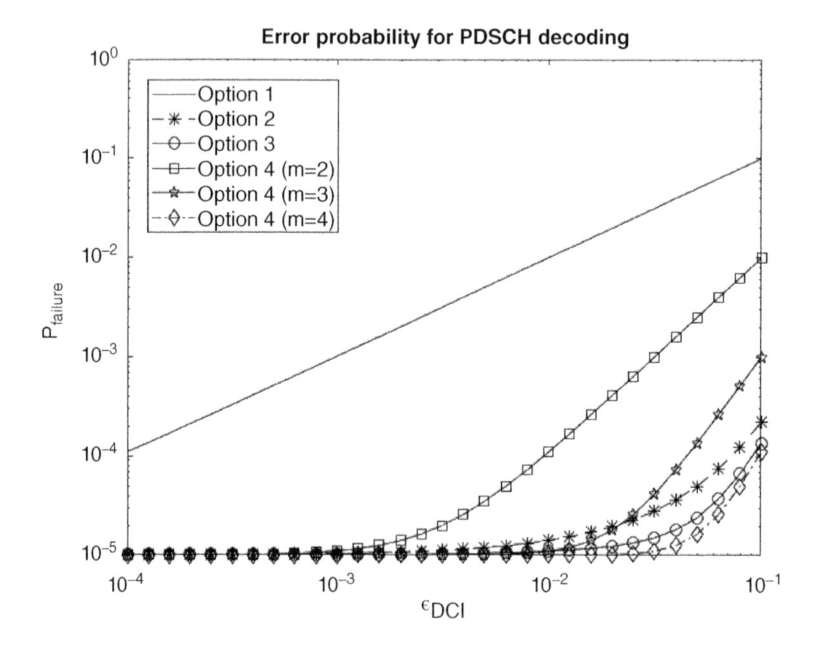

Figure 8.28 Error probability for PDSCH decoding.

have similar performance, while Option 2 and Option 4 with m = 2 are not on the same level any more. If the error probability of DCI further increases, Option 4 with m = 4 gives the best performance followed by Option 3. From the results illustrated in the Figure 8.28 it can be concluded that in case the repetition number is the same, Option 4 performs best, followed by Option 3. However, in case the maximum number of repetitions cannot be achieved with Option 4 due to the limited number of CCEs which can be monitored by a UE within a slot as specified in 3GPP, then Option 3 becomes the best candidate solution.

HARQ Enhancements As in all previous cellular systems, scheduling based transmission will be supported in both DL and UL in 5G NR. With dynamic scheduling, the network can assign resources to the UE in a very flexible manner according to the amount of data in the buffer and hence optimize radio resource utilization. Furthermore, URLLC traffic can be flexibly multiplexed together with eMBB traffic. URLLC traffic can be offered with a higher priority than eMBB traffic to minimize queuing latency. Considering URLLC traffic, one potential concern is the additional latency in UL due to the SR and resource grant before UL data transmission occurs. This delay is prolonged by potential HARQ retransmissions using dynamic scheduling. As discussed earlier, non-slot based transmission can reduce the overall latency compared to slot based processing.

From latency point of view, TDD is more challenging than FDD. Non-slot based scheduling based HARQ operation is discussed below. As an example, we consider now a 7-symbol mini-slot (GAP symbol is included in the DL mini-slot just for illustration purposes) and only one switching point in the 14-symbol slot.

The above Figure 8.29 illustrates one example with dynamic scheduling based UL transmission where the data packet arrives at a time instant in the 1st slot (UL mini-slot) and the SR can be conveyed in the 2nd slot. Under the assumption that there is sufficient time for the gNB to process SR and scheduling UL resources, the first data transmission can take place in the 3rd slot. Due to very limited processing time between the 1st data transmission in slot #3 and the DL control in slot #4, slot #4 must be used for processing and preparing for feedback and resource allocation for potential retransmissions. Hence, slot #4 cannot be used for retransmission and the second transmission can only take place in slot #5 leading to additional latency. As a result, within a 1 ms time window, there could be no retransmission opportunity. This might not be sufficient for URLLC UEs at the cell edge, especially in case time duration for the transmissions is important to achieve high reliability within a very short time window, e.g. with limited UE Tx power in UL. It should be noted that with the agreement of allowing more than one DL/UL switching point within a 14-symbol slot, there is a chance that latency can be reduced further, especially if processing time is not an issue.

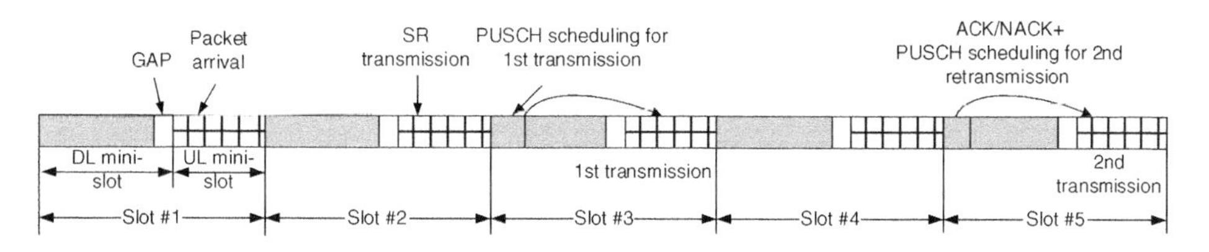

Figure 8.29 Example of UL HARQ processing.

Different enhancements are under investigation to reduce retransmission latency, e.g. proactive repetition (aggregated retransmission or blind retransmission for URLLC). This is also referred to as the support of K repetitions for an UL transmission scheme with grant access. In case of proactive repetition, multiple resources in adjacent slots can be allocated to the same TB as illustrated in Figure 8.28 where two different scheduling options are shown. According to Option 1, after receiving SR in the second slot, the gNB can allocate resources for both initial transmission and potential retransmissions (for example up to K repetitions). Comparing to Figure 8.29, slot #4 can be utilized for the 2nd transmission. In the example given in Figure 8.30, the 1st transmission is sufficient and ACK is sent in slot #5. ACK can also serve the purpose of requesting the UE to stop all following transmissions of the same TB, and the pre-allocated resources are released. The multi-slot scheduling method shown in Figure 8.30 can be understood as one type of slot aggregation scheduling options, i.e. the same TB in multiple slots (with the same or different RVs). Multi-slot scheduling can not only be used for reducing signaling overhead, but also for reducing the overall latency.

Another alternative (Option 2 in Figure 8.28) to achieve the goal of blind retransmission is sending UL resource information in each slot. Compared to Option 1, multiple independent DL signals (up to K under the assumption that one packet can be sent up to K times) are needed which can lead to signaling overhead and negatively impacts the transmission reliability.

In case of DL transmission, the proactive retransmission can be implemented as shown in the following Figure 8.31 (Option 1 only) with the help of multi-slot scheduling. It is assumed that the 1st transmission is successfully received.

Interference Mitigation Either network-based or terminal-based techniques can be used for mitigating interference, which has been identified as a promising complementary solution to improve the SINR. Reducing the received interference from neighboring

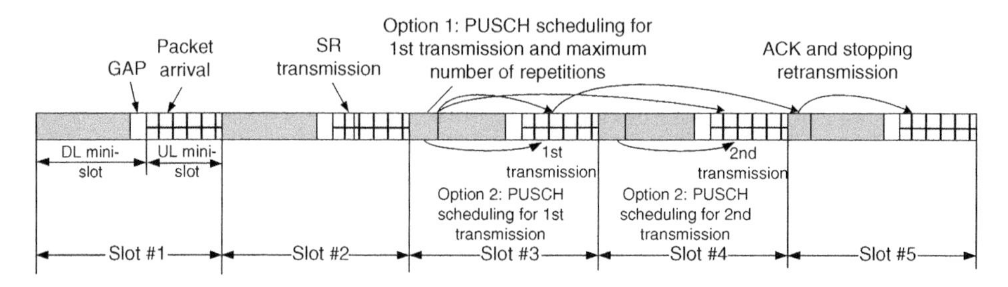

Figure 8.30 Example of UL HARQ processing with multi-slot scheduling.

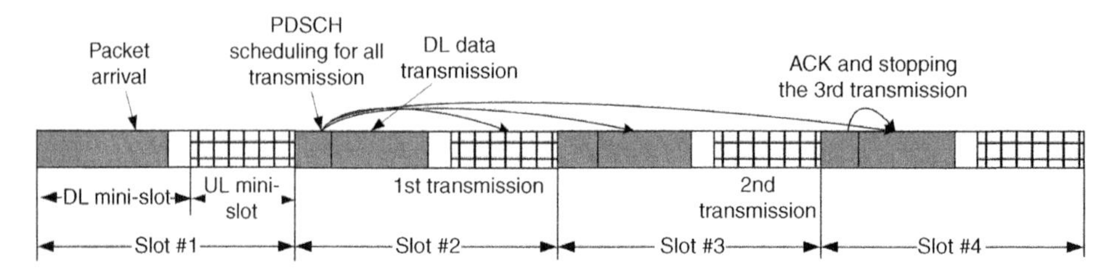

Figure 8.31 Example of DL HARQ processing with multi-slot scheduling.

cells or terminals improves SINR. As a rule of thumb, canceling the strongest or the two strongest interference sources is usually sufficient to achieve most of the potential gain. In LTE, for example with Inter-Cell Interference Coordination (ICIC) and Enhanced Inter-Cell Interference Coordination (eICIC), interference cancelation methods are proactive and semi-statically configured. Beamforming is another efficient way to reduce interference, especially considering high frequency bands. In 5G, more dynamic solutions tailored for increasing the reliability are recommended instead increasing the average capacity as done in LTE.

Seamless (Make-Before-Break) Handover The requirement of 0 ms handover interruption time is considered for 3GPP Rel-15 as part of the mobility enhancements for dual-connectivity, but the simultaneous transmission to or reception from two intra-frequency cells in either synchronous or asynchronous network mode may not be supported in Rel-15.

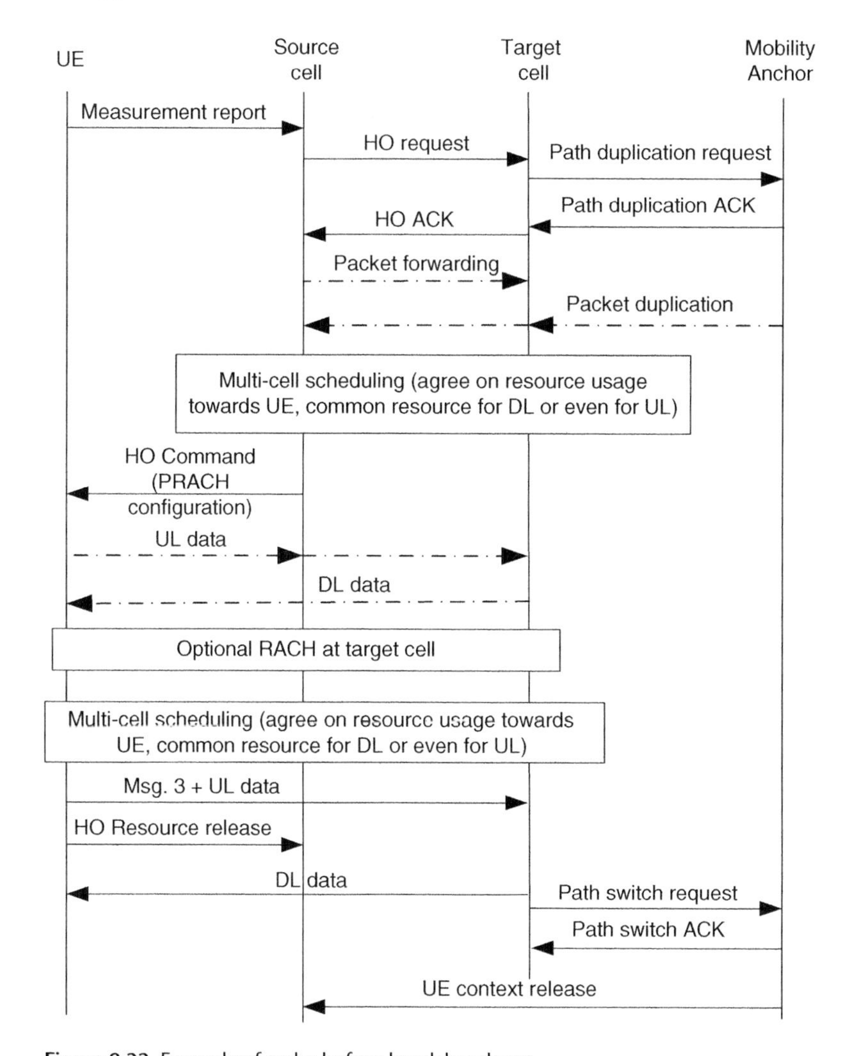

Figure 8.32 Example of make-before-break handover.

From concept perspective, make-before-break type of handover can be considered on the base of path duplication operation. In DL, the same data packet will be transferred from both source and target cell during handover. In UL, objective is to have both source and target cell receiving the same data packet from the UE with one UL transmission only. The cell from which the UE receives a correct timing advance (TA) command is referred to as master cell, the other cell as slave cell. Before RACH (Random Access Channel) in the target cell, the source cell can be the master. After UE gets TA from the target cell, the target cell becomes the master cell. Since TA parameter is not available from both cells, extra guard is needed to protect in case of potential interference. As the same resource transmission cannot happen twice in UL, the UE does not need to have multiple Transmitter Exchange (TX) chains supporting reliable handover, possibly adding delay. The received UL data can be combined at either source cell, target cell or even at the mobility anchor in the core network (Serving Gateway (S-GW)) depending on the operation mode. The following Figure 8.32 illustrates one example signaling diagram for enhanced handover to achieve 0 ms user plane interruption time.

References

All 3GPP specifications can be found under http://www.3gpp.org/ftp/Specs/latest. The acronym "TS" stands for Technical Specification, "TR" for Technical Report.

1 3GPP TR 38.913: "Technical Specification Group Radio Access Network; Study on Scenarios and Requirements for Next Generation Access Technologies", 2017.
2 3GPP TR 22.804: "Study on Communication for Automation in Vertical domains (CAV)", 2018.
3 H. Shariatmadari, Z. Li, S. Iraji, M. Uusitalo, and Riku Jantti: "Control channel enhancements for ultra-reliable low-latency communications", Proc. IEEE International Conference on Communications Workshops (ICC), May 2017, pp 1-6.
4 G. Pocovi, B. Soret, K. I. Pedersen, and P. Mogensen: "MAC Layer Enhancements for Ultra-Reliable Low-Latency Communications in Cellular Networks", Proc. IEEE International Conference on Communications Workshops (ICC), May 2017, pp 1-6.
5 K.I. Pedersen, G. Pocovi, and J. Steiner: "Punctured Scheduling for Critical Low Latency Data on a Shared Channel with Mobile Broadband", Proc. IEEE Vehicular Technology Conference, September 2017, pp 1-6.
6 3GPP TR 38.802: "Study on New Radio (NR) Access Technology; Physical Layer Aspects", 2017.
7 3GPP R1-1804618: "On UL multiplexing between eMBB and URLLC", 2018.
8 3GPP R1-1710992: "Discussion on multiple SR configuration", 2017.
9 3GPP TS 38.214: "Physical layer procedures for data", 2018.
10 Singh, B., Tirkkonen, O., Li, Z., and Uusitalo, M.A. (2018). Contention-based access for ultra-reliable low latency uplink transmissions. *IEEE Wireless Communications Letters* 7 (2): 182–185.
11 3GPP TS 23.501: "Technical Specification Group Services and System Aspects; System Architecture for the 5G System", 2018.

12 G. Pocovi, M. Lauridsen, B. Soret, K. I. Pedersen, and P. Mogensen: "Signal Quality Outage Analysis for Ultra-Reliable Communications in Cellular Networks", Proc. IEEE Globecom Workshops (GC Workshops), December 2015, pp 1-6.
13 3GPP TR 38.323: "Packet Data Convergence Protocol (PDCP) specification", 2018.
14 V. Hytönen, Z. Li, B. Soret, and V. Nurmela: "Coordinated multi-cell resource allocation for 5G ultra-reliable low latency communications", 2017 European Conference on Networks and Communications (EuCNC), June 2017.
15 3GPP R1-1804616: "On further study of compact DCI for URLLC", 2018.
16 H. Shariatmadari et al: "Control channel enhancements for ultra-reliable low-latency communications", Proc. IEEE ICC Workshops, 2017.
17 Shariatmadari, H. et al. (2017). Resource allocations for ultra-reliable low-latency communications. *International Journal of Wireless Information Networks* 24 (3): 317–327.
18 H. Shariatmadari et al.: "Asymmetric ACK/NACK detection for Ultra-Reliable Low-Latency Communications", to be published at EuCNC'18, June 2018.
19 3GPP TS 36.104: "Base Station (BS) radio transmission and reception", 2018.

9

Massive Machine Type Communication and the Internet of Things

Devaki Chandramouli[1], *Betsy Covell*[2], *Volker Held*[3], *Hannu Hietalahti*[4], *Jürgen Hofmann*[3] *and Rapeepat Ratasuk*[2]

[1] *Nokia, Irving, TX, USA*
[2] *Nokia, Naperville, USA*
[3] *Nokia, Munich, Germany*
[4] *Nokia, Oulu, Finland*

9.1 Massive M2M Versus IoT

'What is machine to machine (M2M)?', 'What is the Internet of Things (IoT)?', 'how do they relate to each other?' are some of the commonly asked questions when it comes to machine type communication (MTC) and IoT. In this section, we will attempt to answer these questions in simple terms and also provide an overview. In simple terms, IoT refers to 'anything and everything that is connected', which includes machines and humans. M2M refers to 'communication between devices using any communications channel, including wired and wireless'. In this sense, M2M is a subset of IoT.

IoT, as the name implies, can be considered more like an Internet-like service, where a lot of cross-cultural data and control points are available to a variety of applications or users. Consumers may use various parts of the data and control capabilities for their own purposes.

The classical M2M application is a remote measurement or control application with a dedicated purpose. Measurement and control operations can be performed at well-defined control points, and the resulting data and control procedures are meaningful only within the application they are designed for.

Each M2M measurement application owns the data related to the application, but IoT allows mining of data from multiple sources. The M2M measurement and control data also have a unique meaning in the context of the application. These data may be private and not disclosed, not useful to anyone else except for its owner. An M2M network and traffic grows by installed applications and sensors. An IoT network grows along with the same principle, but the traffic grows based on applications that can benefit from the IoT data that are collected over the network. In this sense the grow of an IoT network might not be linear with the number of sensors. Thus, the data collected by a simple telemetry-type remote control M2M system may attract new users when the coverage of the service reaches volumes that become attractive for new consumers of the same data. Wider use of the same data also involves synthesis of data from multiple sources, related with multiple applications. For instance, the information collected by smart cars

5G for the Connected World, First Edition.
Edited by Devaki Chandramouli, Rainer Liebhart and Juho Pirskanen.

can initially help the individual drivers of those new cars, but when there are more cars acting as 'sensors' for the smart car community, the collected information can be used for traffic control. Combined with the weather data, it can improve road safety, when the risks observed by one car become available to other cars that are approaching the risky spot.

Because IoT is a mesh network with many users, unlike in M2M, the traffic volume of IoT does not directly depend on the size of the network in terms of sensors and actuators and the activity of the related application. For IoT, more data means more potential consumers for the same data when an available IoT service exceeds volumes and coverage that are critical for a service that has not been able to benefit from scarce information in a patchy network.

A practical example can be found in the evolution of the modern car, where automation has progressed from human driver operated controls to automated control processes via several phases of integration. In the first step, the automated processes, like adjustment of the ignition advance remained an enclosed system with no interface to the other sub-systems of the same car, and especially not interfacing with any other system outside of the car. All measurement and control data are meant for an individual use case, and they are application specific.

The second step introduced higher inter-system awareness and shared data. Both applications have started to interoperate with each other and the measurement data from each sensor is shared by many applications. Automatic transmissions started their life as standalone units but have now been integrated with the other sub-systems, considering also the status info from the engine, cruise control, temperature, and so on. The measurement data of a wheel sensor that was initially just driving the speed is now used also for anti-lock brakes, traction and stability control, turning up the volume of the radio at high speeds, and covering the blind spots of satellite navigation when driving in a tunnel without a satellite signal. This is still not yet IoT, even though the characteristics of the application interaction and shared data are met, if the connectivity is still missing.

IoT characteristics of shared data and interaction of applications are met when we connect such an automated vehicle. Not only can the vehicle benefit from the weather, navigation, traffic load, and possible warning information, but also the other vehicles can mutually re-use the sensor information provided by other vehicles on the same road, and the traffic environment can be adjusted for traffic light settings, speed limits, and so on. IoT at its best allows new control logic simply by the addition of the application to re-use the already existing IoT data in a new way. When everything is already connected with everything else, it is up to each application how they use the provided data, and installation of new sensors for new control logic might not be necessary at all.

In summary, M2M and IoT have the following characteristics:

1. *M2M Characteristics*
   - Connected Devices and associated applications;
   - Fixed Solution parameters and rigid architecture;
   - 'Speed' designed in where necessary;
   - Applications in the context of verticals and niches;
   - Data is meaningful in context;
   - Structured data;
   - Predictable growth (in connection with generated data); and
   - Data ownership often clear.

2. *IoT Characteristics*
   - Complex applications and data analysis;
   - Heterogeneity of components and distributed and federated processing/storage/query;
   - 'Speed' needs to be supported as and when requirements emerge;
   - Cross-vertical and cross-functions;
   - Semantic richness, shared context and ontologies;
   - Semi-structured and unstructured data;
   - Unpredictable growth driven by network effects; and
   - Data ownership often unclear.

9.2 Requirements and Challenges

There are many different use cases for IoT such as home automation, car to car communication, industry automation, health monitoring, smart cities connectivity, and the vast area of connected wearables. The requirements, found in TS 22.261 [30], enabling a communication network to support these diverse use cases fall into following categories:

- Long battery life;
- Mobility patterns;
- Bursty data;
- Bulk device management;
- Flexible subscription management; and
- Security.

Each use case can be fulfilled using a combination of requirements from among these categories.

9.2.1 Long Battery Life

Many IoT devices, e.g. smart meters, can be expected to have a very long life-cycle. For many sensor applications, it can further be expected that the device has no other power supply than the original battery. Imagine a sensor embedded in a bridge to detect signs of wear and tear over the life of the bridge. This sensor must be able to work for many years in a situation where it is not connected to an external power supply and where it is not easily possible to replace the battery. Progress in battery development is one part solving the problem, more efficient use of radio and network resources can also extend the device's battery lifetime. There are trade-offs between reachability, data rates, bandwidth, and power consumption, that can be managed through architectural and protocol enhancements in a Third Generation Partnership Project (3GPP) system that will extend device battery life for IoT devices while enabling their full functionality.

9.2.2 Mobility Patterns

Typical wireless networks are optimized for handling the mobility of devices that move quickly (e.g. on a car or train), frequently (e.g. on a delivery truck), or cover large distances (e.g. a global traveller). IoT includes the additional cases that connected devices are stationary (e.g. the embedded sensor), geographically limited (e.g. items in a warehouse), and nomadic (e.g. stationary while in use, but can change location when not in use). Providing the same kind of mobility for these different types of devices is not efficient and results in unnecessary power consumption both in the network and in the device. Adaptable mobility management requirements allow optimal support for each type of device.

One basic requirement is to identify the type of mobility an IoT device requires: stationary, geographically limited, nomadic, fully mobile. From there, different types of mobility can be applied to most efficiently support each device type. Depending on the type of mobility, handover, tracking area updates, and location updates may not need to be supported for this device, providing significant savings in resource usage and extending device battery life for low and no mobility devices.

9.2.3 Bursty Data

Wireless systems were originally designed for smart phones and PC dongles, therefore aimed at maximizing the bandwidth for the users within the capabilities provided by the radio network. This was done under the assumption of a continuous data flow for voice, downloading (e.g. web surfing), or streaming applications. IoT introduces the concept of bursty data, which places different requirements on the wireless system. Bursty data can consist of either a small payload or a large payload, the key criteria being the infrequent, yet regular, transmission of data. Consider a camera installed on a street corner which periodically sends a brief report, (e.g. text or snapshot) that 'all is well'. In case of an accident, the camera is equipped to send streaming video for a certain period, after which it reverts to the brief report format. Traditional user plane setup overhead becomes much more impacting for these bursty data use cases than for voice or lengthy but episodic data transmissions. Session setup delay is also a restrictive factor for some bursty applications. For example, in the corner camera scenario, the video must be sent immediately to fully capture details of the accident. 3GPP has made several enhancements to improve the ratio of payload, session setup, and message overhead to better meet the needs of IoT devices with bursty data.

9.2.4 Bulk Device Management

In 2G, 3G, and 4G systems, it is expected that a user has only a few devices (e.g. one smartphone and one tablet) registering in the network. In an IoT marketplace, a user may range from a person with a dozen devices at home (e.g. smartphone, tablet, thermostat, printer, lights) to an enterprise with thousands of devices (e.g. smart parking meters). The 'one at a time' device management tools provided in 2G, 3G, and 4G for subscription management, activation/deactivation, billing, etc. are not efficient for managing hundreds and thousands of devices with common characteristics belonging to a single owner. 5G will specify bulk provisioning and management tools to more efficiently apply common treatment such as activation/deactivation and subscription updates to a huge number of devices.

9.2.5 Flexible Subscription Management

In 2G, 3G, and 4G systems, user equipments (UEs) are typically expected to belong to one subscriber with no need to change the subscription from the time a UE is activated until it is deactivated. IoT comes with a new business model, where a device, such as a UE embedded in rental equipment, may be passed from one 'user' to another. This model introduces the need to associate a UE in 5G with a new user and subscription in an easy and resource efficient manner.

As IoT continues to grow, it is expected that devices will be manufactured in a more generic manner than UEs today. This means, IoT devices will not be designed for use in a specific operator's network, rather they will be designed and built to be used anywhere in the world, with any mobile network. Consider a manufacturer of smart flower pots. The manufacturer does not want to build different pots for sale in different parts of the world that work with only one operator. From the flower pot manufacturer point of view, it would be better to build smart flower pots that can be sold anywhere and work with any network. It can also be expected that IoT devices will need the flexibility to update and change subscription data as the owner moves. For example, if the owner moves to another country, they will likely find it more desirable to set up a subscription with the new local operator rather than operate the flower pot in a roaming mode.

The 2G, 3G, and 4G paradigm still assumes a human user will insert a Universal Integrated Circuit Card (UICC) into a UE, something not feasible in many IoT cases and by far not cost effective when it comes to small IoT devices that regularly change ownership. To address this issue, the Global System For Mobile Communications Association (GSMA) has defined the E-SIM (Embedded Subscriber Identity Module) specification that partly addresses this shortcoming. The E-SIM is a reprogrammable chip on the device that comes in different shapes and sizes and can be updated over the air, allowing simplified change of the mobile operator by just updating the software. Additionally, 5G specifies more resource efficient mechanisms to administer devices, thus reducing overhead processing.

9.2.6 Security

While the E-SIM solves some problems with device portability, still other solutions are needed to meet the needs of providing security for IoT devices. IoT devices come in a variety of sizes ranging from huge mining trucks to tiny sensors, have different physical characteristics such as a smart T-shirt that will endure multiple washings, and have varying privacy requirements from the high privacy needed for medical device communications to low or no privacy for advertisements sent from a shop to passers-by. 5G will fulfil requirements to support new security mechanisms. Additional work is needed in this area to ensure that the diversity of security requirements can be met in a manner maintaining the security of the mobile operator's network.

9.2.7 Others

Support for the IoT involves many diverse scenarios including:

- large numbers of devices requiring highly reliable instantaneous communication (e.g. V2X);
- smart wearables and other Personal Area Networks (e.g. home, office); and
- sensor networks (many devices, bursty data, static location, reusable, long battery life, low cost devices).

A key challenge is meeting these diverse, and often conflicting, requirements in a network. Fortunately, 5G also includes many new capabilities such as network slicing allowing mobile network operators (MNOs) operating different logical networks simultaneously, tailored to most efficiently meet the needs of a specific type of application.

Based on 2G, 3G, and 4G networks there are already many deployed IoT use cases, e.g. monitoring patients while they are at home or familiar use cases to support portable credit card readers at the point of sale. Another example is Amazon's E-book allowing readers to connect to the bookstore via communication networks, without end user intervention. Common to all of these use cases is their 'vertical' M2M nature, i.e. the company providing such services builds the whole service – or has it built by a specialized company, provides the end user a telemetry device, arranges the connectivity with an operator, monitors the data received and so on, which on one side perfectly fits the needs of the company, but can have downsides when this 'vertical silo' M2M application needs to be extended to support new functionality, fulfil capacity, reliability, or latency requirements. While the silo nature of M2M communication is being addressed by industry for one such as oneM2M which specifies how data collected via IoT devices can be made available to other services when needed, the capacity growth and reliability aspects are being addressed by 3GPP and the new 5G standard.

While 2G, 3G, and 4G systems already support many IoT use cases there are several shortcomings of these systems that 5G needs to tackle. First, considering the expected huge number of IoT devices connected in future, ranging in the order of tens of billion devices in the next decade and growing into the trillions beyond, it becomes obvious that 3G and 4G technology cannot handle this growth rate as they were designed to support much smaller numbers of devices. One limiting factor is, e.g. the maximum number of one billion International Mobile Subscriber Identities (IMSIs) that can be supported in a single cellular public land mobile network (PLMN), another one the ability of a network to keep billions of devices connected at the same time.

9.3 Technology Evolution

Due to the vast number of machines resulting in high signalling load while average revenue per machine is rather low, MNOs are looking for ways to increase revenue with M2M applications by limiting the network resources allocated for such devices and optimizing network procedures for special use cases highlighted in the previous sections.

3GPP has been working on improvements necessary for M2M devices and applications over a period of time (starting in 2009 with Release 10, see Table 9.1 and [1], [2]). Figure 9.1 shows the evolution of M2M features in 3GPP. Until 3GPP Rel-14, M2M features were focused on enhancements to Long Term Evolution (LTE), Narrow Band Internet of Things (NB-IoT), GSM/EDGE RAN (GERAN) radio access technologies, Evolved Packet System (EPS), Cellular Internet of Things (CIoT) and General Packet Radio Service (GPRS) system architecture, respectively. The 5G era starts with 3GPP Rel-15. Since NB-IoT was introduced in 3GPP Rel-13 and the new NB-IoT technology adoption in the market for M2M/IoT devices is expected to take several years (use of a certain technology for use cases such as smart meters is expected to be in the order of 10 years, considering the expected battery lifetime), 3GPP didn't prioritize M2M/IoT specific features for the new 5G System in its first release. However, the 5G System

Table 9.1 Radio and system architecture feature evolution for M2M.

3GPP release	Radio	Architecture
Release 10	LTE, GERAN eWaitTimer	EPS, GPRS Congestion and overload control, Signalling reduction features
Release 11	LTE, GERAN Extended Access Barring	EPS, GPRS Device triggering, support for SMS with PS only subscription, support MT-SMS without MSISDN (with an external identifier)
Release 12	LTE	EPS, GPRS Small data transmission, Low power consumption, dedicated CN node selection
Release 13	NB-IoT, LTE-M enhancements, EC-GSM-IoT	Dedicated Core Network (DECOR), GROUP related enhancements, Monitoring, Support for capability exposure, CIoT enhancements for NB-IoT radio
Release 14	NB-IoT enhancements LTE-M further enhancements EC-GSM-IoT radio interface enhancements	CIoT enhancements, Non-IP GPRS
Release 15		North bound APIs
Release 16		Study item on CIoT enhancements for 5G Study item on massive IoT messaging in 5G

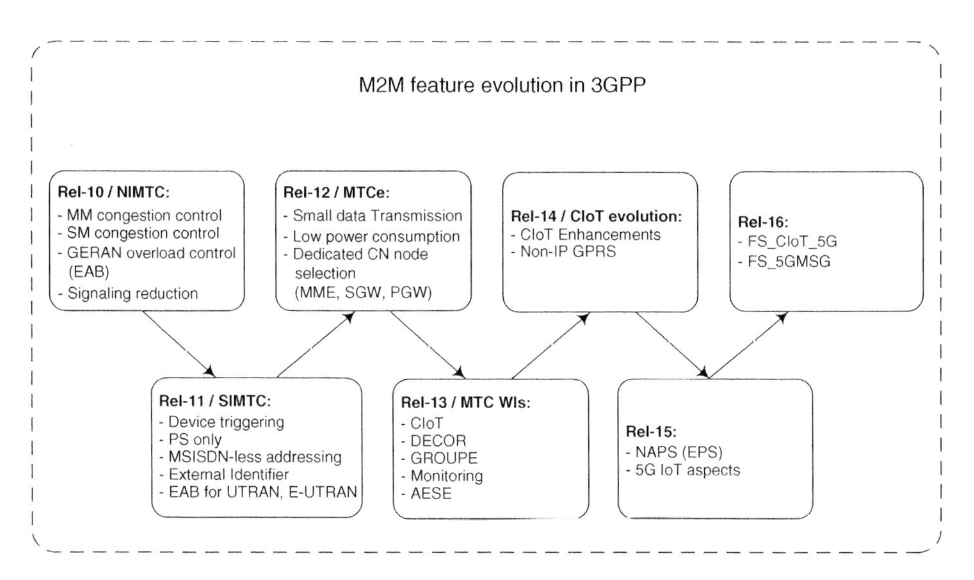

Figure 9.1 System architecture feature evolution for M2M.

introduces features that are considered as 'enablers' to support M2M/IoT devices. Such enablers are explained later in this chapter.

At the time of writing, two interesting IoT related study items have been agreed for 3GPP Rel-16, with the intention to continue this evolutionary 3GPP system development. The topics of the new study items are 'Study on Cellular IoT support and evolution for the 5G system' (FS_CIoT_5G) and 'Study 5G message services for Massive IoT' (FS_5GMSG).

9.4 EPS Architecture Evolution

In the Rel-10 timeframe, 3GPP studied several M2M application scenarios to generate requirements for 3GPP network system improvements for MTC (see also [1], [2]). The objective was to identify 3GPP network enhancements required to support a large number of MTC devices in the network and to provide necessary network enablers for MTC services. Specifically, transport services for MTC as provided by the 3GPP system and the related optimizations were considered, but also aspects for ensuring that data and signalling traffic related to MTC devices does not cause network congestion or system overload. It is also important to enable network operators to offer MTC services at a low-cost level, to match the expectations of mass market machine-type services and applications.

The main MTC functionality specified by 3GPP in Rel-10 provides overload and congestion control. Considering that some networks already experienced congestion caused by M2M traffic, overload and congestion control was considered as high priority. A set of functions and features was specified. It includes introduction of low priority configuration, mobility management congestion control, session management congestion control, radio resource control (RRC, see [9]) connection reject, signalling reduction features and extended access barring for MTC devices. A low priority indicator is sent by the UE to the network so that radio access network (RAN) and Core can take it into account in case of congestion or overload situations (e.g. reject a higher percentage of connection requests from low access priority devices). Some of these functions are also available for terminals that are not specifically considered as low priority access terminals (e.g. smart phones). Furthermore, some already deployed M2M devices are generally using 'normal' access (i.e. do not provide the low priority access indicator). The full range of MTC congestion and overload control means becomes available when terminals are specifically configured for MTC use.

Features and requirements such as device triggering, packet switched (PS) only subscription, E.164 number shortage was addressed under the work item 'System Improvements for Machine Type Communication (SIMTC)' in 3GPP Release 11.

Following are the key features introduced as part of this work and documented in TS 23.682 [27]:

- Enhanced architecture including new functional entities called Machine-Type Communication Interworking Function (MTC-IWF) and MTC-AAA;
- Identifiers (Mobile Subscriber ISDN number, MSISDN, less operation) – Usage of Internet-like identifiers at the external interface between PLMN and service provider domain to replace MSISDN;
- Addressing – IPv6 was recommended for usage with MTC devices;

- Device Triggering – MT-SMS (mobile terminating-short message service) with a standardized interface to the SMS Service Center (SMSC);
- Optimizations for devices with PS only subscription;
- Dual-priority devices – certain applications can override low access priority configuration;
- Extended Access Barring (EAB);
- SMS in mobility management entity (MME) configuration (architecture option for networks with no universal terrestrial radio access network [UTRAN] or GERAN circuit switched [CS] domain where a direct interface from SMSC to MME for SMS delivery is deployed).

9.4.1 MTC Architecture

In Rel-11, 3GPP mainly introduced a new interworking function (MTC-IWF) (Figure 9.2) for service providers to interconnect with the mobile operator network. The MTC-IWF enables device triggering via control plane, identifier translation and

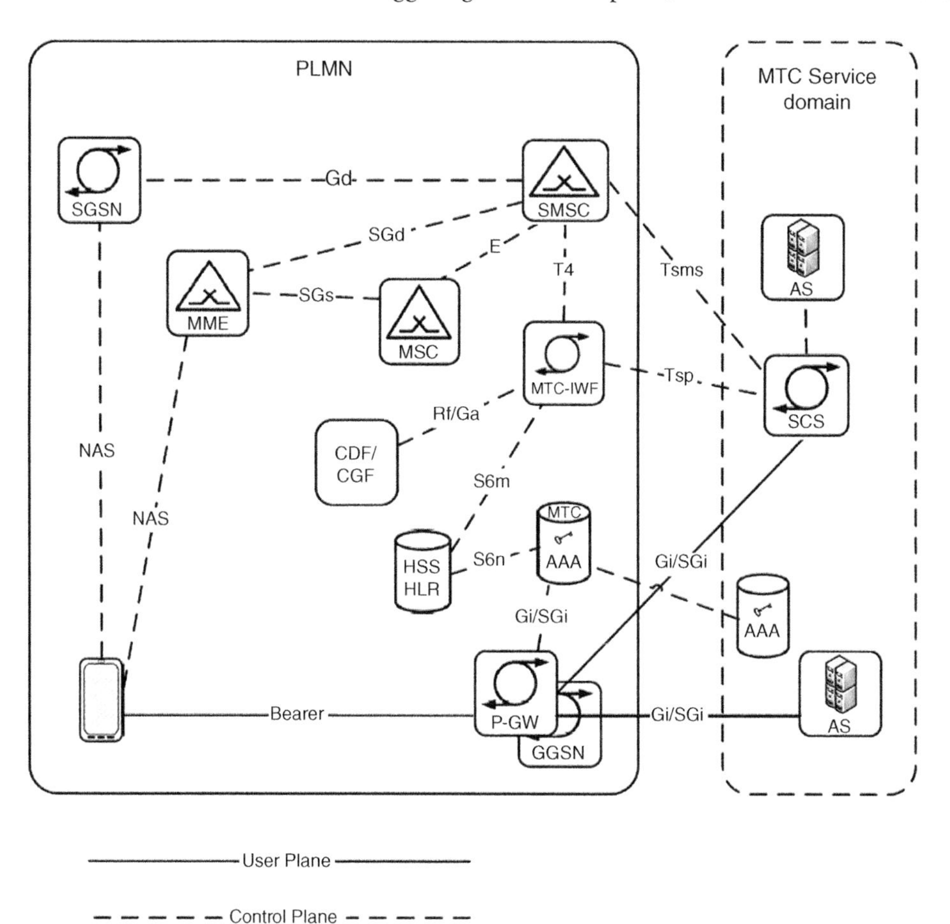

Figure 9.2 MTC architecture.

other features that might be introduced in future. The end-to-end communication between the MTC application in the UE and the MTC application server (AS) may use services provided by the 3GPP system, and optionally services provided by a Services Capability Server (SCS). The MTC Application in the external network is typically hosted by an AS. The SCS can be in the service provider domain (as shown in Figure 9.2) but can also be hosted by the mobile operator as a kind of Service Delivery Platform. In the latter scenario, the SCS can implement charging and security functions.

While the MTC-IWF serves as first contact point for requests coming from the SCS and provides security, charging and identifier translation (external to internal identifier) at the ingress of the PLMN, the newly introduced MTC-AAA function translates the internal identifier (IMSI) at the network egress to the external identifier(s) before forwarding authentication, authorization, and accounting (AAA) requests to an AAA server in the service domain. This avoids exposure of the IMSI outside the operator domain. MTC-AAA can work in server or proxy mode and has interfaces to the PDN Gateway (PGW)/Gateway GPRS Support Node (GGSN) (Gi/SGi) (where AAA requests originate from), Home Subscriber Server (HSS) (S6n) (to retrieve external identifier(s) for a given IMSI and vice versa) and external AAA servers. MTC-IWF receives a device trigger request from the SCS over the Tsp interface and forwards it to the SMSC via T4. It receives subscription data including the IMSI from the HSS via S6m and provides charging data via the existing interfaces Rf/Ga to the charging gateway.

Different deployment models are possible for MTC allowing support for different service level agreements between MNO and service provider:

- *Direct model.* The AS connects directly to the operator network for direct user plane communication with the UE without the use of any SCS; this allows for simple implementation of OTT (over-the-top) applications; OTT deployments are transparent to the PLMN;
- *Indirect model.* The AS connects indirectly to the operator network through the services of a SCS to perform indirect user plane communication with the UE and to utilize additional value-added services (e.g. control plane device triggering). The SCS is either:
  - o MTC Service Provider controlled: The SCS is an entity outside of the operator domain and Tsp is an external interface (i.e. to a third party MTC Service Provider) or
  - o 3GPP network operator controlled: The SCS is an entity inside the operator domain and Tsp is an internal interface to the PLMN;
- *Hybrid model.* The AS uses the direct and indirect models simultaneously to connect directly to the operator's network for direct user plane communication with the UE while also using SCS based services. From 3GPP network perspective, the direct user plane communication from the AS and any value-added control plane related communication from the SCS are independent and have no correlation to each other even though they may be servicing the same MTC Application hosted by the AS.

Since different models are not mutually exclusive, but just complementary, it is possible for a network operator to combine them for different applications. This may include a combination of both MTC Service Provider and 3GPP network operator controlled SCSs communicating with the same PLMN.

9.4.2 Identifiers

As mentioned earlier, a shortage of E.164 numbers is an additional driver for optimizations and improvements in mobile networks. This urged the need to define Internet-like identifiers such as Fully Qualified Domain Names (FQDN), Uniform Resource Names (URN) or Uniform Resource Identifiers (URI) for subscriptions without MSISDN. Such identifiers are referred to as external identifiers.

One IMSI may have one or more external identifier(s) stored in the HSS assigned to it. Rationale behind one to many mapping is twofold. A single device may have several applications running on the device and each application may use its own external identifier. Alternatively, a single device may have subscriptions with several service providers for different applications and each service provider may assign its own external identifier. Although this approach provides more flexibility for deployments, it comes with some drawbacks. At the border between PLMN and service domain external identifiers are used and the PLMN translates them to one internal identifier like the IMSI for usage within the network. Reverse mapping (e.g. for MO-SMS, at Gi/SGi interface) from internal to external identifier may then cause issues in terms of uniqueness of the reverse translation. Choosing the correct external identifier in such scenarios was not resolved in Release 11.

The External Identifier is required to be globally unique and has the following components:

- *Domain identifier*. Identifies a domain that is under the control of the MNO; SCS/AS use the domain identifier to determine the correct MTC-IWF.
- *Local identifier*. Used to derive and obtain the IMSI. It shall be unique within the applicable domain and is managed by the MNO.

The external identifier will use the format of a network access identifier (NAI), i.e. username@realm, as specified in clause 2.1 of Internet Engineering Task Force (IETF) RFC 4282 [62]. The username part of the External Identifier contains a Local Identifier. The realm part contains a Domain Identifier. As a result, an external identifier will have the form '<Local Identifier>@<Domain Identifier>'. This identifier will mainly be used at Tsp, S6m, S6n, T4, Rf/Ga interfaces.

9.4.3 Addressing

To cope with the potentially very huge number of machines connecting to the network, IPv6 is recommended by 3GPP as the preferred addressing format for devices subscribed for MTC. Addressing is a deployment topic, thus use of a certain IP version is an operator decision.

9.4.4 Device Triggering

Device Triggering is needed for delivering a message towards a UE and a specific application on that UE, if the IP address of the UE is not known by the SCS/AS. MTC devices can be triggered when they are attached to the network, with and without an existing packet data protocol (PDP)/packet data network (PDN) connection. In current deployments SMS is used to trigger attached devices but in the early releases before the introduction of the external identifier concept, this required an MSISDN allocated to each

MTC subscription. As MSISDNs are limited in some regions (e.g. in the US and China) solutions need to be found that do not need a require MSISDN per MTC user. In addition, solutions using Internet-like identifiers like NAI are more flexible as such identifiers can be allocated more freely on a per need basis. It must be noted that devices with an established PDP/PDN connection can register their IP address OTT at the AS by application layer means. Thus, the server can trigger the device by sending an application layer trigger without the need to use 3GPP network capabilities. However, when the SCS requests the 3GPP network to trigger a MTC device it can provide the appropriate identifier in the request and the network translates this external identifier into an internal one (e.g. the IMSI) used to trigger the device. The device could be triggered by different means such as SMS, Cell Broadcast messages, session initiation protocol (SIP) messages (Instant Messaging or SMS over IP), or via some new path traversing the MME/SGSN (Serving GPRS Support Node) and/or HSS/HLR (home location register) (e.g. using hypertext transfer protocol (HTTP), DIAMETER/MAP, and non-access stratum (NAS, see [15]) as transport means). However, in Release 11 SMS is the only standardized mechanism that was adopted for device triggering. Cell broadcast messages are used by some operators for triggering groups of devices, but this is a proprietary solution. The external identifier must be stored in the HSS/HLR to allow the network to translate the external request coming from the SCS into an internal trigger request using the proper internal identifier. One device may be assigned multiple external identifiers, thus the HSS/HLR needs to store one IMSI with many external identifiers. Figure 9.2 shows the MTC architecture for device triggering with the Tsp (sp = service provider) interface between SCS and 3GPP network. Tsp is used by the SCS to send a trigger request to the PLMN using the External Identifier for identifying the target device. Tsp is based on DIAMETER and terminates at the MTC-IWF within the PLMN. The MTC-IWF sends the trigger request to the SMSC using the T4 interface, which is also based on DIAMETER and described in TS 29.337 [32]. Device triggering over Tsp/T4 is the only standardized method for triggering in Release 11 (besides application layer triggering). Optionally, the SCS/AS can send a device trigger SMS via the Tsms interface directly to the SMSC. Tsms is the existing legacy interface between a Short Message Entity (SME), e.g. the SCS/AS and the SMSC to send and receive short messages.

9.4.5 PS-only Service Provision

PS-only service provision is providing a UE with all subscribed services via the packet (PS) domain of a mobile network. PS-only service provision implies a subscription that allows only for services exclusively provided by the PS domain, i.e. data bearer services and SMS. Support of SMS via PS domain NAS is a network deployment option and may also depend on roaming agreements. Therefore, a subscription intended for PS-only service provision may also allow for SMS services via CS domain to provide a UE with SMS services in situations when serving node or network doesn't support SMS via the PS domain. The functionality that enables PS-only service provision is described in TS 23.060 [6] and TS 23.272 [13].

9.4.6 Dual Priority Devices

As mentioned above, low priority (or delay tolerant) access configuration was introduced in Release 10 to aid with congestion and overload control when millions or billions of M2M devices are trying to connect to the network. There may however be circumstances when such devices need to access the network for higher priority services. Following are some example scenarios:

- Electricity meters sending a daily report (of the per hour usage) using 'low priority' indication, but, may want to send an alarm without 'low priority', if the meter is being tampered with or is being vandalized.
- A road temperature sensor could send daily 'I'm still working' reports using 'low priority', but, when the temperature falls to sub-zero, immediately send a warning to the control centre without 'low priority'.
- A M2M module which hosts multiple hybrid applications; the room temperature application always requires data transmission using 'low priority' and video streaming application requires data transmission without using 'low priority'.

As a result, it is possible that an application overrides the default low priority setting on rare occasions for establishing normal connections. To accomplish this, a new configuration parameter called 'override low priority access' was introduced. Devices with both low priority access and override low priority access configurations are dual priority devices. Override low priority access indicates to the UE that an application can connect to the network without setting the low priority indicator (e.g. in PDN connection request messages). PDN connections marked as low priority and not marked as low priority may co-exist. When the UE has PDN connections established with low priority and without low priority, it can establish mobility management procedure and RRC connections without low priority ([9]).

9.4.7 Extended Access Barring

EAB is a mechanism to restrict network access for low priority devices. It is activated by the RAN. A network operator can restrict network access for UEs configured for EAB in addition to the common access control and domain specific access control. The UE can be configured for EAB in the Universal Subscriber Identity Module (USIM) or in the mobile equipment (ME). When EAB is activated in the radio base station (e.g. via OA&M) and the UE is configured for EAB, it is not allowed to access the network. When the UE is accessing the network with access classes AC 11–15, and that access class is not barred, the UE can ignore EAB. Also, if it is initiating an emergency call and emergency calls are allowed in the cell, it can ignore EAB. The UE can also respond to paging when barring is active, this is under the assumption that the network will initiate paging only when there is no congestion situation. Dual priority devices may also be configured with override EAB configuration.

9.4.8 Short Message Service in MME

SMS in MME was introduced mainly to address requirements from operators who do not deploy a 3GPP mobile switching centre (MSC) (thus no SGs interface is available) and do not want to support MAP in their network. SMS over IP (i.e. SMS over Internet Protocol Multimedia Subsystem [IMS]) could be one solution, however the concern with this solution was the need for an IMS/SIP client in the devices and not all devices (e.g. machine type devices, dongles) will have such a client implemented. Furthermore,

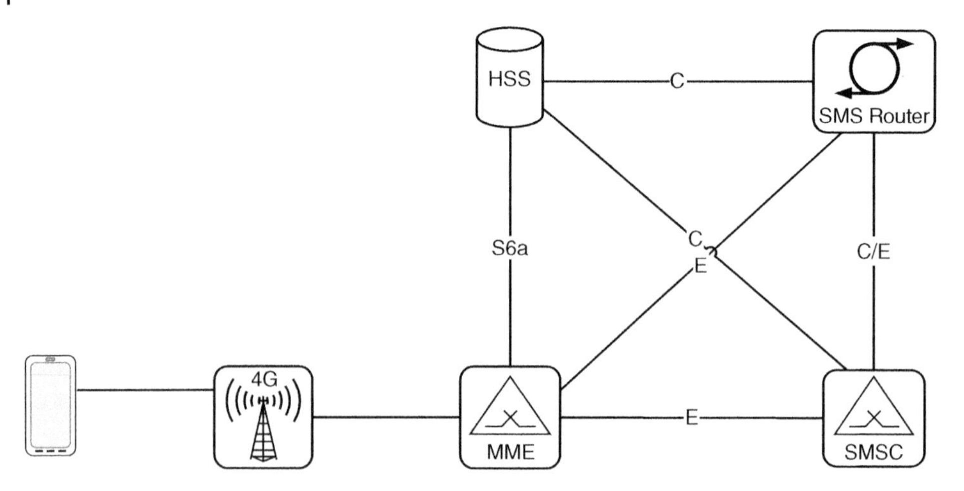

Figure 9.3 SMS in MME architecture.

inbound roamers whose home operators do not support IMS cannot be offered SMS over IMS and thus they will need support for SMS over NAS. These factors resulted in the need to introduce a new architecture for supporting SMS services in evolved packet core (EPC) defined in TS 23.272 [13] (Figure 9.3). This feature can be enabled or disabled in the MME via configuration.

From UE perspective it is transparent whether SMS in MME or SMS over SGs is offered by the network. The UE will perform combined EPS/IMSI attach (or combined tracking area updating [TAU]) to obtain SMS services. The network can decide to offer SMS over SGs or SMS in MME depending on several factors such as user's subscription (PS only, PS, and CS), user requested service (SMS only or SMS and voice), support for the feature in general and local policies. If the UE is performing 'combined attach' to request for SMS services only and the network supports SMS in MME, the network need not establish a SGs association between MME and MSC. The network will then indicate 'SMS only' in the accept message to inform the UE that it has been attached only for SMS services. To keep it transparent to the UE, MME will include a non-broadcast location area identity (LAI) ('dummy LAI') and a reserved temporary mobile subscriber identity (TMSI) in the combined attach accept or combined TAU accept. This ensures backward compatibility so that legacy UEs consider the attach procedure to be successful. Between UE and MME, SMS messages as defined in 3GPP TS 23.040 [31] are encapsulated and transferred within NAS messages.

A third release of improvements was developed for MTC devices and mobile data applications running in smart phones. This work was covered by the Rel-12 feature 'Machine-Type and other mobile data applications Communications Enhancements (MTCe)'. Two main features are part of Rel-12:

- Device triggering enhancements and Small Data Transmission (infrequent and frequent)
- UE power consumption optimizations.

Power consumption is important for UEs using a battery and for UEs using an external power supply. Its importance increases with the continued need for energy savings.

Following are the solution options that were considered in 3GPP to achieve power savings:

- Introducing Extended discontinuous reception (DRX) cycle in idle mode
- Introducing Power saving mode for devices
- Introducing long DRX cycles in connected mode
- Keeping UEs in detached state when not communicating.

Finally, 3GPP agreed to introduce Power Saving Mode (PSM) feature for devices in Release 12.

9.5 Cellular Internet of Things

9.5.1 General

IoT applications are in general assumed to be access agnostic. The application layer needs connectivity (e.g. IP connectivity) to transport application specific payload between IoT device and AS. However, especially the connectivity and mobility pattern of typical telemetry type use cases differs substantially from that of mobile smart phones which are usually always online. The EPS design focussed mainly on highly mobile always-on use cases. The IoT use case was not supported in an optimal way, until the MTC related work items improved the efficiency of infrequent and small data usage patterns. The main stage 2 architecture specifications for CIoT are TS 23.401 [14] and TS 23.682 [27].

Up to Rel-12, the 3GPP M2M work introduced compatible and evolutionary improvements to serve MTC devices more efficiently within the existing GPRS and EPS architectures. CIoT in Rel-13 took a revolutionary approach by introducing a new architecture supporting small data transfer with the 'CIoT EPS Optimization' feature set that is not fully backwards compatible with the Rel-12 EPS architecture.

The Service Capability Exposure Function, SCEF was introduced in the earlier 3GPP releases for monitoring the UE mobility and registration status. It is re-used in CIoT for user data transport. The Rel-14 work item 'Non-IP_GPRS' imported some selected Rel-13 CIoT EPS Optimizations to GPRS, allowing the support of Non-IP Small Data and the user plane Small Data via SGSN and SCEF. The PS domain and the triggering part of the architecture is shown in Figure 9.4.

The SCEF handles Non-IP PDN connections over the control plane. The PGW handles all types of PDN connections over control and user plane.

Also, other energy saving methods, such as UE PSM and extended idle mode DRX (eDRX) can be used in conjunction with CIoT EPS Optimizations. A UE that needs extremely long battery life benefits from eDRX or PSM. Both methods lead to extended but predictable UE unreachability periods, and consequently the need for the network to buffer downlink (DL) data towards UEs that may be unreachable due to their power saving cycle for up to some hours. The maximum eDRX cycle length depends on the hyper frame structure of the used radio. For example, the LTE maximum DRX cycle of 2.56 seconds can be extended via eDRX to more than 42 minutes, and in NB-IoT up to almost three hours.

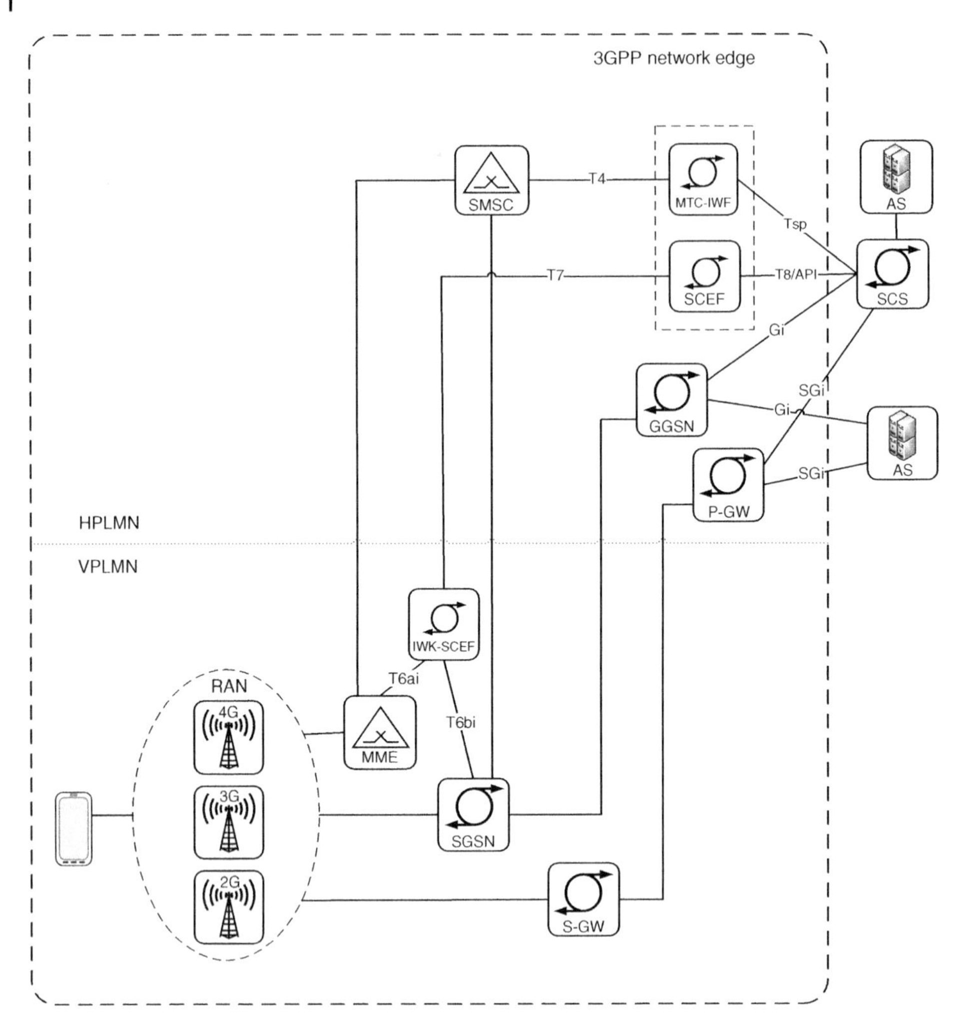

Figure 9.4 CIoT architecture.

9.5.2 CIoT EPS Optimizations

One of the main goals of CIoT EPS optimizations is to allow reduction of the power budget and cost of a CIoT UE. While the legacy 3GPP 2G, 3G, and 4G technologies were optimized for continuous big data, downloading and streaming use cases of smart phones and PC dongles, CIoT is optimized for infrequent one-shot small data transfer. Whether the CIoT EPS Optimizations lead to more compact or more elaborate network design, depends on the deployment. An all-new standalone CIoT network need not support all the legacy network capabilities. Such a standalone CIoT network can be deployed as a Dedicated Core Network using the 3GPP DECOR feature or as an extra PLMN making use of Network Sharing (see [12]). But to support both the traditional EPS UEs and CIoT UEs in the same network elements, an integrated EPS and CIoT network needs to support both sets of capabilities. In integrated deployments, the selected network elements need to be updated to support CIoT EPS Optimization features.

The individual features in CIoT EPS Optimization feature set specified in TS 23.401 [14] can be supported selectively on both UE and network side, and they are not mutually exclusive, even though some conditions exist for the sake of achieving compatibility.

Although a new NB-IoT radio access technology (RAT) was specified by 3GPP, CIoT EPS Optimizations features can use multiple radios, including WB-E-UTRAN (LTE) and GERAN. WB-E-UTRAN stands for wideband E-UTRAN, contrary to NB-IoT which is a narrowband technology based on E-UTRAN.

The CIoT related improvements address the following areas:

1. Simplified UE implementation and power saving;
2. Network attach without PDN connectivity (possibly followed by PDN connectivity on demand);
3. Infrequent and short bursts of data;
4. Infrequent and long bursts of data;
5. SMS transfer for PS only UE without CS domain attach.

In addition to these four main optimization areas, a fallback to traditional EPS user plane connectivity is available.

9.5.3 Attach Without PDN Connection

Until Rel-13, an EPS UE always requested PDN connection when attaching to the network. Rel-13 CIoT allows a UE to request attach without PDN connection. Since this feature is not backwards compatible, and it would be rejected by an MME that has not been upgraded to support CIoT, the network capability to support attach without PDN connection is broadcast to the UE via the radio (in System Information Blocks). If support of this feature is broadcast, the UE can request for attach without PDN connectivity.

The EPS attach procedure including security runs normally, but session management and bearer establishment procedures are omitted. The logic to support attach without PDN connection is to allow a minimalistic UE implementation. Such a UE can still support location procedures and act as a transponder to indicate the present location of the cargo it is included in, and without having any PDN connection, it can communicate with an AS using SMS.

It is rather exceptional for 3GPP to specify a feature that is not backwards compatible with previous releases. The indication of no support for attach without PDN connection in the selected PLMN saves signalling for attach attempts that are doomed to fail. There is no way to obtain compatibility between networks that do not support attach without PDN connection and UEs that cannot support PDN connections. The fallback method for such case is designed in the UE, which needs to select another PLMN in attempt to obtain connectivity.

9.5.4 UE Requested and Network Supported CIoT Capabilities

Since many of the new CIoT EPS Optimizations are not backwards compatible, the network support cannot be taken for granted by the UE. To obtain compatibility between the CIoT EPS Optimizations that are supported on UE and network side, the main capabilities are indicated in System Information Blocks broadcast by the serving cell, and the supported capabilities are negotiated on NAS level between UE and MME ([15]).

The UE indicates in the ATTACH REQUEST the CIoT EPS Optimizations that it would prefer to use. The MME responds in ATTACH ACCEPT with the CIoT EPS Optimizations that it has chosen for this UE.

Network supported CIoT EPS Optimization features are the result of MME and other network elements capabilities, UE subscription and PLMN operator local policy in the MME. Broadcast network CIoT capabilities are:

- NB-IoT and WB-E-UTRAN: Support of attach without PDN connection
- NB-IoT and WB-E-UTRAN: Support of U-plane EPS Optimization
- WB-E-UTRAN only: Support of C-plane EPS optimization is only broadcast by WB-E-UTRAN cells, as in NB-IoT both sides shall support it. Hence there is no need to indicate this capability over NB-IoT.

The following CIoT UE requested and network supported capabilities are negotiated:

- Whether C-Plane is supported or not;
- Whether U-plane is supported or not;
- Whether S1-U data transfer is supported or not;
- Whether SMS transfer without combined attach is supported;
- Whether attach without PDN connection is supported;
- Whether robust header compression (RoHC) is supported for C-plane or not;
- If UE indicates support of both C-plane and U-plane, it also indicates which one it prefers.

9.5.5 Selection of Control or User Plane

One of the difficulties in designing CIoT was the lack of a single reference application using this kind of connectivity. The IoT traffic patterns vary substantially depending on the application. These span from very infrequent small data of tens or hundreds of octets once a day or week, to video streaming capability. It is also foreseen that some CIoT devices may need to change behaviour between pure telemetry-type one-shot messaging to large amount of data streaming, e.g. for updating SW or providing additional information in case of alarms. Thus, both small data and big data optimizations were specified.

Control Plane CIoT EPS Optimization works most efficiently with applications whose data transfer needs area infrequent and small. User Plane CIoT EPS Optimization is an efficient method of sending either a large amount of data or frequent small data. It is not possible to give a single answer which one is more efficient, as the breakeven point in terms of message round trips depends on the traffic pattern of the CIoT application.

The properties of the PDN connection are determined when the connection is established. The UE may request an IPv4, IPv6, dual stack IPv4/IPv6, or Non-IP PDN type. The Non-IP type was added for CIoT EPS Optimizations.

It is possible for the MME to 'pin' a PDN connection to control plane by including the 'Control Plane Only Indicator' in the PDN connection establishment signalling. A pinned PDN connection is permanently locked to the control plane, and it cannot be changed to user plane transport during its life time.

If a PDN connection is not pinned to the control plane, it is possible for either the UE or MME to switch a PDN connection between control and user plane transport.

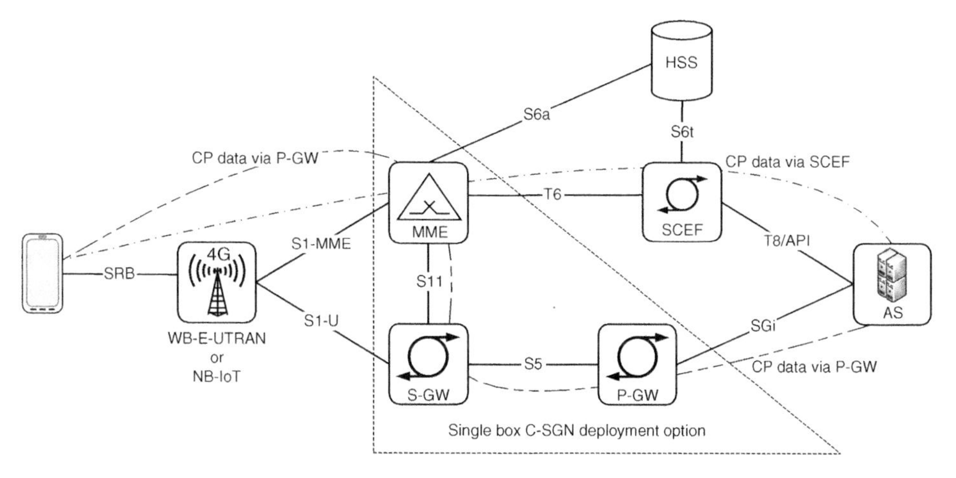

Figure 9.5 Control plane data over SCEF or P-GW.

9.5.6 Control Plane CIoT EPS Optimization

Figure 9.5 shows the possible control plane data paths via SCEF and P-GW.

The network can provide Control Plane CIoT EPS Optimization via two paths, either via the P-GW and its SGi interface, or via the SCEF that was initially introduced to exposure data like UE registration status and reachability to the Application Server (AS) via T8/API. These paths are differentiated by different access point names (APNs) hosted by P-GW or SCEF. The MME determines between use of T6 interface towards SCEF and S11/S5 interfaces towards S-GW/P-GW based on the APN, either requested by the UE or one that is selected by the MME. The APN selected by the MME can be the default APN stored in the HSS as part of the subscription data. In case of P-GW, EPS interfaces S5 and S8 are re-used, in case of SCEF, the T6 interface between MME and SCEF, and the T7 roaming interface towards the Interworking Service Capability Exposure Function (IWK-SCEF) needs to be supported. P-GW and S-GW can support IPv4, IPv6, IPv4v6 dual stack, and Non-IP PDN connection types, but the SCEF supports only Non-IP PDN connections.

User data are encapsulated in NAS signalling messages (NAS Packet Data Unit). Establishment of a Data Radio Bearer (DRB) is omitted, as the Signalling Radio Bearer (SRB) carries also the user data payload. The encapsulated user data messages may be chained to allow application level dialogues by indicating in the NAS signalling that there are more data to send. However, the Control Plane CIoT EPS Optimization yields the highest efficiency on one-shot messages, and if the network sees that the UE is sending an excessive amount of small data using control plane procedures, and if both Control Plane and User Plane CIoT EPS Optimization have been negotiated during UE attach, then the network can assign a DRB and move the transport of user data from control plane to user plane.

A CIoT UE with infrequent and small data traffic pattern may use power saving optimizations such as extended idle mode DRX (eDRX) or PSM to optimize the UE power budget. Both eDRX and PSM allow the UE to negotiate with the MME sleep cycles during which it need not listen to paging. During this sleep cycle, the UE is considered by the network as not reachable, and consequently the network needs to buffer all incoming data for later delivery. The S-GW and the SCEF, based on the MME awareness of the UE sleep cycle, may buffer data until the UE becomes available again. If the UE sleep cycle is excessively long, the AS may be notified about UE unreachability and the estimated reachability time. MME is aware of the negotiated eDRX and PSM cycles, and thus able to predict when the UE might respond to paging again. If MME receives downlink C-plane data 'just before' the next UE Paging Time Window (PTW), then subject to MME local policy and buffering capacity, the MME may buffer the data while waiting for the next paging opportunity. If the MME considers the amount of data or time for buffering excessive, it can request S-GW or SCEF to buffer the message by indicating that the UE is not reachable and giving an estimate of the next paging opportunity.

User identification in Control Plane EPS Optimization follows the same principles irrespective of IP type or P-GW versus SCEF path. Inside the 3GPP system the user is identified by IMSI. IMSI is not disclosed outside of the operator's trust domain, but publicly usable identities are External Identifier or MSISDN. Since the MSISDN has been made optional in earlier releases, a CIoT UE can be addressed without having MSISDN assigned to it.

P-GW and SCEF establish a mapping between IMSI and External ID or MSISDN at connection setup, and they do the mapping of internal (IMSI) and external (External ID or MSISDN) identities of the UE as well as the mapping of the PDN connection related EPS Bearer Identities to the services provided by the ASs.

Volume based rate control would not be practical with infrequent small data, so the Control Plane rate control is based on counting of messages that are sent over a period of time. There are different rate control parameters for uplink (UL) and DL traffic.

In roaming situations (Figure 9.6), both home PLMN (HPLMN) and visited PLMN (VPLMN) may apply rate control on C-plane messages. Serving VPLMN operator cannot dimension the data volumes for the services that are running on roaming CIoT UEs.

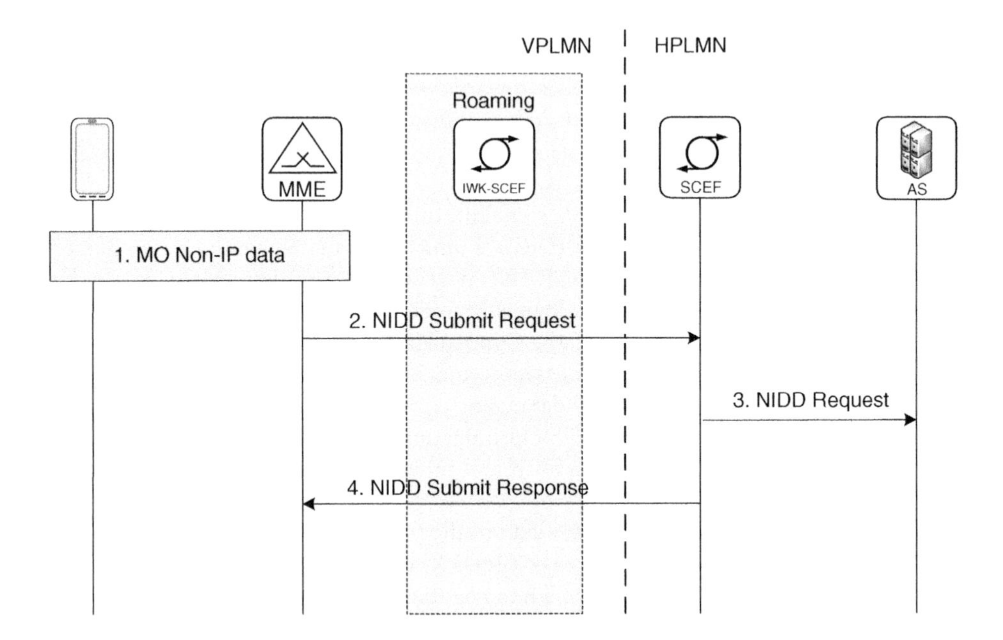

Figure 9.6 C-plane via SCEF.

The Serving PLMN rate control is for safeguarding the serving MME from overload excessive C-plane user data message exchange might cause. APN rate control is set up by the HPLMN operator to dimension the service in terms of how many messages in the agreed time unit a CIoT UE can send or receive.

9.5.7 User Plane CIoT EPS Optimization

User plane connectivity in EPS is always provided via S-GW and P-GW. Normal S1 user plane connection procedures are also supported for a CIoT UEs, i.e. connection establishment and release after data transmission, without the option to suspend and resume connections.

User Plane CIoT EPS Optimization (see Figure 9.7) uses S1-U between eNB and S-GW and DRB on the radio interface between UE and eNB for user data transport (see e.g. [8]). This suits well application level dialogues and occasional big data, e.g. for device configuration or SW updates.

Transport of data via user plane (DRB and S1-U) is by default supported in EPS, but it fails to address the infrequent data transport pattern of IoT device as it incurs signalling overhead to setup and tear down the RRC and User plane connection. User Plane CIoT EPS Optimization adds the capability to suspend the existing RRC connection between UE and eNB when no data are transferred, and to resume it later, to minimize the signalling overhead introduced by RRC connection and UE context setup (see Figure 9.8). In addition, a CIoT UE can fall back to normal S1-U EPS user plane procedures.

The eNB, based on detected UE inactivity, requests the MME to suspend the connection. The MME will trigger bearer release on the user plane but keeps the UE context for later use. The suspend is indicated to the UE in an RRC message, including the Resume ID that the UE stores for later use when it needs to resume the connection (Figure 9.9).

The UE identifies the connection to be resumed with the Resume ID, and if the UE context is available in eNB, it is resumed. If the previously suspended connection cannot be resumed, e.g. due to UE mobility, then normal RRC connection establishment is performed.

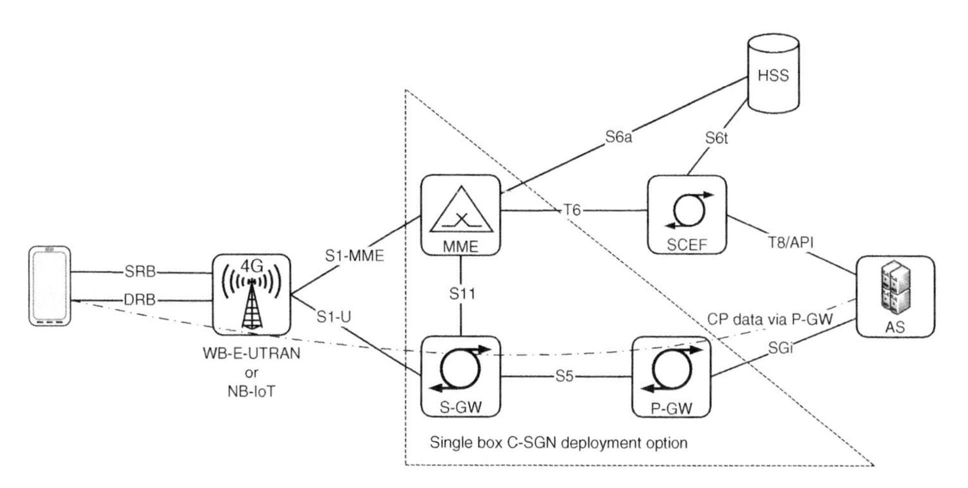

Figure 9.7 User plane CIoT EPS optimization.

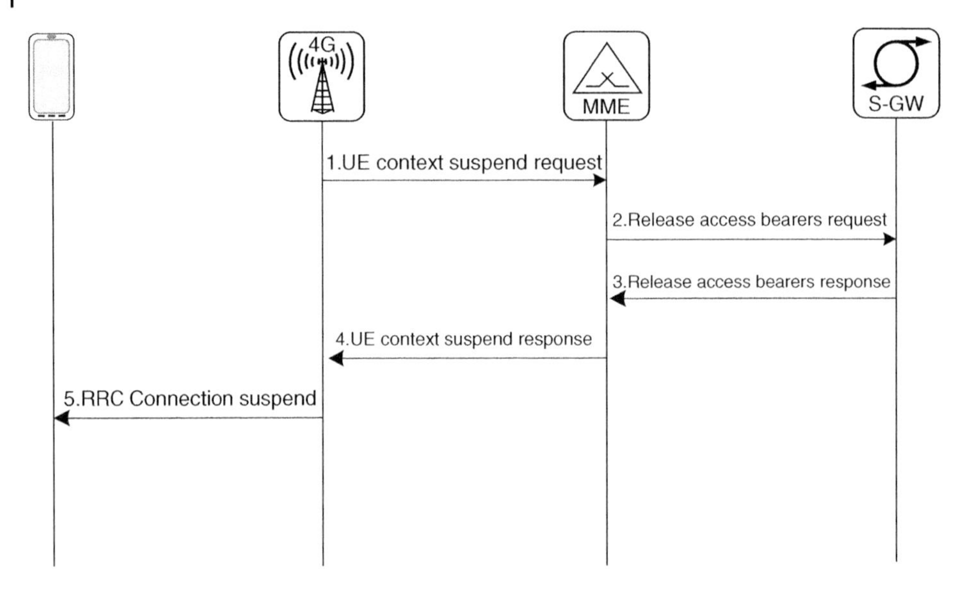

Figure 9.8 Connection suspend procedure according 3GPP TS 23.401.

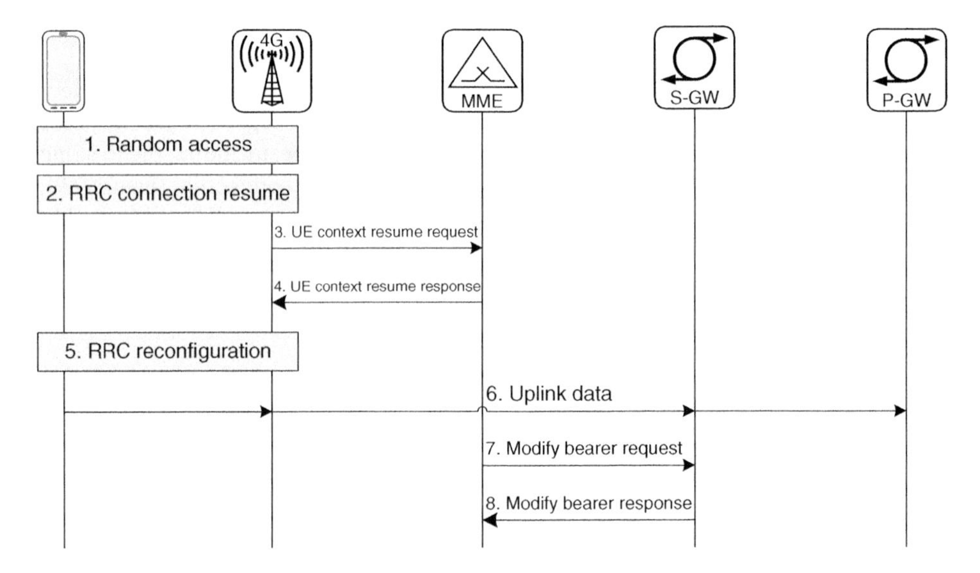

Figure 9.9 Connection resume procedure according 3GPP TS 23.401.

9.5.8 Non-IP PDN Connection

The 3GPP system already supports IP types IPv4, IPv6, and IPv4v6 dual stack. The fourth IP type 'Non-IP' was added for CIoT to minimize the IP header overhead. The support of header compression is one of the optional features that are negotiated in UE requested and network supported CIoT EPS Optimizations during the attach procedure. IP header compression is obviously not useful for Non-IP data as IP header is not included.

Non-IP data refers to data without IP header sent over Non-IP PDN connection. Since there is no IP header, header compression (e.g. RoHC) techniques are unnecessary for such data.

The P-GW can support Non-IP data as one of its four IP types, but the SCEF is restricted to Non-IP control plane data only.

Non-IP PDP type and T6b interface between SGSN and SCEF was added in Rel-14 to allow the GPRS system to support Non-IP Data Delivery (NIDD) via SCEF.

9.5.9 SMS Transport

A UE that has attached without PDN connection cannot send or receive any Control Plane data or User Plane data. This leaves SMS as the only means of data transport available for a CIoT UE that has attached without PDN connection. Such a UE need not support CS domain procedures, PS only registration is possible and indicated in the attach procedure. SMS is transported using the same NAS message encapsulation principle as in NIDD but using a different NAS protocol data unit (PDU). The data path is shown in Figure 9.10.

SMS for CIoT only requires PS SMS support for a UE. As the CIoT UE is likely to use long eDRX or PSM cycles, thus the same challenges with long UE unreachability periods like in MT NIDD will arise.

The SMS architecture differs from the one used for NIDD, but the principle for handling extended UE sleep cycles is the same: the network elements need to buffer a MT SMS towards the UE that has just started its extended power saving cycle. The maximum eDRX time is (depending on RAT) in the order of hours, and the maximum predictable PSM cycle is more than a year. This is not necessarily a practical value applicable in a network, but it shows why network elements do have the option to reject downlink messages with an indication that the UE is not reachable instead of buffering it.

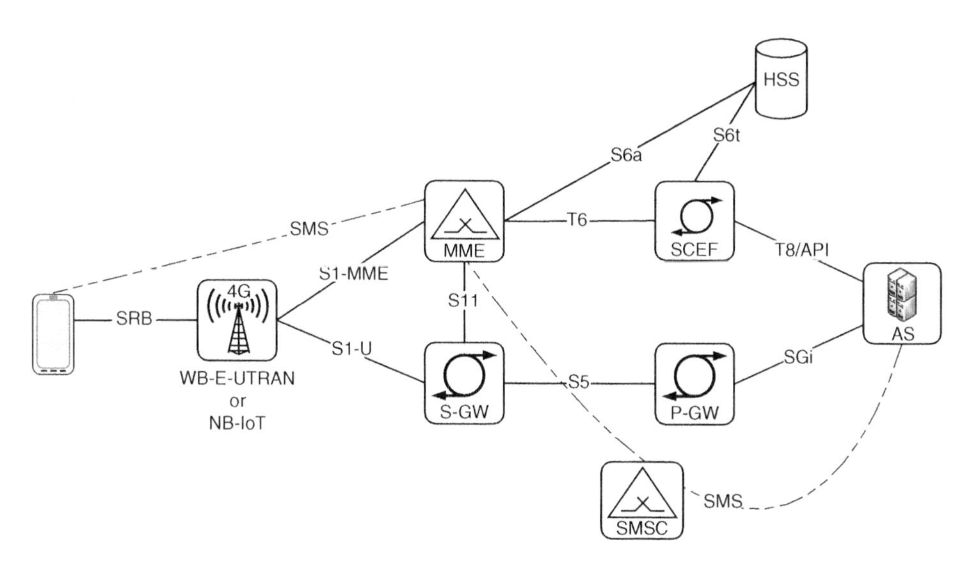

Figure 9.10 SMS data path.

9.5.10 Rel-14 CIoT Extensions

CIoT functionality was further enhanced under two 3GPP work items in Rel-14, 'Non-IP GPRS' and 'CIoT Extensions'.

9.5.10.1 Non-IP GPRS

The scope of Non-IP GPRS was to enhance the GPRS system with the selected CIoT optimizations introduced in Rel-13 for EPS. The T6a interface between MME and SCEF was mirrored to T6b interface between SGSN and SCEF. This enables Non-IP data delivery between AS and a GPRS UE. Since SCEF only supports Non-IP data type, support of Non-IP data was added to GPRS radio.

With Rel-13, there is no interoperability between Non-IP EPS PDN connections and GPRS, since Non-IP PDP context type is not specified in GPRS. It is the responsibility of MME and SGSN to ensure that only supported bearers are transferred during handover. If no compatible bearers exist, the handover fails. Rel-14 Non-IP GPRS removes this limitation by allowing those Non-IP bearers to be transferred.

9.5.10.2 Coverage Enhancement

Coverage Enhancement (CE) allows a UE in a weak coverage area to access the network even with low signal strength. CE techniques (e.g. using higher transmit power at the base station) allow for compensating coverage loss due to reduced number of antennas in low-cost devices and for LTE coverage extensions to static devices in locations prone to weak coverage such as the basement. CE was already a 3GPP Rel-13 feature, but since CE consumes more network resources than non-CE operation, Rel-14 added the network capability to control use of CE based on subscription information. The HSS subscriber data indicate per PLMN whether CE is allowed for the user or not. For backwards

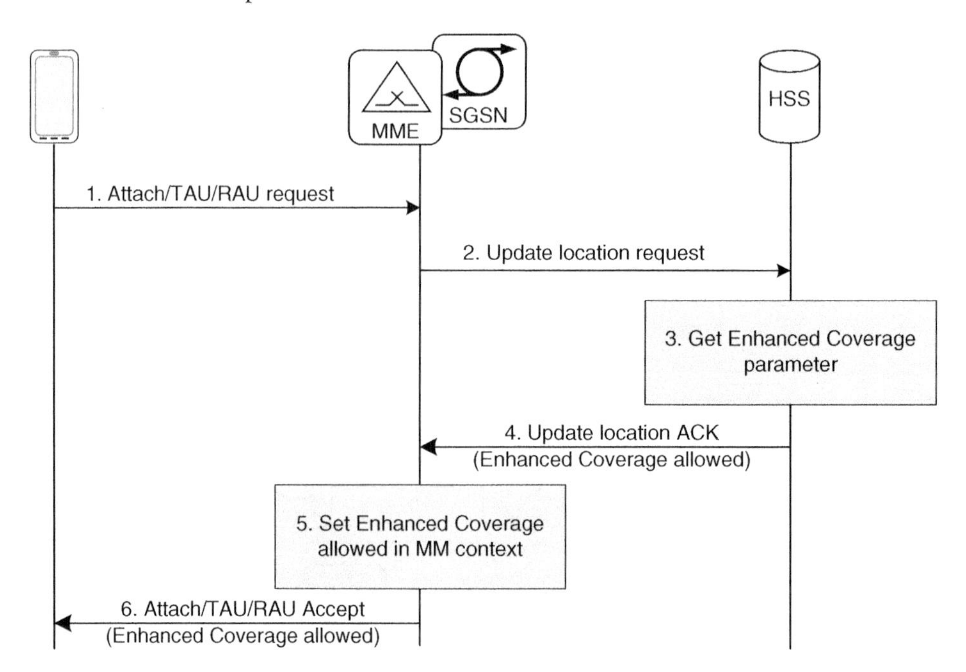

Figure 9.11 UE requesting coverage enhancement.

compatibility reasons, the UE assumes that CE is allowed, unless it is explicitly restricted by the serving PLMN.

The CE setting is negotiated during attach procedure (Figure 9.11), where the MME informs the eNB of CE authorization. The eNB can then determine the CE level based on the UE capabilities and the subscriber data it receives from the MME. When paging the UE, MME includes the CE information in S1 paging message. It is also possible for the AS to request CE, if the subscription allows it.

9.5.10.3 Reliable Communication Service Between UE and SCEF

Use of reliable transport protocols on the path between UE and SCEF provides reliable delivery at each hop, but it cannot guarantee successful processing on the application layer. A message that is successfully transported may still contain syntactical or semantical issues causing it to be ignored or rejected by the receiver.

Reliable communication service between UE and SCEF (Figure 9.12) overcomes this problem for the chargeable NIDD information in two alternative ways. The full end-to-end solution exists of an overlay protocol between UE and SCEF managing the acknowledgements each way to ensure error handling and successful delivery. An

RDS between UE and SCEF
Reliable and fault tolerant NIDD from UE to network edge

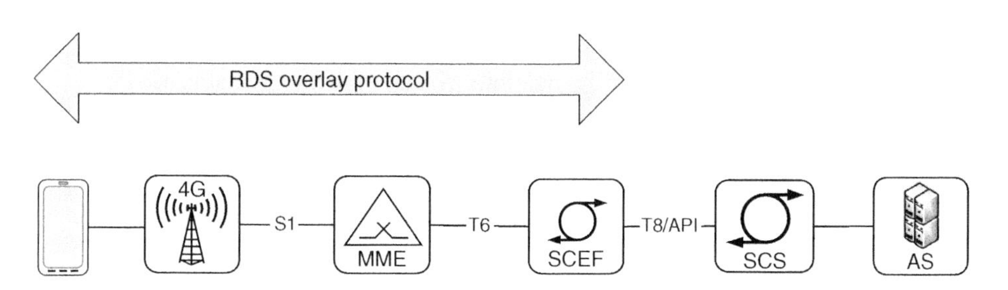

Additional improvement by using hop-by-hop acknowledgements
- Use of RLC Acknowledged mode at the radio interface
- Addition of S1-AP acknowledgements
- Addition of T6 Acknowledgements

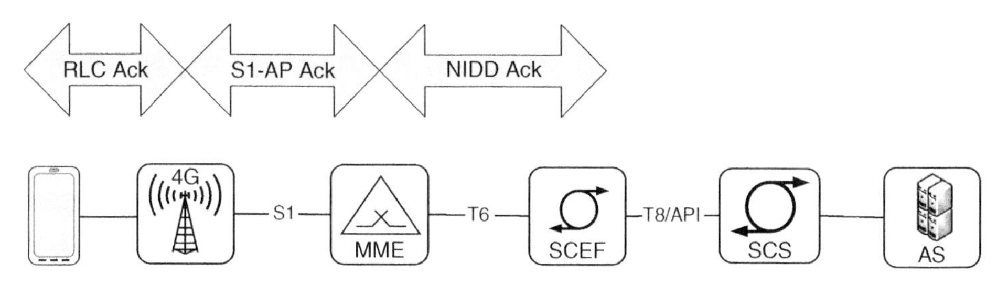

Figure 9.12 Reliable data service.

alternative solution is based on hop-by-hop security, where each interface is secured by acknowledgements.

9.5.10.4 Idle Mode Mobility Between NB-IoT and Other RAT

Rel-13 CIoT EPS Optimizations suffer the limitation that idle mode mobility between NB-IoT and other 3GPP RATs is not supported, as the MME supporting CIoT shall actively prevent it. This forces a UE supporting NB-IoT and other 3GPP RAT to detach and re-attach in the new RAT for inter-RAT mobility.

Rel-14 CIoT MME may allow inter-RAT idle mode mobility between NB-IoT and other RATs (see [7]). To force a Tracking Area Update of the UE, the MME indicates a single-RAT TAI list to the UE.

As usual in inter-RAT mobility cases, the MME needs to remove all PDN connections and bearers that are not compatible with the target RAT. The inter-RAT mobility may also be offered by the operator as a service, as the HSS can indicate per APN which bearers are maintained in inter-RAT TAU, and which are dropped.

9.5.10.5 MBMS Service for CIoT

CIoT EPS Optimizations do not prevent multimedia broadcast/multicast service (MBMS) operation, but the foreseen extremely long eDRX or PSM UE power saving cycles pose the usual problem of UE reachability also for data delivery via MBMS (see [11]). In the case of multicast and broadcast services, the problem is slightly different from the standalone UE reachability for NIDD, as the service would need to be offered to multiple UEs whose non-synchronous PTW reachability periods between sleep cycles are scattered randomly over the maximum sleep cycle range that is used.

The basis of the solution builds on synchronizing the receiving UEs to wake up for receiving broadcast data at the same time. Depending on the application, the solution can be AS or UE centric.

The AS may issue a trigger to the UEs to wake up for MBMS service announcement (see [11]) well ahead of the data delivery so that all affected UEs can receive the service announcement timely before data delivery, as shown in the Figure 9.13 (reference 3GPP TR 23.730 [28]). The figure shows UE1 using a shorter eDRX cycle than UE2.

The UEs may also be configured to wake up at a certain pre-configured time to check a potential MBMS service announcement. Once the service announcement is issued, both methods are aligned, but with this UE synchronization the lead time needed before the data transfer is shorter, as all UEs are synchronously reachable at the same time of the day.

9.5.10.6 Location Services for CIoT

Several methods were designed in Rel-14 to overcome the problems caused by long UE power saving cycles to location services.

The last known location can be used to estimate the location of UE with long sleep cycles. For this purpose, the MME maintains the last serving cell of the UE, and the time stamp when the UE was detected in connected mode. The serving cell ID can be converted to geo-location information.

Deferred location request can be used, if the location need is very precise, and if it can wait until the UE becomes reachable again.

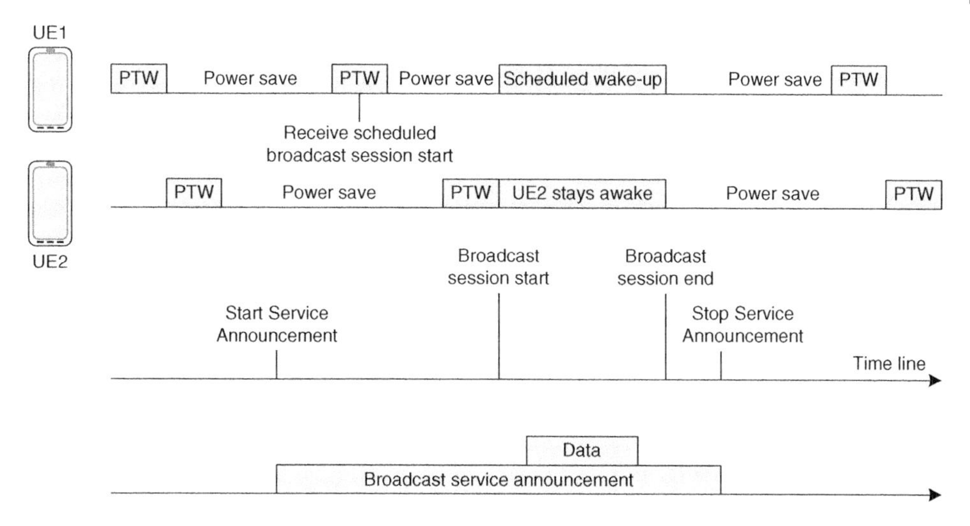

Figure 9.13 MBMS delivery to UEs with non-synchronous power saving cycles.

Triggered or periodic location request can be used, if some of the benefits gained by power saving cycles on UE power budget can be sacrificed for location accuracy. Additional location procedure triggers, whether based on mobility or time, will sacrifice some of the gains in UE power budget for the sake of location accuracy and timeliness.

Further measures to decrease the implicit UE processing requirements were added by allowing the UE to indicate its capability for active and idle mode measurements and indication of NB-IoT cell to reduce the message size and thus transport delay.

9.5.10.7 QoS Differentiation Between UEs Using Control Plane CIoT EPS Optimization

CIoT applications are expected to span a wide range of services, ranging from toys and hobbies to critical application involving safety, security and money. Prioritization mechanisms are needed to ensure that critical IoT services do not suffer when a large number of low priority applications are connected in the same place at the same time.

Serving PLMN Rate Control protecting the serving MME from NIDD overload should ideally never need to be enforced, as that throttles the traffic that otherwise would fit in the APN Rate Control quota that dimensions the service in terms of number of messages in time unit. Priority mechanism is designed for the case when a dense population of CIoT devices operating within their rate control limits would risk congesting the control plane.

Figure 9.14 shows an RRC connection establishment, where the eNB retrieves the UE context using the temporary UE identity from the MME, and may apply prioritization between UEs when establishing the RRC connection.

9.5.10.8 CN Overload Protection in Control Plane

General overload protection mechanisms already exist, but in case of control plane overload (see Figure 9.15), it would be an overkill to deny other services unnecessarily. Part of the deal is that a UE blocked for control plane access is encouraged to use user plane services instead.

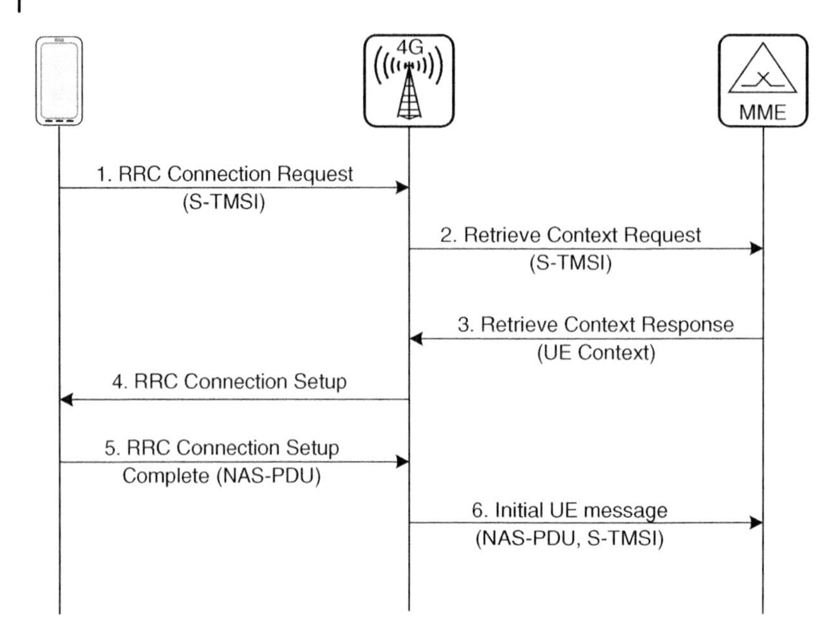

Figure 9.14 QoS differentiation.

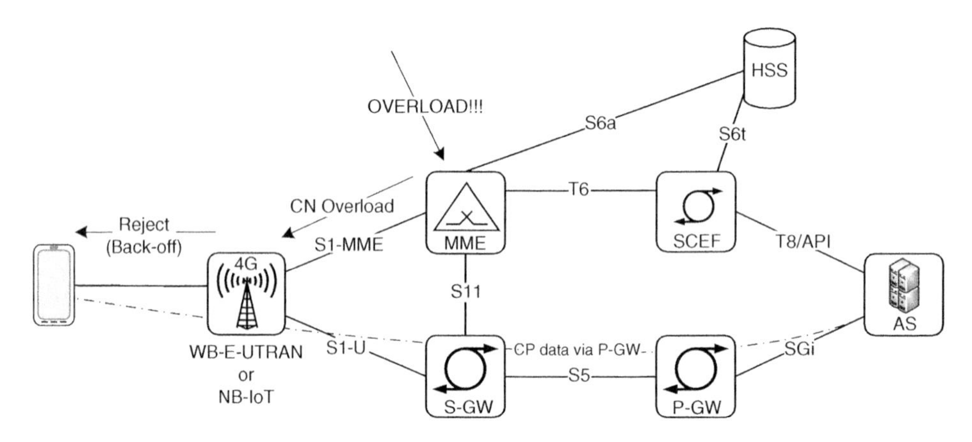

Figure 9.15 Control plane overload control.

Already before CIoT, an MME may request eNB to reject connection requests on its behalf to avoid congestion. If the MME load is mainly due to processing control plane NIDD traffic, then the MME may send S1 OVERLOAD START indicating control plane overload. This instructs the eNB to reject control plane data transmission requests from the UE with a control plane back-off timer.

Following the traditional back-off timer principle, the UE processing of control plane back-off timer prevents new control plane requests, but as the control plane back-off does not affect the user plane data transfer, the UE is allowed using user plane connection, if this was negotiated during attach.

9.5.11 Northbound API for SCEF (NAPS)

The T8/API interface between SCEF and SCS/AS was out of scope of 3GPP specifications until Rel-14. In Rel-15, 3GPP specified the T8 interface in detail. As the main goal of this work was to satisfy the need providing a single application programming interface (API) to manage MTC UEs, also the option for co-located MTC-IWF – SCEF deployments was added so that triggering services available via MTC-IWF can be provided over T8 using a common API.

Some modifications and improvements of the existing 3GPP services were also made under the Northbound API for SCEF – SCS/AS Interworking (NAPS) work item. Depending on network authorization and local policy, the AS may request certain UE reachability and buffering parameters. In earlier 3GPP releases, the AS could set network parameters for monitoring and buffering only in a monitoring request, but a standalone 'Network parameter configuration' procedure was added under NAPS work item.

The parameters the AS may request from the serving 3GPP network are Maximum Latency that guides the setting of periodic update timer by indicating how frequently the AS wants to communicate with the UE, the Maximum Response Time controlling the period of time that the UE remains reachable when it has woken up from power saving, and the Suggested Number of Downlink Packets that gives a hint of the foreseen buffer size.

The other enhancements include the capability for the AS to request CE to be used (subject to authorization), and the addition of source and destination port numbers to identify applications.

9.6 GERAN

9.6.1 Introduction

The continued growth of MTC and IoT provide new business scenarios for MNOs. To serve this demand, different technologies were standardized in 3GPP Release 13, Release 14, and Release 15, such as LTE based enhanced machine type communication (eMTC) and NB-IoT as well as GSM based Extended Coverage-GSM-IoT (EC-GSM-IoT), the latter being dealt with in this section. It is shown how key radio aspects of a Cellular IoT system – such as enhanced coverage (EC), efficient small data transmission, high system capacity, low device complexity, energy efficient device operation and flexible usage in existing network deployments – are realized in the EC-GSM-IoT system. For this purpose, the objectives defined as part of the Cellular IoT study item for 3GPP Rel-13 are depicted, followed by an overview of the design aspects of the selected EC-GSM-IoT solution. Then logical channel design and basic concepts of EC-GSM-IoT such as Coverage Class design, Overlaid Code Division Multiple Access (CDMA) and the description of idle and connected mode operation are presented. Finally, improvements to operation in tight frequency reuse networks and enhancements standardized in 3GPP Rel-14 and Rel-15 are briefly summarized.

9.6.2 Objectives

Objectives for EC-GSM-IoT were set during the Feasibility Study on Cellular Support for IoT, carried out in 3GPP TSG GERAN in 2014 and 2015, which investigated both backward compatible and clean slate solutions and generated the Technical Report 3GPP TR 45.820 [16]. These relate to:

- *Enhanced coverage.* CE versus GPRS by 20 dB to cover devices in extended coverage (e.g. indoor, basements, shielded in moving vehicles) with support of both stationary devices and moving devices (speed up to 50 km h^{-1}).
- *Efficient small data transmission.* Allowed latency of 10 s for high priority reports such as alarms; minimum supported data rate on downlink and uplink of 160 bps on the SubNetwork Dependent Convergence Protocol (SNDCP) layer.
- *High system capacity.* Capacity enhancement versus GPRS; support of massive number of devices (~52 500 per cell) based on the assumptions of 40 devices per household (metropolitan area).
- *Low device complexity.* Low device cost with considerable complexity reduction versus GPRS legacy mobiles.
- *Energy efficient device operation.* Long battery life time of targeted 10 years with 5 Wh battery to minimize operational efforts for most battery-operated devices.
- *Flexible usage in existing network deployments.* No HW impact to GSM/EDGE legacy networks due to introduction of this technology and support of combined operation with GSM/EDGE legacy networks in the same spectrum as well as of standalone operation in separate spectrum.
- *Non-radio related aspects.* Improved security compared to GSM/EDGE achieved by the introduction of integrity protection and improved ciphering algorithms as output of the feasibility study undertaken in 3GPP TSG SA3 [17].

9.6.3 Feature Design Overview

Among the investigated candidates, EC-GSM-IoT has been selected as the backward compatible GSM based solution for CIoT and been standardized in 3GPP Rel-13. To satisfy the feasibility study objectives, the following design aspects are aligned to the key radio aspects of a CIoT network.

Enhanced Coverage

Introduction of four coverage classes to support devices on different coverage levels up to Maximum Coupling Loss (MCL) of 164 dB, supporting physical layer repetitions both in normal and extended coverage, more robust channel encoding for some logical channels and Hybrid Automatic Repeat Request (ARQ) for retransmitted traffic data.

Efficient Small Data Transmission

- Handling of device exception reports (indicating, e.g. alarms) with higher priority.
- Handling of small data payloads of few bytes in few 20 ms radio blocks (normal coverage).

High System Capacity

- Usage of Coverage Class specific training sequence for random access to lower failure due to collisions.

- Coherent repetition of bursts in same Time Division Multiple Access (TDMA) frame, thus improving co-channel and adjacent channel interference performance in connected mode.
- Introduction of Overlaid CDMA allowing to multiplex up to four devices in extended coverage for data transfer on the same uplink resources.

Low Device Complexity

- Simplified half duplex operation with relaxed switching times: either transmit or receive operation per each TDMA frame to lower device complexity.
- Reduction of puncturing patterns compared to enhanced general packet radio service (EGPRS) to lower device complexity.
- Use of Fixed Uplink Allocation (FUA) to achieve resource efficiency for delay tolerant services and to omit Uplink State Flag (USF) monitoring yielding energy efficient operation in connected mode.

Energy Efficient Device Operation

- Support of extended DRX (eDRX) and/or PSM in idle mode reducing paging availability of the device to ensure long battery lifetime.
- Support of Power Efficient Operation (PEO), indicated per cell, to support relaxation for measurement requirements for the device.
- Energy efficient downlink common control channel (CCCH) operation to reduce paging and access grant monitoring time.
- Removal of continuous timing advance procedure in connected mode to adapt to the range of device speeds and to lower device complexity.

Flexible Usage in Existing Network Deployment

- Support of existing infrastructure: Radio frequency (RF) and baseband processing performance requirements are defined in a way to support legacy base stations in the field.
- Support of both combined/standalone system operation through CCCHs for EC-GSM-IoT operated on different timeslots of the Broadcast Control Channel (BCCH) carrier (i.e. TN 1, TN 3, TN 5, and TN 7) than used for legacy GSM/EDGE CCCHs (i.e. TN 0, TN 2, TN 4, and TN 6) and support of tighter frequency reuse by extended base station identity code (BSIC).
- Scalable resources for EC-GSM-IoT, such that at minimum one additional timeslot is needed for EC-GSM-IoT CCCHs in a combined deployment with legacy GSM/EDGE.

9.6.4 Logical Channel Design

Out of the logical channels for GSM/(E)GPRS, only the frequency correction channel (FCCH) is reused due to its suitable power spectral density (PSD) characteristics, while all other logical channels for EC-GSM-IoT have a new design compared to (E)GPRS. The set of logical channels for EC-GSM-IoT is depicted in the Table 9.2.

9.6.5 Coverage Class Concept

To enable operation of devices in normal and in extended coverage, the concept of Coverage Classes (CC), applicable to the operation on the radio interface only, was introduced in 3GPP. For EC-GSM-IoT, each of the defined four Coverage Classes serves for

Table 9.2 Set of logical channels for EC-GSM-IoT.

Logical channel	Purpose	Mapping onto radio interface
FCCH (DL)	Frequency adjustment as part of network synchronization	BCCH carrier, TN 0
EC-SCH (DL)	Timing adjustment (frame and symbol) as part of network synchronization	BCCH carrier, TN 1
EC-BCCH (DL)	Broadcast of EC-GSM-IoT relevant system information	BCCH carrier, TN 1
EC-PCH (DL)	Conveys paging requests for sending network commands (MT calls) or routing area update requests	BCCH carrier, TN 1 (also TN 3, TN 5, TN 7 in case of multiple EC-CCCH cell configuration)
EC-RACH (UL)	Conveys channel access requests initiating EC-GSM-IoT data transfers (MO calls) or for responding to paging requests on EC-PCH	BCCH carrier, TN 1 (also TN 3, TN 5, TN 7 in case of multiple EC-CCCH cell configuration)
EC-AGCH (DL)	Conveys access grant for responding to channel access request or paging responses on EC-RACH	BCCH carrier, TN 1 (also TN 3, TN 5, TN 7 in case of multiple EC-CCCH cell configuration)
EC-PDTCH (DL/UL)	Used for Packet data transfer (payload) on DL (network commands)/UL (Mobile autonomous/exception reporting)	one timeslot (Coverage Class 1) or four timeslots (Coverage Class 2–4), either on BCCH carrier or on TCH carrier, optionally using FH
EC-PACCH (DL/UL)	Conveys control signalling associated with the packet data transfer in DL or UL	Same mapping as for EC-PDTCH in the same link direction; according to assigned coverage class in the reverse link direction

operating a device according to its coverage condition, up to the CE target of 20 dB versus GPRS. The coverage gain is obtained through:

- The use of blind physical layer repetitions, i.e. consecutive repetitions of radio blocks over a larger number of TDMA frames thus increasing the effective transmission time interval (TTI) of the basic radio block. At the receiver the repetitions across TDMA frames are soft combined.
- The coherent transmission of four consecutive bursts within a TDMA frame by applying the same initial modulator phase and amplitude for the start of each burst (or in case of Overlaid CDMA a sequence of predefined burst phase shifts, see Section 9.6.6). At the receiver, the IQ samples of the four bursts are accumulated providing ideally a signal power gain of 12 dB and an signal-to-noise ratio (SNR) gain of 6 dB.
- The use of Hybrid ARQ after the receiver has sent feedback information on the received blocks. At the receiver side, the initial transmission and retransmissions are soft combined.

- The use of more robust encoding for CCCHs in downlink and for the packet associated control channel in downlink and uplink conveying smaller payload sizes compared to (E)GPRS.

Figure 9.16 illustrates the assignment of the defined four Coverage Classes to devices located outdoor or indoor with increasing building penetration loss requiring an increase of the applied Coverage Class.

The details of the coverage classes are summarized in the following Tables 9.3–9.5 for the packet traffic channel and for the CCCHs. The MCL figures are taken from 3GPP TS 43.064 [18] and confirm a maximum gain of 20 dB versus GPRS with a reference MCL of 144 dB [16].

Depending on the traffic load, the network may request the MS using Coverage Class 1 to perform the channel access either on TN 0 using legacy channels or on TN 1 using EC channels, which is signalled in the synchronization channel (EC-SCH). In addition, as an option the random access channel (EC-RACH) can be operated on dual timeslots (TN 0 and TN 1). For this deployment option, indicated in the broadcasted EC System Information, the device transmits the same number of repeated access bursts for Coverage Classes CC2 to CC4, as given above, both on TN 0 and TN 1 and uses coherent transmission between TN 0 and TN 1 to enable IQ combining at the base station, which yields further improved sensitivity and interference performance compared to the single timeslot EC-RACH configuration. Each Coverage Class on EC-RACH is assigned a distinct new synchronization sequence to enable the Base Transceiver Station (BTS) distinguishing these channel requests from those originated by legacy GSM/(E)GPRS users.

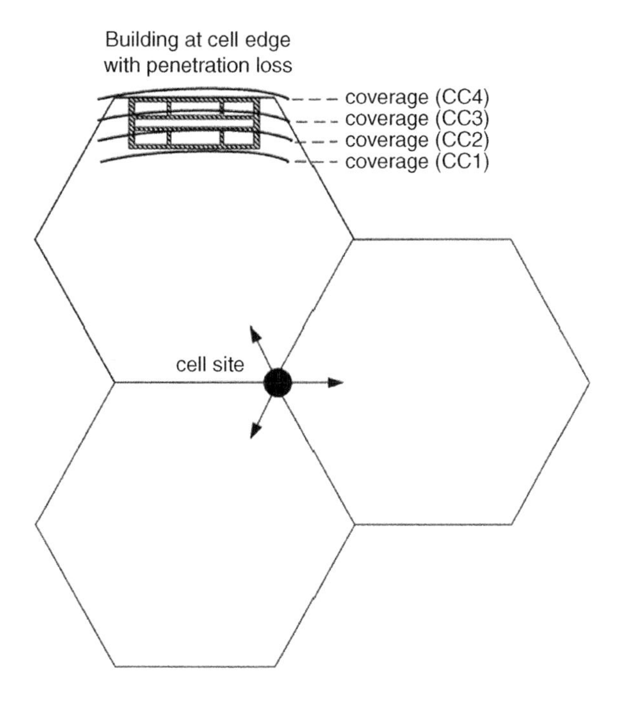

Figure 9.16 Illustration of the application of coverage classes in a sensitivity limited deployment scenario.

Table 9.3 CC parameters for EC-GSM-IoT packet traffic channel.

Coverage class	MCL (dB)	TTI of radio block (ms)	Total burst repetitions	Mapping of packet traffic channel (EC-PDTCH and EC-PACCH) to radio interface
CC1	150	20	1	Single burst occupies one timeslot per TDMA frame. Radio block is mapped onto four consecutive TDMA frames
CC2	156	20	4	Single burst is repeated over four consecutive timeslots per TDMA frame. Radio block is mapped onto four consecutive TDMA frames
CC3	159	40	8	Single burst is repeated over four consecutive timeslots per TDMA frame. Radio block is mapped onto four consecutive TDMA frames, and is repeated once in another four TDMA frames
CC4	164	80	16	Single burst is repeated over four consecutive timeslots per TDMA frame. Radio block is mapped onto 4 consecutive TDMA frames, and is repeated three times in another 12 TDMA frames

Table 9.4 CC parameters for EC-GSM-IoT DL common control channels (sent in TN 1, TN 3, TN 5, or TN 7 according to multiple EC-CCCH configuration).

Coverage class	MCL (dB)	TTI of radio block (ms)	Total burst repetitions	Mapping of common control channel (EC-PCH/ EC-AGCH) to radio interface
CC1	150	5.2	2	Burst in one timeslot of TDMA frame is repeated in next TDMA frame. Radio block is mapped onto two consecutive TDMA frames
CC2	156	268.3	16	Radio block consists of burst in one timeslot repeated seven times in the consecutive seven TDMA frames and these eight bursts are repeated at the same position once in the next 51-multiframe
CC3	159	305.2	32	Radio block consists of burst in 1 timeslot repeated 15 times in the consecutive 15 TDMA frames and these 16 bursts are repeated at the same position once in the next 51-multiframe
CC4	164	776	64	Radio block consists of burst in 1 timeslot repeated 15 times in the consecutive 15 TDMA frames. These 16 bursts are repeated at the same position in the next three 51-multiframes

Table 9.5 CC parameters for EC-GSM-IoT UL common control channel (single timeslot; only TN 1 considered, same applies for TN 3, TN 5, TN 7 in case of multiple EC-CCCH cell configuration).

Coverage class	MCL (dB)	TTI of radio block (ms)	Total burst repetitions	Mapping of common control channel (EC-RACH) to radio interface
CC1	150	0.55	1	One access burst is sent, either on RACH (TN 0) or EC-RACH (TN 1) in one TDMA frame of the 51-multiframe.
CC2	156	14.4	4	A sequence of four access bursts is sent on EC-RACH (TN 1) in four consecutive TDMA frames.
CC3	159	69.8	16	A sequence of 16 access bursts is sent on EC-RACH (TN 1) in 16 consecutive TDMA frames.
CC4	164	342.1	48	A sequence of 48 access bursts is sent on EC-RACH (TN 1) in consecutive 24 TDMA frames, and is repeated at the same position once in the next 51-multiframe.

Table 9.6 Radio parameters of DL channels for EC-GSM-IoT network synchronization.

Coverage class	MCL (dB)	TTI of radio block	Total burst repetitions	Mapping of common control channel to radio interface
FCCH	164	0.55 ms	none (periodic)	One burst sent in TN 0 of FN 0, 10, 20, 30, 40 in each 51-multiframe (identical to legacy FCCH).
EC-SCH	164	0.734 s	28	28 bursts sent in TN 1 of FN 0, 1, 2, 3, 4, 5, 6 in 4 consecutive 51-multiframes.
EC-BCCH	164	1.68 s	64	64 bursts sent in TN 1 of FN 7, 8, 9, 10, 11, 12, 13, 14 in 8 consecutive 51-multiframes.

The design of logical channels for network synchronization to EC-GSM-IoT cells, depicted in Table 9.6, was made sufficiently robust to allow operation in the entire targeted range for extended coverage.

9.6.6 Overlaid CDMA

To improve system capacity for mobile originated data transfer, Overlaid CDMA may be applied on the uplink packet data traffic channel (EC-PDTCH/U) for users in a higher coverage class (CC2 to CC4) using four consecutive timeslots per TDMA frame. This allows the network to multiplex up to four mobile stations on the same uplink physical resources for simultaneous transmission of packet data.

Orthogonality between mobile stations is achieved through orthogonal codes, corresponding to a set of phase shifts to be applied on a per transmitted burst basis. The phase shifts shall, in sequence, be applied to the transmitted bursts per TDMA frame. For this purpose, an individual Hadamard sequence is used by each device multiplying

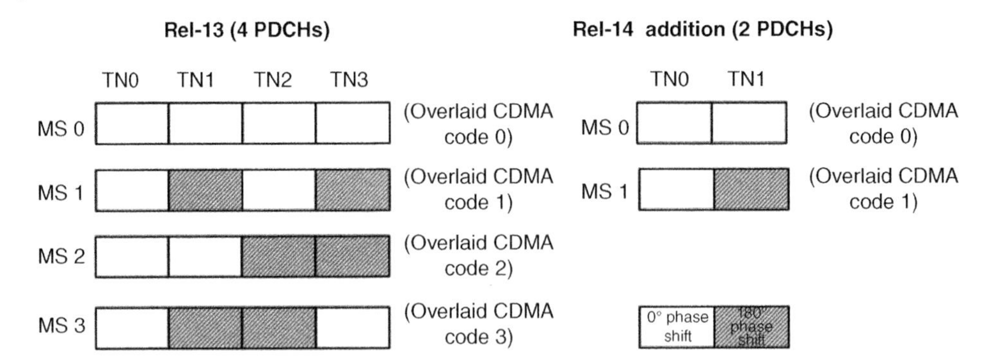

Figure 9.17 Burst phase shift on UL for overlaid CDMA (examples of multiplexing four users on TN 0 to TN 3 or two users on TN 0 and TN 1 on a traffic carrier).

the modulation symbols with the predefined phase shifts, being multiples of π, for the four consecutive bursts in the TDMA frame. By knowing the Hadamard sequence of each multiplexed device, the base station can distinguish the transmissions of the multiplexed devices, since based on orthogonal codes and provided the amplitude and phase coherency requirements between consecutive bursts are satisfied by each device.

Overlaid CDMA is used on the packet data channel and its packet associated control channel (EC-PDTCH and EC-PACCH) and the code to be used is included in the channel assignment command. Figure 9.17 depicts the principle of Overlaid CDMA for achieving orthogonality between uplink transmissions (for details see [19]).

Either two, three, or four users may be multiplexed in 3GPP Rel-13 on 4 consecutive time slots; additionally in 3GPP Rel-14 two users on two consecutive timeslots can be multiplexed. Support of Overlaid CDMA is mandatory for the mobile station and optional for the base station.

9.6.7 MS Support Level for EC-GSM-IoT

Different support levels are specified for the device. Aside non-support, the device may support only EC-GSM-IoT, in which case it cannot camp on a legacy GPRS/EGPRS cell, or it may support EC-GSM-IoT in combination with GPRS/EGPRS, in which case this restriction does not apply, i.e. the device may (re-)select to a legacy GPRS/EGPRS cell or a EC-GSM-IoT cell depending on the received signal level.

9.6.8 Operation in Idle Mode

9.6.8.1 Network Synchronization

In idle mode (see [7]), the device searches for cells supporting EC-GSM-IoT by detecting the FCCH on time slot 0 (TN 0) of the BCCH carrier and the synchronization channel (EC-SCH) on timeslot 1 (TN 1) of the BCCH carrier, which carries an orthogonal synchronization sequence to that of the legacy SCH for allowing fast detection of an EC-GSM-IoT capable cell. It then performs cell selection based on the received power level and received quality of FCCH and EC-SCH for candidate cells and camps on the best available cell. The synchronization channel contains aside assisted time synchronization to the cell to construct the TDMA frame number, information on the base station identity code, on barring of the cell for MTC devices, the broadcast change mark indication and on options for sending the channel access request on TN 0 or TN 1. If the cell is not barred for MTC devices, the device continues to read the broadcast channel (EC-BCCH) to acquire the EC-GSM-IoT specific system information, sent on up to four messages (EC SI 1 to EC SI 4). In any case the device must read the system information at least once every 24 hours or based on the broadcast change mark indication sent on the synchronization channel.

9.6.8.2 Reading the System Information

The EC System Information, sent on the EC-BCCH contains information on cell specific settings, such as supported Coverage Classes on DL and UL and related thresholds for their selection, parameters for system access on RACH/EC-RACH, on network sharing ([12]), the change mark indication to support updates of the EC SI content, on supported frequencies in the cell and neighbour cell information, such that EC-GSM-IoT devices are not required to read any BCCH information. A device can only access the EC-GSM-IoT cell once it has read its EC System Information [20].

9.6.8.3 Paging

The device is paged on the paging channel (EC-PCH). A paging block spans over two TDMA frames for CC1 and several TDMA frames for CC2 to CC4, as outlined in Table 9.4; it is dedicated either to one device (using IMSI paging) or to two devices (using two P-TMSIs). When sending a paging response, the device performs a channel request on the random access channel using RACH or EC-RACH as indicated in EC-SCH.

9.6.8.4 Energy Efficient Operation on Downlink Common Control Channels

Coverage Class detection is enabled for reception of EC-PCH and access grant channel (EC-AGCH) blocks by using a training sequence code (TSC) based discrimination, such that devices in the lower coverage class (CC1) are paged/acknowledged using a TSC from TSC Set 1, i.e. a TSC of the legacy training sequence set and devices in the higher coverage class (CC2, CC3, or CC4) are paged/acknowledged using the paired TSC from TSC Set 2, as defined in 3GPP Rel-9 for VAMOS (Voice services over Adaptive Multi-user channels on One Slot) subchannel 2. Multiplexing pages to two users in the lower and higher coverage class is then performed by utilizing a TSC from Set 2. This allows an early CC detection for a device in a higher coverage class, listening to a longer paging block due to blind physical layer transmissions, since paging requests are always using a TSC from TSC Set 2.

9.6.8.5 Extended Idle Mode Discontinuous Reception (eDRX)

The device is paged according to its negotiated DRX cycle. If extended DRX was negotiated on NAS level, the device is paged with the negotiated eDRX cycle (ranging between 1.92 s and 52 min), which can be considerable larger than the broadcast DRX cycle (0.5–15 s in case of split cycle). To synchronize paging requests to devices using eDRX, the network configuration must consider the following restrictions:

- All cells in the same Routing Area (RA) need to support eDRX.
- The timing of paging groups for devices using eDRX in all cells of a Routing Area needs to stay in a four seconds time window.

9.6.8.6 Power Saving Mode (PSM)

Like eDRX the PSM (3GPP Rel-12) is negotiated on NAS level. After leaving connected mode an active timer is started, its duration being commanded by the network, and if this expires the device enters the PSM mode, not performing AS level based operations except monitoring of paging requests. The PSM mode is terminated once the device needs to send data or receives a paging request.

9.6.8.7 Power Efficient Operation (PEO)

Relaxations to neighbour cell monitoring and system information acquisition in idle and connected mode were introduced in 3GPP Rel-13. Also, introduction of a fast one phase-based channel request initiated by a device supporting PEO or EC-GSM-IoT was specified. For an EC-GSM-IoT device, PEO support is mandatory.

9.6.9 Operation in Connected Mode

The EC-GSM-IoT device leaves packet idle mode and enters connected mode (packet transfer mode) when it has received a matching page or when there is a trigger due to data on higher layers to be sent from the device to the network.

9.6.9.1 Channel Request

In case packet data need to be sent by the device or a matching paging request was received, the device performs a channel request sending one or a sequence of short access bursts on the random access channel, i.e. EC-RACH (TN 1) or on the legacy RACH (TN 0), if so indicated by the network in the synchronization channel (EC-SCH). For sending the channel request the device selects the appropriate UL Coverage Class according to UL Coverage Class thresholds broadcast in the EC SI and includes in the message the preferred DL Coverage Class for the access grant message on EC-AGCH as well as receive power information, derived from the DL Coverage Class thresholds broadcast in the EC SI to assist the network to select the appropriate DL Coverage class for its response.

For EC-GSM-IoT combined system operation, if TN 1 is used for sending the channel request on EC-RACH, the device adopts one of the four defined new synchronization sequences for CC1 to CC4 depending on its selected uplink Coverage Class. Upon detection of the synchronization sequence, the base station derives the repetition pattern for the respective Coverage Class (see Table 9.5) and combines access bursts with the same synchronization sequence for decoding the information on EC-RACH. If the device in CC1 is commanded by the network to transmit on RACH (TN 0), it uses the same synchronization sequence as it would use on TN 1. Legacy devices may use one of three defined synchronization sequences, and PEO capable devices, not supporting EC-GSM-IoT, use a separate synchronization sequence. Thus, the base station needs to be capable distinguishing five synchronization sequences on TN 0 and four on TN 1 in this scenario.

For a EC-GSM-IoT standalone system, where TN 0 is not used for legacy device access, the device in a higher coverage class (CC2 to CC4) may transmit on EC-RACH in two consecutive timeslots (TN 0 and TN 1) and the device in CC1 only on TN 1, such that access requests with only three different synchronization sequences in TN 0 and four in TN 1, respectively, need to be distinguished by the base station.

9.6.9.2 Access Grant

After decoding and processing the channel request or the paging response, the network responds with an EC Immediate Assignment message on the EC-AGCH containing the Fixed Uplink Allocation for UL data transfer or the Dynamic Allocation for DL data transfer and commands the device to use specific coverage classes on DL and UL. Like for the EC-PCH, the same TSC discrimination is used for devices in the lower coverage class (CC1) and in one of the higher coverage classes (CC2–CC4) to shorten the access grant monitoring time.

9.6.9.3 Data Transfer

After receiving the channel assignment message, the mobile switches to the assigned packet traffic channel and starts to listen or transmit in the indicated starting frame.

Uplink Data Transfer This is used for Mobile Autonomous Reporting (MAR) or Exception Reporting (e.g. alarms). In this case the device transmits radio blocks including blind physical layer transmissions, according to the commanded UL CC, on the granted UL resource. The device thereafter listens to the base station's feedback on received radio blocks (Packet Uplink Ack/Nack), sent in the associated packet control channel (EC-PACCH) identified by the assigned Temporary Block Identifier (TFI) and containing also the uplink resource for FUA related to the hybrid automatic repeat request (HARQ) retransmission, and reconfirms reception of this control message via the EC Packet Control Acknowledgement (EC PCA). The next data blocks are sent or resent until all radio blocks are received by the base station or the maximum number of retransmissions is reached for a given radio block.

Downlink Data Transfer This is used for network commands (e.g. software downloads). In this case the device listens to the assigned packet data traffic channel (EC-PDTCH) identified by the TFI and attempts to decode the radio blocks, eventually sent using blind physical layer repetitions (according to the commanded DL CC). The network polls the device to send its feedback on received radio blocks (Packet Downlink Ack/Nack) in the associated packet control channel (EC-PACCH) including the channel description for the UL resource to be used. Upon reception, the base station performs HARQ retransmission of reported non-received blocks and continues to send new radio blocks until all radio blocks are received by the device or the maximum number of retransmissions is reached for a given radio block.

After data transfer the device stays for a certain duration in packet transfer mode where it can be reached on the dedicated channel until it switches back to idle mode.

9.6.10 Location Services

EC-GSM-IoT in 3GPP Rel-13 supports cell ID TA procedure. Both the timing advance to the serving cell, the cell ID and optionally received quality parameters are sent to the Serving Mobile Location Centre (SMLC) determining the location estimate, which is fed back to the SGSN in the Core Network (CN). An enhanced location method is specified in 3GPP Rel-14 (see Section 9.6.14).

9.6.11 Core Network Support

To achieve energy efficient operation in idle and connected mode, the device attaches to the network indicating EC operation and negotiates its eDRX cycle or its PSM active timer with the Serving GPRS Support Node (SGSN) in the CN. The CN takes the restricted paging availability of the device into account and eventually postpones the paging request based on feedback from the Base Station Subsystem (BSS), which has knowledge of the paging group occurrence for the device under interest. The BSS also forwards DL and UL coverage class as well as cell ID information in connected mode to the SGSN which is used for subsequent paging requests in idle mode. In case the device selects a higher DL coverage class in idle mode due to degraded radio conditions, a cell update is sent to the SGSN including the new DL CC to be used for subsequent paging requests. These messages are exchanged on the Gb interface between BSS and SGSN. Due to longer radio transmissions, the CN uses increased NAS timer values for EC-GSM-IoT devices compared to legacy GPRS/EGPRS devices to avoid NAS layer timeouts when the device is in extended coverage.

9.6.12 Security Enhancements

Several enhancements are introduced for EC-GSM-IoT to protect the user data from undesired interception or mitigate operation of faked base stations or EC-GSM-IoT devices. Weak ciphering algorithms on the user plane such as GEA1 to GEA3 were removed for EC-GSM-IoT and only the usage of GEA4 [21] and the newly introduced GEA5 [23] are allowed, while there is no ciphering on the control plane. In addition, integrity protection has been introduced with the definition of GIA4 and GIA5 algorithms for NAS layer messages [22, 23].

9.6.13 Operation in Reduced Spectrum Allocation

For reduced spectrum allocation due to tight frequency reuse networks, the system operation has been optimized by introducing following measures:

1. Usage of a longer 9-bit BSIC, consisting of the Network Colour Code (NCC, 3 bit), the Base Station Colour Code (BCC, 3 bit), and the Radio Frequency Colour Code (RCC, 3 bit). The new RCC part is used to distinguish cells operating on the same BCCH carrier frequency. For instance, in a 1/3 tighter reuse network, the same BCCH carrier frequency is used in one cell of each three-sectored cell site and hence the device synchronizing to the network and selecting the best available cell needs to have means to distinguish between these sector cells. The 9-bit BSIC is contained in the EC-SCH. The 9-bit BSIC is also supported by PEO devices, not supporting EC-GSM-IoT but eDRX and/or PSM on NAS level, composed by the 6-bit long BSIC being signalled in the SCH and the 3-bit RCC broadcast in the BCCH as well as in PCH/AGCH rest octets.
2. Signal power measurement mandated for the device: The mandated usage of measurements on FCCH and EC-SCH channels, further the removal of interference and noise contributions to determine the signal power of the cell, and the definition of new cell reselection criteria yield improved performance in particular in tighter reuse networks with larger interference power levels compared to the legacy measurement method, requiring the mobile to measure the total power

including noise and interference for cell selection purposes, and hence degrading the estimation of the received quality of a cell in this scenario.
3. Signal to interference plus noise ratio (SINR) based Coverage Class Selection: In addition to signal level based DL and UL coverage class selection, an SINR based DL coverage class selection is specified. The network signalling SINR based DL CC thresholds requires the device to estimate the appropriate DL CC based on the experienced SINR yielding improved performance in tighter frequency reuse networks, typically being interference limited.

Operation of EC-GSM-IoT in tighter frequency reuse networks is described in more detail in 3GPP TR 45.050 [24].

9.6.14 Rel-14 Enhancements

In 3GPP Rel-14 following enhancements for EC-GSM-IoT were standardized:

- *Positioning enhancements.* Position accuracy of EC-GSM-IoT devices is improved with a target of 100 m by using Multilateration Timing Advance (MTA). The device being paged in idle or connected mode, reselects to the strongest cell and performs channel access to this cell as well as to other suitable, non-co-sited cells to enable the BSS to acquire the timing advance to multiple cell sites and to report these measurements to its assigned SMLC which performs a Timing Advance multilateration for determining the position of the device. SMLC forwards this information via the CN to the requesting application. In addition, Multilateration Observed Time Difference (MOTD) is introduced, where the SMLC after the mobile station's reselection to the strongest cell, provides it with assistance data for selecting appropriate neighbour cells for measuring observed time differences of arrival between the strongest cell and any of these neighbour cells at its location and reporting them via the BSS to the SMLC. MOTD provides better device energy efficiency since minimizing transmissions, but exhibits lower position accuracy compared to MTA. It may be used in synchronized networks in standalone mode and in non-synchronized networks in combination with MTA. The functional description for MTA and MOTD is contained in 3GPP TS 43.059 [25].
- *Support of Dedicated Core Networks (DCN).* The support of a DCN for EC-GSM-IoT requires to (re-)route data and control information from the BSS to the correct CN node. This (re-)route capability is controlled by new information elements: UE Usage Type (fetched by the SGSN from HSS), SGSN Group Identity and additional P-TMSI.
- *Radio Interface Enhancements.* Alternative mappings for higher Coverage Classes (CC2 to CC4) based on two consecutive PDCHs (i.e. time slots), rather than four consecutive PDCHs as used in 3GPP Rel-13, are introduced in 3GPP Rel-14 both for downlink and uplink packet traffic channels and associated signalling channels to allow for EC-GSM-IoT data transfer in conditions with constraint radio resources, i.e. few carriers in the cell, or temporary high traffic load due to other services. In this case Overlaid CDMA is supported on two consecutive time slots, see Figure 9.17. In addition, a more robust Coverage Class (CC5) for uplink is designed to achieve a target MCL at least 3 dB higher compared to 3GPP Rel-13, this enabling an uplink MCL beyond 164 dB for devices with 33 dBm and equally improving uplink MCL for devices with low output power (i.e. 23 dBm) towards 157 dB, thus getting closer to downlink MCL. The functional description is contained in 3GPP TS 43.064 [18].

9.6.15 Rel-15 Enhancements

In 3GPP Rel-15 following enhancements for EC-GSM-IoT were standardized:

- *Further enhancements.* With the increase of CIoT deployments, new use cases requiring reduced latency for network commands are becoming more relevant. Use cases, such as the unlocking of smart-bikes based on payment reception at server or network command based tracking of devices for fleet management purposes, require low latency for the network command which in turn requires the MS to operate a lower eDRX cycle, such as a few minutes rather than tens of minutes, while the frequency of network commands to the MS in these use cases is still expected to be rather low. However, the lower eDRX cycle will yield a substantial decrease of the battery life time; a reduction from beyond 10 years to 2.5 years is observed calling for further MS energy savings [26]. Two enhancements are specified in 3GPP Rel-15 which are expected to yield significant MS energy savings for these scenarios:
  - Introduction of a paging indication channel (EC-PICH): using up to two bursts, informing the EC-GSM-IoT device in a higher coverage class (CC3 or CC4) on the presence of a paging request in upcoming paging blocks and thus allowing for an early 'go-back-to-sleep' mode, in case no paging is indicated. This enhancement particularly reduces energy consumption of the device in extreme coverage; a reduction of up to around 25% for a stationary device in CC4 was evaluated [26].
  - Introduction of deferred System Information acquisition: mitigating the need to read system information messages on EC-BCCH, for cell reselection as well as after cell reselection to a specific neighbour cell prior to receiving a matching page in that cell. This is achieved through the definition of common cell parameters for idle mode mobility, such as broadcast carrier allocation, PCH organization, cell reselection parameters and Coverage Class threshold parameters, for a group of geographically contiguous cells, which is known to the MS in the serving cell prior to performing cell reselection to another cell. The MS will thus defer the reading of EC System Information for these cells after receiving a matching paging request. This enhancement is observed to yield an energy consumption benefit of up to around 35% [26] and will particularly serve well for devices with low or moderate velocity. At the time of writing, the specification of this feature is being finalized, the functional description being contained in 3GPP TS 43.064 [18].

9.7 LTE-M

LTE-M (LTE for M2M, also known as eMTC in 3GPP) was standardized in Rel-13 to provide low-power, wide-area cellular access for MTC devices. It reuses existing LTE channels (see [36], [37], [38], [39]) and signals with some restrictions to support low-complexity devices and enhancements for extended coverage. The design objectives for LTE-M include (see [29]):

- Support of low complexity and low-cost devices for MTC operation in any LTE duplex mode (i.e. full Frequency Division Duplex [FDD], half duplex FDD, and Time Division Duplex [TDD]). Both low-cost MTC and legacy UEs can operate on the same carrier.
- Improved coverage of 15 dB in FDD compared to nominal LTE cell coverage footprint. Based on [29], this corresponds to a MCL of 155.6 dB.

- Power consumption reduction to target ultra-long battery life.

These design objectives can be met via the following techniques:

- Reducing device complexity by supporting bandwidth of 1.08 MHz, having only one receiver antenna at the device, reduced peak data rates to below 1 Mbps, optimized to support half-duplex operation.
- Reuse of almost all legacy signals and channels (only a new downlink control channel is introduced) with some restrictions. This allows low-complexity devices to operate in any system bandwidth up to 20 MHz.
- Improving coverage by using repetition, power boosting, and frequency hopping.
- Improving power efficiency by minimizing downlink and uplink transmissions using extended discontinuous transmission/reception (eDRX), PSM, and signalling reduction for small data transmission.

LTE-M occupies 1.08 MHz of spectrum, which is equivalent to the bandwidth of six Physical Resource Blocks (PRBs) in LTE. It can be deployed in-band in any of the LTE system bandwidths from 1.4 to 20 MHz. This is done by introducing a narrowband concept to LTE. The wideband LTE system is divided into narrow bands. Each narrowband comprises six PRBs and spans over 1.08 MHz, which is within the radio frequency bandwidth of narrowband LTE-M UEs. UEs can retune from one narrowband to another as configured or scheduled by the eNB.

In the downlink, LTE-M reuses the following legacy signals and channels – Primary Synchronization Signal/Secondary Synchronization Signal (PSS/SSS), Common Reference Signals (CRS), Demodulation Reference Signal (DMRS), Physical Broadcast Channel (PBCH), Physical Downlink Shared Channel (PDSCH). These signals and channels are either already confined to or can be restricted to six PRBs, requiring no changes in the specification to the physical layer structure. One new control channel, the MTC Physical Downlink Control Channel (MPDCCH), based on the legacy Enhanced Physical Downlink Control Channel (EPDCCH) was introduced. In addition, the following legacy LTE channels are not supported: Physical Hybrid-ARQ Indicator Channel (PHICH), Physical Control Format Indicator Channel (PCFICH) and Physical Downlink Control Channel (PDCCH). To support EC, repetitions and frequency hopping were introduced to the supported channels.

For cell access, the UE can use legacy synchronization signals and PBCH with new information added to the Master Information Block (MIB) that are specific to LTE-M. The enhanced MIB contains scheduling information for the LTE-M system information block. From decoding the MIB, MTC devices can determine whether the system supports LTE-M operations or not. New system information blocks for LTE-M are also transmitted. Figure 9.18 illustrates system access for a narrowband LTE-M UE operating in the wideband LTE system.

In the uplink, LTE-M reuses the following legacy signals and channels: Physical Random Access Channel (PRACH), Physical Uplink Shared Channel (PUSCH), DMRS and Sounding Reference Signal (SRS). The Physical Uplink Control Channel (PUCCH) was reused but without intra-subframe inter-slot frequency hopping like in legacy LTE. No new channel was added, although enhancements (repetitions and frequency hopping) were introduced to the uplink channels to support CEs.

In LTE-M, two CE modes were defined: CE Mode A and B. CE Mode A is intended to provide small to medium CEs, while CE Mode B is intended to provide large CEs.

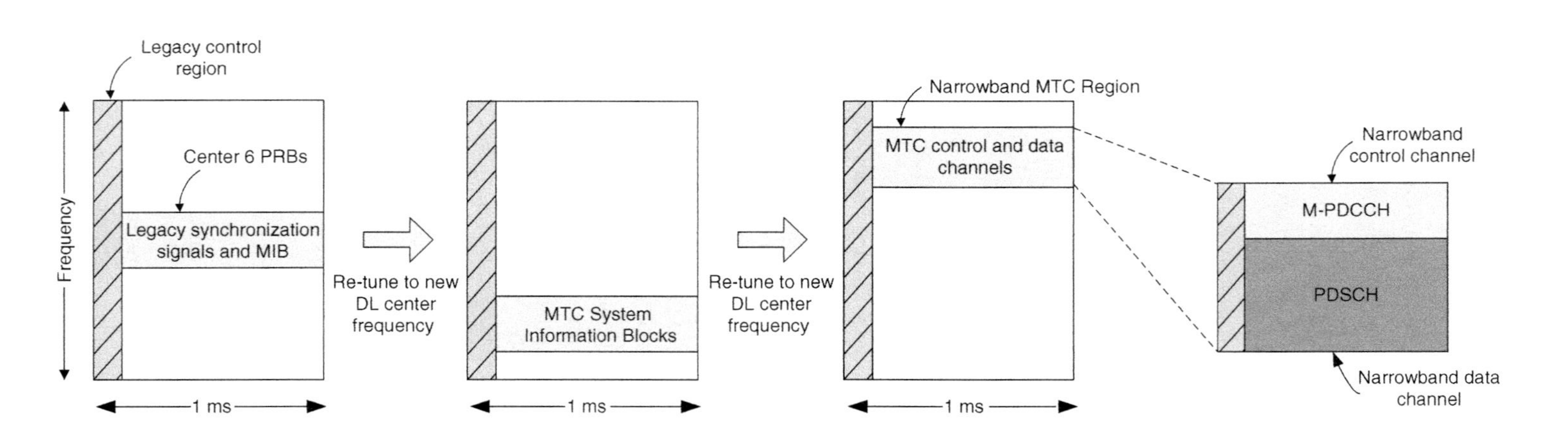

Figure 9.18 LTE-M downlink operation.

Table 9.7 Coverage enhancement modes A and B.

Channel		CE Mode A	CE Mode B
MPDCCH	Aggregation levels	2, 4, 8, 16, 24	8, 16, 24
	Maximum number of repetitions	256	256
PDSCH	Supported modulation levels	QPSK, 16-QAM	QPSK
	Maximum number of repetitions	32	2048
PUSCH	Supported modulation levels	QPSK, 16-QAM	QPSK
	Maximum number of repetitions	32	2048
	Power control	Power control on	Max power is always used
PUCCH	Supported formats	ACK/NACK, CQI, SR	ACK/NACK, SR
	Maximum number of repetitions	8	32

Table 9.8 LTE-M link budget.

Channel	PBCH	MPDCCH	PDSCH	PRACH	PUSCH	PUCCH
TBS/Format			152 bits	Format 0	104 bits	Format 1a
Number of repetitions			32	32	256	32
Transmitter						
Max Tx power (dBm)	46	46	46	23	23	23
(1) Actual Tx power (dBm)	36.8	36.8	36.8	23	23	23
Receiver						
(2) Thermal noise density (dBm/Hz)	−174	−174	−174	−174	−174	−174
(3) Receiver noise figure (dB)	9	9	9	5	5	5
(4) Interference margin (dB)	0	0	0	0	0	0
(5) Occupied channel bandwidth (Hz)	1 080 000	1 080 000	1 080 000	1 080 000	180 000	180 000
(6) Effective noise power = (2) + (3) + (4) + 10 log ((5)) (dBm)	−104.7	−104.7	−104.7	−108.7	−116.4	−116.4
(7) Required SINR (dB)	−15.0	−16.0	−15.6	−24.6	−19.9	−19.3
(8) Receiver sensitivity = (6) + (7) (dBm)	−119.7	−120.7	−120.3	−133.3	−136.3	−135.7
(9) Rx processing gain	0	0	0	0	0	0
(10) MCL = (1)–(8) + (9) (dB)	156.5	157.5	157.1	156.3	159.3	158.7

Support for CE Mode A is mandatory for all LTE-M UEs, while CE Mode B is an optional capability. Table 9.7 summarizes key differences between CE Mode A and B.

Coverage evaluation is provided via link budget analysis as shown in Table 9.8. From [29], the MCL corresponding to nominal LTE cell coverage footprint was determined to be 140.6 dB. Table 9.8 shows that 15 dB CE (up to MCL of 155.6 dB) is feasible.

In Rel-14, the following enhancements were introduced to LTE-M:

- Positioning enhancements including defining measurement performance requirements and introducing enhanced packet radio services (PRSs) resource configurations for Observed Time Difference of Arrival (OTDOA) localization. Up to three PRS time-frequency configurations can be supported in the cell, and frequency hopping and repetitions are also supported on the PRS to help improve measurement accuracy.
- Multicast support based on single-cell point-to-multipoint (SC-PTM) transmission. SC-PTM uses the logical control channel SC-MCCH to convey SC-PTM configuration messages and the transport channel SC-MTCH to carry MBMS sessions (see [11]). Data rate of up to 1 Mbps using 1.4 MHz or 4 Mbps using 5 MHz can be supported on the SC-MTCH. To keep complexity low, SC-PTM is only supported in RRC_IDLE mode ([9]) and the UE is not required to receive the SC-MCCH and SC-MTCH at the same time.
- Support for higher data rates. A new UE category (Cat-M2) supporting 5-MHz bandwidth was introduced. This new UE category can support peak rates of approximately 4 Mbps in the downlink and 7 Mbps in the uplink. In addition, the uplink peak rate for Cat-M1 UE category was increased from 1 Mbps to approximately 3 Mbps. Non-bandwidth limited UE in CE mode can now support up to 20 MHz for the downlink and 5 MHz for the uplink.
- Furthermore, the number of downlink HARQ processes in FDD was increased from 8 to 10 to allow continuous downlink data transmission. HARQ-ACK bundling was introduced in HD-FDD to allow a single HARQ-ACK feedback for multiple downlink transport blocks.
- Support of Voice over LTE (VoLTE) enhancements to improve coverage for VoLTE and other delay sensitive services. They included supporting additional PUSCH repetition numbers, dynamically adjusting the HARQ-ACK delay, enabling modulation step down, and enhancing SRS coverage.

Release 15 enhancements for LTE-M are ongoing and enhancements are being considered to improve latency, battery life, and spectral efficiency.

9.8 NB-IoT

9.8.1 Introduction

NB-IoT was standardized in Rel-13 to provide low-power, wide-area cellular access for IoT devices. It is a new radio access system built from existing LTE functionalities with essential simplifications and optimizations. The design objectives for NB-IoT include (see [16]):

- Support of ultra-low complexity and low-cost devices.
- Improved coverage of 20 dB compared to legacy GPRS, corresponding to a MCL of 164 dB. At this coverage level, a data rate of at least 160 Bps should be supported at the application layer.
- Support of massive number of low-throughput devices, at least 52 547 devices within a cell-site sector.

- Improved power efficiency, battery life of 10 years with battery capacity of 5 Watt/hour.
- Exception report latency of 10 seconds or less for 99% of the devices.

These design objectives can be met with the following techniques:

- Reducing device complexity by supporting only narrow channel bandwidth of 200 kHz, having only one receiver antenna at the device, half-duplex operation only, reduced peak data rates, and supporting relaxed processing time.
- Improving coverage via narrowband transmission, using repetition, power boosting, and using single-tone transmission with low power backoff requirement in the uplink.
- Supporting a large number of devices by using single-tone/multi-tone transmission, signalling reduction for small data transmission, and using multiple carriers.
- Improving power efficiency by minimizing downlink and uplink transmissions using extended discontinuous transmission/reception (eDRX), PSM, signalling reduction for small data transmission, and low peak to average power ratio modulation schemes in the uplink.
- Reducing delay for exception report by signalling reduction for small data transmission and adding an establishment cause for exception reporting.

NB-IoT occupies 180 kHz of spectrum, which is equivalent to the bandwidth of one PRB in LTE. It can be deployed in three operation modes, in-band within an LTE carrier, in the guard-band of an LTE carrier, and as a stand-alone carrier. In in-band operation mode, each NB-IoT carrier uses one LTE PRB. This allows LTE and NB-IoT to seamlessly share valuable spectrum between the two systems. In addition, broadband traffic served by LTE is downlink centric, while IoT traffic served by NB-IoT is uplink centric. Thus, the two systems can share spectrum efficiently.

In guard-band operation mode, NB-IoT will be deployed in the guard-band of an LTE system. This allows IoT services to be supported without eating up LTE spectrum. This operation mode is most suitable for wideband LTE with bandwidth greater than or equal 5 MHz. This is because there is sufficiently large amount of guard-band available allowing NB-IoT to be deployed without co-existence issues.

In stand-alone operation mode, each NB-IoT carrier can be deployed as a replacement for one GSM carrier. This provides the ability to efficiently refarm spectrum from older GSM technology to support IoT services.

NB-IoT design is based on LTE and supports most LTE functionalities with many simplifications and some optimizations to support low-power, wide-area, and low data-rate IoT services. Reusing LTE design where possible has been vital in achieving rapid specification of NB IoT in Rel-13.

Table 9.9 lists downlink (DL) and uplink (UL) NB-IoT channels and signals and their functions. A summary of each channel and signal is provided in this section (see also [35], [36], [37], [38], [39]).

9.8.1.1 Downlink

In the downlink, the following channels and signals are transmitted: Narrowband Physical Downlink Control Channel (NPDCCH), Narrowband Physical Downlink Shared Channel (NPDSCH), Narrowband Physical Broadcast Channel (NPBCH) and Narrowband Primary Synchronization Signal (NPSS/Narrowband Secondary Synchronization Signal (NSSS)).

Table 9.9 NB-IoT channels and signals.

Channel			Purpose
Downlink	NPSS/NSSS	Narrowband Synchronization Signal (Primary and Secondary)	Time and frequency synchronization, cell identification
	NPBCH	Narrowband Physical Broadcast Channel	Master information for system access
	NPDCCH	Narrowband Physical Downlink Control Channel	Uplink and downlink scheduling information
	NPDSCH	Narrowband Physical Downlink Shared Channel	Downlink dedicated and common data
Uplink	NPRACH	Narrowband Physical Random Access Channel	Random access
	NPUSCH	Narrowband Physical Uplink Shared Channel	Uplink dedicated data

Synchronization signals are used by the UE to obtain time and frequency synchronization with the network as well as to obtain the cell identification. Two synchronization signals are present in NB-IoT: NPSS and NSSS. NPSS provides symbol level timing while NSSS provides 80 ms timing boundary needed to decode the NPBCH. The NSSS also provides the cell identification determined from the sequence used for NSSS.

NPBCH transmits the Narrowband Master Information Block (MIB-NB). The MIB-NB conveys information needed by the UE to access the system. It includes essential information such as system frame number, operation mode, system value tag to indicate changes in the system information, access barring information, and scheduling information of other system information blocks.

NPDCCH is used to transmit Downlink Control Information (DCI). Three new DCI formats are defined: format N0 is used for scheduling uplink data, format N1 is used for scheduling downlink data and for requesting the UE to perform random access procedure, and format N2 is used for scheduling paging as well as to indicate a change in the system information.

NPDSCH is used to transmit downlink data to the UE and occupies the entire downlink bandwidth of 12 subcarriers. A single transport block may be mapped to multiple subframes from the set {1, 2, 3, 4, 5, 6, 8, 10}. The maximum transport block size (TBS) that can be transmitted on the NPDSCH is limited to 680 bits.

9.8.1.2 Uplink

In the uplink, only two channels are defined: Narrowband Physical Random Access Channel (NPRACH) and Narrowband Physical Uplink Shared Channel (NPUSCH).

NPRACH is used for random access. Only preambles are transmitted on NPRACH. The preamble uses single-tone transmission with frequency hopping. To provide large coverage, the NPRACH uses narrowband transmission. This allows the UE to concentrate its power onto a narrowband, improving the SNR. The NPRACH uses subcarrier spacing of 3.75 kHz. Two cyclic prefix lengths are provided to support two different cell sizes, 10 and 35 km. Up to 48 preambles can be supported in one random access opportunity. In addition, the NPRACH can be repeated up to 128 times to improve coverage.

NPUSCH is used to transmit uplink data to the eNB. The channel was designed to provide extended coverage and massive capacity via the following features:

- Single-tone and multi-tone transmission. Multi-tone transmission can be used by the UE in good condition while single-tone transmission can be used by the UE in poor condition. Single-tone transmission supports both 3.75 and 15 kHz subcarrier spacing.
- Support of phase shifted modulation schemes for single-tone transmission to support efficient power amplifier operation. These schemes have low power amplifier back-off requirements, which results in higher average transmission power and therefore higher coverage.

In addition, NPUSCH supports repetition up to 128 times and larger transport blocks that can be mapped to up to 10 resource units.

In NB-IoT, multiple carriers can be supported. Each carrier takes up one LTE PRB in the in-band deployment mode, or 200 kHz in the guard-band/stand-alone mode. However, there is only one anchor carrier on which the UE can access the system. Once the UE is in RRC connected state, it can be configured via RRC signalling to move to a non-anchor carrier.

On the higher layer, only essential features for small data transmission are supported in Rel-13. Hence, NB-IoT does not support mobility, handover, and measurement reports. Two optimizations for small data transmission have been specified: RRC connection suspend/resume and data transmission using control plane signalling (see also Section 9.5). In addition, multi-carrier operation is supported.

RRC connection suspend/resume ([9]) allows an RRC connection to be suspended at RRC connection release and then resumed later. A Resume ID is provided to the UE and the eNB stores the UE context together with Resume ID upon connection suspension. During the resume procedure, the UE provides the stored Resume ID to the eNB to resume the RRC connection. Security is continued and there is no need for a Service Request procedure.

Data transmission via the control plane allows data to be encapsulated in control plane messages and transmitted via the Signalling Radio Bearer (SRB). An uplink NAS signalling message or uplink NAS message carrying data can be transmitted in an uplink RRC container.

Table 9.10 summarizes the number of signalling messages required among the three methods. As the table shows, the number of messages can be reduced significantly using the two new procedures.

9.8.2 Coverage

Coverage evaluation is provided via link budget analysis. Table 9.11 illustrates NB-IoT link budget for in-band deployment. In this operation mode, eNB transmission power must be shared between NB-IoT and LTE. In this analysis, the total eNB transmit power is 46 dBm, resulting in PSD per PRB of 29 dB. To improve coverage, 6 dB PSD boosting is used for NB-IoT, bringing the total transmission power for NB-IoT to 35 dBm. At the

Table 9.10 Number of signalling messages for NB-IoT data transmission.

Direction	Normal service request procedure	RRC connection resume	Control plane data transmission
UL	Preamble		
DL	Random access response		
UL	RRC connection request	RRC connection resume request	RRC connection request
DL	RRC connection setup	RRC connection resume	RRC connection setup
UL	RRC connection setup complete	RRC connection resume complete	RRC connection setup complete
DL	Security mode command		
UL	Security mode complete		
DL	RRC connection reconfiguration		
UL	RRC connection reconfiguration complete		
Total number of messages	9	5	5

Table 9.11 NB-IoT link budget for in-band deployment mode.

Channel	NPBCH	NPDCCH	NPDSCH	NPUSCH	NPUSCH	NPRACH
Data rate (kbps)			0.44	0.31	0.36	
Transmitter						
Max Tx power (dBm)	46	46	46	23	23	23
(1) Actual Tx power (dBm)	35	35	35	23	23	23
Receiver						
(2) Thermal noise density (dBm/Hz)	−174	−174	−174	−174	−174	−174
(3) Receiver noise figure (dB)	5	5	5	3	3	3
(4) Interference margin (dB)	0	0	0	0	0	0
(5) Occupied channel bandwidth (Hz)	180 000	180 000	180 000	15 000	3750	3750
(6) Effective noise power = (2) + (3) + (4) + 10 log ((5)) (dBm)	−116.4	−116.4	−116.4	−129.2	−135.3	−135.3
(7) Required SINR (dB)	−12.6	−13.0	−13.7	−12.8	−6.6	−5.8
(8) Receiver sensitivity = (6) + (7) (dBm)	−129.0	−129.4	−130.1	−142.0	−141.9	−141.1
(9) Rx processing gain	0	0	0	0	0	0
(10) MCL = (1) − (8) + (9) (dB)	164.0	164.4	165.1	165.0	164.9	164.1

UE side, 23 dBm of power is available. Table 9.11 shows that the target MCL of 164 dB can be achieved.

9.8.3 Capacity

Capacity evaluation is determined using system simulations. A traditional macro-cell system simulation scenario is used with NB-IoT deployed within an LTE system. The cell layout is of a traditional 19-site, 57-cell system setup with wrap-around. The inter-site distance between cells is 1732 m, which represents a suburban macro-cell deployment scenario. LTE system bandwidth is 10 MHz and NB-IoT occupies one LTE PRB. NB-IoT transmit power is 35 dBm and thermal noise level of −174 dBm/Hz is assumed. The carrier frequency is 900 MHz.

The traffic model is based on MAR [16]. In this model, devices wake up periodically to transmit a report, then go back to sleep. The inter-arrival time for the reports are as follows: 40% of the devices transmit a report once a day, 40% of the devices transmit a report once every two hours, 15% of the devices transmit a report once every hour and remaining 5% of the devices transmit a report once every thirty minutes. The size of each report is random and follows a Pareto distribution with a minimum packet size of 20 bytes and a maximum packet size of 200 bytes. The mean packet size is 32.6 bytes. In addition, to each packet a total protocol overhead of either 65 bytes (without IP header compression) or 29 bytes (with IP header compression) is added.

NB-IoT objective is to support at least 52 547 devices within a cell-site sector. System-level simulation results show that at least 250 000 devices can be supported within a cell site sector per NB-IoT carrier (in this case, one NB-IoT carrier equals one LTE PRB). Additional capacity, if needed, can be acquired through using multiple carriers.

9.8.4 Extended Battery Life

Battery life analysis is based on MAR traffic model described previously and battery capacity of 5 Wh. The target is to achieve battery life of more than ten years at extreme coverage (164 dB maximum coupling loss). In the analysis four power states are used: transmit, receive, idle and power saving. The current consumptions corresponding to the four states are 543, 90, 2.4, and 0.015 mW, respectively.

Based on these assumptions, the estimated battery lifetime at maximum coverage level are 10.5 years to send a daily report of 200 bytes and 16.8 years to send a daily report of 50 bytes. This shows that the NB-IoT 10-year battery life target can be met or exceeded for daily reporting.

9.8.5 Rel-14 NB-IoT Enhancements

In Rel-14, the following enhancements were introduced to NB-IoT:

- Positioning based on Enhanced Cell-ID (E-CID) and OTDOA. Narrowband Positioning Reference Signal (NPRS) was defined allowing the UE to measure the time of arrival (ToA) of reference signals received from multiple transmission points and report the Reference Signal Time Difference (RSTD) to the location server. In addition, performance requirements were defined for E-CID.

- Multicast support based on single-cell point-to-multipoint (SC-PTM) transmission.
- Support for higher data rates. A new UE category (Cat-NB2) was introduced. This new UE category can achieve peak data rates of approximately 127 kbps in the downlink and 159 kbps in the uplink.
- Support of paging and random access on non-anchor carriers. This increases paging and random access capacity and allows the network to perform load balancing for paging and random access among carriers.
- A new UE power class with a maximum output power of 14 dBm has been introduced with relevant RF requirements defined. This new power class is suitable for small form-factor batteries supporting use cases such as wearables.

Release 15 enhancements for NB-IoT are ongoing and enhancements are being considered to improve latency and battery life. Rel-15 will also introduce TDD support for NB-IoT and enhance the range and reliability of the random access channel.

9.9 5G for M2M

9.9.1 Requirements

Although M2M support in 5G will be specified by 3GPP only in Rel-16, some high-level assumptions and key targets for the 5G radio to fulfil M2M/IoT device requirements are known today:

- 5G wide area coverage is a necessity for many applications.
- Local area support for low cost IoT devices is also envisioned.
- Use cases for local area, low cost IoT devices may include connected homes (light, heating, ventilation, etc.).
- Having 5G IoT devices supporting both local area and wide area networks requires a solution to enable multiband support in 5G IoT devices.
- Native support for low cost and low power IoT, both local area and wide area.
- Lower energy consumption compared to LTE-M.
- Optimized architecture for massive IoT support.

Like for other radio technologies, M2M/IoT devices using 5G radio are supposed to support all or a subset of following features:

- Power savings and paging optimizations.
- Efficient infrequent and frequent small data transmission.
- Optimizing signalling and latency for idle to active mode transition.
- Support of Embedded SIM/Soft SIM and SIM-less authentication.
- Automated subscription and provisioning.
- Group based features, e.g. group-based messaging.
- Mobility and session on demand.

9.9.2 Challenges

The 5G New Radio (NR) as specified in Rel-15 provides a unified, flexible radio interface capable of supporting various verticals such as automotive, healthcare, industry, smart city, utility, smart home, etc. The 5G radio interface can be tailored to meet requirements coming from different services, including enhanced Mobile Broadband (eMBB), massive Machine-Type-Communication (mMTC) and Ultra Reliable and Low Latency Communication (URLLC):

- Scalable orthogonal frequency-division multiplexing (OFDM) numerology to support different bandwidth, subcarrier spacing, and frequency bands including mmWave.
- Flexible slot structure supporting low latency services and dynamic TDD.
- Massive multiple input multiple output (mMIMO) support to utilize large number of antennas for capacity and Coverage Enhancements (CE).
- Beamforming based operation for capacity and CEs.
- Advanced coding schemes for both control and data packets.

With respect to M2M, the 5G radio must support two use cases:

- mMTC with the following performance objectives: ultra-low complexity and low-cost IoT devices and networks, MCL of 164 dB for a data rate of 160 Bps at the application layer, connection density of one million devices per square km in an urban environment, battery life in extreme coverage beyond 10 years (15 years is desirable), and latency of 10 seconds or less on the uplink to deliver a 20-byte application layer packet measured at 164 dB MCL.
- URLLC with the following performance objectives: user plane latency of 1 ms (0.5 ms for downlink and 0.5 ms for uplink), and packet transmission reliability of 10e−5 with a user plane latency of 1 ms (for a packet size of 32 bytes).

For mMTC, it has been agreed in 3GPP that no 5G radio solution will be specified for low-power, wide-area use cases. Instead, these use cases will continue to be addressed by evolving LTE-M and NB-IoT as those two radio technologies can satisfy the 5G requirements for mMTC. Note that peak data rates up to 27 Mbps in DL and 7 Mbps in UL are possible with LTE-M. Further analysis of LTE-M/NB-IoT deployment within 5G carriers might be required to ensure optimal co-existence between these technologies.

For URLLC, key radio techniques have been specified for 5G in Rel-15 including short slot length, mini-slot allocation, data pre-emption, advanced coding, massive MIMO and beamforming. Further 5G radio enhancements are being studied including grant-free transmission, compact scheduling grant, control channel repetition and defining new modulation and coding table.

In addition, 3GPP is studying M2M use cases that fall in between mMTC and URLLC. Examples include CCTV cameras, mass transit train control and building automation. These use cases require higher data rates, e.g. peak data rates of 100 Mbps, than can be delivered by mMTC, but do not require URLLC level of latency and reliability. A low-complexity 5G UE category may be defined to support these use cases.

9.9.3 Architecture Enablers

9.9.3.1 General

The Rel-15 5G System feature set did not include MTC functionality on complete feature parity basis compared to EPS (as explained in Section 9.3), but it introduces some 'enablers' for future support of M2M/IoT devices. Similar requirements as addressed by earlier technologies are solved using similar or different solutions in 5G. Consequently, some EPS MTC optimizations are adopted to 5G, some others are not.

9.9.3.2 Mobile Initiated Communication Only (MICO)

As the name implies, only the UE can initiate communication in the Mobile Initiated Communication Only (MICO) mode. This mode of operation is optimized for telemetry type of devices sending data event driven or on scheduled basis. MICO mode is always negotiated between UE and access and mobility management function (AMF), to avoid waste of paging resources, as the UE in MICO mode need not listen to paging. The UE may request MICO mode during the Registration procedure, and if the network authorizes it based on subscription and network local policy for this UE, then the AMF acknowledges use of MICO in the Registration Accept message.

The AMF assigns a Periodic Registration Update timer also to MICO mode UEs, and to use MICO, the UE needs to re-confirm it explicitly at every Registration Update. The new 'All PLMN' registration area concept allows the AMF assigning a PLMN-wide Tracking Area List to the UE, which means that the UE will not meet a registration area border if it remains registered to the same Registered PLMN. Thus, the only Registration Updates the UE needs to send are the timer based Periodic Registration Updates.

Since the UE in MICO mode cannot be paged, the AMF will consider a UE in this mode as unreachable for mobile terminated data. A MICO mode UE is reachable for DL data only in CM-CONNECTED state, once the UE has initiated the connection establishment.

MICO mode offers an extreme power saving method for the UE, as the only system-based responsibility for such a UE is the Periodic Registration Update. If there are no application data to send, the modem in the device can be switched off completely between Periodic Registration Updates.

9.9.3.3 RRC Inactive

RRC Inactive is a new RRC state that was introduced for 5G (see [3] and [4]). This state is under RRC control, and normally the CN need not be aware of it, even though some information exchange may occur between radio network and CN.

The Core provides RRC Inactive assistance information to the RAN over the N2 interface (see [5]). This assistance information comprises the following data:

- Registration Area;
- Periodic Registration Update timer;
- DRX values;
- MICO mode indication;
- (Partial) information of the UE permanent identifier to allow RAN calculating the UE paging occasions.

RAN can decide to save radio resources by moving some UEs from RRC Connected to RRC Inactive state. These RRC state transitions are invisible for the Core (AMF), for which the UE remains in CM-CONNECTED state.

At the time of writing this book it is open whether support of RRC Inactive is indicated in RRC signalling.

UE mobility in RRC Inactive state is managed by RAN. When moving the UE to RRC Inactive, RAN need to consider the NAS Registration Area and the Periodic Registration Update timer, to avoid exceeding either of them. The RAN Notification Area issued to the UE is completely inside the NAS Registration Area, or at the maximum, the same as the NAS Registration Area. The RAN Notification Area Update timer is set considering the value of the NAS Periodic Registration Area Update timer and the DRX parameters. When the UE in RRC Inactive state re-selects to a cell outside of its RAN Notification Area, or the RAN Notification Area Update timer expires, the UE performs RAN Notification Area Update, which is an RRC procedure that is not visible to the AMF.

To send uplink data, a UE in RRC Inactive state resumes the inactive RRC connection instead establishing a new one. Downlink data is just sent towards the UE over the existing N3 connection as the CN considers the UE to be in CM-CONNECTED state. If the UE is in RRC Inactive state, the RAN pages the UE to resume the inactive connection to deliver the pending data.

UE mobility during RRC Inactive may lead to context retrieval between RAN nodes, to enable the UE resuming the inactive connection at different RAN nodes. If the resume fails, the RAN paging fails, or the UE fails to perform Periodic RAN Notification Area Update, the RAN releases the existing connection, which moves the UE state in AMF to CM-IDLE. Also cell re-selection to another RAT (GERAN, UTRAN) leads to transition of the UE state to CM-IDLE.

The UE behaviour in RRC Inactive state uses some aspects from the Idle state, as the UE in CM-CONNECTED/RRC Inactive can perform Idle mode PLMN selection procedure.

9.9.3.4 Idle Mode DRX and High Latency Communication

Idle mode DRX is supported by 5G, but there is no support of Extended Idle mode DRX or High Latency Communication (HLCom) in the 5G CN. Consequently, the DRX capabilities are limited by the buffering and scheduling capabilities of the radio network, as the Core does not offer any help via extended buffering. Before DRX can be used, the DRX cycle must be negotiated between UE and Core, and the Core makes the RAN aware of the negotiated DRX parameters.

9.10 Comparison of EPS and 5GS

9.10.1 Goals of MTC and IoT Optimizations

Main drivers for the architectural design to support MTC and IoT use cases were the following:

- UE simplicity and power saving capability;
- Registration without PDN connection (and possible connectivity on demand);
- Infrequent and frequent small data support.

Efficient use of radio frequencies and MTC optimized radios is by no means less important than these architectural requirements, but it is covered in other sections of the book. In the following sections we will compare EPS (see [14]) and 5GS (see [33], [34]) concepts to cope with the above-mentioned goals and drivers.

9.10.2 UE Simplicity and Power Saving

CIoT EPS Optimizations aim at UE complexity reduction via omitting capabilities that are not required for MTC use cases. One major advantage is removal of CS domain support. With 5G, a UE gains the same advantage, as native interworking between 4G and 5G is specified, and the 5G system does not contain a CS part. Thus, CIoT and 5G UE need not support combined attach/registration to use SMS service.

The UE needs to communicate with the network due to periodic updates and the need to monitor possible paging causes heavy impact on the UE power budget.

The protocol details for Periodic Registration Update have not been specified at the time of writing this book. If the extended timer values in GPRS are also used for the extended 5G timer values, it would bring 5G on par in terms of periodic updates that can be set to maximum value of 31 times 320 hours, which is more than one year's update period. Such values should be long enough for any application.

The longest possible eDRX period in EPS is determined by the radio hyperframe structure, which means the maximum eDRX cycle depends on the layer 1 frame structure of each radio technology. The longest possible eDRX cycle in NB-IoT (close to three hours) and in WB-E-UTRAN (almost 44 minutes) are far longer than the maximum DRX cycle of 5G. Extended Idle mode DRX and HLcom are not supported for 5G in Rel-15.

The EPS concept of UE PSM was not copied over to 5G as it stands, but MICO mode serves the same purpose. Considering the very long periodic update timers, both MICO and PSM allow the UE to switch off the modem until the next uplink data transfer or the next periodic update. MICO allows to assign an 'All PLMN' Registration Area to the UE, reducing the number of Registration Updates from UE side.

In terms of simplicity, supported procedures, and power saving features, a 5G UE with uplink only application is comparable with CIoT EPS Optimizations. If the UE needs to be paged, then the 5G UE suffers from lack of extended idle mode DRX feature support in 5G.

9.10.3 Connectionless Registration

Attach without PDN connectivity is supported for SMS-only CIoT UEs. With 5G, the registration procedure in AMF is independent from session management function (SMF) controlled PDU session establishment. Connectionless registration is supported natively from the first release of 5G onwards.

Connectivity on demand is supported for UEs initially attached/registered to the network without PDN connection. The UE may remain registered without PDU/PDN connections, but it may request for PDN connection establishment or PDU session establishment later, if this is required by the application.

Connectionless network attachment is supported in 5G with less restrictions than in CIoT EPS Optimizations.

9.10.4 Small and Infrequent Data

For frequent and infrequent data with bursty nature, CIoT offers the User Plane CIoT EPS Optimizations with the capability to suspend the RRC connection to resume it later with an expedited signalling procedure. Introduction of RRC Inactive state offers the same option for a 5G UE. Thus, efficiency of the user plane communication depends mainly on the implementation of radio layers.

CIoT supports also control plane user data using the Control Plane CIoT EPS Optimizations. Control plane data was left out of 5G in 3GPP Rel-15, but it is one of the

Rel-16 topics to be studied. Even without dedicated method for control plane data, 5G can compete with Control Plane CIoT EPS Optimizations offering very low latency and minimal overhead with RRC Inactive state applicable to one-shot messages.

9.11 Future Enhancements

9.11.1 General Rel-16 IoT Aspects

The first 5G Rel-15 specification version follows the 3GPP tradition of establishing the first full 5G release which builds the baseline for further work and enhancements in future releases. At the time of writing this book, several Rel-16 studies are ongoing. One of the agreed study items is 'Study on Cellular IoT support and evolution for the 5G system'. This work considers alignment with the CIoT EPS Optimizations, but it is not restricted to the already specified EPS procedures, even though the goals are mostly common.

The Rel-16 aims for CIoT and 5G evolution in the MTC area and will focus on the following main topics:

- Enable CIoT/MTC functionalities in 5G CN;
- Co-existence and migration from EPS based eMTC/NB-IoT to 5G CN;
- Enhancements to address 5G service requirements.

TR 23.724 [10] documents the goals of the study and candidate solutions. The following key issues were already identified. The study will consider possible solutions for these topics:

1. Support of frequent and infrequent small data transmission;
2. High Latency communication;
3. Power Saving functions;
4. UE TX power saving functions;
5. Management of EC;
6. Overload control for small data;
7. Support of Reliable Data Service;
8. Support of common North-bound APIs for EPC-5GC interworking;
9. Network parameter configuration API provided by network exposure function (NEF);
10. Monitoring support;
11. Inter-RAT mobility support to and from NB-IoT;
12. Support for expected UE behaviour;
13. Quality of Service (QoS) support for NB-IoT,
14. CN selection for Cellular IoT.

As can be seen in the above list, the 5G CIoT study has a wider scope than CIoT EPS Optimizations. In addition to the Rel-13 CIoT work scope, 5G CIoT comprises also the Rel-14 CIoT enhancements, the Rel-15 Northbound API for SCEF and network and UE parameter provisioning and monitoring capabilities. Even though the goals are mostly common, the 5GC architecture shown in Chapter 4 requires some re-design. It is expected that the agreed candidate solutions will be specified as part of 3GPP Rel-16, but the final conclusions on the candidate solutions that will move forward have not been made at the time of writing this book.

9.11.2 Small Data

Both infrequent and frequent small data will be supported in 5G, but the architectural differences between 5GC and EPC do not guarantee that all CIoT EPS Optimizations can be re-used without change in 5GC. The main differences possibly requiring a re-design are the functional split between AMF and SMF and the control plane – user plane split between SMF and user plane function (UPF) in 5GC. Both could make the transfer of user data over the control plane more complicated in 5G.

The candidate solutions for small data include enhanced versions of both control plane and user plane solutions. Among control plane solutions, the main candidate solutions include routing of mobile originating (MO) small data on the path

- UE-RAN-AMF-SMF-UPF-DN or
- UE-RAN-AMF-UPF-DN or
- UE-RAN-AMF-NEF-DN.

The user plane solutions can lead to possible optimizations in suspend and resume procedures. The new candidate solutions for user plane small data include Connectionless and Fast Path solutions, allowing transfer of single packets even without establishment of an RRC connection. This requires the existence of a pre-configured path between RAN and UPF. RAN may either maintain the UE state to know where to route packets, but can also be stateless and retrieve path information from the control plane node when needed. In this case, UPF needs the UE related security parameters for ciphering and integrity protection.

9.11.3 UE Power Saving and High Latency Communication

The initial version of the 5G radio does not support extended idle mode DRX, but various interworking scenarios also allow wide-band LTE and NB-IoT to interwork with the 5G Core over N2 and N3 interfaces, which enables eDRX at least in LTE RAT. UE PSM is not copied to 5GC as such, but MICO mode also requires the network to either drop DL packets targeted for a UE during its sleep cycle, or to buffer them. This requires the addition of High Latency Communication to 5GC. When this is specified in 3GPP, it needs to be determined which node (NEF, UPF, SMF) buffers DL packets during the UE sleep cycle.

Also, MICO mode enhancements are considered in 3GPP, i.e. the addition of an Active Timer to keep the UE pageable for a known period after moving to idle state and a timer to keep the UE connected until the pending DL data can be delivered.

9.11.4 Management of Enhanced Coverage

Management of EC copies the EPC principles where the UE may request for EC, which the serving CN may grant after checking subscription information.

An additional aspect for EPS-5GC interworking and mobility is the capability of MME and AMF to exchange the EC restriction information as part of the UE context transfer during inter-system mobility.

9.11.5 Overload Control for Small Data

User plane overload controls are mostly already in place, but if user data transfer on control plane is supported, then it also requires additional CP-related overload control mechanisms.

Small Data Rate control corresponds to the APN Rate Control in CIoT EPS Optimizations. This is the operator's tool to dimension CIoT subscription in terms of number of messages that are allowed in time unit (hour, day, week). The operator may delay or drop the packets that exceed the rate control quota, or may pass them on and possibly over-charge.

Service Gap control was added in Rel-15 EPS to discourage the CIoT UEs from excessive frequent detach and re-attach cycles. A UE that detaches after every data packet generates very high signalling overhead, so typically it is more efficient to keep such a UE registered all the time between transmit intervals. CIoT optimizations provided in EPS and in 5GS ensure UE power budget optimizations by means of signalling efficiency.

AMF Overload control corresponds to the Serving PLMN Overload control, which protects the serving CN node (AMF or MME) against signalling overload that could be caused by excessive control plane small data signalling. It is expected that the AMF in the serving PLMN allows at least a reasonable Small Data Rate control quota to pass, without having to intercept small data that is within the subscribed Small Data Rate control quota.

Control Plane Back-Off timer protects the serving CN nodes against overload caused by control plane small data signalling by means of allowing the AMF to assign a back-off timer to some or all UEs, if it is running very high load. If the UE supports both user plane and control plane signalling, it can bypass the Control Plane Back-Off by transfer data via user plane.

9.11.6 Reliable Data Service

Reliable Data Service (RDS) was initially introduced between UE and SCEF, but the capability to support RDS overlay protocol has been broadened to apply also between UE and P-GW on the user plane.

To support the same service in 5GC, both NEF and UPF need to terminate the RDS protocol.

9.11.7 Northbound API

In the 5GC architecture, NEF takes the role of the network exposure function. An AS, whether it is modelled as an EPS SCS/AS or 5GC AF, should not need to worry about the UE camping on a 4G or 5G cell, thus the NEF needs to expose the Northound API like the SCEF in EPS. For privacy reasons, it is also important to hide the 3GPP identities of the user, so the 3GPP network edge should look the same from the outside irrespective of what lies southbound of the NEF or SCEF.

Using just a single exposure function towards the applications is ideal for application designers, but it fits the substantially different southbound architectures inside the 3GPP network very badly and forcing an update of all related EPC legacy interfaces is not an attractive option to build the interworking between EPC and 5GC.

The 3GPP architecture is a logical model that does not bind the mapping of the logical entities to physical deployments. Merging multiple logical nodes into a single box implementation is always a viable option. In this case the combined SCEF and NEF seems an attractive alternative that solves most of the interworking challenges. Such combined nodes must support the sum of SCEF and NEF interfaces. The SCS/AS or AF can use the common T8 API without the need to distinguish between SCEF and NEF, but inside the 3GPP system, a UE roaming in a 4G or 5G system is addressed via the procedures that are specified for each system. More importantly, AMF and SMF need not support T6 interface towards SCEF and the MME need not support the 5GC interfaces, but the combined SCEF and NEF node exposes both EPC and 5GC interfaces towards all affected (EPS and 5GS) nodes. However, the combined SCEF and NEF node exposes only one API interface towards the AS.

This admittedly leaves two main points open for the implementation. These are the interaction between SCEF and NEF roles inside the combined node and handling of the subscriber data. The user specific subscriber data are considered as a single set of data managed jointly by the HSS and unified data management (UDM)/unified data repository (UDR). The beauty of this design is that duplication of updates is avoided, as either HSS or UDM/UDR can update the same subscription data record, depending on whether the update was received from EPC or 5GC. For example, a UE monitoring request received from the application is inserted by HSS/UDM into the subscription data, and then updated to the affected MMEs and AMFs. Whichever detects a monitoring event will send the notification towards the exposure node they interface with (either SCEF or NEF).

The API offered to the AS is expected to be the same in all cases when a service is provided via EPC and 5GC. However, this does not mandate all EPC and 5GC services to be the same, e.g. if for some reason a service can be provided only in one system.

If small data is supported over user plane, then the user plane nodes must offer the API services corresponding to EPS NIDD. Candidate solutions allowing the SMF or UPF to play the role of the NEF are documented in TR 23.724 [10]

9.11.8 Network Parameter Configuration

The 3GPP network may open an interface for authorized third party applications to update certain parts of the user subscription based on predicted UE behaviour. A generic subscription data provisioning framework is specified for an external party to provision UE communication pattern parameters that are called Expected UE behaviour. These parameters give an educated guess on the UE's behaviour, such as:

- *Expected UE moving trajectory.* The foreseen or planned UE movement in terms of geo-location;
- *Stationary indication.* Indicates whether the UE is static or mobile;
- *Communication duration time.* The time that the UE usually stays in CM-CONNECTED state;
- *Periodic Time.* Time interval for periodic communication;
- *Scheduled communication time.* Weekday and time of the day when the UE is expected to be available for communication;
- *Scheduled communication type.* Downlink only, Uplink only, or bi-directional communication.

If the AS can predict the above Expected UE behaviour accurately, then the 3GPP network can benefit from the provisioned values and provide a better service to the application by configuring the UE mobility parameters and UE power saving cycles accordingly.

9.11.9 Monitoring

The monitoring framework is already supported in Rel-15 5GC, but UE mobility between EPS and 5GS is still an open topic. As already mentioned in section 9.11.7, ideally the external AS should not need to worry about whether the UE is camping on LTE or 5G, it only needs to request monitoring of a certain UE identified by an External Identifier or External Group Identifier. It is the network responsibility notifying the AS on the reachability of the UE irrespective of whether the UE becomes available for communication over LTE or 5G.

In EPS, the HSS updates the monitoring request in the MMEs that can notify on the UE availability for communication. In a multi-system network environment this requires a monitoring request entered in the subscription data by HSS or UDM is forked to all affected MMEs, AMFs, and SMFs that can process the monitoring notification part. The common subscription data stored in UDR is key for a combined monitoring solution.

9.11.10 NB-IoT Inter-RAT Mobility

The goal is to provide UE mobility without having to re-attach to the network at RAT change. This requires inter-RAT TAU and Mobility Registration each way, however the various 5G deployment scenarios (see Chapter 1) are adding complexity to potential solutions. The main point is whether NB-IoT radio network is connected to EPC over S1 or to 5GC over N2/N3.

Inter-RAT mobility requires the maintenance of the UE Radio Capabilities for 5G and NB-IoT, which implies handling of those capabilities on each RAT separately. This is the result of the large size of the multi-RAT, multi-band UE Radio Capability information data in relation with the limited data rate of NB-IoT. Signalling over 3GPP protocols must be optimized to avoid excessive signalling overhead over narrow-band radio.

An aspect related with the different capabilities supported in each system is the way how the active 5G PDU Sessions and 4G PDN Connections can follow in the UE's inter-RAT mobility. The capabilities of the serving network elements are of course considered, but additionally a subscription-based indication of which connections can be maintained during mobility may be introduced. The CN transfers only those PDN Connections or PDU Sessions that can be supported after mobility, considering the network capabilities, the subscription data and the network local policy.

9.11.11 QoS Support for NB-IoT

The goal is to enable QoS differentiation between NB-IoT UEs mainly for control plane small data.

As a possible solution the AMF can assign the subscribed QoS index received from the UDM during registration to the UE. The UE includes this QoS index in subsequent

Table 9.12 Other technologies for M2M/IoT.

	SIGFOX	LoRa	Weightless	On-ramp wireless	Short-range wireless
Range MCL L	<12 km 160 dB	<10 km 157 dB	<10 km 156 dB	<1 km	10 cm to 200 m
Spectrum	Unlicensed 900 MHz	Unlicensed 900 MHz	Unlicensed 470–900 MHz	Unlicensed 2.4 GHz	Unlicensed 2.4 GHz
Battery lifetime	>10 years	>10 years	>10 years	>10 years	>10 years
Data rate	<100 bps	<37 kbps	1–500 kbps	<20 kbps	<100 s Mbps
Use case	Smart Grid/ City/ Monitoring	Smart Grid/ City/ Monitoring	Smart Grid/ City/ Monitoring	Smart Grid/ City/ Monitoring	Smart home/ factory
Chipset costs	1.5$	1.5$	1.5$	tbd	0.5$

control plane data messages. This allows the RAN to evaluate the UE's priority compared to other UEs. The QoS Index can be policed by the AMF by comparing the QoS index value provided by the UE against the one from UDM. While this service is considered Quality of Service, the technical implementation has got similarities with overload control.

9.12 Other Technologies

With the increased interest in the IoT, many companies and engineers are focusing on developing wireless connectivity solutions for IoT use cases.

Table 9.12 provides a summary of other (proprietary) technologies that are used today for M2M/IoT communication.

References

All 3GPP specifications can be found under http://www.3gpp.org/ftp/Specs/latest. The acronym 'TS' stands for Technical Specification, 'TR' for Technical Report.

1 3GPP TR 23.887: "Machine-Type and other Mobile Data Applications Communications Enhancements".
2 3GPP TR 23.888: "System Improvements for Machine-Type Communications".
3 3GPP TS 38.331: "NR; Radio Resource Control (RRC); Protocol specification".
4 3GPP TS 38.300: "NR; Overall description".
5 3GPP TS 38.413: "NG-RAN; NG Application Protocol (NGAP)".
6 3GPP TS 23.060: "General Packet Radio Service (GPRS); Service description".
7 3GPP TS 23.122: "Non-Access-Stratum (NAS) functions related to Mobile Station (MS) in idle mode".
8 3GPP TS 36.321: "Medium Access Control (MAC) protocol specification".
9 3GPP TS 36.331: "Radio Resource Control (RRC); protocol specification".

10 3GPP TR 23.724: "Study on Cellular IoT support and evolution for the 5G system".
11 3GPP TS 23.246: "Multimedia Broadcast/Multicast Service (MBMS); Architecture and functional description".
12 3GPP TS 23.251: "Network Sharing; Architecture and functional description".
13 3GPP TS 23.272: "Circuit Switched (CS) fallback in Evolved Packet System (EPS)".
14 3GPP TS 23.401: "GPRS enhancements for E-UTRAN access".
15 3GPP TS 24.301: "Non-Access-Stratum (NAS) protocol for Evolved Packet System (EPS)".
16 3GPP TR 45.820: "Cellular System Support for Ultra Low Complexity and Low Throughput Internet of Things".
17 3GPP TR 33.860: " Study on Enhanced General Packet Radio Service (EGPRS) access security enhancements with relation to cellular Internet of Things (IoT)".
18 3GPP TS 43.064: "Overall description of the GPRS radio interface".
19 3GPP TS 45.002: "GSM/EDGE Multiplexing and multiple access on the radio path".
20 3GPP TS 44.018: "Mobile radio interface layer specification; Radio Resource Control (RRC) protocol".
21 3GPP TS 55.226: "3G Security; Specification of the A5/4 Encryption Algorithms for GSM and ECSD, and the GEA4 Encryption Algorithm for GPRS".
22 3GPP TS 55.241: "Specification of the GIA4 integrity algorithm for General Packet Radio Service (GPRS); GIA4 specification".
23 3GPP TS 55.251: "Specification of the GEA5 and GIA5 encryption algorithms for General Packet Radio Service (GPRS); GEA5 and GIA5 algorithm specification".
24 3GPP TR 45.050: "Background for Radio Frequency (RF) requirements".
25 3GPP TS 43.059: "Functional stage 2 description of Location Services (LCS) in GER-AN".
26 R6-180020: "On Further Energy Saving for EC-GSM-IoT", source: Nokia, Nokia Shanghai Bell, 3GPP RAN6#7, February 2018.
27 3GPP TS 23.682: "Architecture enhancements to facilitate communications with packet data networks and applications".
28 3GPP TS 23.730: "Study on extended architecture support for Cellular Internet of Things (CIoT)".
29 3GPP TR 36.888: "Study on provision of low-cost Machine-Type Communications (MTC) User Equipments (UEs)".
30 3GPP TS 22.261: "Service requirements for the 5G system".
31 3GPP TS 23.040: "Technical realization of Short Message Service (SMS)".
32 3GPP TS 29.337: "Diameter-based T4 Interface for communications with packet data networks and applications".
33 3GPP TS 23.501: "System Architecture for the 5G system; Stage 2".
34 3GPP TS 23.502: "Procedures for the 5G system; Stage 2".
35 3GPP TS 36.300: "Overall description; Stage 2".
36 3GPP TS 36.211: "Physical Channels and Modulation".
37 3GPP TS 36.212: "Multiplexing and channel coding".
38 3GPP TS 36.213: "Physical layer procedures".
39 3GPP TS 36.214: "Physical layer – Measurements".

10

Summary and Outlook

Rainer Liebhart[1] *and Devaki Chandramouli*[2]

[1] *Nokia, Munich, Germany*
[2] *Nokia, Irving, TX, USA*

10.1 Summary

5G is not just another radio technology, (i.e. not just an evolution of long term evolution, LTE), providing more capacity and lower latency to the end consumer, rather it's design and architecture principles follow a revolutionary concept, as for the first time this new system allows building open networks with open interfaces where in principle any level of network control can be offered to the outside world. The network can even offer itself as a service to tenants. The system architecture is designed to use network function virtualization (NFV) and software defined networking (SDN) concepts natively and exploit these technologies. The 5G technical enablers allow cost-efficient support of diverse use cases with very different demands for throughput, latency, reliability, security, as described in Chapter 8 for ultra-reliable low latency communication (URLLC) and Chapter 9 for Internet of Things (IoT). This will provide new opportunities to mobile operators, industries and end consumers. It will help people in their daily life, at their work places and in emergency situations. It allows connecting factories, which helps increasing the level of automation in various industries, i.e. less interruption time when designing and manufacturing goods of any kind. Connecting machines, things, and people can make life easier but also safer. The steadily growing data demand (a big US operator expects data traffic grows 10 times by 2020) will be addressed by 5G as well. Especially in dense urban areas and hot spots big chunk of available bandwidth, massive multiple-input-multiple-output (MIMO) and 3D beamforming will boost the capacity for consumers. Together with very short reaction times of the network this will increase user satisfaction and willingness to pay more for a 5G subscription as indicated by some recent opinion polls.

However, creating new revenue streams does not come along the lines of higher throughput for end consumers or few Dollars or Euros more per monthly bill. Competitive pressure in the end consumer market segment is too high and will not decrease over time. A new operator on the Indian market gives a good example how a new business idea can change a market, however not in terms of revenue per subscription. New revenue streams can only be generated, if operators invest consequently in cost-effective, optimized connectivity solutions, which can be offered to verticals and

5G for the Connected World, First Edition.
Edited by Devaki Chandramouli, Rainer Liebhart and Juho Pirskanen.

service providers. These solutions must be fine-tuned to fit the needs of the customer, they must be flexible, open and must allow a tenant to control its connectivity solution (e.g. a network slice) whenever there is a need. These business models are new to the telecom industry. Current mobile networks are huge and complex machines, designed mainly for voice and Internet access. Adopting existing networks to the needs of special verticals like IoT service providers, factory owners, small and medium enterprises is not an easy task. LTE/Evolved Packet System (EPS) was not designed from the beginning to cope with the requirements coming from connecting millions or billions of IoT devices. It took several 3rd Generation Partnership Project (3GPP) releases to introduce new concepts into the overall EPS system, together with enhancements on the radio side such as Long Term Evolution category M1 (LTE-M) and Narrow Band Internet of Things (NB-IoT), which will continue to play an important role as low power wide area (LPWA) radio technology in future, to offer a solution that is competitive to Sigfox and LoRa or other low power and wide area wireless systems. With 5G, this should not happen as 5G provides support for all use cases inherently, enhanced mobile broadband (eMBB), critical machine type communication (cMTC) and massive machine type communication (mMTC). Cloud-based architecture and SDN support is part of the 5G system design natively. Flexibility comes with features like service based architecture (SBA), micro services, enablers for network function resiliency and mobile edge computing, the new Quality of Service (QoS) model and network slicing. The 5G Core is designed to work as a converged core applicable for multiple access systems: New Radio (NR), Evolved Universal Mobile Telecommunications System Terrestrial Radio (E-UTRA), Wireless Fidelity (Wi-Fi) and fixed. This allows usage of the same services via different accesses, seamless mobility without the need for re-authentication, common protocols, thus helps lowering capital expenditure (CAPEX) and operating expenses (OPEX). Whether these capabilities offered by 5G are successfully used to open new businesses and connect all kind of machines in the interest of mankind is an open question. This requires clearly an open and critical mind from the (telecom) industry and the whole society, close partnership with other industries as now undertaken with 5G Automotive Association (5GAA) and 5G for Connected Industries and Automation (5GACIA), and the willingness to change internal and external processes to exploit the full capabilities of the 5G System. It is possible that verticals will build their own private networks, i.e. these new revenue streams mentioned above could bypass mobile operators, at least to a significant extent. What is the right choice for a given vertical depends on the specific needs and on the offer an operator or infrastructure vendor can make and is willing to make.

10.2 Outlook

Will 5G be the last “G” or is a 6th generation of mobile networks already on the way in the minds of people? This question sounds strange as there was always a next “G” after completion of one “G.” In addition, work on new wave forms and new techniques has already started for the next “G.” The Academy of Finland has selected the University of Oulu’s proposal for 6G-enabled Wireless Smart Society & Ecosystem as one of the first two flagships in the new national research funding program. Japanese DoCoMo announced recently successful tests achieving 100 Gbps wireless transmission using a new principle, Orbital Angular Momentum (OAM) multiplexing, with the aim of achieving terabit-class wireless transmission in the 2030s. This may be the starting point of 6G, and we can also define a new Core for this 6G radio technology, if there is a Core needed at all or 6G will just follow the Wi-Fi network architecture paradigm.

On the other hand, 5G promises to cover potentially all use cases, eMBB, cMTC, mMTC. Even if a new radio technology achieves higher data rates and smaller latency in future, is it then really 6G? Progress in technology will not stop, but 5G in its current form could be an umbrella (radio and core) system and architecture, flexible enough to integrate any new radio and core technology for the next decade and longer.

3GPP will progress on 5G in its Release 16, completion date is end of 2019 or beginning of 2020. The new features (list is not exhaustive) targeted for the New Radio in Release 16 can be grouped under the following themes:

- Efficiency:
  - Enhanced MIMO
  - Mobility enhancements
  - user equipment (UE) power consumption
  - Remote interference management
  - Positioning methods
  - Waveforms above 52.6 GHz
- Verticals:
  - Industry 4.0, Smart city, Private networks: Industrial IoT, NR in unlicensed bands with focus on bands below 7 GHz
  - Industry 4.0, Smart city: URLLC enhancements
  - NR Vehicle-to-X (V2X)
- Deployment and Operability:
  - NR-NR Dual-Connectivity
- NR for non-terrestrial networks

The new features (not exhaustive list) targeted for the overall 5G System Architecture in Release 16 can be grouped under the following themes:

- Verticals:
  - Cellular IoT support in 5G
  - Enhancements for URLLC
  - Vertical and local area network (LAN) Services including Private Networks and Time Sensitive Networking (TSN)
  - Location Services
  - V2X Services
- General System Architecture enhancements:
  - Service Based 5G System Architecture enhancements
  - Network Slicing enhancements
  - Enhancements to session management function (SMF) and user plane function (UPF) topology
  - User data migration
- Enablers for Network Automation
- Support for additional access types

- o Wireless and Wireline Convergence
- o Access Traffic Steering, Switching and Splitting
- o Architecture aspects for using Satellite access in 5G

Following the international telecommunications union radio communication sector (ITU-R) Workshop on international mobile telecommunication (IMT) 2020 terrestrial radio interfaces in October 2017, 3GPP is working on the initial description of its 5G solution (including results achieved in 2017 and 2018). This initial description has been provided to the international telecommunications union (ITU) in accordance with the IMT 2020 submission and evaluation process to make 3GPP's 5G a candidate for inclusion in IMT 2020. The summary is not final but gives a good overview of the content of 3GPP Release 15 (see http://www.3gpp.org/NEWS-EVENTS/3GPP-NEWS/1937-5G_DESCRIPTION). The description will be enhanced in future based on 3GPP progress.

As for previous generations interoperability between different vendors based on standardized interfaces and features is also key for 5G. 3GPP plays a key role when defining the new radio and core network architecture. Due to the very diverse nature of use cases covered by 5G, other organizations play also a fundamental role for defining an overall 5G framework. This includes activities in European Telecommunications Standards Institute Multi-Access Edge Computing (ETSI MEC) to define a platform providing standardized functions and data via well-defined APIs to applications. This allows application developers making use of data and information on network and user level. In addition, standardized ways to manage and orchestrate network slices in an efficient way is extremely important to roll-out services in a short time and without intensive manual input. For that purpose, the European Telecommunications Standards Institute (ETSI) created the Zero touch network and Service Management Industry Specification Group (ZSM ISG). As mentioned in Chapter 1 associations such as 5GAA and 5GACIA are considering the specific requirements of certain industries and provide input to 3GPP or other groups. This has never happened before in this extensive manner.

Big data analytics, predictive analytics, edge computing, a service oriented architecture for core network (CN) and radio access network (RAN), Machine Learning (ML) and Artificial Intelligence (AI) will play a key role in the future. Exposing data from the network via northbound interfaces will be essential to monetize these data and allow automatic and fast network (re-)configurations. Using AI and ML to make intelligent decisions, e.g. in Radio Resource Management of one or several adjacent cells influenced by a radio controller (a topic the Open RAN initiative is studying), optimizing network resources in radio, core and transport locally, regionally and globally based on predictive QoS schema, Self-Optimized Networks (SON) and cognitive analytics are some of the exciting innovative new ideas that may come in future. One trend that is already visible today is using predictive data analytics and ML techniques with respect to IoT traffic. Automation is key to network efficiency. With the introduction of virtualization, the total cost of operation could increase initially and it can increase even more with manual support of network slicing. The real reduction in total cost of operation is expected mainly with introduction of "zero touch network and service automation," enabling also support for fully automated slicing, especially for dynamic and large-scale networks. For instance, when considering potentially billions of connected IoT devices, accurate predictive analysis of their behavior, optimizing operations and processes based on that is essential to support extreme network automation and reduce operational costs. This kind of predictive analysis can also be executed in Edge Clouds and requires potentially huge computing capabilities at the edge. Innovative ideas and foreseen trends will radically change the network as we know it today.

The New Radio and 5G System Architecture will give operators and verticals the right tools to harvest new business potentials and at the same time, 5G will blend into everyone's life. Every industry will be affected by 5G. 5G will transform our individual lives, economy and society.

Appendix of 3GPP Reference Points

The list contains reference points and service-based interfaces applicable to the 5G System.

Cx	Reference point between CSCF and HSS/UDM based on Diameter.
E1	Reference point between CU-CP and CU-UP (Central Unit) to control user plane resources. E1 is based on E1AP (E1 Application Part).
F1	Reference point between RAN DU (Distributed Unit) and RAN CU (Central Unit). It consists of F1-C and F1-U to exchange control and user plane traffic respectively. F1 is based on F1AP (F1 Application Part).
N1	Reference point between UE and AMF to exchange NAS (Non-Access Stratum) messages.
N2	Reference point between 5G-RAN and AMF based on NGAP (Next Generation Application Part).
N3	Reference point between 5G RAN and UPF based on GTPv1-U (GPRS Tunneling Protocol User Plane).
N4	Reference point between SMF and UPF to manage data sessions at the user plane. N4 is based on PFCP (Packet Forwarding Control Protocol).
N5	Reference point between PCF and AF.
N6	Reference point between UPF and packet data networks. It can, e.g. transport IP and Ethernet packets.
N7	Reference point between SMF and PCF.
N8	Reference point between AMF and UDM.
N9	Reference point between two UPF to transmit user plane data. It is based on GTPv1-U.
N10	Reference point between SMF and UDM.
N11	Reference point between AMF and SMF.
N12	Reference point between AMF and AUSF.
N13	Reference point between AUSF and UDM.
N14	Reference point between two AMF.
N15	Reference point between AMF and PCF.
N16	Reference point between visited SMF (vSMF) and home SMF (hSMF).
N17	Reference point between AMF and 5G-EIR.

5G for the Connected World, First Edition.
Edited by Devaki Chandramouli, Rainer Liebhart and Juho Pirskanen.

N18	Reference point between NF and UDSF.
N22	Reference point between AMF and NSSF.
N23	Reference point between PCF and NWDAF.
N24	Reference point between visited PCF (vPCF) and home PCF (hPCF).
N27	Reference point between visited NRF (vNRF) and home NRF (hNRF).
N31	Reference point between visited NSSF (vNSSF) and home NSSF (hNSSF).
N32	Reference point between V-SEPP (visited SEPP) and H-SEPP (home SEPP) in roaming cases.
N33	Reference point between NEF and AF.
N40	Reference point between SMF and CHF.
N50	Reference point between AMF and CBCF.
N5g-eir	Service based interface to offer 5G-EIR services.
Naf	Service based interface to offer AF services.
Namf	Service based interface to offer AMF services.
Nausf	Service based interface to offer AUSF services.
Nnef	Service based interface to offer NEF services.
Nnrf	Service based interface to offer NRF services.
Nnssf	Service based interface to offer NSSF services.
Npcf	Service based interface to offer PCF services.
Nsmf	Service based interface to offer SMF services.
Nudm	Service based interface to offer UDM services.
Nudr	Service based interface to offer UDR services.
Nudsf	Service based interface to offer UDSF services.
Nnwdaf	Service based interface to offer NWDAF services.
Nwu	Reference point between UE and N3IWF for establishing secure tunnels. Nwu is based on IPsec/IKEv2.
Rx	The Rx reference point provides application layer information to the PCF, e.g. to establish new multimedia sessions. Rx is based on Diameter.
S1-MME	Reference point between eNB and MME based on S1AP (S1 Application Protocol).
S1-U	Reference point between eNB and S-GW based on GTPv1-U.
S5	This reference point is used to tunnel user plane packets and manage the user plane tunnels between S-GW and P-GW. S5 is based on GTPv2-C or alternatively on PMIPv6.
S6a	Reference point between MME and HSS to enable transfer of subscription and authentication data. It is based on Diameter.
S10	Reference point between MMEs for MME relocation and information transfer, based on GTPv2-C.
S11	Reference point between MME and S-GW, used to manage new or existing sessions, to relocate S-GW during handover, establish direct or indirect forwarding tunnels and trigger paging. S11 is based on GTPv2-C.
SBc	Reference point between CBC and MME for warning message delivery and control functions. The used protocol on this interface is the SBc Application Protocol (SBc-AP).

SGi	This is the reference point between the P-GW and a packet data network. Protocols on this interface are e.g. IPv4, IPv6, RADIUS, DIAMETER and DHCP.
X2	Reference point between master eNB and secondary gNB in dual connectivity mode used for floor control and data forwarding.
Xn	Reference point between gNB and gNB, gNB and ng-eNB. It consists of Xn-C and Xn-U.
Y1	Reference point between UE and non-3GPP access. Y1 is access technology dependent and not defined by 3GPP.
Y2	Reference point between untrusted non-3GPP access and N3IWF for the transport of NWu traffic.

Following reference points are realized as service-based interfaces: N5, N7, N8, N10, N11, N12, N13, N14, N15, N16, N17, N18, N22, N24, N27, N31, N33, N40 and N50.

Index

5G for the Connected World, First Edition.
Edited by Devaki Chandramouli, Rainer Liebhart and Juho Pirskanen.

b

d

m

n

o

q

t

V

W

TJI88027-9781119247081-13-06-19